Innovation in Concrete Structures: Design and Construction

Proceedings of the International Conference
held at the University of Dundee, Scotland, UK
on 8-10 September 1999

Edited by

Ravindra K. Dhir
Director, Concrete Technology Unit
University of Dundee

and

M. Roderick Jones
Senior Lecturer, Concrete Technology Unit
University of Dundee

Published by Thomas Telford Publishing, Thomas Telford Limited, 1 Heron Quay, London E14 4JD.

URL: http://www.t-telford.co.uk

Distributors for Thomas Telford books are
USA: ASCE Press, 1801 Alexander Bell Drive, Reston, VA 20191-4400, USA
Japan: Maruzen Co. Ltd, Book Department, 3–10 Nihonbashi 2-chome, Chuo-ku, Tokyo 103
Australia: DA Books and Journals, 648 Whitehorse Road, Mitcham 3132, Victoria

First published 1999

The full list of titles from the 1999 International Congress 'Creating with Concrete' and available from Thomas Telford is as follows

- *Creating with concrete*
- *Radical design and concrete practices*
- *Role of interfaces in concrete*
- *Controlling concrete degradation*
- *Extending performance of concrete structures*
- *Exploiting wastes in concrete*
- *Modern concrete materials: binders, additions and admixtures*
- *Utilizing ready-mixed concrete and mortar*
- *Innovation in concrete structures: design and construction*
- *Specialist techniques and materials in concrete construction*
- *Concrete durability and repair technology*

A catalogue record for this book is available from the British Library

ISBN: 0 7277 2824 5

Printed and bound in Great Britain by MPG Books, Bodmin, Cornwall

PREFACE

Concrete is the key material for Mankind to create the built environment, the requirements for which are both demanding in terms of technical performance and economy and yet greatly varied from architectural masterpieces to the simplest of utilities. This presents the greatest challenge and the question is how best to advance concrete and create imaginatively.

In response, the Concrete Technology Unit (CTU) of the University of Dundee organised this Congress following on from its established series of events, namely, Concrete in the Service of Mankind in 1996, Concrete 2000: Economic and Durable Concrete Construction Through Excellence in 1993 and Protection of Concrete in 1990.

Under the theme of Creating with Concrete, the Congress consisted of five Seminars: (i) Radical Design and Concrete Practices, (ii) Role of Interfaces in Concrete, (iii) Controlling Concrete Degradation, (iv) Extending Performance of Concrete Structures and (v) Exploiting Wastes in Concrete, and five Conferences: (i) Modern Concrete Materials: Binders, Additions and Admixtures, (ii) Utilising Ready-Mixed Concrete and Mortar, (iii) Innovation in Concrete Structures: Design and Construction, (iv) Specialist Techniques and Materials for Concrete and Construction and (v) Concrete Durability and Repair Technology. In all, a total of 421 papers were presented from 67 countries.

The Opening Addresses were given by Mr Henry McLeish, MP, MSP, Minister for Enterprise and Lifelong Learning, Scotland, Dr Ian Graham-Bryce, Principal and Vice Chancellor of Dundee University, Mrs Helen Wright, Lord Provost, City of Dundee and Professor Peter Hewlett, President of the Concrete Society. This was followed by four Opening Papers by leading international experts; Dr Bryant Mather, US Army Corps of Engineers, Professor Charles F Hendriks, Delft University of Technology, Netherlands, Dr Bjørn Jensen, Danish Technological Institute, Dr Oliver Kornadt, Philipp Holzmann AG, Germany, Professor Jurek Tolloczko, Concrete Society, UK, Mr Michael Téménidès, CIMBÉTON, France and Professor Yves Malier, Ecole Normale Superieure de Cachan, France. The Closing Address was given by Professor John Morris, University of the Witwatersrand, South Africa.

The support of 20 International Professional Institutions and 31 sponsors was a major contribution to the success of the Congress. An extensive Trade Fair, participated in by 50 organisations, formed an integral part of the Congress. The work of the Congress was an immense undertaking and all of those involved are gratefully acknowledged, in particular, the members of the Organising Committee for managing the event from start to finish; members of the International Advisory and National Technical Committees for advising on the selection and reviewing of papers; the Authors and the Chairmen of Technical Sessions for their invaluable contributions to the proceedings.

All of the proceedings have been prepared directly from the camera-ready manuscripts submitted by the authors and editing has been restricted to minor changes where it was considered absolutely necessary.

Dundee
September 1999

Ravindra K Dhir
Chairman, Congress Organising Committee

INTRODUCTION

Defining exactly what is 'innovation' can be difficult. The dictionary typical says "to invent or begin to apply methods or ideas". Peter Drucker, the highly regarded management guru, states that "business has only two functions - marketing and innovation". It can be argued, therefore, that the degree to which an industry innovates is vital to its success. It is almost axiomatic that companies which do not innovate are the most likely to fail.

Given this, the concrete construction industry is often regarded by many analysts as being slow to innovate and is, too often, only driven to examine current practices and methods, by revisions to codes of practice, specifications, regulations etc. This situation can lead to complacency, as many traditional engineering industries in western countries have found to their cost. There is no doubt that this reactive form of innovation must give way to a more proactive approach of continuous self-examination of materials, technology, design and construction processes.

In some ways, the concrete industry's tardiness in undertaking and adopting innovations is understandable. Concrete is both a mature and a highly successful material which is required in virtually all forms of construction. In addition, it can be difficult to protect the necessary financial investment for research and development given that the generic nature of concrete means establishing patents can prove either uneconomic or impossible.

The key to successful innovation in the concrete industry appears to lie in a close relationship between users and researchers at an early stage and a willingness by industry to commit the necessary resources. The role of central and local government in this process will always be important, not least, as it is the key regulatory authority for the built environment as legislators and planners and, in the majority of cases, the direct or indirect client. The role of government in this process will probably always be somewhat nebulous, but for innovation to be successful, the driving force must come from industry directly.

The concrete construction industry worldwide has become increasingly alert to the need for innovation and the competitive advantage it can bring. This Conference sought to review, discuss and report new practices and methods in the design and construction of concrete structures from around the world.

The Proceedings for the Conference; *Innovation in Concrete Structures: Design and Construction* dealt with issues in six themes, namely (i) Civil Engineering Structures, (ii) Sub-Structures, (iii) High-Rise Structures, (iv) Precast Concrete Construction and (v) Housing. Each of these themes was opened by a Leader Paper presented by an expert in that field. There were a total of 64 papers presented during the Conference which have been compiled into these Proceedings.

Dundee
May 1999

Ravindra K Dhir
M Roderick Jones

ORGANISING COMMITTEE
Concrete Technology Unit

Professor R K Dhir, OBE (Chairman)

Dr M R Jones (Secretary)

Mr M D Newlands (Joint Secretary)

Professor P C Hewlett
British Board of Agrément

Dr N A Henderson
Mott MacDonald Ltd

Professor V K Rigopoulou
National Technical University of Athens, Greece

Dr S Y N Chan
Hong Kong Polytechnic University

Dr N Y Ho
L & M Structural Systems, Singapore

Dr M J McCarthy

Dr M C Limbachiya

Dr T D Dyer

Dr K A Paine

Dr T G Jappy

Mr P A J Tittle

Mr J C Knights

Mr S R Scott (Unit Assistant)

Miss A M Duncan (Unit Secretary)

INTERNATIONAL ADVISORY COMMITTEE

INTERNATIONAL ADVISORY COMMITTEE (CONTINUED)

NATIONAL TECHNICAL COMMITTEE

Mr P Barber
Manager of the Scheme, The Quality Scheme for Ready Mixed Concrete

Professor A W Beeby
Professor of Structural Design, University of Leeds

Mr B V Brown
Divisional Technical Executive, Readymix (UK) Ltd.

Dr T W Broyd
Technology Development Director, W S Atkins Ltd.

Professor J H Bungey
Professor of Civil Engineering, University of Liverpool

Dr P S Chana
Director, CRIC, Imperial College of Science, Technology & Medicine

Professor J L Clarke
Principal Engineer, The Concrete Society

Dr P C Das
Group Manager, Structures Management, Highways Agency

Dr S B Desai, OBE
Principal Civil Engineer, Department of the Environment, Transport and the Regions

Professor R K Dhir, OBE (Chairman)
Director, Concrete Technology Unit, University of Dundee

Mr C R Ecob
Director Special Services Division, Mott MacDonald Ltd.

Professor F P Glasser
University of Aberdeen

Professor T A Harrison
Technical Consultant, Quarry Products Association

Professor P C Hewlett
Director, British Board of Agrément

Professor J Innes
Director of Roads, Scottish Office

NATIONAL TECHNICAL COMMITTEE (CONTINUED)

Mr K A L Johnson
Director, AMEC Civil Engineering Ltd.

Dr M R Jones
Senior Lecturer, Concrete Technology Unit, University of Dundee

Mr P Livesey
National Technical Services Manager, Castle Cement Ltd.

Professor A E Long
Director of School, Queens University of Belfast

Professor P S Mangat
Head of Research, Sheffield Hallam University

Mr G Masterton
Director, Babtie Group Ltd.

Professor G C Mays
Director of Civil Engineering, Cranfield University

Mr L H McCurrich
Technology Development Consultant, Fosroc Construction

Professor R S Narayanan
Partner, SB Tietz & Partners Consulting Engineers

Dr P J Nixon
Head, Centre for Concrete Construction, Building Research Establishment Ltd.

Dr W F Price
Senior Associate, Messrs Sandberg

Professor G Somerville, OBE
Director of Engineering, British Cement Association

Professor D C Spooner
Director, Materials and Standards, British Cement Association

Dr H P J Taylor
Director, Tarmac Precast Concrete Ltd.

Mr M Walker
Technical Manager, The Concrete Society

Dr R J Woodward
Senior Project Manager, Transport Research Laboratory

SUPPORTING INSTITUTIONS

American Concrete Institute, USA

American Society of Civil Engineers, USA

Australian Concrete Institute

Concrete Association of Finland

Concrete Society of Southern Africa

Concrete Society, UK

Danish Concrete Society, Denmark

Fédération de l'Industrie du Beton, France

German Concrete Association (DBV)

Hong Kong Institution of Engineers

Indian Concrete Institute

Institute of Concrete Technology, UK

Institution of Civil Engineers, UK

Instituto Brasileiro Do Concreto, Brazil

Japan Concrete Institute

Netherlands Concrete Society

New Zealand Concrete Society

Norwegian Concrete Association

Singapore Concrete Institute

Spanish Association for Structural Concrete

Swedish Concrete Association

SPONSORING ORGANISATIONS WITH EXHIBITION

AMEC Civil Engineering Ltd.

Babtie Group Ltd.

Bardon Aggregates

Blue Circle Cement

Blyth & Blyth

British Board of Agrément

British Cement Association

Building Research Establishment

Castle Cement Ltd.

Cementitious Slag Makers Association

CIMBÉTON, France

Du Pont de Nemours International S.A., Switzerland

ECC International Ltd.

Elkem Ltd. (Materials)

Fosroc International Ltd.

Grace Construction Products

HERACLES General Cement Co., Greece

John Doyle Group

Lafarge Aluminates

L M Scofield Europe Ltd.

Minelco Ltd.

Mott MacDonald Ltd.

O'Rourke Group

Ove Arup and Partners

SPONSORING ORGANISATIONS WITH EXHIBITION (CONTINUED)

Readymix (UK) Ltd.

Rugby Cement

Scottish Enterprise Tayside

Sika Ltd.

SKW - MBT Construction Chemicals

Thomas Telford Publishing Ltd.

United Kingdom Quality Ash Association

W A Fairhurst & Partners

ADDITIONAL EXHIBITORS

Christison Scientific Equipment Ltd.

CMS Pozament Limited

The Concrete Society

David Ball Group plc.

E & FN Spon

Flexcrete Ltd.

Germann Instruments A/S, Denmark

Natural Cement Distribution Limited

Palladian Publications Ltd.

Quality Scheme for Ready Mixed Concrete

UK Certification Authority for Reinforcing Steel

Wacker-Chemie GmbH, Germany

Wexham Developments

CONTENTS

THEME 3 HIGH-RISE STRUCTURES

THEME 5 PRECAST CONCRETE CONSTRUCTION

THEME 6 HOUSING

LEADER PAPERS

INTERFACING INNOVATION WITH BEST PRACTICE

P C Hewlett
British Board of Agrément
United Kingdom

ABSTRACT. Product quality and client satisfaction governs success. Change and innovation are perceived as carrying risk and their adoption is usually approached with caution. Nevertheless, innovation within the construction and building sector is essential if the business activity is to progress and benefit from new ideas and competition.

However, innovation based on research and development is not supported as an activity within this sector as in other industrial sectors such as electronics and pharmaceuticals. The reasons for this disparity are discussed and the relationship between innovation and invention suggested and the distinction between Technical Approval and Standardisation made.

Those factors that may encourage or hinder innovation are identified and the roles of regulation and certification/performance assessment are explored since the adoption of innovation into reliable and best practice will govern whether commercial benefit and client well being are the ultimate results.

Keywords: Innovation, Quality, Performance, Testing, Research, Standards, Approvals, Regulation.

Professor Peter C Hewlett is the Director of the British Board of Agrément and Visiting Industrial Professor to the Department of Civil Engineering at the University of Dundee.

He specialises in materials for construction and building both inorganic and organic and in particular the enhanced performance of concrete by way of chemical modifications.

He is President of the European Organisation for Technical Approvals and the UK Concrete Society.

INTRODUCTION

For a progressive future change is inevitable. This truism applies to all matters and construction and building products and systems are no exception. To change we have to innovate but like so many commonly used words their meaning becomes blurred. Therefore I will start with a definition.

Innovation is the adverb derived from the verb to innovate and both relate to the noun innovation. Since we are concerned with building products, the noun is probably more appropriate and may be defined as alteration, change, new idea, departure, introduction, newness, novelty and variation.

Innovation is above invention since it is more deliberate and less random being based upon knowledge rather than flair. Additionally, innovation does not stop at the product itself but is also concerned with the adoptive step of incorporating a product or process into practice. In this regard invention is part of the innovation process.

Two impressions are obvious, namely that innovation is dynamic and results in change and secondly, departure and variation are recognised. This is sometimes referred to as deviation meaning departure from what is regarded as normal or in the case of products, as conventional and established. Because innovation is dynamic it sometimes creates uncertainty and carries with it a perceived risk. Indeed, lack of innovation may well reflect a lack of personal and/or national confidence.

A number of factors work against accepting innovative products. For instance,

- the product is harder to sell
- indifference to benefits of change
- ignorance
- attachment to old ideas
- fear of remedial costs
- risk and assumed liability
- lack of confidence

Problems of accommodating innovation are global and in the construction and building sector, some would argue against innovation. However, and by way of example in the USA, the value of buildings and infrastructure is put at $20 trillion ($20^9$) and getting the best from such investment would seem to make sense [1]. The objective of innovating is to improve the product and process of construction.

Some industries rely on innovation to survive. For instance, pharmaceuticals, electronics, aeronautical engineering and plastics technologies. However the building and construction sector accept change only gradually and it is suspicious of new things.

Influencing Factors

Within the field of concrete there have been relatively few fundamental innovations, e.g. inclusion of reinforcement as rods, bars or fibres, alternative hydraulic binders to lime and then Portland type cements although there have been numerous variations on established themes.

This attitude has its roots in the permanency and empiricism associated with buildings and constructions and these related activities. Over many years confidence and familiarity have been built up with certain types of material. For instance, brick, wood, iron and masonry. The alternatives of concrete, plastics and steel do not have the same pedigree of age although they have become accepted because they solved a real need and had measurable benefits.

It is perhaps more the current range of innovations such as coatings, new generation composite plastics, alloys and modified cements that raise questions and doubts.

In this regard many professions are involved in helping or hindering innovation and its adoption. For instance, corporate executives, designers, project managers, craftsmen, legislators, insurance underwriters, government departments/officers, regulators, lawyers, inventors, manufacturers and contractors [1]. This is a very disparate collection of disciplines with different motivations that may help or hinder innovation.

However, there is only one creator of materials wealth and that is industry. Construction and building product manufacture are part of industry and hence wealth creation. If new products and systems are not accepted and brought into active use then building practice and designs would stagnate.

It is usually intended that most innovations should be changes for the better, more efficient, longer lasting, lighter, cheaper or less dangerous. In other words, innovation helps to achieve functional improvement. Such a drive cannot be resisted and in that regard innovation is inevitable and has to be accommodated.

STAGES IN INNOVATION

There are four readily recognisable stages in innovation

1. Research or discovery

2. Development

3. Production and application

4. Exploitation/commercialisation

Basic research produces the majority of discoveries from which all other progress flows [2][3]. The motives behind basic research are often curiosity and understanding. As such they may seem indulgent and a little out of fashion with the current preoccupation with focussing effort, managing programmes and meeting targets.

These aspects are more appropriate to development than research and within construction and concrete research and development, often become synonymous and result in reduced quality research and/or poorly disciplined and irresolute development.

In this regard innovation is contemplative rather than haphazard. The drive is not altruistic but the creation of wealth and material wellbeing. This sequence is shown in Figure 1.

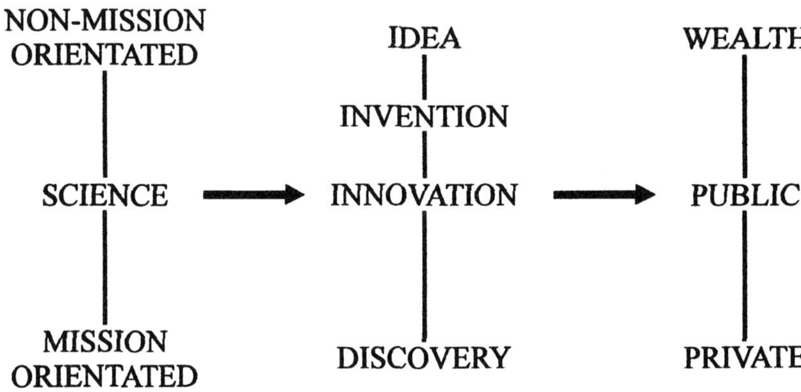

Figure 1 Innovation and wealth creation

There are levels of innovation, some minor, some not. For instance, a new concept might be regarded as having more potential importance than a new way of doing the same thing. The principle of using a high modulus reinforcement in a matrix having low tensile strength such as the inclusion of steel in concrete was at the turn of the century a new concept, This concept opened up design opportunities and there are now many variations on the same basic concept but these changes cannot compare with the initial idea.

According to Bernstein and Lemer [1] there have been 117 innovations in permanent residential structures over the period 1945 – 1990. [Table 1 – reproduced with kind permission of the authors].

Innovation would nevertheless appear to be regional reflecting different priorities, one country to another. This is evident from Table 2 also taken from reference [1] comparing USA, Europe and Japan. Such a situation is probably best since it addresses national priorities and needs but also reduces duplication and offers trade opportunity between different countries.

Transposing scientific and technological potential into viable innovations has become recognised in the last decade as an important step in the creation of wealth. Within the building and construction sectors in the UK at least, it is competing with the funding of research and development within the building and construction sectors. Perhaps to the disadvantage of the latter. Clearly a balance has to be struck.

Table 1 Sample of innovations in permanent residential structures 1945 – 1990 [1]

FUNCTIONAL AREA	NO. OF INNOVATIONS
Structural exterior wall framing	07
Enclosure and insulation	08
Openings	13
Interior wall framing	07
Foundation	12
Floor framing	10
Roof framing	07
Roof covering	07
Plumbing	12
Electrical wiring	04
Heating/ventilation/air conditioning	12
Interior finish	18
Total	117

Table 2 Regional variability of innovation[1]

US LEADS	EUROPE LEADS	JAPAN LEADS
High-performance concrete	High-performance asphalt	High-performance steel
Waste/wastewater treatment	High-speed rail/Magnetic levitation vehicles (Mag-lev)	Automated equipment
Computer Aided Design (CAD)/ Computer Aided Engineering (CAE)	Tunnelling	Field computer use
Solid/hazardous waste disposal	Real-time site positioning systems	High-speed pavement assessment
Environment	Restoration	Safety
Geographic Information and Positioning Systems (GIS/GPS)	Marine construction	Intelligent Buildings
Integrated database	Energy conservation	Building systems

Relative levels of technology adoption in design and construction industries, as assessed in reconnaissance trips sponsored by the US National Science Foundation and the Civil Engineering Research Foundation. Source: Civil Engineering Research Foundation European Research Task Force. *Constructed Civil Infrastructure Systems R&D: A European Perspective, Washington, DC: Civil Engineering Research Foundation, 1994, p5*

Innovation and Research and Development

Research and development in the construction and building sectors is not a substantial percentage of total market value. Various estimates put the figure within the range 0.1 – 0.5% of turnover. Some sectors are above this level and others below. Notwithstanding this low level of reinvestment, the industry is inventive and new ideas and ways of doing things often result from experience and commonsense rather than scientific understanding. On occasions this 'pioneering' approach is often found wanting when a new innovation allegedly fails. If the product or system has not been based on a clear technical pedigree it is often difficult to substantiate the role played by the invention or innovation and to give it a credible defence. Such a background reinforces the need for independent performance-based assessments to establish the fitness and worth of an innovation. That is the role of Technical Approval and Agrément. It is not really the role of product conformity or compliance to standards. National and international standards are usually based on accumulated experience and reflect what has been and is being done. In this respect they are archival but serve the useful purpose of establishing comparative if not absolute performance levels. Approvals on the other hand depend on performance-based bespoke assessments reflecting function rather than historical practice.The advantages of this approach are clear and there is a growing trend for standards to become more performance-based. If this trend is maintained then assessment and compliance will become coincident. However the present situation is far from that point.

It is a characteristic of the building and construction sectors that appreciation of the need for sound science is not so obvious as in other activities such as medicine, space exploration and electronics. The reasons for this are complex having their root in the education of civil engineers and architects and the mechanics of specifying for construction and building work. Prescription by way of codes and standards pervades the activity and such a situation is likely to continue. This being so, science and its outturn through innovation have to be assisted in the transposition from scientific concept to new practice. In this regard standards, approvals, conformity and quality management certification all have a part to play. These interactions are shown schematically in Figure 2.

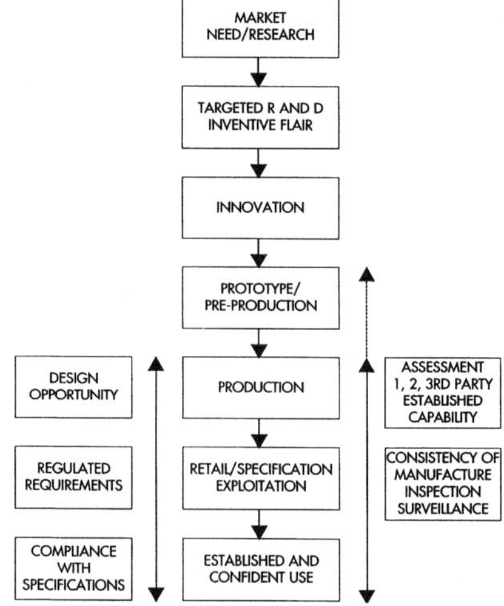

Figure 2 Transposing innovation into adopted practice

Not withstanding this situation there have been notable innovations in the construction and building sectors. For instance,

- high performance steels
- robotics
- high performance concrete
- high performance / engineered wood
- anti static flooring tiles
- silent pile driving equipment
- green technologies

In order to evaluate a new product it is important to have available test methods that can be used to measure the performance both in isolation and as incorporated in a building. Within the activity of standards making these are the so-called horizontal standards and are dealt with in the next section.

Test Methods

Many test methods are simulations of what happens in practice. They are useful for making comparisons, one product or system to another, but often do not permit cause and effect judgements to be made. For instance one may measure the vapour permeability of a protective membrane and it is a property of the material. Alternatively one may measure vapour transmission under a set of fixed conditions. Changing the conditions changes the value. Relating the value to a real event can be difficult if there is no absolute basis.

Much standardisation and certification work is product oriented with test procedures based on past experience with particular product types. In addition the measured responses also reflect the product types and not the actual performance required. For instance it is feasible to test an insulation in such a way as to reflect let us say, its fibrous nature, and to give ranges of values for thermal conductivity and transmittance that would typify such material. These values can be built into specifications so that only those materials that conform to such values would be chosen and used. This approach creates problems for new products that may impart their insulation properties (staying with the example) in quite a different way.

An alternative approach is to put emphasis on the performance required by way of characteristics and relate these to a measurement/test methodology and recording how the product responds. The designer can then used these measured characteristics in whatever way is appropriate. This approach can accommodate innovation.

Whatever approach is used they both depend upon having available agreed test methods. To design such tests it is necessary to understand the physics and perhaps the chemistry of the performance characteristics, e.g. freeze-thaw and durability testing of concrete.

The scale of test is another aspect that has to be considered. The interrelationship between testing a product, for instance a window, in the laboratory and relating that to the performance of the same window when installed in a brick wall, fitted with insulation and consisting of aerated concrete blocks for the inner leaf and a brick wall outer leaf can be very complex. Full scale testing is very desirable. For instance, windows and doors can be examined using the guarded hot box [5] [Figure 3]. Similar scaled up testing of roofing and external wall insulation can be assessed using BBA's Tempest rig [6] [Figure 4].

Figure 3 Guarded Hot Box apparatus

The BBA Guarded Hot Box is accredited for the determination of the thermal transmittance and conductance of building elements up to a size of 2.3m x 2.8m. It can also be used to test glazing and complex wall panels. The two main components of the apparatus, the hot chamber and the cold chamber, are mounted together in a frame which allows them to be rotated to change their orientation and permit the testing of horizontal elements such as ceiling panels or rooflights.

The use of a large number of calibrated thermocouples with computerised date acquisition enables the identification of cold bridges and the estimation of the potential condensation risk.

The BBA Tempest Rig comprises a computer that runs from specially software, a valve, an electrically drive radial fan, connecting flexible trunking and an enclosure (to suit the sample under test). The rig allows large building components to be subjected to simulated wind loading. A typical sample size is 3 x 3 metres. This apparatus has been recently adapted so that other tests may be run on external wall insulation systems using a vertical enclosure that can beclamped to the insulation system installed on a masonry wall.

These tests are described in EOTA (European Organisation for Technical Approvals) draft ETAG [7].

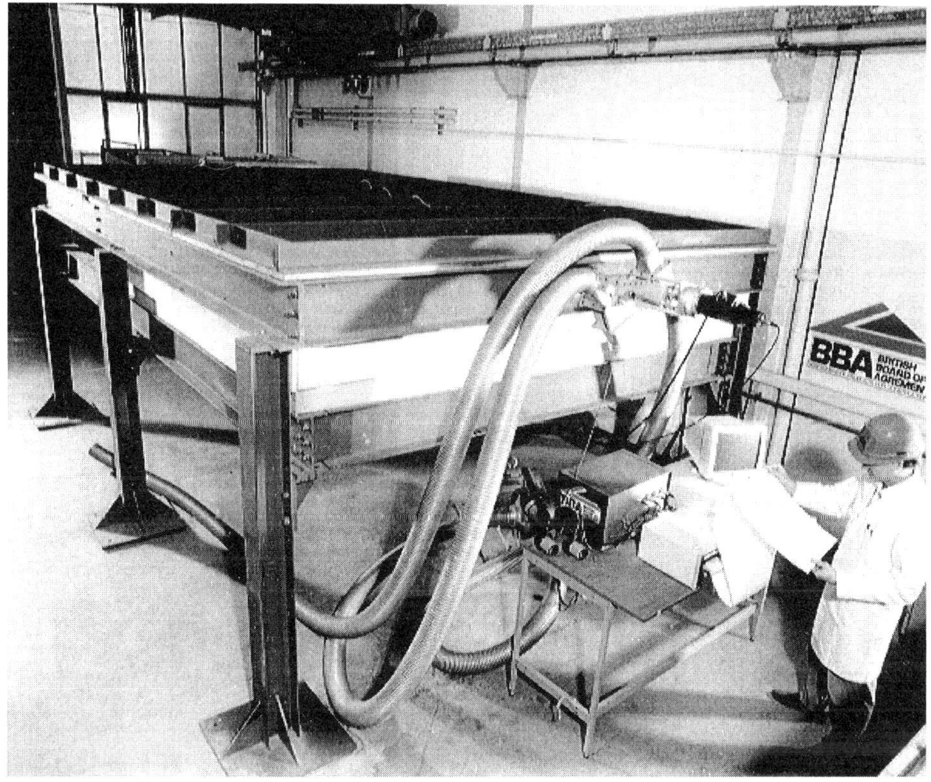

Figure 4 BBA Tempest Wind Loading apparatus

Such equipments are expensive representing considerable capital investment. In this respect some form of national facility having a greater chance of continuous use is to be preferred.

There is scope for international co-operation in the field of special facility provision and testing. This already exists on a limited basis within the UEAtc (European Union of Agrément)

Innovation and Regulation

Regulation generally and building regulations in particular provide a tempering of commercial enthusiasm that may otherwise only be concerned with profit, that on occasion may lead to excesses resulting in the health and safety of the population at large being put at risk. Regulation these days is also concerned with quality, fairness and the expectations of the purchaser/user whilst encouraging the manufacturers, suppliers/contractors to adopt good practices.

It is perhaps easier to realise such objectives by adopting well established routines and specifying products of good pedigree. Such an approach is 'safe' but acts against the use of new products and systems that may offer benefit.

Compliance with a national standard, preferably by a third party/independent accredited body is one way but if the variant product cannot comply with the recognised specification some alternative has to be available. That alternative is Agrément or national/European Technical Approvals.

The concept of approvals has now become worldwide with developing reciprocal arrangements coming into play between the national approvals bodies. The WFTAO (World Federation of Technical Approval Organisations) oversees such global cooperation. Very relevant if one exports as well as trading nationally.

In the UK the link between regulated requirements and national standards/Agrément is via what we call the Approved Documents. Recognition is given to approvals on a deemed to satisfy basis. However, there is a third option of an equivalent alternative. It is relevant to note that mutual recognition between various countries on compliance with Building Regulations has not developed very far. Harmonisation of regulations, even within Europe is more a matter of discussion that practice. Present attempts to evolve performance-based regulations (CIB TG 11) [8] could well ease if not entirely solve this problem.

At the moment Articles 16 and 17 of the CPD [4] require that country of destination requirements can be addressed in the country of manufacture but do not go so far as to accept the role of mutual recognition, as far as regulated requirements are concerned. This omission shows
itself from time to time. For instance, a manufacturer with a CE marked product will deem to satisfy regulated requirements in all member states of the EU*. The specification (with attestation) against which the CE mark is given, be it on the harmonised standards or ETAs, should therefore meet the regulated requirements in all member states. To achieve this tends to add requirements to the specification resulting in a 'maximalist' approach that becomes progressively more expensive. The alternative is to have a 'minimalist' approach keeping costs of compliance low and also not increasing the regulated level in any member country. This is the lowest common factor approach but has the disadvantage of not necessarily satisfying the needs in some member states, that in turn can lead to arbitrary additional requirements resulting in possible technical barriers to trade. The very objective the CPD was designed to prevent.

Innovative products, it may be argued should be subject to more scrutiny than those that are established and can be covered by standards. There is some truth in this view and would justify a more maximalist approach to approvals. If that was the agreed position it would have to be accepted by all parties and regarded as appropriate by manufacturers and their collective trade body representatives so that the technical underwriting of innovative products by way of approvals, at whatever level, was accepted.

The support for technical approvals should be in proportion to that for standards making if innovation is to be encouraged and exploited.

Regulators and specifiers should encourage responsible innovation and therefore also encourage the use and development of Technical Approvals and Agrément.

A balance has to be struck between caution where there is risk and yet recognising the continuum between idea/innovation, approvals, regulation and ultimately standardisation. This sequence is shown in Figure 5[9].

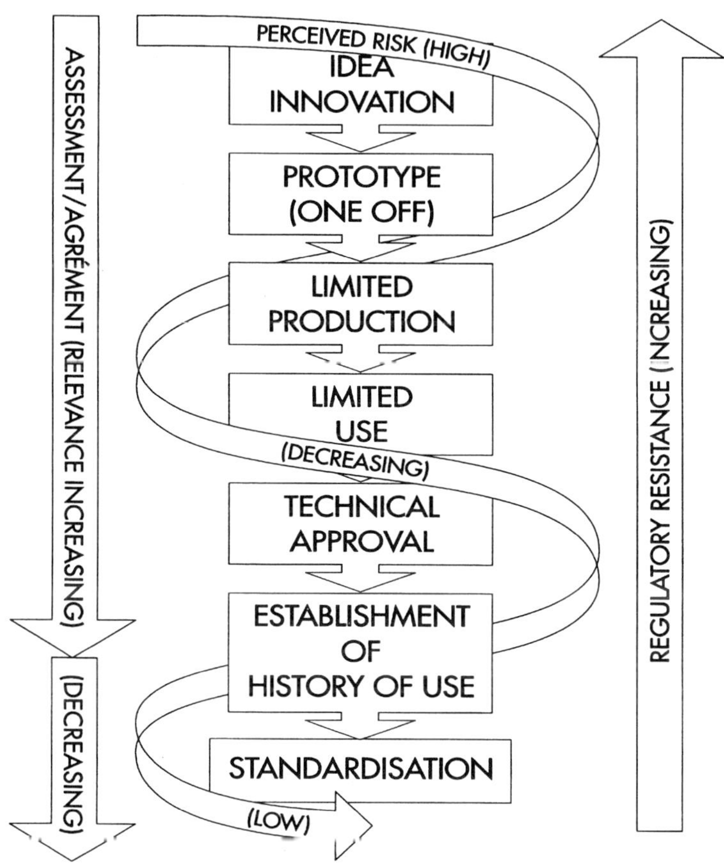

Figure 5 Perception of risk from innovation to established use

Innovation is both inevitable and desirable. It should be encouraged and the products of innovation used so long as their fitness for purpose has been established.

Innovation has its roots in the science of building pathology and functional performance. Only if we understand can we innovate in an objective and logical way. Innovation can, however, be confused with invention. The latter is unpredictable and haphazard originating from flair and practical awareness rather than scientific understanding. However, in the building and construction sectors invention can be significant. Even so, the results of invention should still be subject to rigorous performance assessment.

At the present time there are no clear and agreed criteria by which organisations may be recognised as competent to compile a performance-based assessment specification and deliver the approval against it. There are however such criteria for approved testing, inspection and certification bodies at least in the European Union [10]. This situation may have to be considered during the next decade.

CONCLUSIONS

1. Innovative building products are essential for a vigorous and competent building and construction activity. They should not be regarded with suspicion and considered second best if fitness for purpose has been established by a credible third party assessment/approvals body.

2. Many countries are fortunate in having this service available to its product manufacturing industry or if not already established has the latent capability to create it. This service should be used to the full ultimately carrying forward and helping to create better national and international standards. It may also offer a prospect in demonstrating regulatory compliance one country to another.

3. Only in this way will innovative concrete products be readily used justifying the expenditure on the research and development to begin with.

REFERENCES

1. BERNSTEIN, H M and LEMER, A C. “Solving the Innovation Puzzle” ASCE Press, New York 1996 pp 127

2. BULLETIN OF THE EUROPEAN UNION SUPPLEMENT. “Green Paper on Innovation” 5/95, European Commission, 1996, pp 102

3. LANGRION, J , GIBBONS M , EVANS W G and JERONS, F R. “Wealth from knowledge’ a study of innovation in industry, Macmillan 1972, pp 477

4. The Construction Products Directive 89/106/EEC, Official Journal of the European Communities, 21 December 1988

5. BRITISH STANDARDS INSTITUTION British Standard 874: Part 3: Section 3.1, 1987,

6. BRITISH BOARD OF AGRÉMENT:
BBA Method Of Assessment And Testing, No 50, ‘UEAtc Technical Guideline for the Assessment of thermal insulation systems intended for supporting waterproof coverings on flat and sloping roofs’. November 1992,
BBA MOAT No 55, Supplementary Guide for the assessment of mechanically fastened roof waterproofing systems’ April 1991

7. EUROPEAN TECHNICAL APPROVAL GUIDELINE 'Guideline for Technical Approval for external thermal insulation composite systems with rendering', private communication EOTA (not published)

8. CIB TG11 'Performance-based building codes' May 1998. Prepared by The Inter-jurisdictional Regulatory Collaboration Committee.

9. HEWLETT, P C and DE NORMANN, J. 'Innovation and related matters: Technical Approvals and Standards'. BSI Committee B/-/2 Working Group report (unpublished document) August 1995

10. EUROPEAN STANDARDS ORGANISATION:
EN45001: 1989 "General criteria for the operation of testing laboratories"
EN45004: 1995 "General criteria for the operation of various types of bodies performing inspection"
EN45011: 1989 "General criteria for certification bodies operating product certification"
EN45012: 1989 "General criteria for certification bodies operating quality system certification"

EUROCODE 2 - INNOVATIONS IN DESIGN AND CONSTRUCTION OF CONCRETE STRUCTURES

H-U Litzner

German Concrete Society

Germany

ABSTRACT. The Structural Eurocodes of the European Communities establish requirements for building and civil engineering works in terms of reliability, adequate performance in service conditions and durability. For the achievement of durability, several steps are necessary in the design process. They are subject of Eurocode 2 and the European Standard EN 206 for concrete technology. The basic elements of this integrated design concept are described in the present paper.

Keywords: European Standards for the Design of Concrete Structures, Design Working Life, Durability, Environmental Actions, Concrete Technology, Cover to Reinforcement, Crack control, Execution, Curing of concrete.

Dr H-U Litzner is managing director of the German Concrete Society. He specialises in design of concrete structures and concrete technology. Since 1980, he is involved in the Eurocode programme. Since 1990, he is Chairman of CEN/TC250/SC2 which is responsible for Eurocode 2 "Design of Concrete Structures“.

STRUCTURAL EUROCODES AND THEIR OBJECTIVES

For the realisation of the European single market, the Commission of the European Communities (CEC) has initiated the work of establishing a set of unified technical rules for the design of building and civil engineering works which will gradually replace the different rules in force in the various EC-Member States. These technical rules which became known as the *Structural Eurocodes* shall lead to structures which fulfil the following fundamental requirements established in [1]:

1. With acceptable probability, they will remain fit for the use for which they are required, having due regard to their intended life and their cost.

2. With appropriate degrees of reliability, they will sustain all actions and influences likely to occur during execution and use and have adequate durability in relation to maintenance costs.

In other words, the fundamental requirements which shall be met are adequate *performance* in use, appropriate degree of *reliability* and adequate *durability* during the design working life (Table 1). The relationship between durability and *maintenance costs* should be noted.

The *Structural Eurocodes* provide the technical tools for the achievement of these requirements. The corresponding elements of the design concept are described in the following. They are related to Classes 3 and 4 in Table 1, where the design working life is defined as follows [1]:

"... The design working life is the assumed period for which a structure is to be used for its intended purpose with anticipated maintenance but without major repair being necessary.“

Table 1 Indication of the design working life required in [1]

CLASS	REQUIRED DESIGN WORKING LIFE IN YEARS	EXAMPLE
1	1-5	Temporary structures
2	25	Replaceable structural parts, e.g. gantry girders, bearings
3	50	Building structures and other common structures
4	100	Monumental building structures, bridges and other civil engineering structures

EUROPEAN STANDARDS SYSTEM FOR CONCRETE STRUCTURES

Figure 1 presents the actual European Standards System for building and civil engineering works in concrete, which consists mainly of European Prestandards (ENV). They are actually converted to European Standards, which will replace the corresponding national standards in force in the various CE-Member States.

In this European Standards System which provides all elements for structural and durability design four levels can be distinguished:

- Level 1 comprises standards for structural safety [1] and actions on structures; in particular, in [1] basic durability requirements are established.
- Level 2 consists of Eurocode 2 [2] for the design and detailing of concrete structures.
- Level 3 gives data for structural materials, in particular for concrete [3], [4], and the execution of concrete structures [5].
- Level 4 consists of standards for the testing of materials.

It should be noted, however, that the standards shown in Figure 1 will only lead to the performance required in [1], [2], if they are applied in appropriate combination.

VERIFICATION OF THE DURABILITY OF CONCRETE STRUCTURES

According to [1], [2] it shall be verified that a concrete structure satisfies the following condition:

$$S_d \leq R_d \qquad (1)$$

where:

S_d denotes the design value of an action effect

R_d is the corresponding design resistance associating all structural properties with the respective design values.

Equation (1) has initially been derived for *direct* actions such as, for example, permanent or imposed loads, and for *indirect* actions such as, for example, imposed constraint or deformation (e.g. due to hydration heat).

In the context of [1], [2] and [4], the format given by Equation (1) can also be used for environmental actions ([6] and Figure. 2). For example, R_d can be interpreted as the actual concrete cover c and the corresponding design value S_d as the carbonation depth. This will be illustrated in a later section.

Such methods for durability design require, however, well defined performance characteristics, precise tests methods, reliable models for material behaviour and good knowledge of the environmental conditions. The values of S_d are strongly related to climatic parameters and to other deterioration factors as described later.

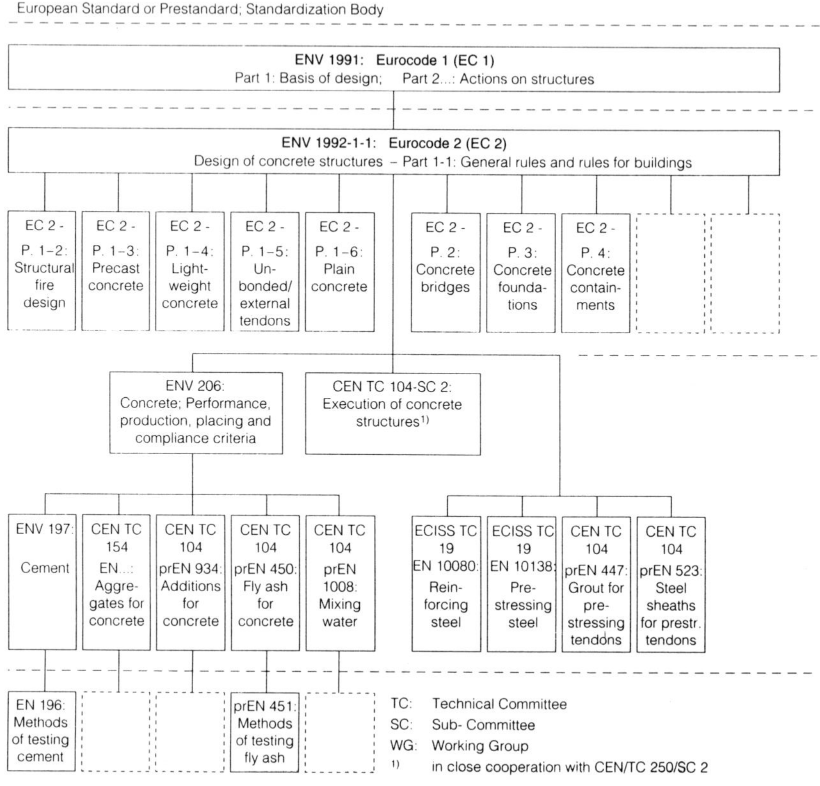

Figure 1 Structure of the actual European Standards System for building and civil engineering works in concrete

The design resistance R_d of concrete structures depends on several parameters. The most important of them are:

- Permeability and structure of the concrete
- Crack pattern and crack width
- Type of reinforcement (steel reinforcement, prestressing steel)
- Cover to reinforcement
- Quality of workmanship during execution (e.g. curing)

The possible influence of those parameters on R_d will be considered in the subsequent sections.

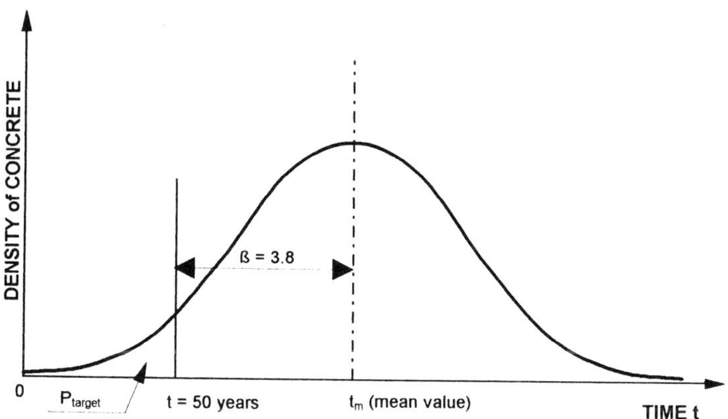

Figure 2 Example of a service life distribution with a reliability index $\beta = 3.8$ and an intended service life of $t = 50$ years [6]. The probability p_f of a failure (i.e. $S_d > R_d$) during the intended lifetime of 50 years shall be smaller than an assumed target probability ($p_{f,\,tar} = 7 \cdot 10^{-5}$ in [1])

DEFINITION OF THE ENVIRONMENTAL ACTIONS S_d

According to [1], a concrete structure shall be designed in such a way that deterioration of concrete and/or steel (see Figure 3) should not impair the durability and performance of the structure, having due regard to the anticipated level of maintenance. In other words, an adequate maintenance strategy is part of the design concept of the Structural Eurocodes.

The above requirements to be met by concrete structures depend mainly on the environment to which the concrete is exposed. Environment in this context implies chemical and physical actions resulting in effects (see Figure 3) which are not considered as loads in structural design. The environmental actions defined in [4] are shown in Tables 2 and 3 where rough distinction is made between six deterioration mechanisms for concrete and steel respectively

The actions in Tables 2 and 3 may, where relevant, be considered as *local* or *micro* conditions. Local conditions are those around the structure after having been built, taking into account the specific actions where the structure or the structural element is located (e.g. realitve humidity RH, CO_2-content).

However, in some circumstances, micro conditions need to be considered. These denote environmental actions on a specific surface of a structural element. This may, for example, apply to the following circumstances:

- Exposition to driving rain
- Exposition to sun radiation
- Contact with earth, ground water, sea water etc.

Table 2 Exposure classes defined in [4]

DETERIORATION MECHANISM	CLASS DESIGNATION	DESCRIPTION OF THE ENVIRONMENT	INFORMATIVE EXAMPLES WHERE EXPOSURE CLASSES MAY OCCUR
1 No risk of corrosion or attack	XO	Very dry	Concrete inside buildings with very low humidity (RH < 45 %)
2 Steel corrosion induced by carbonation	XC1	Dry	Concrete inside buildings with low humidity (RH < 65 %)
	XC2	Wet, rarely dry	Parts of water retaining structures, many foundations
	XC3	Moderate humidity (RH < 80 %)	Concrete inside buildings with moderate or high air RH; external concrete sheltered from rain
	XC4	Cyclic wet and dry	Surfaces subject to water contact, not within class XC2
3 Steel corrosion induced by chlorides	XD1	Moderate humidity	Concrete surfaces exposed to direct spray containing chlorides
	XD2	Wet, rarely dry	Swimming pools; concrete exposed to industrial water containing chlorides
	XD3	Cyclic wet and dry	Parts of bridges; pavements; car park slabs
4 Steel corrosion induced by chlorides from sea water	XS1	Exposed to airborne salt, not in direct contact with sea water	Structures near to or on the coast
	XS2	Submerged	Parts of marine structures
	XS3	Tidal, splash and spray zones	Parts of marine structures
5 Freeze/thaw attack on concrete	XF1	Moderate water saturation, no de-icing agents	Vertical concrete surfaces exposed to rain and freezing
	XF2	Moderate water saturation, with de-icing agents	Vertical concrete surfaces of road structures exposed to freezing and airborne de-icing agents
	XF3	High water saturation, no de-icing agents	Horizontal concrete surfaces exposed to rain and freezing
	XF 4	High water saturation, with de-icing agents	Road and bridge decks exposed to de-icing agents and vertical concrete surfaces exposed to direct spray containing de-icing agents and freezing
6 Chemical attack on concrete	XA1, XA2, XA3	Aggressive chemical environment	See Table 3

It becomes evident that the approach in the design for durability is similar to structural design where normally *global* verification (e.g. structural analysis) and *local* checks (e.g. stress limitation) are performed.

Table 3 Limiting values for exposure classes XA for chemical attack in [4]

CHEMICAL CHARACTERISTIC	XA1	XA2	XA3	TEST METHOD
SO_4^{2-} mg/l in water	≥ 200 and ≤ 600	> 600 and ≤ 3000	> 3000 and ≤ 6000	EN 196-2[7]
SO_4^{2-} mg/kg in soil[1] total amount	≥ 2000 and ≤ 3000[2]	> 3000[2] and ≤ 12000	> 12000 and ≤ 24000	EN 196-2[3]
ph of water	≤ 6.5 and ≥ 5,5	< 5.5 and ≥ 4.5	< 4.5 and ≥ 4.0	DIN 4030-2 [8]
Acidity of soil	> 20 ° Baumann Gully			DIN 4030-2
CO_2 mg/l aggressive in water	≥ 15 and ≤ 40	> 40 and ≤ 100	> 100	
$_4NH^+$ mg/l in water	≥ 15 and ≤ 30	> 30 and ≤ 60	> 60 and ≤ 100	ISO 7150-1[9] ISO 7150-2[10]
Mg^{2+} mg/l in water	≥ 300 and ≤ 1000	> 1000 and ≤ 3000	< 3000	ISO 7980 [11]

Footnotes:

1. Clay soils with a permeability below 10^5 m/s may be moved into a lower class.
2. The 3000 mg/kg limit is reduced to 2000 mg/kg, where there is a risk of accumulation of sulphate ions in the concrete due to drying and wetting cycles or capillary suction.
3. The test method prescribes the extraction of SO_4^{2-} by hydrochloric acid; alternatively, water extraction may be used, if experience is available in the place of use of the concrete.

Figure 3 Deterioration of concrete due to environmental actions

RESISTANCE OF CONCRETE AGAINST ENVIRONMENTAL ACTIONS

EN 206 [4] provides two general methods for the assessment of the design resistance R_d in Equation (1). The standard method ("macro-level design" in [6]) consists of the provision of limiting values for concrete composition in terms of maximum water/cement-ratio, minimum cement or air content and, where relevant, of additional requirements for cement and/or aggregates (see Table 4). Alternatively, performance - related methods with respect to durability may be used. These may be based on refinements of the standard method, on approved and proven tests or on analytical models.

When using the latter approach, for example, for environmental class XC in Table 4, i.e. deterioration of steel by carbonation , it shall be verified that

$$d_c \leq c_{act}, \qquad (2)$$

where d_c denotes the carbonation depth and c_{act} the actual cover to reinforcement of the member considered.

From Equation (3) it can be concluded that d_c - or, in more general terms, the physical model - depends on several parameters with a statistical distribution. These parameters may be characterized as follows:

- the standard deviation of their statistical distribution is high;
- they are significantly time - dependent;
- the „scale effect" should be considered, i.e. it should be checked whether results from lab tests can be applied in insitu conditions.

It should be noted, however, that the actual development in the field of durability is characterized by world-wide activities with the objective to define design values of all relevant parameters which can be introduced in the verification format described above (see Figure 5 and the references in [6]).

The carbonation depth d_c may be calculated from:

$$d_c = \sqrt{2 \cdot k_1 \cdot k_2 \cdot k_3 \cdot \Delta_{con}} \quad \sqrt{\frac{D_{nom}}{a} \quad t} \quad \left(\frac{t_0}{t}\right)^n \leq c_{act} \qquad (3)$$

where:

d_c the carbonation depth

D_{nom} the diffusion coefficient of dry concrete for carbon dioxide in defined environment (20°C, 65 % rel. humidity)

a the amount of CO_2 for complete carbonation

Δ_{con} the concentration difference of carbon dioxide at the carbonation front and in the air

k_1, k_2, k_3 parameters for micro climatic conditions, to describe curing conditions and the effect of water separation (local w/c-ratio) respectively

n parameter for micro climatic conditions, describing wetting and drying

$n = 0$ for interior conditions; $n \leq 0.3$ for outdoor conditions

t_0 reference period, $\sqrt{t}$ -law valid (e.g. 1 year)

t time

c_{act} actual cover to reinforcement

Table 4 Limiting values for composition and properties of concrete made with cement type CEM I conforming to ENV 1997-1 [12]

EXPOSURE CLASS	LIMITING VALUES FOR CONCRETE COMPOSITION			
	w/c-Ratio	Strength Class	Cement Content kg/m³	Air Content [%]
XO	—	C12/15	—	—
XC1	0,65	C20/25	260	—
XC2	0,60	C25/30	280	—
XC3	0,55	C30/37	280	—
XC4	0,50	C30/37	300	—
XS1	0,50	C30/37	300	—
XS2	0,45	C35/45	320	—
XS3	0,45	C35/45	340	—
XD1	0,55	C30/37	300	—
XD2	0,55	C30/37	300	—
XD3	0,45	C35/45	320	—
XF1	0,55	C30/37	300	— 1),2)
XF2	0,55	C25/30	300	4,0 1),2)
XF3	0,50	C30/37	320	4,0 1),2)
XF4	0,45	C30/37	340	4,0 1),2)
XA1	0,55	C30/37	300	—
XA2	0,50	C30/37	320	— 2)
XA3	0,45	C35/45	360	— 2)

1) freeze / thaw resisting aggregates

2) Sulfate resisting cement

CONCRETE COVER TO REINFORCEMENT

According to Eurocode 2 [2], a *nominal* concrete cover to reinforcement shall be introduced in the design calculations. It is given by:

$$\text{nom c} = \text{min c} + \Delta h \tag{4}$$

where:

nom c	denotes the nominal cover
min c	is the minimum cover
Δh	is an allowance for tolerances.

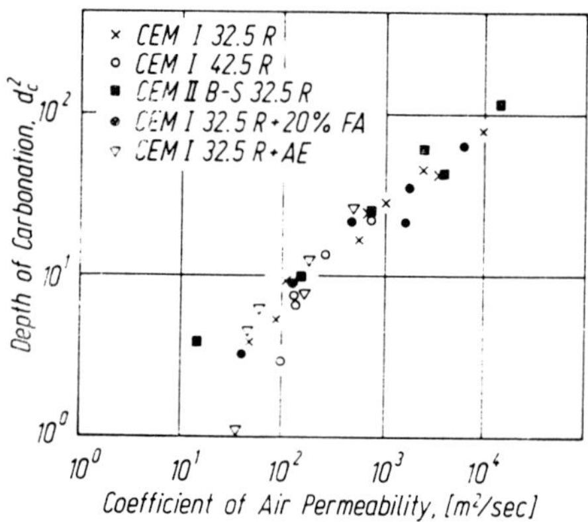

Figure 4 Depth of carbonation of concrete after 1 year of storage at 20°C, 65 % RH, as a function of the air permeability coefficient at an age of 56 days for concretes made of various types of cements [13]

For the determination of the *minimum* concrete cover, *min c*, the following criteria apply:

- Safe transmission of bond forces
- Avoidance of spalling
- Adequate fire resistance
- The protection of the steel against corrosion

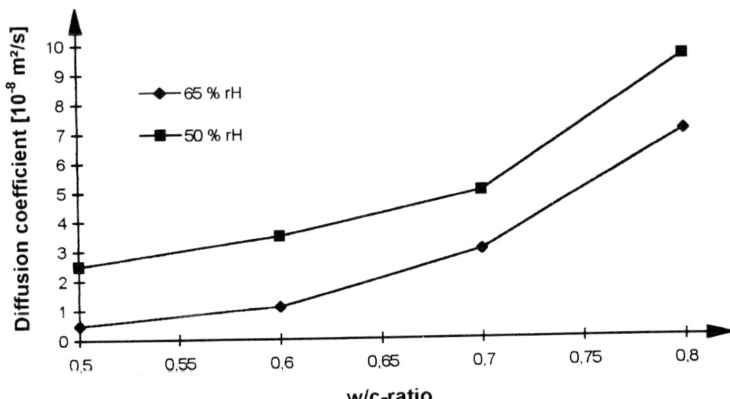

Figure 5 Example for the diffusion coefficient of concrete as function of the w/c-ratio for two relative humidities

In the latter case, the protection against corrosion depends upon the continuing presence of a surrounding alkaline environment provided by an adequate thickness of good quality, well-cured concrete (see later section). In the absence of other provisions, adequate thickness may be assumed if the values of *min c* given in Table 5 are used [15]. Except for exposure class XC1 they may be reduced by 5 mm for slab elements.

A further point concerns the trade-off between *min c* and the actual concrete grade used (see Table 6). Generally, a reduction of the values in Table 5 is allowed provided that the actual concrete grade is higher than the minimum grade given in Table 4 for the relevant exposure class. For example, in [15] which is based on [2], the trade-off is restricted to prefabricated concrete elements and a reduction of 5 mm of the values in Table 5 is permitted for normal strength concrete ($f_{ck,cylinder} \leq 55$ N/mm^2) in the following cases:

- Exposure Class XC1 : no reduction permitted
- Classes XC2, XC3 : reduction permitted if the actual concrete grade corresponds to the minimum strength class in Table 4
- Class XC4 : reduction permitted for C35/45 and higher
- Classes XS, XD : reduction permitted if the concrete grade is at least two strength classes higher (see Table 6) than the minimum strength class in Table 4

Table 5 Minimum cover, *minc*, to reinforcement for corrosion protection [mm]

TYPE OF STEEL	EXPOSURE CLASS IN RELATION TO STEEL CORROSION									
	XC1	XC2	XC3	XC4	XS1	XS2	XS3	XD1	XD2	XD3
Reinforcing steel	10	20	20	25	40	40	40	40	40	40
Prestressing steel	20	35	35	35	50	50	50	50	50	50

For high strength concrete (see Table 6), additional consideration may be necessary. The allowance for tolerances, Δh, will usually be in the range of 0 mm < Δh < 5 mm for precast concrete elements if the production control can guarantee these values. The allowance will be in the range of 5 mm < Δh < 10 mm for in-situ reinforced concrete construction. Higher or lower tolerances, Δh, may be used if this can be justified by the construction method used. These rules underline that for durability design a close relationship between concrete technology and construction on site exists.

CONTROL OF CRACKING

The durability of concrete structures may adversely be affected by excessive cracking. Besides that, cracking shall be limited to a level that will not impair the proper functioning of the structure or cause its appearance to be unacceptable.

Cracks in concrete structures have two primary causes:

- Cracks caused by the rheological properties of the fresh or hardening concrete
- Cracks caused by loading and/or imposed deformations

The first type of cracks can be controlled by appropriate measures of concrete technology, in particular by the composition of the concrete mix, proper placing and curing. Corresponding rules are provided in [3 to 5]

For the control of cracks caused by loading and/or imposed deformation, the design concept in Eurocode 2 provides two basic tools:

- The requirement of a minimum bonded steel reinforcement
- The limitation of crack width

The minimum steel reinforcement has two functions: it should ensure an equilibrium at the time when cracks may first be expected. In addition, the area of the minimum reinforcement should be such that crack widths with an unacceptable value are avoided. In most cases, the minimum reinforcement is calculated for imposed deformations due to the dissipation of the hydration heat, i.e. for a concrete age between 3 to 5 days after casting. It depends mainly on the actual concrete tensile strength, f_{ct}

Table 6 Strength classes for normal-weight and heavy-weight concrete in prEN 206 [4]

STRENGTH CLASS	$f_{ck, cylinder}$ [N/mm²]	$f_{ck, cube}$ [N/mm²]	DEFINITION
C12/15	12	15	Normal strength concrete
C16/20	16	20	
C20/25	20	25	
C25/30	25	30	
C30/37	30	37	
C35/45	35	45	
C40/50	40	50	
C45/55	45	55	
C50/60	50	60	
C55/67	55	67	
C60/75	60	75	High strength cconcrete
C70/85	70	85	
C80/95	80	95	
C90/105	90	105	
C100/115	100	115	

For the limitation of crack width, Eurocode 2 provides a classification of verification criteria which is presented in Table 7. The principle is that for a certain load level (infrequent, frequent, quasi-permanent) either the limit state of decompression or the limit state of crack width shall not be exceeded. The load levels in Table 7 have been derived from investigations and from experience. At the limit state of decompression, no tensile stresses in the concrete are allowed under the relevant combination of actions. It is relevant for prestressed members or for those subjected to significant compression forces.

For crack width control, the deemed to satisfy criteria given in Table 7 are based on the following design crack width w_k:

- For members with internal bonded prestressing tendons: w_k = 0.2 mm
- For members with reinforcing steel: w_k = 0.3 mm.

The categories A to E in Table 7 are chosen in relation to the environmental actions, the risk of deterioration and the design load level. The latter is mainly a function of the variable load, Q_k, which is defined in [1] as an upper value of a statistical distribution with an intended probability of 98 % of not being exceeded within a reference period of one year.

Values of Q_k can be found in Eurocode 1 (see Figure 1). The infrequent value of a variable action corresponds (approximately) to Q_k.

Table 7 Classification of verification criteria

CATEGORY	COMBINATION OF ACTIONS FOR THE VERIFICATION OF	
	Decompression	Crack width
A	infrequent	—
B	frequent	infrequent
C	quasi-permanent	frequent
D	—	frequent
E	—	quasi-permanent

The frequent value of a variable action, $\Psi_1 \cdot Q_k$, corresponds according to [1] to a value which is exceeded either 5 % of the reference time or 300 times per year. The highest value should be chosen. The corresponding combination coefficient, Ψ_1, will vary between 0.5 and 0.9, according to the variable action considered.

The quasi-permanent value of a variable action, $\Psi_2 \cdot Q_k$, which is commonly used for crack control in reinforced (i.e. not prestressed) concrete members, corresponds to the time average or to the value with a probability of being exceeded of 50 %. The corresponding value Ψ_2 will vary between 0.3 and 0.8.

Besides the above provisions which require numerical checks, a couple of deemed to satisfy rules are provided in Eurocode 2. They concern bar spacing, spacing of stirrups for the control of inclined cracks due to shear and/or torsion, a limitation of bond stresses and minimum reinforcement areas along the surface of concrete members. The latter are intended to resist self-equilibrating stresses and, thus, to ensure an adequate quality of the concrete at the surface of concrete members.

CURING

The durability of a concrete surface zone depends on several parameters, in particular on an adequate resistance against carbonation and a low-permeability. Both of these two parameters are a function of the degree of hydration which is a time-dependent process which depends mainly on the type of binder (cement) and on the ambient temperature during reaction. On the other hand, the strength development of the concrete is also a function of the hydration. An analogy between permeability *decrease* and strength *increase* exists. For this reason and in order to provide a practice-oriented engineering model, the concrete strength was used in [5] for the characterization of the degree of hydration after the curing period.

Therefore, according to [5] a concrete surface exposed to environmental conditions other than XO and XC1 in Table 2 shall be cured until the surface has achieved a certain percentage of the specified compressive strength. This percentage depends mainly on the moisture supply *after* curing, i.e. on the continuation of the hydration process after the curing period. In [16], corresponding values are given which vary between 60 % (very dry environment) and 10 % (humid environment). In [5], however, a constant percentage of 50 % is required for reasons of simplification.

This value may be considered to be achieved if the minimum curing periods recommended in Table 8 are applied. The Table applies to all environmental conditions in Table 4 other than XO and XC1 and distinguishes between different ambient temperatures t and the strength development of the concrete. This development is expressed by the ratio

$$r = f_{cm,2} / f_{cm,28} \qquad (5)$$

where:

$f_{cm,2}$ is the mean value of compressive strength after 2 days

$f_{cm,28}$ denotes the mean value of compressive strength after 28 days determined from initial tests or based on known performance of concrete of comparable composition.

Table 8 Minimum curing period for EN 206 exposure classes other than XO and XC1

SURFACE CONCRETE TEMPERATURE (t), °C	MINIMUM CURING PERIOD, DAYS [1),2)] FOR A CONCRETE STRENGTH DEVELOPMENT $r = f_{cm,2}/f_{cm,28}$			
	$r \geq 0.50$ rapid	$r \geq 0.30$ medium	$r \geq 0.15$ slow	$r < 0.15$ very slow
$t \geq 25$	1.0	1.5	2.0	3.0
$25 > t \geq 15$	1.0	2.0	3.0	5
$15 > t \geq 10$	2.0	4.0	7	10
$10 > t \geq 5$[3)]	3.0	6	10	15

Note:
1. Plus any period of set exceeding 5 hours.
2. Linear interpolation between values in the rows is acceptable.
3. For temperatures below 5°C, the duration should be extended for a period equal to the time below 5°C.

For concrete surfaces to be exposed to exposure classes XO and XC1 in Table 2 only, the minimum curing period should be 0.5 day, provided the set does not exceed 5 hours and the surface temperature is equal to or above 5°C.

However, [5] does not exclude the application of other minimum curing periods provided that this can be justified by the conrete used and the curing method applied.

REFERENCES

1. EUROPEAN COMMITTEE FOR STANDARDIZATION (CEN). (1994) Eurocode 1: Basis of Design and Actions on Structures. Part 1: Basis of Design. CEN, Brussels. ENV 1991-1.

2. EUROPEAN COMMITTEE FOR STANDARDIZATION (CEN). (1991) Eurocode 2: Design of Concrete Structures. Part 1: General Rules and Rules for Buildings. European Prestandard. CEN, Brussels. ENV 1992-1-1.

3. EUROPEAN COMMITTEE FOR STANDARDIZATION (CEN). (1990) Concrete. Performance, Production, Placing and Compliance Criteria. European Prestandard. CEN, Brussels. ENV 206.

4. EUROPEAN COMMITTEE FOR STANDARDIZATION (CEN). (1997) Concrete. Performance, Production and Conformity. Draft. CEN, Brussels. prEN 206.

5. EUROPEAN COMMITTEE FOR STANDARDIZATION (CEN). (1998) Execution of Concrete Structures. Part 1: General Rules and Rules for Buildings. CEN, Brussels. Document CEN/TC104/SC2-N126.

6. COMITÉ EURO-INTERNATIONAL DU BÉTON (CEB). (1997) New Approach to Durability Design. An example for carbonation induced corrosion. CEB, Lausanne. CEB-Bulletin d'Information No. 238.

7. EUROPEAN COMMITTEE FOR STANDARDIZATION (CEN). (1994) Methods of testing cement - Part 2: Chemical analysis of cement. CEN, Brussels. EN 196-2.

8. DEUTSCHES INSTITUT FÜR NORMUNG E.V. (DIN). (1991) Assessment of water, soil and gases for their aggressiveness to concrete - Part 2: Collection and examination of water and soil samples. DIN, Berlin. DIN 4030-2.

9. INTERNATIONAL STANDARDS ORGANIZATION (ISO). (1984) Water quality; Determination of amonium - Part 1: Manual spectrometric method. ISO, Geneva. ISO 7150-1.

10. INTERNATIONAL STANDARDS ORGANIZATION (ISO). (1986) Water quality; Determination of amonium - Part 2: Automated spectrometric method. ISO, Geneva. ISO 7150-2.

11. INTERNATIONAL STANDARDS ORGANIZATION (ISO). (1986) Water quality; Determination of calcium and magnesium; Atomic absorption spectrometric method. ISO, Geneva. ISO 7980.

12. EUROPEAN COMMITTEE FOR STANDARDIZATION (CEN). (1992) Cement; Composition, specifications and conformity criteria - Part 1: Common cements. CEN, Brussels. ENV 1997-1.

13. HILSDORF, H.K. (1995) Concrete, published in Concrete Structures - Euro-Design Handbook, Ernst & Sohn, Berlin, pp. 1-103.

14. VISSERS, J.L.J. (1998) k-value for powder coal fly ash. Brussels, Internal paper of CEN/TC104/SC1.

15. DEUTSCHES INSTITUT FÜR NORMUNG E.V. (DIN). (1997) Concrete, reinforced and prestressed concrete structures - Part 1: Design. DIN, Berlin. prDIN 1045-1.

16. GRÜBL, P. (1996) European Concept on the Curing of Concrete; published in Concrete Precasting Plant and Technology, Vol. 62, Bauverlag GmbH, Wiesbaden, pp. 82-91.

THEME ONE:

CIVIL ENGINEERING STRUCTURES

EVALUATION OF A REINFORCED CONCRETE ROAD USING SURFACE STRESS WAVES

W Haegeman

E De Winne

Ghent University

Belgium

ABSTRACT. The Spectral-Analysis-of-Surface-Waves (SASW) method is a non-destructive seismic method which has been used in situ to determine the elastic moduli of soils and pavements at low level of strain and the variation of these moduli with depth. The test is based on the dispersion of Rayleigh-waves which means that Rayleigh waves of different wavelengths propagate at relatively different depths. If the medium of propagation is vertically inhomogeneous, then the different wavelengths propagate at different phase velocities. This variation of phase velocity with wavelength is called a dispersion curve and is related to the structural stiffness of the medium of wave propagation. This paper presents the results of an evaluation of the stiffness of every construction phase of a reinforced concrete road, starting with the embankment, surface compaction, foundation and top layers.

Keywords: Reinforced concrete road, Non-destructive testing, Rayleigh-waves, Stiffness parameters.

Dr Wim Haegeman is a research/teaching fellow at the laboratory for soil mechanics of the Ghent University, Belgium. His main research topics are small strain parameters of materials and soils and non-destructive (geophysical) testing

Professor Etienne De Winne is head of the road department, Ghent University, Belgium, and Inspector-General at the Road Administration of the Flemish Community. He specialises in the testing, durability and protection of roads.

INTRODUCTION

Seismic methods are most often used today to profile near-surface soils. These methods involve body wave measurements and thus require the installation of one or more boreholes. Borehole installation is generally time consuming and costly. The SASW method, on the other hand, involves measurement of surface waves of the Rayleigh type to evaluate shear wave velocity and shear modulus profiles and can be used on stiffer materials like pavements. In the SASW method, both the source and receivers are placed on the surface. Rayleigh waves are generated by applying vertical loading to the surface. The propagation of these waves along the surface is then monitored. From the velocities of propagation, the stiffness profile of the site is typically calculated through a forward modelling or inversion process.

Any significant stiffness change due to for instance a soil improvement or a construction phase can be directly determined without recourse to empirical correlations. The non-intrusive nature of the SASW test and the fact that it is based on stress wave propagation makes it ideal for evaluating purposes. Besides an introduction about the fundamental principles of the SASW technique, this paper will present the results of an evaluation of the elastic moduli of every layer in a reinforced concrete road during construction.

THE SASW METHOD

Theoretical Background

Surface waves used in the SASW method are the vertically polarised Rayleigh waves. These are seismic waves that travel along the exposed surface of any solid system. These waves have particle motion that decreases with depth into the system.

The depth of wave motion is determined by the wavelength (or frequency) of the wave. Low frequency, hence long wavelength, waves extend deeper into the system than high frequency, hence short wavelength, waves. This property is illustrated in Figure 1.

Rayleigh waves with short wavelengths propagate through the surface layer ; their velocity will only be determined by the properties of that layer. On the other hand, longer wavelength Rayleigh waves propagate through the top several layers, and their velocities will be determined by the combined properties of the layers through which they propagate. Conclusion, in layered media, the velocity of propagation of a surface wave depends on the frequency (or wavelength) of the wave. This variation of velocity with frequency is called the dispersion. Therefore, all layers in the profile can be sampled simply by generating surface waves over a wide range of wavelengths (i.e. a wide range in frequencies) and the velocities will vary with the stiffness and thickness of the layers in the system. The objective in the field testing of the SASW method is to measure this surface wave dispersion.

Surface wave velocity (V_R) of a material is closely related to the shear wave velocity (V_s) of the material.

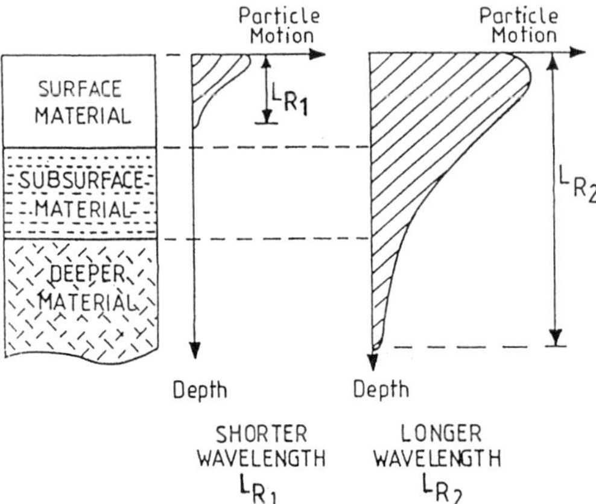

Figure 1 Distribution of vertical particle motion with depth for two surface waves of different wavelengths

The surface wave propagates at a velocity slightly less than the shear wave velocity. The relationship between surface wave velocity and shear wave velocity depends on Poisson's ratio (ν).

For values of Poisson's ratio between 0.1 and 0.3, surface wave velocity can be approximated by :

$$V_R \approx 0.9\ V_s \qquad \text{for } 0.1 < \nu < 0.3$$

From the theory of elasticity, values of the shear modulus can be calculated from shear wave velocity and mass density (ρ) :

$$G = \rho\ V_s^2 \qquad (1)$$

$$E = 2\ G\ (1+\nu) \qquad (2)$$

where G is shear modulus and E is Young's modulus.

Testing Procedure

The general configuration of the source, receivers, and recording equipment is shown in Figure 2. Surface waves are generated by applying a dynamic vertical load to the ground surface. The propagation of these waves along the surface is monitored with two receivers placed at distances of d_1 and d_2 from the source.

The most common types of sources are either simple hammers (small, hand-held hammers or sledge hammers) or dropped weights weighing from 200 to 1500 N.

Electromagnetic vibrators in conjunction with sinusoidal or random input motion can also be used as sources.

A dual channel Fast Fourier Transform (FFT) dynamic signal analyser is used to record and analyse the motions at any two transducers. The ability to calculate transforms rapidly in the field, is an essential part of the SASW method, allowing operators to immediately assess the quality of the data being collected and, if necessary, modify the arrangement of source and receivers or other test parameters accordingly. This data can be easily transferred to a computer for further analysis as desired.

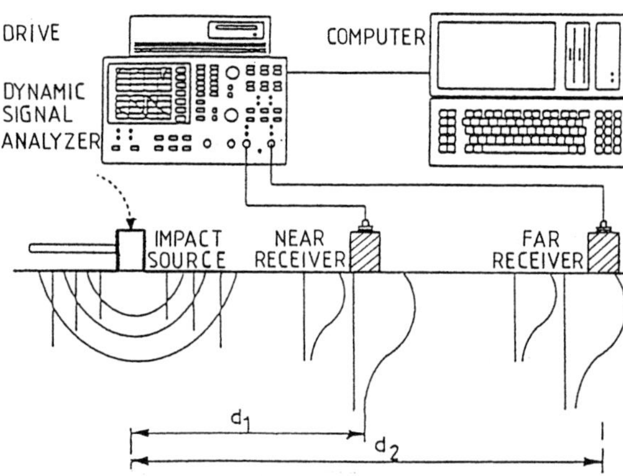

Figure 2 Source-receiver configuration

Analysis Procedure

For each source/receiver spacing, the time histories recorded by the two receivers, x(t) and y(t), are transformed to the frequency domain resulting in the linear spectra of the two signals, X(f) and Y(f). The cross power spectrum of the signals, $G_{xy}(f)$ is then calculated by multiplying Y(f) by the complex conjugate of X(f). In addition to the cross power spectrum, the coherence function and auto power spectrum of each signal are also calculated. It must be emphasized that all of these frequency domain quantities are calculated in real time by the waveform analyzer.

The key data are the phase of the cross power spectrum and the coherence function. The coherence function represents a signal-to-noise ratio and should be nearly one in the range of acceptable data.

The phase of the cross spectrum represents the phase difference of the motion at the two transducers. The surface wave velocity (V_R) and the wavelength (L_R) can be determined from the phase of the cross spectrum ($\theta_{xy}(f)$) using the following expressions :

$$t(f) = \theta_{xy}(f)/2\pi f \qquad (3)$$

where the phase angle is in radians and the frequency, f, is in Hertz. The surface wave phase velocity, V_R, is determined using :

$$V_R(f) = (d_2 - d_1)/t(f) \tag{4}$$

and the corresponding wavelength of the surface wave is calculated from :

$$L_R = V_R/f \tag{5}$$

The result of these calculations is a dispersion curve (V_R versus L_R) for a given receiver spacing. Individual dispersion curves for all receiver spacings are assembled together to form the composite dispersion curve for the site. For a layered system in which stiffness changes with depth, an inversion process is required to obtain the stiffness profile from the measured dispersion curve. This requires that a velocity profile be assumed, and a theoretical dispersion curve be calculated for that profile. The theoretical dispersion curve is then compared to the measured curve, and the assumed profile is adjusted in an attempt to improve the match. This procedure is repeated until the theoretical and measured dispersion curves closely match at which time the assumed profile is taken to represent the stiffness profile in the material system.

Application of inverse theory to surface wave testing has increased the accuracy of resulting wave velocity profiles and has significantly expanded the variety of sites of which the SASW method can be successfully used.

EVALUATION OF THE CONSTRUCTION PHASES OF A ROAD EMBANKMENT

The evaluation of a road construction must ideally be fast to reduce equipment downtime, to produce results in the field for immediate assessment, to be customizable for investigating any zone of interest and to directly measure the road properties as a function of depth without recourse to empirical correlations. The Spectral Analysis of Surface Waves (SASW) is an emerging in-situ testing technique that potentially offers all of the above advantages.

The site test consisted of a 3 kilometres reinforced road construction, included an embankment leading to a bridge. This sand embankment was placed hydraulically with a maximum height of 10 m above the natural soil level.

The profile of the road is shown in Figure 3 and consisted of 20 cm drainage layer, 20 cm lean concrete, 5 cm asphalt and 20 cm reinforced concrete. During construction it was decided to put an extra 4 cm coverlayer of “whisper” asphalt. This complex profile gave us an excellent opportunity to validate the usefulness of the SASW method in every construction phase of this road embankment.

Therefore six measurement profiles were choosen along the road and by repeating the SASW test on this places after every construction of a layer, the stiffness changes of the layers underneath were evaluated.

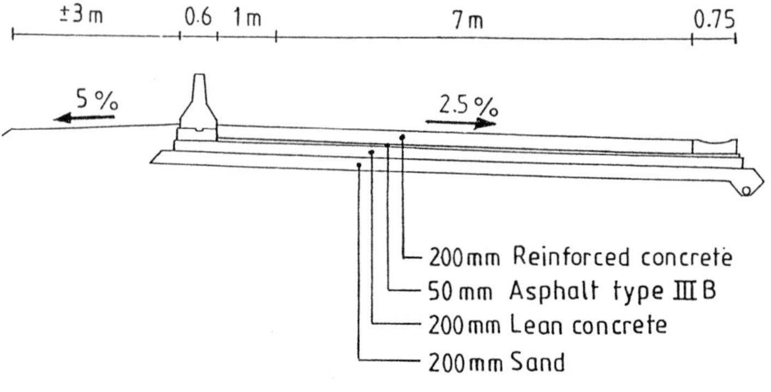

Figure 3 Road profile

On the soil material two seismic geophones were consecutively placed at spacings of 0.5, 1, 2 and 4 m while on the concrete or asphalt material two accelerometers were used at spacing of 0.125, 0.25, 0.5 and 1 m. Elastic stress waves were then generated by the impact of a (sledge) hammer on an equal distance for the first receiver as the corresponding receiver spacing.

For each receiver spacing, five or more signals were averaged in the frequency domain where the phase of the cross-spectrum and coherence were also calculated. The phase delays and receiver spacing were then used to calculated the Rayleigh wave velocities as a function of frequency using equations (3), (4) and (5). The procedure was repeated for all the spacings listed above. At the end of testing, dispersion points from the individual spacings are combined into a single dispersion curve for that well choosen location.

To evaluate the stiffness of the profile, the dynamic shear modulus and young modulus were calculated using equations (1) and (2).

Some of the field results of the SASW tests are shown in Figures 4 to 8. In each case the combined and theoretical dispersion curves are shown together with the corresponding layer thickness and stiffness.

Figure 4 shows the SASW test on the embankment after the hydraulic fill. This is a typical soil profile with a shear wave velocity increasing with depth.

The maximum E modulus of the top layer is 62.9 MPa. On the same spot a plate load test was performed which gave a plate modulus of 25.45 MPa. This result is 30 % of the constrained modulus (E_{oed}) calculated with the SASW test due to the much larger strains which occur in a plate load test and proves the statement made in [1] that a plate modulus is about 1/3 of the SASW oedometer modulus.

Some weeks later a surface compaction was performed on the embankment and the results of this SASW test on the same place are shown in Figure 5. One clearly sees the higher velocity at the surface which gives an elasticity modulus of 229.3 MPa and a compacted layer of 0.4 m thickness.

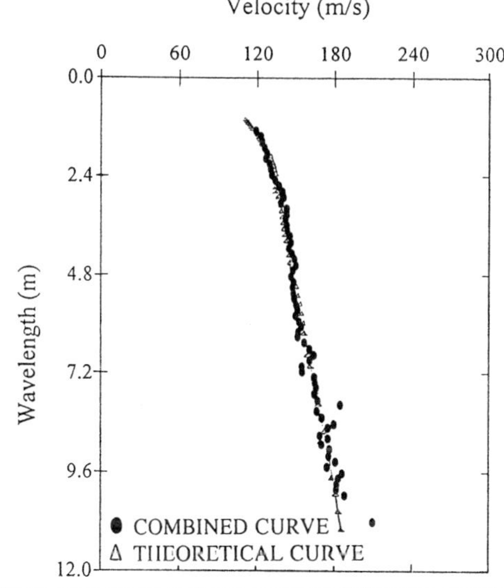

D(m)	Vs(m/s)	ν	ρ(t/m^3)	G(MPa)	E(MPa)	E_{oed}(MPa)
0.5	110	0.3	2	24.2	62.9	84.7
2	160	0.3	2	51.2	133.1	179.2
2	200	0.3	2	80.0	208.0	280.0
	260	0.3	2	135.2	351.5	473.2

Figure 4 SASW test on the embankment after hydraulic fill

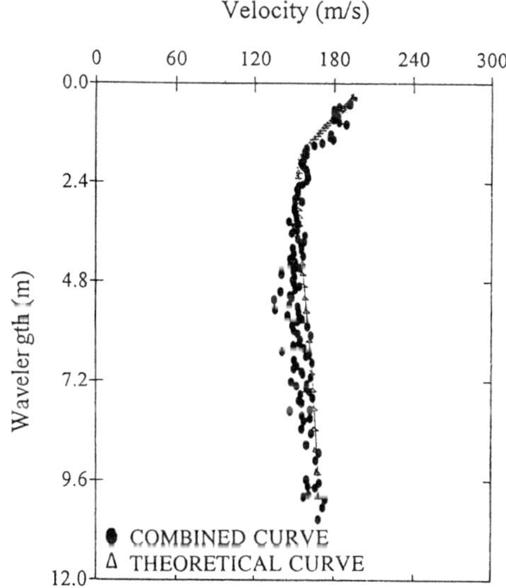

D(m)	Vs(m/s)	ν	ρ(t/m^3)	G(MPa)	E(MPa)	E_{oed}(MPa)
0.4	210	0.3	2	88.2	229.3	308.7
2	155	0.3	2	48.1	124.9	168.2
	190	0.3	2	72.2	187.7	252.7

Figure 5 SASW test on the embankment after surface compaction

The soil underneath this compacted layer became a little bit weaker because of the heavy rainfall in the days before the performing of the test.

The test performed on the partial constructed road on top of the asphalt layer is shown on Figure 6. One sees a 5 cm thick asphalt layer with a maximum elasticity modulus of 18785 MPa on top of a lean concrete layer of 20 cm thickness. The drainage layer and natural soil underneath show a shear wave velocity of 200 m/s. This is higher than the velocity in the natural soil without the road because of the surface compaction and the weight of the road resting on this layers.

In profiles with big differences in stiffness between the layers stiffness contrast peaks may occur in the dispersion curves because of reflecting waves on the layer interfaces. This is seen in Figure 6 at a wavelength of about 25 cm in the theoretical curve but not yet measured in the experimental curve.

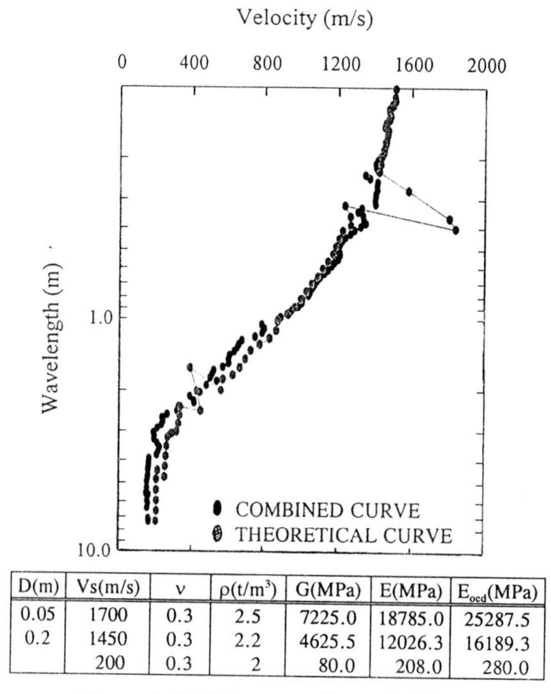

D(m)	Vs(m/s)	ν	ρ(t/m³)	G(MPa)	E(MPa)	E_{oed}(MPa)
0.05	1700	0.3	2.5	7225.0	18785.0	25287.5
0.2	1450	0.3	2.2	4625.5	12026.3	16189.3
	200	0.3	2	80.0	208.0	280.0

Figure 6 SASW test on the asphalt layer

Figure 7 however shows the experimental dispersion curves of a SASW test on the reinforced concrete. At a wavelength of 20 cm one clearly observes the stiffness contrast peak caused by the big difference in stiffness between the reinforced concrete and the asphalt layer. The transition of the lean concrete to the drainage layer is again causing a peak at a wavelength of about 60 cm. Inversion shows that because of the existance of this peak it is difficult to determine the parameters of the this weaker asphalt layer underneath the stiff reinforced concrete.

For the tip and bottom concrete layers however a maximum elasticity modulus of 43740 MPa and 20649 MPa is found. The value of the modulus of the lean concrete increased in comparison with the measurement of Figure 6 because of the overburden of the reinforced concrete.

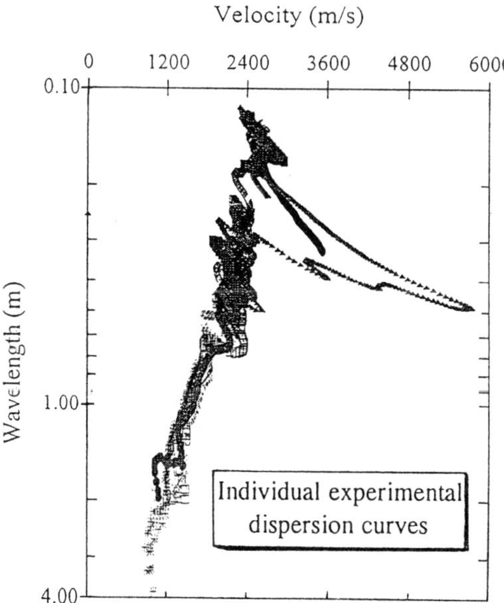

Figure 7 SASW test on the reinforced concrete

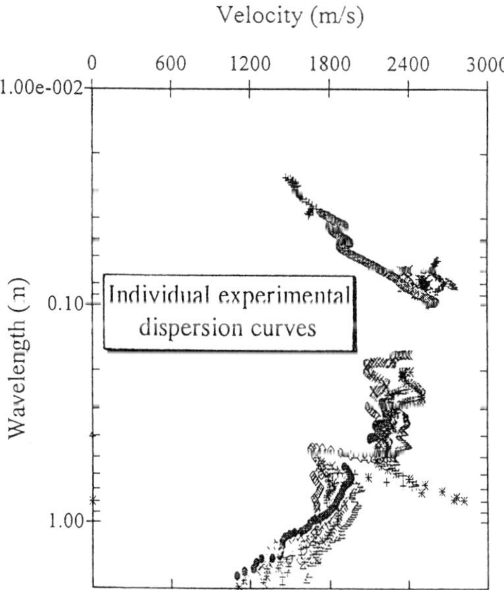

Figure 8 SASW test on the whisper asphalt

Finally Figure 8 is showing the experimental dispersion curves measured after completion of the road. Although the variation of the phase velocity looks rather complex, this Figure is easy to interprete. In the top four centimetres the phase velocity is increasing because of the decrease of the porosity with depth in the whisperasphalt. The phase velocity is ± 2500 m/s in the reinforced concrete and ± 2200 in the lean concrete. Because of elimination of the stiffness contrast peak the curves are showing a gap at a wavelength of 10 to 20 cm. At greater depths the second peak is noticed again and the phase velocity is decreasing fast in the drainage layer and the natural soil profile. The inversion gives us a 20 cm thick reinforced concrete layer (E = 43740 MPa), a 22 cm thick lean concrete (E = 20649 MPa) and a soil deposit with G = 80 MPa. Parallel to this in-situ measurement small scale tests were performed in the laboratory to investigate the influence of a gravelbed, reinforcement, vertical boundaries, cracks...etc. on the wave propagations. Detailed data of these tests can be found in [2].

Finally some Falling Weight Deflectometer (FWD) tests were performed in comparison with SASW tests on the same spot. Moduli determined from the FWD and SASW tests differed significantly for the top layer materials primarily because of the different strain levels at these shallow depths associated with the two testing methods. Moduli of the top layers from the FWD tests were significantly lower. However moduli from the foundation layers and soil compared well for both testing methods since strain levels at these depths can be regarded as equal.

CONCLUSIONS

The SASW method is a valuable tool for evaluation of soil improvement techniques and road constructions. Sample data of this case study presented herein illustrates the testing and analysis procedures at a site with very different material properties and stiffnesses.

Shear wave velocities of each layer of the pavement system are directly converted into a stiffness modulus. The easy handling of the test and the simplicity of the required field equipment moreover assure that the SASW test is quite cost-effective.

Improvements in theoretical analysis and inversion procedures promise to continue to improve the accuracy of the method and allow more automation of the analysis procedures.

REFERENCES

1. HAEGEMAN, W, and VAN IMPE, W F. Developments in SASW Testing, Proceeding XIth ECSMFE, Vol. 1, May 1995, pp. 133-140.

2. HAEGEMAN, W. Spectrale analyse van oppervlaktegolven ter onderkenning van de dynamische stijfheidskarakteristieken van gelaagde media, PhD, May 1996, Ghent University, Belgium.

RAPID HARDENING CONCRETE FOR HIGHWAY PAVEMENT REPAIR

L Vandewalle

Katholieke Universiteit Leuven

Belgium

ABSTRACT The repair of concrete highway pavements requires the use of high-early strength concrete to reduce the traffic jams and the consequential economical damage. At the Department of Civil Engineering of the K.U.Leuven, an extensive research programme has been executed to design a concrete mix that meets among other performances the following compressive strength requirements : after 2 days, the compressive strength, determined on cores (ϕ = 113 mm, h = 100 mm) drilled out of a small plate (thickness $\approx$ 250 mm), which is stored at a temperature of +7 °C, has to reach 40 N/mm²; after 7 days the strength must at least be equal to 50 N/mm². The sides of the small plate are not isolated in order to simulate the behaviour of a real pavement. The most important parameters in the investigation are the cement type, the cement quantity and the ambient air temperature. From the test results, it follows that for repair at low ambient air temperature, using a blend of blast furnace slag cement and normal Portland cement is very beneficial. Substituting blast furnace slag cement partly with normal Portland cement results in a much higher strength at early ages. The use of CEM I 52,5 instead of CEM I 42,5 for the substitution hardly improves the strength. For mixes stored at laboratory conditions (+20 °C), the substitution of blast furnace slag cement by normal Portland cement not always results in an increase of the compressive strength.

Keywords: Ambient air temperature, Blended, Cement, High early strength concrete, Mix proportions.

Professor Lucie Vandewalle is an assistant professor in the Civil Eng. Dept., Katholieke Universiteit Leuven, Belgium. She received her engineering degree in 1981 and Ph.D. in 1988. Her field of interest mainly concerns reinforced and prestressed concrete, high strength concrete and steel fibre concrete. She is president of RILEM TC162-TDF "Test and Design Methods for Steel Fibre Reinforced Concrete" and convenor of FIB TG 8.3 "Fibre Concrete".

INTRODUCTION

Due to the ever increasing traffic volumes, the rapid repair of highways has become one of the most difficult problems. The main objective in any pavement reconstruction operation is to provide a new, high quality pavement to the public for the least cost and inconvenience. The use of concrete with a "high strength at a very early age" makes it possible to pave and open the work quickly to traffic. The result is that alternatives can be provided for projects that in the past were not considered feasible with concrete due to the lengthy cure-times [1-3].

To achieve an adequate concrete strength at a very early age, material selection and concrete design do require careful considerations. The early strength of any mixture is controlled by a lot of parameters such as W/C-ratio, cement content, cement type, aggregate particle distribution, ambient air temperature, etc. It is recommended that prior to specifying the mixture proportions of a high early strength concrete, a thorough laboratory analysis is conducted to determine the properties of concrete developed with "local materials".

The objective of this research work was to develop high early strength concrete for highway pavement repair applications. The requirements for the concrete included:

- low alkali content (alkali-aggregate reaction) : The maximum allowable Eq.Na_2O is dependent on the used cement type [5] :
  100 % CEM I (normal Portland cement): $(Eq.Na_2O)_{max} = 3$ kg/m^3
  100 % CEM III/A (blast furnace slag cement) : $(Eq.Na_2O)_{max} = 4,5$ kg/m^3;

- enough workability for placement operations, this means that among other things also truck travel time between the plant and the job site has to be taken into account;

- achieving an in-situ compressive strength in excess of :
  40 N/mm^2 after 2 days hardening at +7°C
  50 N/mm^2 after 7 days hardening at +7°C.
  The strength is determined on drilled cores (ϕ=113 mm, h=100 mm);

- use of locally available materials, i.e. materials normally available at the concrete plant.

The main parameters in this investigation [4], the results of which are reported in this paper, are: cement content, cement type and ambient air temperature.

EXPERIMENTAL PROGRAMME

Properties of Materials Used

Eleven mixes have been proportioned and tested in the laboratory with three types of cement were used, i.e. one blast furnace slag cement (CEM III/A 42,5 LA) and two types of normal Portland cement (CEM I 42,5 R and CEM I 52,5 R). The physical properties and chemical analysis of those cements are shown in Table 1. The admixture, a polycondensate of melamine and formaldehyde, is not only a high-range water reducer but also an accelerator. Natural river sand 0/4 and minus 32-mm crushed porphyry were used as fine and coarse aggregates, respectively.

Table 1 Physical properties and chemical analysis of cement according to EN 196 "Methods of Testing Cement" [1991]

	CEM III/A 42,5 LA	CEM I 42,5 R	CEM I 52,5 R
Physical tests			
Fineness : Blaine (m^2/kg)	471	327	468
Compressive strength (N/mm^2)			
2 days	20,8	31,3	39,5
28 days	62,8	55,3	59,8
Setting time (minutes)			
initial	231	219	220
final	275	235	240
Chemical analysis (%)			
SO_3	2,91	2,72	3,18
Na_2O	0,27	0,14	0,12
K_2O	0,61	0,91	0,88
Eq.Na_2O	0,67	0,74	0,70
Cl^-	0,033	0,054	0,047
Al_2O_3	8,12	4,45	4,43
Fe_2O_3	2,13	2,84	2,81
MgO	4,37	2,05	2,06
SiO_2	26,02	20,43	20,41
CaO	52,29	64,35	64,13
Loss on ignition	1,79	1,49	1,17
Insoluble residue	0,77	0,30	0,02

Mix Proportions

To proportion a concrete mix that achieves the required compressive strength at 2 and 7 days, the following variables were considered :

- cement content : 450 kg/m^3 - 475 kg/m^3 - 500 kg/m^3
- cement type : the "reference mixes" (A, B, C) contain 100 % blast furnace slag cement (CEM III/A 42,5 LA). To accelerate gain in strength a part (20 - 25 %) of the blast furnace slag cement was substituted by normal Portland cement CEM I 42,5R, CEM I 52,5R respectively: "composite-mixes" D to K.

The dosage of the admixture was equal to 3 % by weight of the total cement content and W/C-ratio amounted to 0,33. Addition of the admixture was subdivided into two equal dosages with an interval of 60 minutes. This corresponds with a first addition at the batch plant and a second one at the point of placement. Mixes A and B, however, are exceptions: the total quantity of the admixture was added in one time during mixing operations. The requirement with respect to the maximum allowable alkali-content was fulfilled for all mixes.

The mix proportions used are summarized in Table 2.

Casting and Curing

Two types of specimens were used, i.e. 150 mm cubes and concrete slabs with dimensions 450x350x250 mm^3. A vibrating table was used during placing of concrete to ensure full compaction.

Immediately after finishing the specimens, they were stored at their respective curing conditions:

- T = +20°C and R.H. >= 95 % : simulation of laboratory conditions;
- T = +7°C and R.H. = 86 % during the first 7 days, afterwards gradual increase of temperature from +7°C till +9,5°C at 28 days, +17°C at 90 days respectively and gradual decrease of R.H. from 86 % to 81 % at 28 days, 69 % at 90 days respectively: simulation of ambient weather conditions in Belgium during winter and spring.

At the age of 2, 3 and 7 days respectively, the compressive strength was determined on cylinders (ϕ = 113 mm) which were drilled out of the concrete slab. These cores were sawn into two pieces (h = 100 mm) in order to measure the concrete strength of both the upper half and the lower part of the plate. To simulate as well as possible the real behaviour of concrete in a pavement at low ambient temperature conditions,the sides of the concrete slab were isolated. So cooling of concrete was mainly the result of heat emission by the top surface of the slab. Only the slabs, stored at low ambient temperature conditions, were isolated.
Table 2 Mix proportions

The compressive tests at the age of 7, 28 and 90 days respectively, were conducted on cubes. The sides of the cubes were not isolated. No tests were performed at 28 and 90 days on specimens stored at laboratory conditions. In total compressive strength was determined on 144 cubes and 288 cylinders.

TEST RESULTS - WORKABILITY

The workability of the fresh concrete was measured by means of the slump and the Vebe test. To simulate the behaviour of the concrete during transportation by a truck mixer from the batch plant to the job site as well as possible, the mixer in the laboratory was set in action for a moment every ten minutes. This procedure was executed during the hour, following the termination of mixing. Slump and Vebe test were performed 30 and 60 minutes after completion of mixing operations. The results are given in Table 3. It can be observed that the workability in 60 minutes after termination of mixing was still sufficient.

TEST RESULTS - COMPRESSIVE STRENGTH

As already mentioned in the introduction, the objective of this research was to proportion a concrete mix that achieves at least 40 N/mm^2 after 2 days hardening at +7°C, 50 N/mm^2 after 7 days respectively.

Table 2 Mix proportions

MATERIALS, kg/m^3	MIX										
	A	B	C	D	E	F	G	H	I	J	K
CEM III/A 42,5 LA	450	475	500	360	360	338	338	380	380	356	356
CEM I 42,5 R	-	-	-	90	-	113	-	95	-	119	-
CEM I 52,5 R	-	-	-	-	90	-	113	-	95	-	119
Cement content	450	475	500	450	450	450	450	475	475	475	475
Porphyry 20/32	600	600	600	600	600	600	600	600	600	600	600
Porphyry 7/20	388	388	388	388	388	388	388	388	388	388	388
Porphyry 2/7	350	334	316	353	353	353	353	334	334	334	334
Sand 0/4	521	489	461	516	516	516	516	489	489	489	489
Water	150	158	167	150	150	150	150	158	158	158	158
Superplasticizer	13,50	14,25	15,00	13,50	13,50	13,50	13,50	14,25	14,25	14,25	14,25

Table 3 Workability

WORKABILITY	MIX										
	A	B	C	D	E	F	G	H	I	J	K
slump 30' (cm)	3,8	11,5	2,5	1,8	2	2	0	1,5	1,5	2	2,5
slump 60' (cm)	1	6,7	22	13,5	9,5	16,5	9	16,5	14	15	17
VeBe 30' (sec)	4	2	6	7	9	6	12	7	6	6	6
VeBe 60' (sec)	6	4	3	4	6	3	4	4	3	3	3

Only the reference mixes A and B didn't satisfy this requirement (Table 4).

Compressive Strength Development at Early Age: 2 - 7 Days

At the age of 2 days the compressive strength, measured on the top-cylinders was always smaller than that measured on the corresponding bottom cylinders (Table 4). The difference between both values is more pronounced for the cylinders, stored at cold weather conditions (+7°C). Moreover, one can see that this "absolute difference in strength between top and bottom" remained almost unchanged during the further strength development.

A possible explanation for this phenomenon is that the temperature increase in the concrete due to the heat of hydration was higher in the bottom part than at the top of the slab because the top of the plate was not isolated.

Table 4 Compressive strength of "Top" and "Bottom" cylinders between 2 and 7 days (N/mm^2)

	7°C						20°C					
	cylinder (top)			cylinder (bottom)			cylinder (top)			cylinder (bottom)		
	2d	3d	7d	2d	3d	7d	2d	3d	7d	2d	3d	7d
A	28,5	32,9	56,0	32,0	45,0	63,8	67,6	73,3	94,1	67,3	78,6	96,3
B	32,2	40,0	60,0	36,3	44,3	59,0	61,4	69,9	87,4	65,5	74,5	87,1
C	40,2	47,1	64,7	50,2	57,5	73,0	72,9	83,3	95,8	78,6	87,4	99,4
D	39,8	52,0	65,8	47,2	57,8	77,0	65,6	74,0	87,7	76,5	78,9	92,6
E	44,4	49,0	65,3	50,0	59,1	74,6	74,6	76,4	85,6	76,1	82,7	87,7
F	43,2	52,9	71,4	56,0	60,1	82,9	71,4	77,3	82,6	79,2	80,1	91,3
G	50,6	60,2	75,2	55,4	65,4	86,7	72,3	80,0	84,9	79,1	81,4	93,1
H	40,9	55,1	67,0	50,9	58,9	72,9	70,2	78,7	92,8	79,1	81,0	95,3
I	46,9	55,1	73,7	51,2	61,7	76,4	75,4	76,3	96,3	82,9	82,2	99,4
J	47,3	52,8	74,1	54,2	57,9	79,7	74,5	77,4	94,7	77,8	84,3	99,3
K	49,9	61,9	71,9	56,3	67,9	81,9	66,4	75,3	87,7	79,5	84,8	97,0

After a few days, the temperature in the whole concrete slab was almost uniform which resulted in a parallel strength development between "top" and "bottom"concrete.

In addition the analysis of the unit weight of the concrete [4] showed that the top-cylinders were about 0,5 to 2 % lighter than the corresponding bottom-cylinders. This difference in density, caused by segregation during compaction, could be an additional reason for the difference in compressive strength. Hereafter, only the mean value of the compressive strength "top"-"bottom" will be used in the further discussion of the test results.

The results of the reference mixes are shown in Figure 1. The compressive strength of the specimens, stored at +7°C, increased with higher amount of blast furnace slag cement in concrete. However, at laboratory conditions, the relation between cement content and corresponding compressive strength was not unequivocal : the compressive strength of mix B (475 kg/m^3 CEM III/A 42,5 LA) was smaller than that of mix A (450 kg/m^3 CEM III/A 42,5 LA) and mix C (500 kg/m^3 CEM III/A 42,5 LA) respectively. It can also be observed that the compressive strength after 2 days at laboratory conditions was higher than that observed after 7 days storage at +7°C.

The strength development of the composite-mixes is presented in Figures 2 and 3.

From Figure 2, it follows that for concrete, stored at +7°C, the substitution of a portion of blast furnace slag cement by normal Portland cement results in a higher compressive strength. The strength gain is the greatest for "young concrete", i.e. at the age of 2 and 3 days.

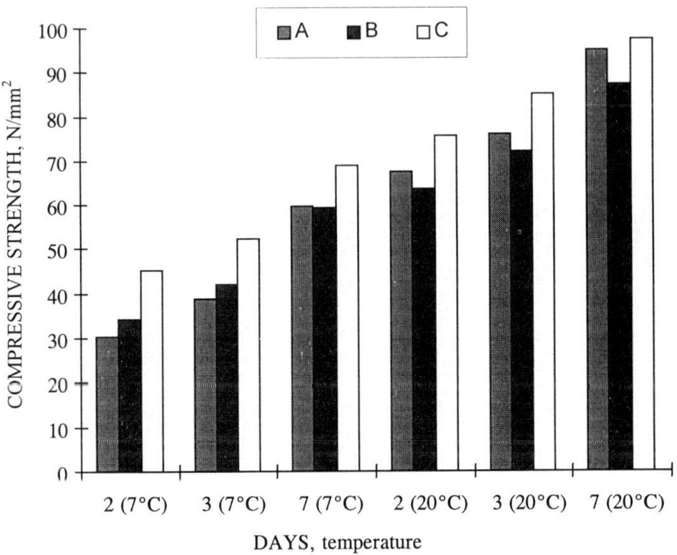

Figure 1 Influence of cement content on compressive strength development

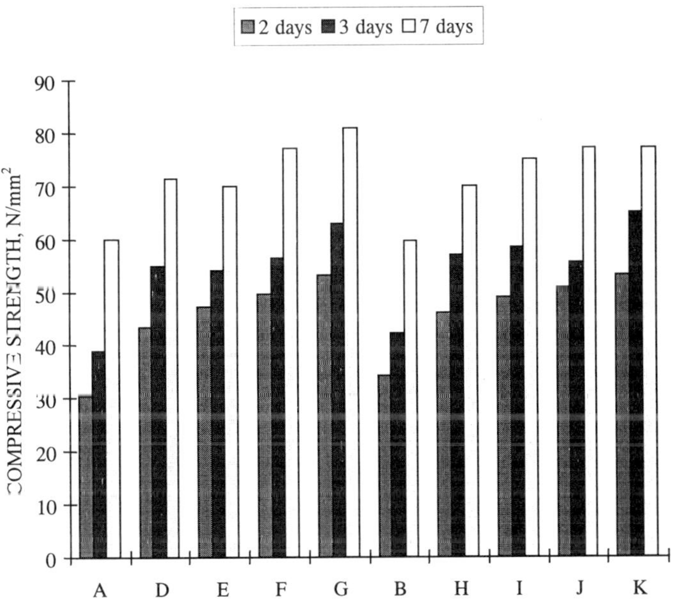

Figure 2 Compressive strength development at low temperature

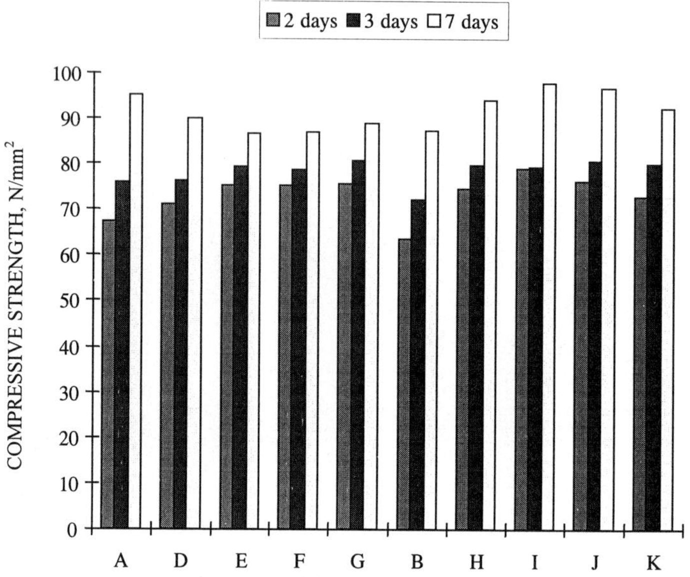

Figure 3 Compressive strength development at laboratory conditions

In most cases it can be said that the more blast furnace slag cement was substituted by normal Portland cement, the larger was the relative increase in compressive strength. The same remark can be made for the quality of the used normal Portland cement: the better the quality of the normal Portland cement, the higher the relative gain in compressive strength.

However, the influence of the total amount of cement (450 - 475 ⇒ mixtures D to K) on compressive strength development seems to be negligible for these composite-mixes. In laboratory conditions (Figure 3), it is not necessary to substitute a portion of blast furnace slag cement by normal Portland cement to obtain a high early strength. It was even found that at the age of 7 days the composite-mixes, containing 450 kg/m^3 cement, had a lower strength than the corresponding reference mix A.

Compressive Strength Development between 7 and 90 Days

Compressive tests on concrete, stored at +20°C, were only performed at the age of 7days (Figure 4). The strength development between 7 and 90 days has been determined on cubes which were not isolated at their sides. Concerning the cold weather conditions, the temperature increased gradually from +7°C at 7 days till +17°C at 90days.

The results of the cubes, stored at +20°C, are analogous to those of the corresponding drilled cores. The difference in compressive strength between the reference and the composite mixes is very small : the total cement content, the substitution percentage of the blast furnace slag cement and the quality of the normal Portland cement respectively, do not play an important part in the compressive strength of the tested concrete.

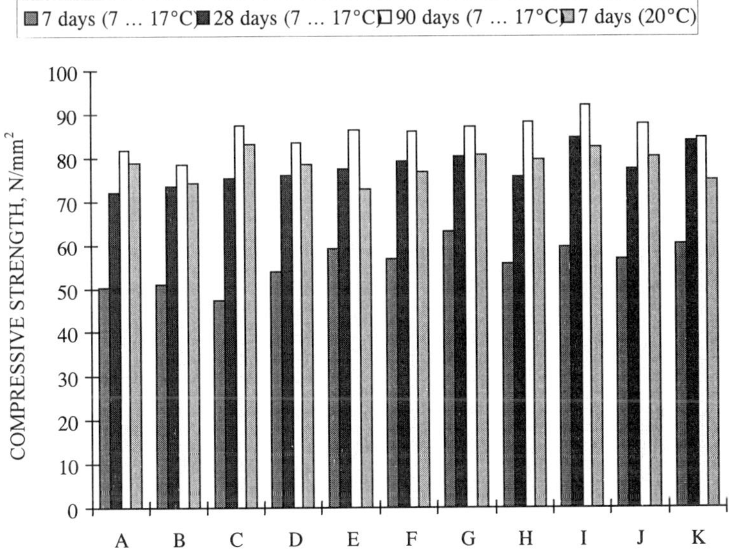

Figure 4 Compressive strength development between 7 and 90 days

This is in contrast with the strength of the cubes, stored at +7°C and tested at the age of 7 days. For these latter cubes the same conclusions can be drawn as for the corresponding cylinders.

At the age of 7 days, the compressive strength of the drilled cores is considerable higher than that of the corresponding cubes (Figures 2, 3 and 4). The mean value of the ratio between cube strength and cylinder strength amounts to :

+7°C => 0,78 (0,69 - 0,85)
+20°C => 0,85 (0,81 - 0,90).

Since the concrete block (cylinders) was isolated but the cubes not, it seems obvious that the above-mentioned ratio is smaller at low ambient temperature conditions than at laboratory conditions. As shown in Figure 4, the strength gain due to the substitution of blast furnace slag cement by normal Portland cement decreased with increasing age of concrete.

CONCLUSIONS

1. At low ambient temperature conditions (+7°C) a substitution of a portion of blast furnace slag cement by normal Portland cement results in a much higher strength at early age (2 - 3 days). However, the mutual difference in strength gain, due to various substitution percentages and cement qualities is rather small.

2. In laboratory conditions, a substitution of blast furnace slag cement by normal Portland cement not always leads up to a compressive strength increase.

Consequently, when repair of highways has to be executed at low ambient temperature conditions, it is recommended to substitute about 20 % of the blast furnace slag cement by normal Portland cement CEM I 42,5R. Once the ambient temperature fluctuates between +15°C and +20°C, the blast furnace slag cement provides almost just the same concrete strength development as a blend of blast furnace slag cement and normal Portland cement does.

AKNOWLEDGEMENTS

The author would like to thank FEBELCEM (Federation of Belgian Cement Industry) for their cooperation and the ready mixed concrete industry for the delivery of materials.

REFERENCES

1. WALKER, B., The Road Ahead is Concrete: A Review of Developments, Concrete, 1997, Volume 31, No.1, January, pp.12 - 14.

2. KNUTSON, M., MIKULANEC, J., SOLBERG, C., Fast Track Overlays, Seminar 29-17, "World of Concrete '91", Las Vegas, January 28 - February 1, 1991, 12 pp.

3. ANSARI, F., LUKE, A., VITILLO, N.P., BLANK, N., TURHAN, I., Developing Fast Track Concrete for Pavement Repair, Concrete International, May 1997, pp.24-29.

4. BOUCKAERT, S., LINDEKENS, K., Evolutie van de Druksterkte van Snelhardend Beton op Jonge Leeftijd, Thesis K.U.Leuven, May 1997,170 pp. (In Dutch, English title: Compressive Strength Development of Concrete at Early Age).

5. Omzendbrief MET met referenties 95-02-5187 van 17.02.1995.

PREDICTION OF CONCRETE STRENGTH USING THE MATURITY AND EQUIVALENT AGE CONCEPTS

N Buch

Michigan State University

United States of America

ABSTRACT. "Fast-track" concrete repair mixtures that cure within 8-24 hours were used to carry out full-depth slab repairs on a section of Interstate highway 496, near Lansing, Michigan. The mixtures included in the study were: a control mixture (C1) with a projected opening time of 7-days, a non-chloride accelerator mix (M2) with a projected opening time of 24-hours, a calcium chloride ($CaCl_2$) accelerated mix (FS), a high early strength mixture (HESC) developed by the Strategic Highway Research Programme (SHRP) (M1) and a modified fiber accelerated mix (MFS) with projected opening time of 8-hours. Before carrying out the repairs, job materials and mixture designs were obtained, and correlations between strength and maturity were developed. Maturity functions based on Nurse-Saul and Arrhenius (equivalent age) approaches were developed. To account for the heat rise in the concrete mixtures in the actual pavement sections, laboratory correlations were carried out by curing test cylinders and beams in specially insulated moulds. Maturity functions were used to predict the in-situ strength gain of concrete in instrumented test repair sections. Temperatures were monitored at slab and specimen mid-depths for the first 24-hours of curing. The average concrete temperatures ranged from 30°C to 70°C depending on the mix ingredients. The 8-hour mixtures exceeded 2MPa in flexural strength and 15MPa in compressive strength, whereas, the 24-hour mixture exceeded 3MPa in flexural strength and 20MPa in compressive strength. The control mixture gained strength at a much slower rate.

Keywords: Maturity, Datum temperature, Equivalent age, Flexural strength, Fast track

Dr Neeraj Buch is an assistant professor in the Department of Civil and Environmental Engineering at Michigan State University in East Lansing, Michigan, U.S.A. His research interests are in the design and analysis of rigid pavements, design and analysis of rehabilitation alternatives for rigid pavements, and fiber reinforced concrete properties. He is a member of several professional societies in the field of transportation and concrete materials

BACKGROUND

A typical fast track concrete mixture is obtained by using high cement content, low water cement ratio and accelerating admixtures. Concrete mixtures used for full-depth repairs must meet minimum strength levels before a pavement can be opened to traffic. Traditionally, concrete cores from the pavements are tested periodically to estimate strength gain. This process is both time consuming and destructive in nature. A variety of methods for in-situ strength measurements are available [1,2]. One such method that is widely used is the maturity method and was chosen for evaluation for this research programme.

Maturity of concrete is a time-temperature relationship that provides an indication of concrete strength (compressive or flexural) at a particular age [2]. The maturity method provides a solution to predict the complicated process of hydration and strength gain. Mathematically it can be expressed as follows [2, 3]:

$$M(t) = \Sigma (T - T_o) \Delta t \tag{1}$$

where:

$M(t)$	=	Maturity at age t, degree-days or degree-hours;
T	=	Temperature of concrete, °C;
T_o	=	Reference/Datum temperature, °C; and
Δt	=	Increment in time, days or hours.

Datum temperature is the temperature at which concrete ceases to gain strength in time i.e. hydration stops. Typically the datum temperatures range from –10.6°C to 0°C (14°F to 32°F) but could fall outside this range depending on the mixture ingredients. Equation 1 is used with the assumption that maturity increases linearly with temperature. However, it is well known from chemical kinetics that the rate of chemical processes increases with temperature exponentially and according to Arrhenius the phenomenon can be quantified as follows [2]:

$$K = A \bullet \exp\left(\frac{-E}{RT}\right) \tag{2}$$

where:

K	=	rate constant, 1/time;
A	=	constant, 1/time;
E	=	activation energy, J/mol;
R	=	gas constant, J/K•mol; and
T	=	temperature, °K.

Based on the above equation, the variation in maturity or the "equivalent age" at specified temperature can be computed as follows [2]:

$$t_e = \Sigma \exp\frac{-E}{R}\left(\frac{1}{T_a} - \frac{1}{T_s}\right)\Delta t \tag{3}$$

where:

t_e	=	equivalent age at a specified temperature, days or hours;
T_a	=	average temperature of concrete during time interval Δt, °K;
T_s	=	specified temperature, °K; and
Δt	=	time interval, days or hours.

The goal of maturity testing is to establish a relationship between strength gain and maturity, which will facilitate in predicting pavement opening time without exhaustive destructive testing. Both in newly cast concrete pavements and full depth repair slabs, environmental and load factors contribute to stress development. Concrete pavements may develop cracks when the total induced stresses exceed the concrete strength. The temperature of the concrete material (especially during early ages) was found to be useful in predicting a window for activities such as saw cutting timing and opening to traffic [4]. Temperature development in concrete can be an indication of not only stress, but also the strength of the concrete.

SCOPE OF THE INVESTIGATION

The basic mixture ingredients were Type I Portland cement (ASTM C150), 6AA (Michigan series) crushed limestone coarse aggregates with a 25mm nominal maximum size, 2NS natural sand with a 4.75mm nominal maximum size, both chloride and non-chloride accelerators, varying dosages of high range water reducers and air entraining admixtures. The target slump range for these mixtures was 25-75mm and the target air content range was 4-7%. The mixture quality was tracked through periodic gradation analysis of coarse and fine aggregates. The mixtures used for this study are presented in Table 1. The opening criteria as established by the state agency for the mixtures used in this study are presented in Table 2.

Table 1 Mixture designs

MIX CODE	WATER kg/m^3	CEMENT kg/m^3	COARSE AGG. kg/m^3	FINE AGG. kg/m^3	AEA ml/m^3	ACCELERATOR	WATER REDUCER
C1	164	290	1136	770	270	0	0
FS-8hrs*	190	500	1040	585	500	16 kg/m^3	0
MFS-8hrs**	166.4	415	1068	617.5	415	13 kg/m^3	0
M1-8hrs	162	540	646	700	980	25 l/m^3	560 ml/m^3
M2-24hrs	150	445	711	806	980	26 l/m^3	560 ml/m^3

* Fast Track mixture w/chloride accelerator.
** Modified fiber mixture. Fiber dosage of 1.2 kg/m^3.

The mixture ingredients were mixed in a horizontal rotary drum mixer with a capacity of 0.1m^3. Three replicate specimens were moulded for each test to track variability within any mixture design. Based on the final setting times of the different high early strength concrete (HESC) mixtures, strength test times were determined. The bulk of the early age testing was completed within 48-hours after molding the specimens.

The balance of the test specimens were tested at 3, 7, 14, and 28 days. The insulated concrete block samples were instrumented with a maturity meter at mid-depth (100mm). The insulated flexural strength beam specimens were 152 × 152 × 460mm. in dimension, the compressive strength cylinders measured 76 mm Φ × 152mm. The compressive strength test specimens were sulfur capped prior to testing (ASTM C617-87). The insulated maturity test blocks were 610 × 610 × 200 mm in dimension.

Table 2 Opening flexural strength criteria

CONCRETE MIXTURE	OPENING STRENGTH CRITERIA
C1	3.8 MPa @ 7-days
FS	2.1 MPa @ 8-hrs
M1	2.1 MPa @ 8-hrs
M2	3.5 MPa @ 24-hrs
MFS	2.1 MPa @ 8-hrs

Field Investigation

The field project involved full-depth patching of I-496 (westbound) between the cities of Lansing and East Lansing, Michigan. Most of the patches used in the study were in the passing lane. Two fast track mixtures, FS and MFS were used in the field investigation. Approximately $40m^3$ of MFS concrete (20-full depth patches) and approximately $5m^3$ of FS concrete (5-full depth patches) were placed as part of this investigation. All the mixture ingredients except calcium chloride were mixed at the batch plant. Calcium chloride was added at the job-site. Temperature development in patches was monitored by placing maturity meters at mid-depth. The instrumentation was placed in the slab interior to minimize edge cooling. Maturity readings were recorded every 30 minutes.

In order to conduct laboratory tests on the field concrete, insulated prismatic specimens 100×100×350mm for flexural testing and cylindrical specimens 100mm Φ × 200mm in height for compression tests were molded. Slab specimens 200mm thick and 750mm square in planar dimensions were cast and instrumented with thermocouples to monitor concrete temperature. The flexural and compressive specimens were cured until test ages.

RESULTS AND DISCUSSION

To estimate the strength gain from maturity data, the concrete temperature of the mixtures and the full-depth patches were monitored during the entire curing period. The ambient temperature during concrete placement ranged from 22°C to 40°C and the relative humidity was about 45%. The maximum concrete temperature for MFS mixture at mid-depth was 57°C and the maximum air temperature was 37°C, both these maximums were recorded at approximately 8 hours after concrete placement. The maximum concrete temperature for the FS mixture at mid-depth was 65°C. The difference in maximum temperatures of MFS and FS is possibly due to the reduced cement and accelerator content in the MFS mixture. The temperature profiles for FS and MFS are presented in Figure 1. The temperature profiles for C1, M1 and M2 are presented in Figure 2. The temperature distribution for mixtures MFS and FS of insulated concrete blocks (58°C and 64°C respectively) is similar to the temperature distribution of the concrete placed in the full-depth patches. The impact of water-cement ratio and accelerators is evident from the temperature distributions presented in Figure 2.

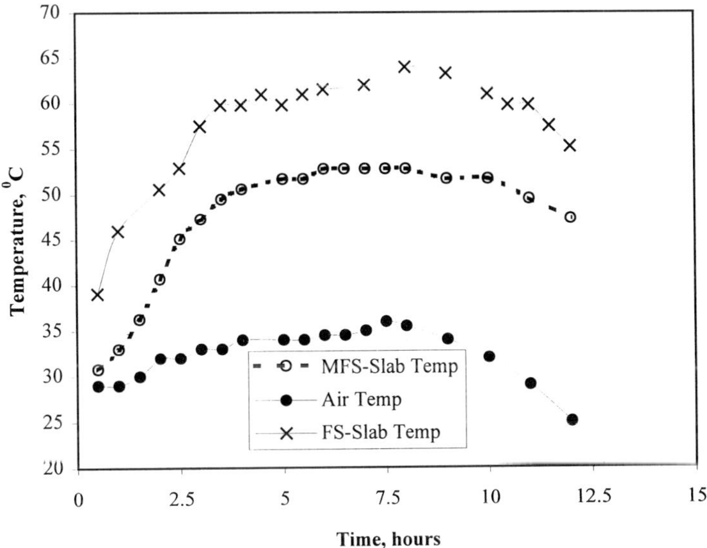

Figure 1 Temperature profiles full-depth patches, I-496(W)

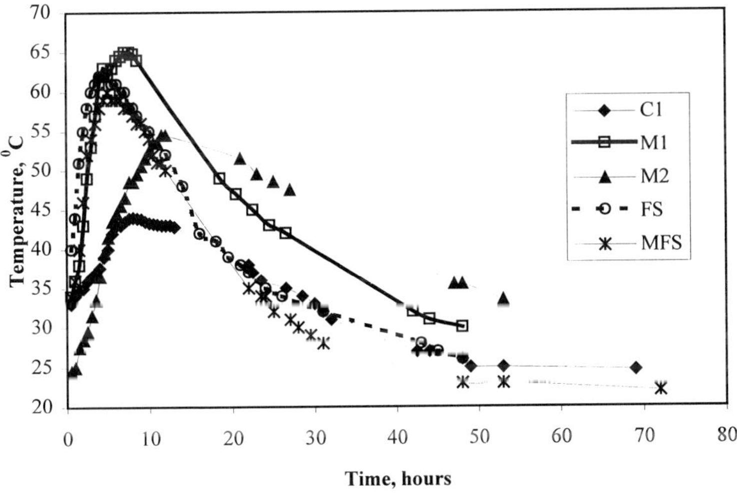

Figure 2 Temperature profiles for the laboratory mixes

The temperature data acquired from the laboratory insulated blocks and full-depth patches were used to develop maturity-strength (flexural) plots for the 5 concrete mixtures under investigation. The results of this analysis are presented in Figures 3 and 4. A summary of opening strength criteria and maturity are presented in Table 3.

Table 3 Summary of Opening Strengths and Maturity (datum temp=-10°C)

CONCRETE MIXTURE	OPENING STRENGTH CRITERIA	MATURITY @ OPENING, °C-hrs
C1	3.8 MPa @ 7-days	-
FS	2.1 MPa @ 8-hrs	300
M1	2.1 MPa @ 8-hrs	275
M2	3.5 MPa @ 24-hrs	525
MFS	2.1 MPa @ 8-hrs	275

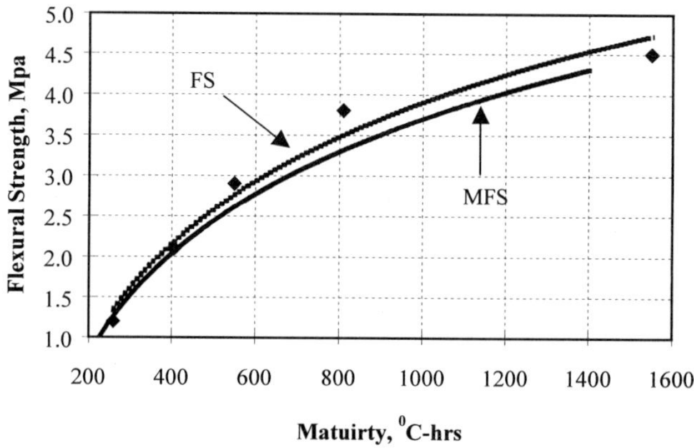

Figure 3 Strength-maturity plot for MFS and FS full-depth patches

The data collected in the laboratory phase of the testing can be used to develop procedures for predicting opening times. It is assumed that the strength-maturity relationship is hyperbolic. A two step procedure is associated with the determination of limiting strength (S_∞) and rate constant k_T and age, which then leads to development of a relative strength versus equivalent age relationship for the mixtures under investigation. Equivalent age is defined as "the time required for a concrete mixture to develop desired strength if it was cured at 23°C and 100% R.H." Table 4 summarizes the computed and measured limiting strength, rate constant and age at 23°C for the mixtures under investigation.

Table 4 Rate Constants and Limiting Strength Data

MIXTURE ID	RATE CONSTANT, DAY^{-1}	AGE, DAYS	MEASURED LIMITING STRENGTH, MPa	PREDICTED LIMITING STRENGTH, MPa
FS	2.4652	0.021	5.6	5.3
M1	2.4486	0.0522	5.1	4.9
M2	1.7217	0.076	5.0	5.3
MFS	4.9443	0.1270	5.4	5.2

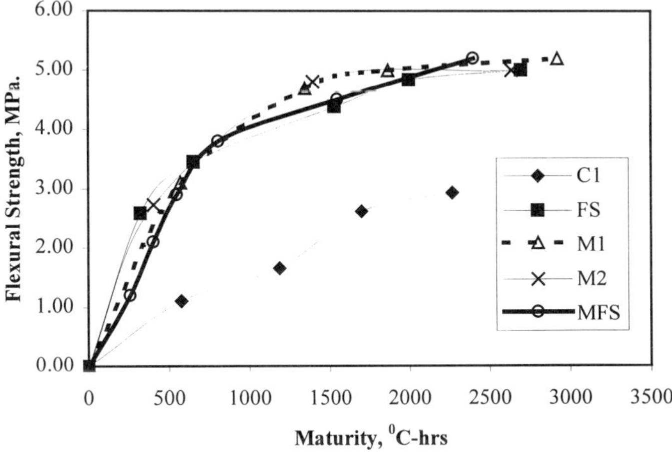

Figure 4 Strength-maturity plot for laboratory mixtures

The standard strength equivalent age relationship (hyperbolic) is expressed as follows:

$$S = S_\infty \left[\frac{k_T (t_e - t_o)}{1 + k_T (t_e - t_o)} \right] \quad \textbf{(4)}$$

Rearranging the above equation

$$\frac{S}{S_\infty} = RS = \left[\frac{k_T (t_e - t_o)}{1 + k_T (t_e - t_o)} \right] \quad \textbf{(5)}$$

where:

S = strength, MPa.;
S_∞ = limiting strength, MPa.;
RS = relative strength;
k_T = rate constant;
t_e = equivalent age;
t_0 = age;
R = universal gas constant; and
E = activation energy

The activation energy of the concrete is sensitive to the cement content, the dependence on temperature changes is explained by the changing values of the rate constant. The above form of the equation assists the pavement engineer in determining opening times for concrete pavement construction as a function of curing temperatures and strength gain. If the curing temperature in the field is other than 23°C, then the following age conversion factor α can be used:

$$\alpha = \exp \frac{-E}{R} \left[\frac{1}{T_a} - \frac{1}{T_s} \right] \quad \textbf{(6)}$$

The number of days needed to reach an equivalent age at 23°C is obtained by dividing the required equivalent age by the age factor. With this information the pavement engineer and the contractor can analyze the economics of different construction schedules. For example, if the full-depth repair were to be performed during cold weather, then the equivalent age concept would assist the contractor in determining duration of traffic closure. Based on the laboratory data for the 4 concrete mixtures under investigation Table 5 summarizes the results from the equivalent age analysis

Table 5 Summary of equivalent ages

TEMPERATURE, °C	α	FS	M1	M2	MFS
5	0.35	21.48	26.35	52.13	23.83
10	0.47	15.82	19.41	38.40	17.55
15	0.64	11.78	14.45	28.59	13.07
20	0.85	8.86	10.87	21.50	9.83
23	1.00	7.50	9.20	18.20	8.32
30	1.46	5.15	6.32	12.50	5.72
35	1.88	3.98	4.88	9.66	4.42
40	2.42	3.10	3.81	7.53	3.44
45	3.08	2.44	2.99	5.91	2.70

Note: All times are in hours.

CONCLUSIONS

Based on the results of this study described in this paper the following conclusions may be drawn:

- NDT readings must be calibrated for strength before construction. Strength as a function of NDT reading must be established using job materials and proposed mixture designs. Changes in mixture designs can lead to significant prediction errors.

- The maturity technique can be used to monitor early strength gain during the curing period in pavement repair. Results obtained from flexural strength specimens agree with the maturity approach.

- Data has shown that the limiting strength of the concrete mixtures under investigation is a function of early-age curing temperatures. This means that no unique strength-maturity relationship can be developed. However, it appears that there is a unique relationship between relative strength and maturity. A combination of the relative strength-maturity relationship and age conversion factor a paving engineer can predict relative strength gains at temperatures other than 23°C.

- The concrete hydration temperature of FS on an average is higher than that of MFS, the coincidence of maximum paving and concrete temperatures may have adverse effects on the long-term performance of the concrete patch.

- The maturity method cannot and should not be used for detecting batch errors or any errors related to construction practices. The maturity method should be supplemented with other tests before performing critical construction operations.

ACKNOWLEDGEMENTS

The Michigan State University Laboratories carried out this work with financial assistance provided by the Michigan Department of Transportation and the Pavement Research Center of Excellence.

REFERENCES

1. WHITING, D., ET.AL, Optimization of Highway Concrete Technology, SHRP C 373, Strategic Highway Research Program, National Research Council, Washington,D.C., 1994.

2. Handbook on Nondestructive Testing of Concrete, edited by MALHOTRA AND CARINO, CRC Press, Boca Raton, Florida, 1991, pp. 101-147.

3. SAUL, A.G.A. Principles Underlying the Steam Curing of Concrete at Atmospheric Pressure, Magazine of Concrete Research, 1951, Vol. 2, No. 127, pp.127-140.

4. NAGI, M. AND WHITING, D. Strength and Durability of Rapid Highway Repair Concrete, Concrete International, Sept. 1994, Vol. 16, pp. 36-41.

5. LEMING, M.L., SCHEMMEL, J.J., ZIA, P. AND AHMAD, S.H. High Performance Concrete; North Carolina Filed Installation Results, TRR 1282, Transportation Research Board, Washington D.C., 1993, pp. 78-81.

DYNAMIC ANALYSIS AND PASSIVE CONTROL SYSTEM FOR CONCRETE BRIDGES

P A Lopez-Yanez

T G de Sousa

Federal University of the Paraíba

Brazil

ABSTRACT In this paper, a new practical application of tendon based passive control system for a concrete grid bridge is proposed. The control system, that is obtained by a group of tendon supports associated with viscoelastic dampers, is used to mitigate the structural vibration of simple supported grids. The dynamic model which considers consistent-mass matrix and finite element discretization is used to determine the stiffness matrix. The traffic loads are represented by moving concentrated loads applied on the nodal points. The analysis is carried out by time-history technique, considering the small deformations theory and linear stress-strain relationship. A numerical example is presented in order to show the efficiency of this technique.

Keywords: Viscoelastic dampers, Passive control system, Concrete grid bridges

Professor Pablo Anibal Lopez-Yanez is Head Professor of Technology Center, Federal University of the Paraíba, Brazil. His main research include dynamic analysis and control of concrete structures.

Judas Tadeu Gomes de Sousa is a PhD student in automation and control Engineering at Federal University of the Paraíba, Brazil. His main research include dynamic analysis and control of bridges.

INTRODUCTION

Now a days, the continuous technological progress in the areas of new materials of civil engineering construction and methods of calculation of structures, allied to the readiness in the market of computers and softwares which are becoming more and more sophisticated every day, helped the civil engineering professionals to develop projects of increased elegance and lightness in its structural elements.

The increase of the lightness in those new constructions, which can result in the decrease of the expenses with the construction. It also brings some problems caused by the introduction of vibrations in the system, which can be stated as: psychological discomfort for its users, decrease in the time of useful life of the structure and in more serious cases, flaws in the structural elements of the construction owed to the continuous use of the material.

The control of the vibrations in bridges, provoked by the action of the traffic of vehicles, has been a subject of several researchers' study. As an example of the works published in this area, the research done by Abdel-Rohman and Leipholz [1-2] and Abdel-Rohman and Nayfeh [3] should be mentioned. All these works demonstrate the efficiency of the control of the traverse oscillations in beams of bridges using tendons and actuators.

In this work, a finite element method for dynamic analysis of bridges is presented, whose shipment corresponds to the traffic of vehicles in the surface of the track. As a second objective of this work a mechanism of passive control for the vibrations of the structure is proposed, originating from the traffic of vehicles, similar to that proposed by Abdel-Rohman and Nayfeh [3], however, with viscoelastic dampers [4].

THE MODEL OF THE PRIMARY SYSTEM

The primary system consists basically of a single span bridge of concrete with 30(m) of length, whose traverse section is presented in Figure 1.

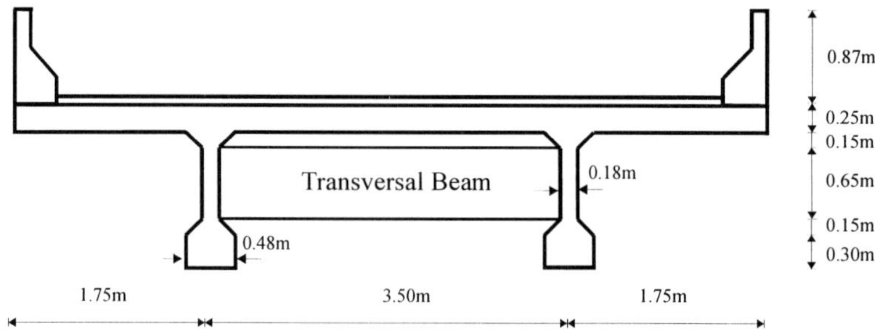

Figure 1 Traverse section of the bridge

Schematic Representation of the Primary System

The primary structure can be schematically represented by the grillage shown in Figure 2

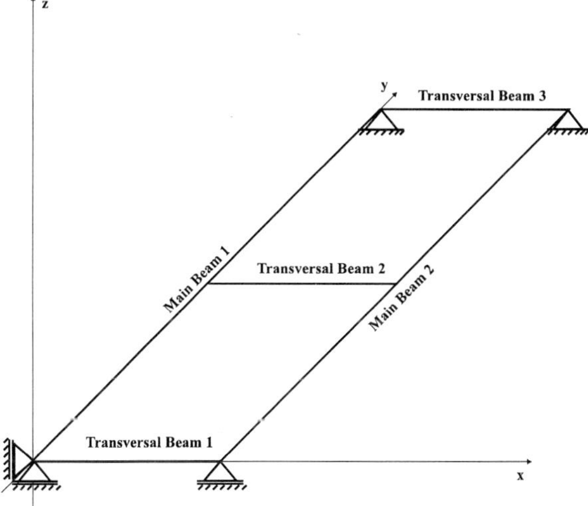

Figure 2 Schematic representation of the bridge

where the traverse section of the main beams and of the concrete traverse beams are shown in Figure 3 and 4, respectively

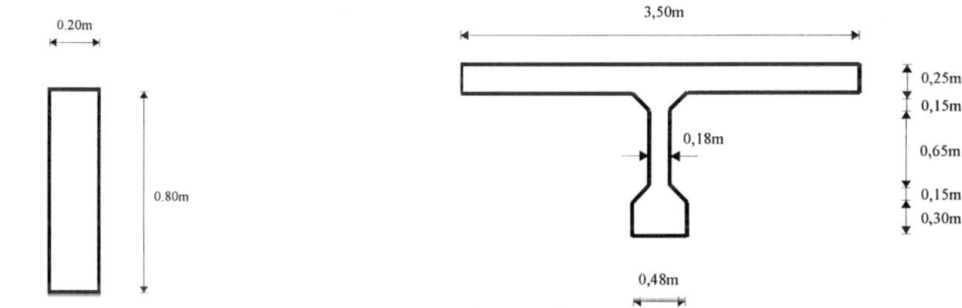

Figure 3 Traverse section of bridge's traverse beam

Figure 4 Traverse section of the main beams

In order to model the system, initially the bridge is decomposed in several grillage elements. Particularly in this example, the traverse beams are considered as one element members, and the main beams are analysed by several finite elements. The displacements of a generic bar are shown in Figure 5.

Where, v_i and v_f are the transverse displacements, θ_i and θ_f are the rotations in the flexion plain, and ϕ_i and ϕ_f are the displacements associated to the torsion reactions.

For an element with length l, area A, moment of inertia to the flexion I, moment of inertia of the torsion J. Being ρ the specific weight, E the Young's modulus and G the shear modulus of the element.

The matricial equations, according to Lopez-Yanez and Sousa [5], that relate each displacement to the forces in the extremities are presented in the continuation.

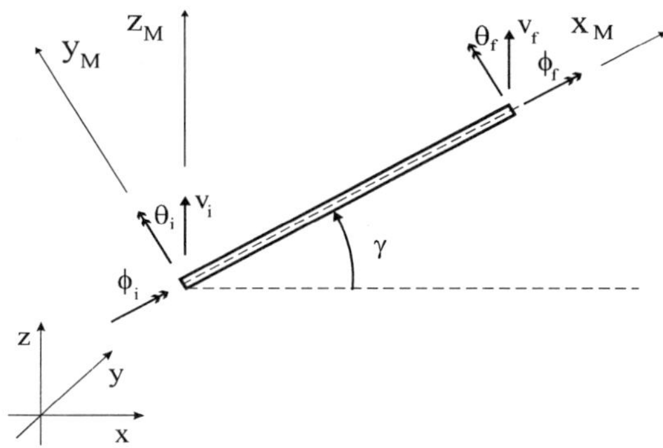

Figure 5 Convention of signs for the displacements of an element of the grillage

$$
\begin{Bmatrix} T_i \\ M_i \\ V_i \\ T_f \\ M_f \\ V_f \end{Bmatrix} = \frac{\rho A l}{420}
\begin{bmatrix}
140 & 0 & 0 & 70 & 0 & 0 \\
0 & 4l^2 & 13l & 0 & -3l^2 & 22l^2 \\
0 & 13l & 156 & 0 & -22l & 54 \\
70 & 0 & 0 & 140 & 0 & 0 \\
0 & -3l^2 & -22l^2 & 0 & 4l^2 & -13l \\
0 & 22l & 54 & 0 & -13l & 156
\end{bmatrix}
\begin{Bmatrix} \ddot{\phi}_i \\ \ddot{\theta}_i \\ \ddot{v}_i \\ \ddot{\phi}_f \\ \ddot{\theta}_f \\ \ddot{v}_f \end{Bmatrix} +
$$

$$
+ \begin{bmatrix}
GJ/l & 0 & 0 & -GJ/l & 0 & 0 \\
0 & 4\,EI/l & -6\,EI/l^2 & 0 & 2\,EI/l & 6\,EI/l^2 \\
0 & -6\,EI/l^2 & 12\,EI/l^3 & 0 & -6\,EI/l^2 & -12\,EI/l^3 \\
-GJ/l & 0 & 0 & GJ/l & 0 & 0 \\
0 & 2\,EI/l & -6\,EI/l^2 & 0 & 4\,EI/l & 6\,EI/l^2 \\
0 & 6\,EI/l^2 & -12\,EI/l^3 & 0 & 6\,EI/l^2 & 12\,EI/l^3
\end{bmatrix}
\begin{Bmatrix} \phi_i \\ \theta_i \\ v_i \\ \phi_f \\ \theta_f \\ v_f \end{Bmatrix} +
\begin{Bmatrix} T_i^F \\ M_i^F \\ V_i^F \\ T_f^F \\ M_f^F \\ V_f^F \end{Bmatrix}
\tag{1}
$$

where, V_i and V_f are the transverse elastic reactions, V_i^F and V_f^F are the reactions of perfect setting, M_i and M_f are the elastic moments, M_i^F and M_f^F are the moments of perfect setting, T_i and T_f are the elastic reactions of torsion and T_i^F and T_f^F are the torsional reactions of perfect setting.

The Model of the Dynamic Loading

The study of vibrations in bridges, owed to the action of the traffic of vehicles, has been the recent objective of many researches, [6] which present an important contribution in this area, so much with relationship to the model of the structure, as of the vehicle. In their work the authors consider problems such as: the interaction of the movements of the structure and of the vehicle, variation in the speed of the vehicle and presence of irregularities in the surface of the track.

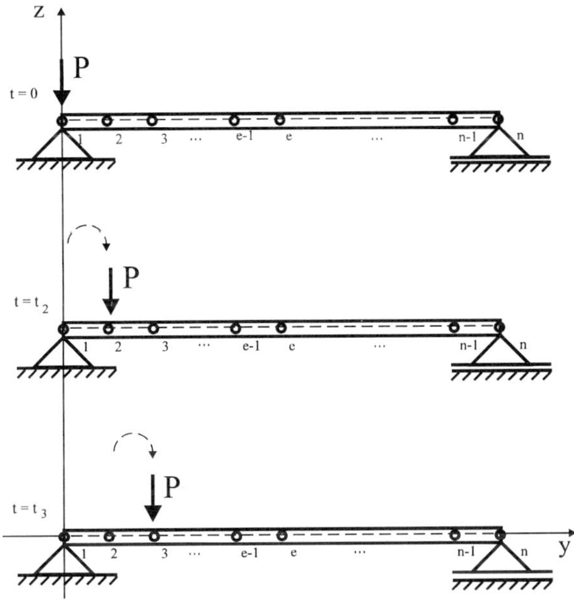

Figure 6 Loading along the time

In the present work, a simpler model for the dynamic loading is used. In this case the vehicle is substituted by a moving impulsive force that jumps on the nodes of the main beam 1, travelling all its extension with constant velocity (Figure 6). This approach can be considered when the traverse dimensions of the structure are very inferior to the longitudinal ones and the main beams of the structure are divided in a large number of elements.

Equations of the Movement of the Primary System

The equations of the not softened movement are obtained by making the analysis of the balance for each knot, by means of Equation (1). This way it is determined by:

$$[M]\{\ddot{x}\} + [K]\{x\} = \{P(t)\} \tag{2}$$

Where [M] is the mass matrix, [K] the rigidity matrix and {x} the vector of the displacements. But the vector {P(t)}, represents the dynamic loading at the time, that consists of an impulsive load with width of 450KN, that travels the extension of the beam with constant speed.

The bridge damping is assumed to be an absolute viscous type, the modal ratio damping are considered to be 0.01% [11]. The solutions for the equations can be obtained through the method of the sum of the normal modes and Duhamel's integral.

CONTROL OF VIBRATIONS IN OF CIVIL ENGINEERING WORKS

In the last decades, several research have been conducted with the objective of not only to study the vibrations in works of civil engineering, but also to attenuate or to eliminate its noxious effects. For example, in the case of the control of vibrations, the works published by Roorda [7] and Yang and Giannopoulos [10]. Talking about the control of the oscillations in bridges, it can be exemplified, by means of the publications elaborated by Abdel-Rohman and Leipholz [1-2], Abdel-Rohman and Nayfeh [3], and Michalopoulos et al [6].

In general, structural elements with varying, but controllable mechanical behaviour, integrated into a structural system may affect the overall behaviour of the system and eventually improve it [6]. This is the primordial idea of the structural control. Depending on whether externally supplied energy is required during the operation of the system or not, the control is characterised as active or passive, respectively.

In this work, the mechanism of passive control used is based on the example presented by Abdel-Rhoman and Nayfeh [3], although in this case it is destined to actively control the structure. Besides that, an example of the work published by Zhang et al [11], that it is destined to reduce the vibrations in buildings due to the action of earthquakes using viscoelastic dampers, a group of viscoelastic dampers serving as an additional source of reduction is also projected for the bridge

The Model of the Complete System

The complete system is formed by the primary system associated with the mechanism of passive control, as shown in Figure 7.

The tendons and dampers are elements submitted exclusively to the extension. The matricial Equation (3) relates the displacements with the forces in the extremities of the bars [5],

$$\begin{Bmatrix} U_i \\ U_f \end{Bmatrix} = \frac{AE}{l}\begin{bmatrix} 1 & -1 \\ -1 & 1 \end{bmatrix}\begin{Bmatrix} u_i \\ u_f \end{Bmatrix} + \rho A l \begin{bmatrix} 1/3 & 1/6 \\ 1/6 & 1/3 \end{bmatrix}\begin{Bmatrix} \ddot{u}_i \\ \ddot{u}_f \end{Bmatrix} + \begin{Bmatrix} U_i^F \\ U_f^F \end{Bmatrix} \qquad (3)$$

where U_i and U_f are the normal elastic reactions, u_i and u_f are the displacements associated to the reactions and U_i^F and U_f^F are the reactions of perfect setting, in this way i and f become indexes of initial and final extremities of the bar.

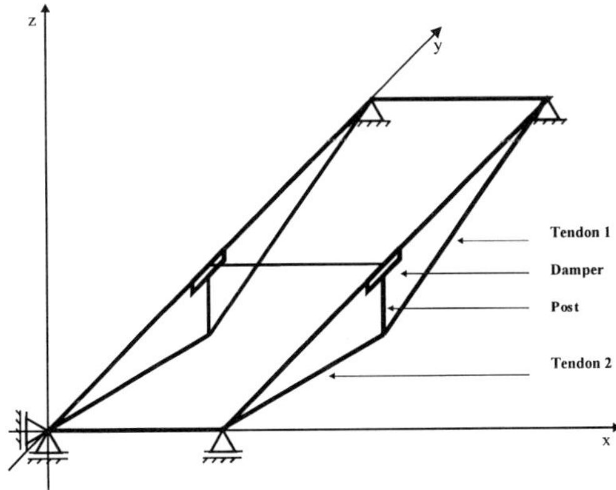

Figure 7 Schematic representation of the complete system

CALCULATION OF THE FACTOR OF ADDITIONAL MODAL REDUCTION

A viscoelastic material when submitted to deformations of harmonic type presents combined characteristics of an elastic solid and of a viscous liquid, that is, after a complete cycle of deformation, the material comes back to its original form, but there is a liberation of energy in form of heat. This effect on the system can be evaluated in terms of equivalent damping, that is proportional to the reason among the energy dissipated in a cycle and the maximum deformation energy [11]. In the case of a system with several degrees of freedom,

$$\xi_{eq}^{n} = \frac{E_{d}^{n}}{4\pi E_{ms}^{n}} \tag{4}$$

where ξ_{eq}^{n}, is the ratio of equivalent modal damping, E_{d}^{n} is the energy dissipated by the dampers, VE and E_{ms}^{n} the maximum deformation energy, in the n*th* mode. The values for energy are determined as proposed by Zhang et al [11]. Thus, the ratio of final damping is a sum of the ratio of the structure's original damping (1%) and the equivalent modal damping.

RESULTS

Setting up the matricial equation of the movement, a model for system is developed using the MATLAB software [9]. The solution found for the different equation show the behaviour of the structure along the time. In the problem are used, for the bridge the following are used: Young's modulus, E_{bridge}, is 29000(MPa), shear modulus, G_{bridge}, is 1210(MPa) and the density, ρ_{bridge}, is 2500(Kg/m^3); for the post: length, l_{post}, is 4.90(m), extensional rigidity, EA_{post},, is 2470.02(MN) and the density, ρ_{post}, is 7850(Kg/m^3); for the tendons: length, l_{tendon}, is 5.81(m), extensional rigidity, EA_{tendon}, is 157.71(MN) and the density, ρ_{tendon}, is 7850(Kg/m^3); for the viscoelastic dampers: length, l_{damper}, is 0.10(m) and the extensional rigidity, EA_{damper}, is 7.04(MN).

The history of the vertical degrees of freedom, correspondent to the middle of the main beams of the bridge, for the structure with and without the mechanism of passive control, is presented in Figures 8, 9, 10 and 11.

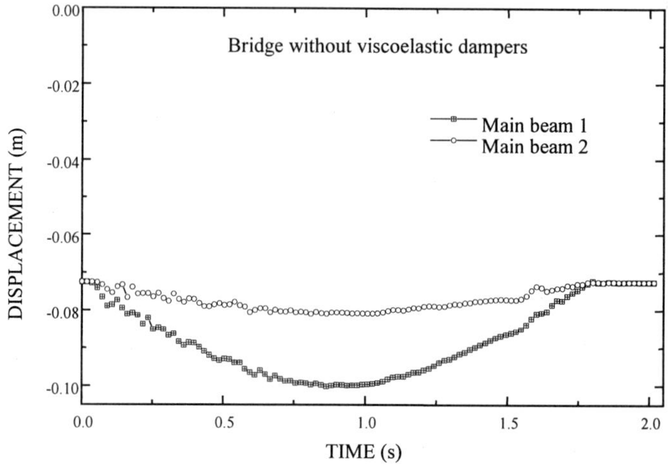

Figure 8 Vertical displacement in the middle of the for a vehicle velocity 60Km/h

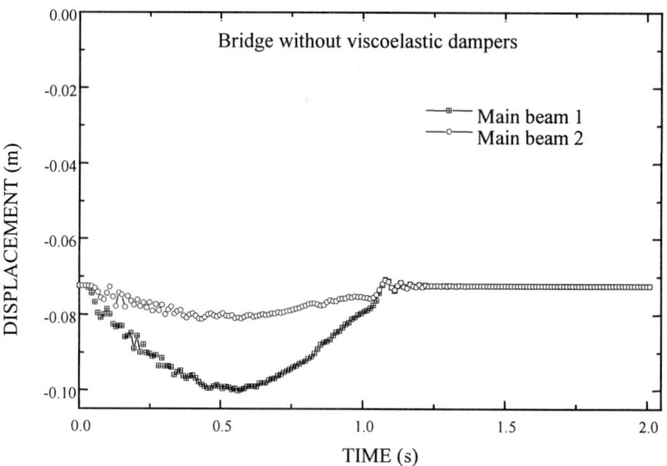

Figure 9 Vertical displacement in the middle of the bridge for a vehicle velocity 100Km/h

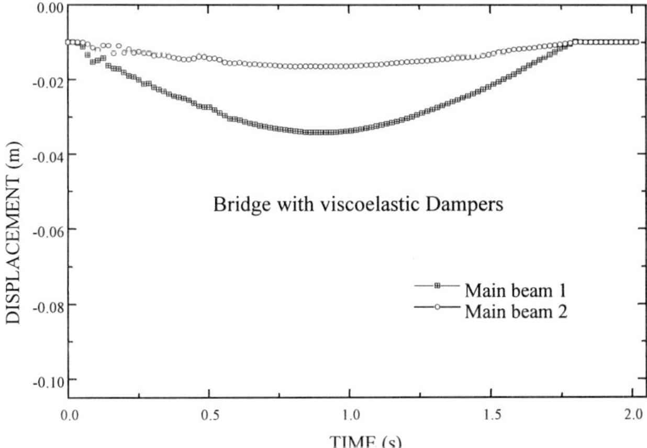

Figure 10 Vertical displacement in the middle of the bridge for a vehicle velocity 60Km/h

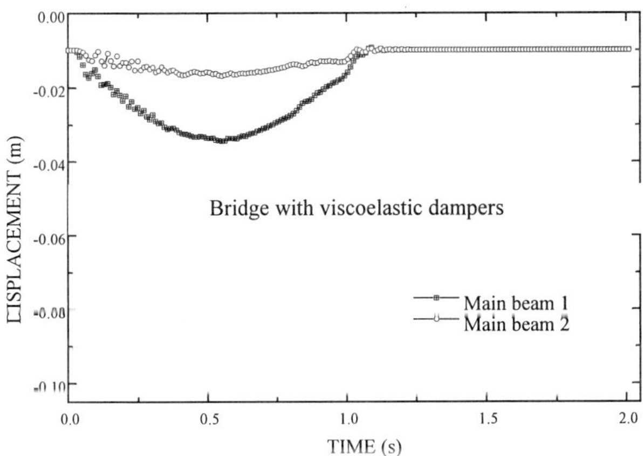

Figure 11 Vertical displacement in the middle of the bridge for a vehicle velocity 100Km/h

CONCLUSIONS

A control mechanism for the traverse vibrations of a structure submitted to a dynamic loading is presented. The main system studied is a single span concrete bridge, in the other hand, the excitation used is an impulsive concentrated load that travels across the bridge with constant speed and the control device consists of an association of a post and steel tendons with a viscoelastic damper. The results presented in the graphs show the variation in the time of the degrees of freedom of the structure, for the cases of the beam with and without the system of dampers. In the analysis of the results, a considerable reduction is verified in the widths of the deformations (static and dynamic) as well as in the accommodation time of the structure. This facilitates an increase in the users' comfort, as well as, a larger economy in the design of the bridge elements.

REFERENCES

1. ABDEL-ROHMAN, M. AND LEIPHOLZ, H H. Active control of flexible structures. Journal of the Structural Division, Vol. 104, No ST8, 1978. pp 1251-1266.

2. ABDEL-ROHMAN, M AND LEIPHOLZ, H H. Automatic active control of structures. Journal of the Structural Division, Vol. 106, No ST3, 1980. pp. 663-677.

3. ABDEL-ROHMAM, M AND NAYFEH, A H. Active control of nonlinear oscillations in bridges. Journal of the Engineering Mechanics Divison, Vol. 113, No EM3, 1987. pp. 303-315.

4. ZHANG, RI-HUI et al. Seismic response of steel frame structures with added viscoelastic dampers. Earthquake Enginering and Structural Dynamics, Vol. 18, 1989. pp 389-396.

5. LOPEZ-YANEZ, P A AND SOUSA, J T G. Análise dinâmica de pontes considerando-se um sistema de controle sob a pista. Proceedings of the XIII CILANCE Congress, Vol IV, Brasília, 1997. pp 2003-2010.

6. CHOMPOOMING, K AND YENER, M. The Influence of Roadway Irregularities and Vehicle Deceleration on Bridge Dynamics using the Method of Lines. Journal of Sound and Vibration, Vol. 183, 1995. pp 567 – 589.

7. MICHALOPOULOS, A. et al. Pretressed tendon based passive control system for bridges. Computers & Structures, Vol 63, No 6, 1997. pp 1165-1175.

8. ROORDA, J. Tendon control in tall structures. Journal of the Structural Division, Vol. 101, No. ST3, 1975. 505-521.

9. SHAHIAN, B AND HASSUL, M. Control systen design using MATLAB. Prentice-Hall International, 1993, pp 44-72.

10. VELETSOS, A S. AND HUANG T. Analysis of dynamic response of highway bridges. Journal of the Engineering Mechanics Division, Vol. 96, No. EM5, 1970. pp 593-620.

11. YANG, J N. AND GIANNOUPOULOS, F. Active tendon control of structures. Journal of the Engineering Mechanics Division, Vol. 104, No. EM3, 1978. pp 551-568.

MODELLING SHEAR LAG IN CONCRETE BOX GIRDER FLANGES BY STOCHASTIC FEM

R Rusina

D Novák **B Teplý**

Technical University of Brno

V Křístek

Technical University Prague

Czech Republic

ABSTRACT. The effect of shear lag on stress distribution across the box girder flange width is analysed using stochastic finite element method. It has been found that the statistical variabilities of material parameters and geometrical characteristics have significant effects on the statistical scatter of the predicted longitudinal normal stresses and on their distribution.

Keywords: Shear lag, Concrete box girder flanges, Stochastic finite element method

Mr Radoslav Rusina is a Ph.D. student at the Institute of Structural Mechanics, Faculty of Civil Engineering, Technical University of Brno, Czech Republic. He specialises in the stochastic finite element method.

Associate Professor Drahomír Novák is a Lecturer at the Institute of Structural Mechanics, Faculty of Civil Engineering, Technical University of Brno, Czech Republic. His main research interest is structural safety and reliability – stochastic computational mechanics.

Professor Břetislav Teplý is the Head of the Institute of Structural Mechanics, Faculty of Civil Engineering, Technical University of Brno, Czech Republic. He is a member of RILEM technical committee TC107 and is also active in National Technical Committee for standardisation "Reliability of Structures".

Professor Vladimír Křístek is the Head of the Department of Concrete Structures and Bridges, Czech Technical University, Prague, Czech Republic. He is a member of RILEM Technical Committee TC107 and is a founding member of the Engineering Academy of the Czech Republic.

INTRODUCTION

When modelling the real behaviour of concrete structures it is inevitable to account for the uncertainties of involved factors and effects. These can be listed in the following groups:

(1) Mechanical characteristics of concrete, reinforcement and prestressing steel,

(2) Technological and compositional aspects of concrete,

(3) Geometry of the structure,

(4) Characteristics of prestressing – human labour,

(5) Environmental characteristics,

(6) Computational model.

The present study is focused on the items (2), (3) and (6) only.

It is well known that shear lag, induced by shear deformations of flanges of box girders, can result in a very non-uniform distribution of longitudinal normal stresses across the flange widths, and may also significantly influence the girder deflections. This shear lag effect, in its *classical meaning*, can result in a significant increase of stress levels close to the edges of the flange. *The negative shear lag* [1], for which is typical that the longitudinal stresses at the middle of the flange exceed those nearer the webs, occur at various important structural arrangements like cantilever and continuous box girders.

The effects of shear lag are particularly severe under the following circumstances: in local regions adjacent to concentrated loads positions and in relatively wide flanged plates. In a continuous bridge girder these circumstances may coincide, and an exceptionally strong shear lag effect in flanges over intermediate support can exist. At prestressed concrete box girder bridges where the cross sectional area of webs represents only a small fraction of the total cross sectional area, also *the shear deformations of webs* become significant.

In design practice, short-term as well as long-term structural analyses of box girder bridges are typically based on simple frame models (like a continuous beam), taking usually only the action of bending moments into account (i.e. disregarding shear deformations and thin-walled effects like the shear lag). This results in underestimation of deflections due to action of vertical loads. The prestressing effects, being accompanied by none or minor shears, are, on the other hand, predicted by the elementary bending theory more satisfactorily. Thus, it is apparent that *the concept of effective widths*, if in the analysis the same effective width is assumed for evaluation of effects of vertical loads as well as of the prestressing, is in case of prestressed concrete bridges completely wrong.

In the reality, all the incoming parameters are of random nature. Among these parameters especially different types of imperfections should be taken into account. Therefore the present paper studies the stress in flanges of concrete box girders having randomly distributed irregularities and imperfections. The irregularities caused by technological imperfections and by the degradation processes are described as random fields.

The influence of uncertainties of various parameters are examined, particularly their effects on the distribution of longitudinal stresses across the widths of flanges of box girder bridges.

STOCHASTIC FINITE ELEMENT ANALYSIS

The theory of random fields and finite element method in computational mechanics resulted during last decade in the stochastic finite element method (SFEM). To solve the stochastic finite element system, several methods have been developed. The crude Monte Carlo simulation method is conceptually the simplest and well-known approach. An improvement in deriving the finite element solution for the response variability is the Neumann expansion technique [2].

Finite element equilibrium equations are written by a well-known formula:

$$\mathbf{Ku} = \mathbf{F}$$

The global stiffness matrix K contains the random parameters that are subjected to spatial variability. The matrix K can be decomposed into two matrices:

$$\mathbf{K} = \mathbf{K}_0 + \Delta\mathbf{K}$$

where K_0 is the deterministic part (calculated with mean values of input parameters) and ΔK is the deviatoric part (based on deviations of random parameters from mean values).

Every uncertainty involved in the analysis is statistically described by the mean value, variance and correlation structure described by the type of auto-correlation function and correlation length. Random fields of parameters are randomly generated - the values of parameters are considered constant within each element.

In case of the different spatial correlation, the procedure has to be performed in two orthogonal directions (here x, y). Auto-correlation function has the form (orthotropic random field)

$$R_{aa}(\xi) = \sigma_0^2 \, exp\left\{-\left[\left(\frac{\xi_x}{d_x}\right)^2 + \left(\frac{\xi_y}{d_y}\right)^2\right]\right\}$$

and is the function of separation vector $\xi = [\xi_x, \xi_y]^T$ between two points $x = [x, y]^T$ and $x + \xi$; σ_0 is a variance of a random field, d_x and d_y are correlation lengths in direction of coordinates x and y.

BOX GIRDER BRIDGES

A box girder is a shell-like spatial system and in its structural performance the decisive role is played by the *membrane* action of its parts – elements. This is why the problem of the in-plane stress distribution in box girders may be simplified and solved as a 2D one utilising the stochastic finite element method.

The goal of the paper is to assess the influence of random variability in 2D space across the flange width in structural response - the membrane stresses. The statistical analysis is performed in order to provide the statistical characteristics of the response. Plane elastic stress/strain problem is solved using 4-node isoparametric quadrilateral FE with linear displacement variation, 2 DOF per node.

The stochastic finite element analysis consists of the following steps: (i) The stiffness characteristic (i.e. the product of Young's modulus E_c and thickness h) is generated as a two-dimensional random field. (ii) The element stiffness matrices are calculated and then assembled into the global deterministic stiffness matrix K_0 (only once) and the deviatoric part ΔK (for every simulation), FEM solution is performed. (iii) The procedure itemised above is repeated N-times (N is the number of simulations). Finally, statistical characteristics of the structural response i.e. deflections and stresses are obtained evaluating the sets gained by simulations.

CASE STUDY

An analysis of the effect of imperfections on stress distribution in the lower flange of the Nusle bridge in Prague, Czech Republic, may be presented as an example. This outstanding bridge was opened for traffic in February 1973. The bridge is a frame structure of prestressed concrete having five spans of lengths 68.25 + 3 x 115.5 + 68.25 m, with the box cell of constant depth and variable thicknesses of webs and lower flange. No diaphragm was located inside the box due to the necessity to maintain free space for transport purposes (the Prague underground system). As an example, the deck of the bridge is loaded by a uniformly distributed load of intensity 10 kN/m^2.

The attention of the study is focused on the part of the lower flange of this bridge adjacent to intermediate supports where, as it has been stated above, an exceptionally high shear lag effect appears resulting in a very non-uniform distribution of the longitudinal stresses across the flange width.

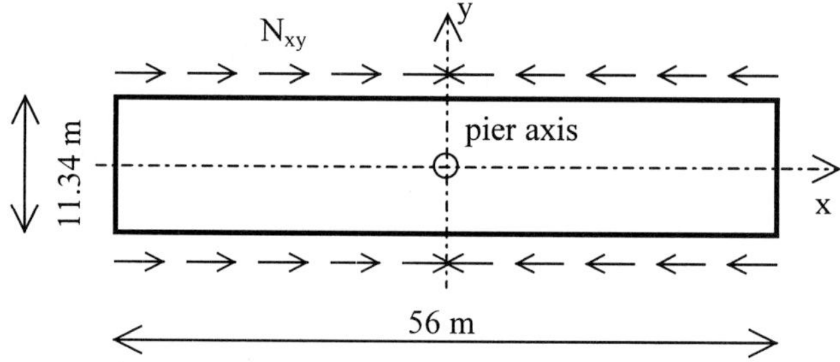

Figure 1 Portion of the lower flange

Figure 1 shows the portion of the lower flange between the points of contraflexure of the box girder bridge. The decisive loading effects in the bottom flange are produced by longitudinal normal stresses σ_x. Such a flange segment thus can be analysed as a separated plane loaded by N_{xy} along the edges. Hence, 2D analysis may be adopted while studying the stress state of the flange. The stress distributions due to this loading, corresponding to *the deterministic analysis*, assuming the dimensions and material properties of the structure without any imperfections are shown in Figure 2. The distribution of the longitudinal normal stresses across the lower flange of the box girder bridge, corresponding to the classic shear lag, is severe in the region close to the pier. On the other hand, the negative shear lag appears along the majority of the lower flange segment length.

The in-plane structural performance of a flange depends on the flange stiffness expressed as a product of the flange thickness and the modulus of elasticity - both these parameters exhibit random variations. This is why in the present study, the above mentioned irregularities are modelled by variations of the flange stiffness.

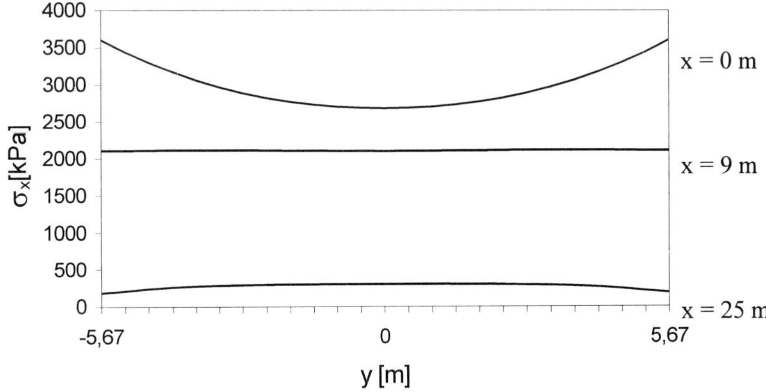

Figure 2 Shear lag effect – deterministic result of longitudinal stresses

While performing the analysis by SFEM, the important step was to select suitable correlation lengths for random fields which should reflect the reality as close as possible. Note that the meaning of the correlation length from practical point of view is that parameter in question observed at a certain point has an insignificant statistical correlation with the same parameter in distance which is equal to correlation length. In case of a very small correlation length – the parameter fluctuates very randomly along the structure (in *x* or *y* direction). In case of a large correlation length (over the size of the structure solved) – the parameter is almost constant along the structure. Then the case is equivalent to modelling of such parameter via simple random variable approach. Random fields most likely follow some shape of waves of medium sizes in reality. In our case one should have in mind technological aspects of construction and curing. We have to emphasise the different selection of correlation lengths in *x* and *y* directions reflecting e.g. the fact that the bridge has been constructed from separately fabricated segments. Results for correlation lengths d_x = 2m, d_y = 2m with mean value $E_c h$ = 1.1 kNm^{-1} and *cov* = 0.10 are presented here.The statistical results for the transverse distribution of longitudinal stresses across the flange width is shown in Figure 3.

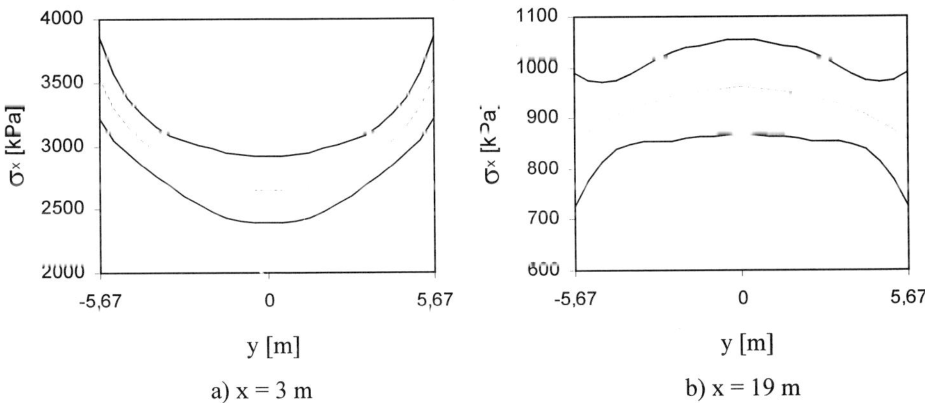

Figure 3 Shear lag effects – results of statistical analysis for selected sections *x* (mean ± 2 standard deviation)

Plotted are the values of the mean +/- 2*s* (*s* = standard deviation). These values represent the 95% confidence limits on the results, i.e. the limits that are exceeded with 2.5% probability on the plus side and 2.5% on the minus side (assuming the normal probability distribution of the response). The mean values are depicted by dotted lines.

Figure 3a shows the stress distribution at the cross section $x = 3$ m (with the classic shear lag). Figure 3b concerns the cross section $x = 19$ m where the negative shear lag occurs. It can be seen that the spread of the results is large in flange parts close to webs and at the middle of the flange, i.e. in the regions decisive for the design.

It has been found that the statistical results are sensitive to values of correlation lengths. This dependency is strong for both classic and negative shear lags. The variation coefficient is large for small correlation lengths in the longitudinal direction of the bottom flange (d_x), generally much larger than that for the correlation lengths in the transverse direction (d_y). Thus, to obtain reliable results, it is necessary to use appropriate values of the correlation lengths in the analysis (corresponding to the investigated case from the point of view of structural arrangement, material characteristics as well as the method of erection and environmental conditions). It seems that at the technology allowing a production process exhibiting long correlation lengths the statistical scatter of stress distribution can be lower than that for small correlation lengths. This finding is particularly important at prestressed concrete box girder bridges erected by the cantilever method where individual segments (typically of short lengths of about 3 m) are made in situ in their full widths in one stage.

CONCLUSIONS

The statistical variability of the technological and compositional aspects of concrete (represented by the modulus of elasticity in this study) and random imperfections in the geometry of cross-section (represented by the variability of the flange thickness) can cause a significant statistical scatter in the predicted stress values as well as in the predicted character of distribution of the stresses across the flange width. The utilisation of an appropriate model, which takes the effects of shear lag into consideration, is strongly recommended. It should be emphasised that the statistical spread of the results strongly depends on the accuracy of the choice of the correlation lengths.

ACKNOWLEDGEMENT

Authors express their thanks for the subsidy to Grant Agency of the Czech Republic - project No. 103/97/0074 and project No. 103/98/1479.

REFERENCES

1. KŘÍSTEK,V., STUDNIČKA, J. Negative shear lag in flanges of plated structures, J. Struct. Division, ASCE, Vol 117, No. 12, 1991, pp 3553-3569.

2. BRENNER, C. E. Stochastic finite element methods (literature review). University of Innsbruck, Working report No. 35-91, 1991.

3. YAMAZAKI, F., SHINOZUKA, M. & DASGUPTA, G. Neumann expansion for stochastic finite element analysis, J. Engrg. Mech., Vol 114, No. 8, 1988, pp 1335-1354.

4. BAŽANT, Z.P. & KŘÍSTEK, V. Shear Lag Effect and Uncertainty in Concrete Box Girder Creep. J. Struct. Engrg., Vol 113, No. 3, 1987, pp 557-574.

MECHANICAL PROPERTIES AND DURABILITY OF HIGH-PERFORMANCE CONCRETE

C Măgureanu

A Popa

Technical University of Cluj Napoca

Romania

ABSTRACT. This paper discusses some of the experimental results of a series of tests carried out on high strength concrete to 65 MPa. Amongst the characteristics evaluated are strength development, shrinkage, creep and elastic modulus. Further study on durability, ductility and long term deformation is currently being undertaken. Similar work on high strength concrete up to 65 MPa are also being undertaken. Based on the characteristics evaluated , it has been shown that high strength concrete at 65 MPa can be produced reliably using materials available locally.

Keywords: Concrete, Compressive strength, Tensile strength, Drying shrinkage, Creep, Modulus of elasticity, Silica fume, Specifications.

Professor Cornelia Măgureanu is at the Technical University of Cluj Napoca, Romania. She specializes in the high strength concrete, reinforced and prestressed concrete and in ferocement as well.

Professor Augustin Popa is at the Technical University of Cluj Napoca. He specializes in durability, rehabilitation, geotechnical and foundation. He is a head of Department of Geotechnical and Foundation at Technical University of Cluj Napoca.

INTRODUCTION

Due to the fast economic growth in Romania as well as the region, buildings are getting higher and more sophisticated. The enable buildings to go higher without structural supports of massive sizes, thus achieving higher rental space, and flexibility in both design and construction, the use of higher strength concrete is inevitable. High strength concrete can also reduce construction cost, a higher durability and accelerate construction cycles. In addition, the much – improved durability of high strength concrete can provide long term benefits to structures.

EXPERIMENTAL PROGRAMME

The experimental programme consisted of production and testing of high performance concrete specimens during a period of approximately 90, 150 and 300 days. The materials and mixture parameters used for preparing the specimens are briefly described below.

Materials

Cement

Ordinary Portland Cement Type II/A – 32.5 R, with fly ash in 20 %. Compressive strengths at 2 and 28 days of cement paste was 10 MPa and 32.5 MPa. The cement content in the specimens was 450 Kg/m^3.

Coarse aggregate

For high strength concrete,the mineral characteristics and strength of coarse aggregate itself controls the strength of concrete. The size of the coarse aggregate affects the strength; in general, the smaller size, the higher the strength. In the present study, for 65 MPa concrete, a 15 mm maximum size siliceous gravel consisting of smooth, round particles was used.

Fine aggregate

The fine aggregate was natural river sand, a 7 mm maximum size.

Silica fume

Silica fume affects the concrete due to its high pozzolanic effect, it reacts with cement and water to produce calcium silicate hydrate gel, that binds the aggregate together and gives the concrete strength. Silica fume acts as a filler occupying the voids between the cement grains reducing the porosity of the cement matrix and resulting in less permeable, more durable concrete. In the present study, the amount of silica fume is 10 percent of the total cement content

Admixtures

A superplasticizing, air entraining admixture (sodium lignosulphonat based) was added with a recommended dosage of 0.1% per 100 Kg of cement.

The water – cementitious ratio was 0.28.

RESULTS AND DISCUSSION

Strength Development

The strength development is shown in Figure 1. At each age, at least three specimens were tested and the average of the test values was taken as the compressive strength. The specimens were moist cured for 28 days and then air dried until the time of testing.

Silica fume concrete has compressive strength development patterns, which are generally different from those of portland cement concretes. Major contributions of silica fume to the strength take place prior to 28 days.

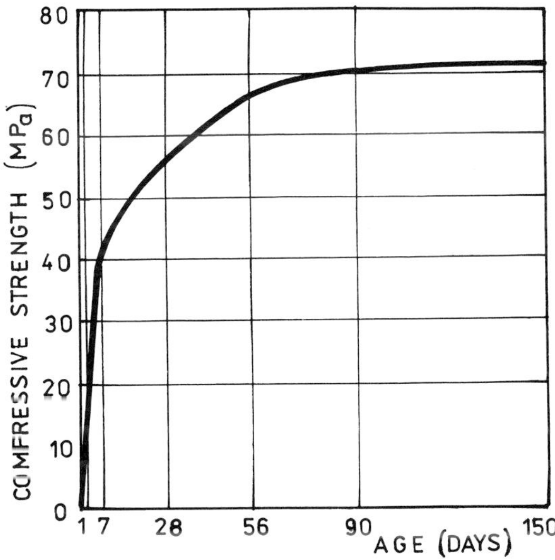

Figure 1 Compressive strength development

Flexural – Tensile Strength (Modulus of rupture)

Figure 2 shows obtained flexural tensile strength, or modulus of rupture at 3, 7, 28, 56 and 150 days. The maximum flexural – tensile strength obtained was 4.3 MPa and was measured by a beam flexural test. Flexural strength of concrete is about 6.5 % of compressive strength.

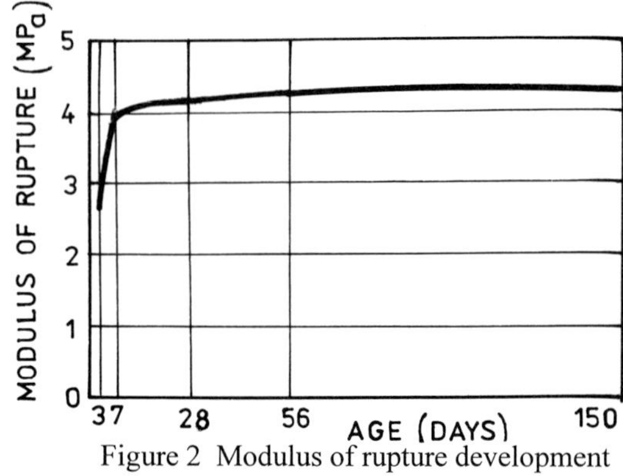

Figure 2 Modulus of rupture development

Static Elastic Modulus of Elasticity "E"

For Portland Cement and silica fume concrete for the compressive strength 65 MPa are of the order of about 40 GPa, Figure 3, and there is no significant difference between the "E" values obtained at 7 and 28 days.

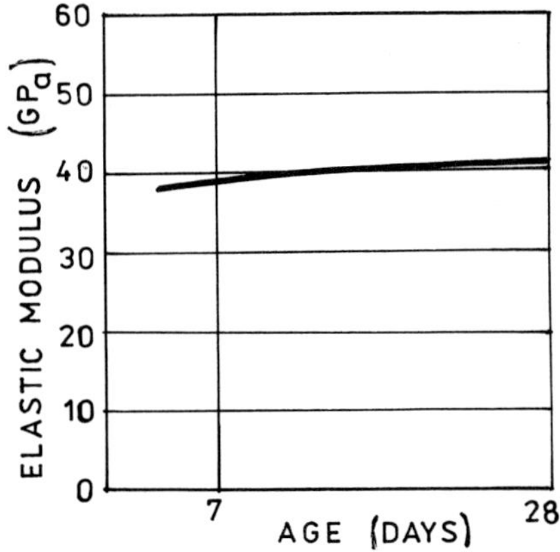

Figure 3 Modulus of elasticity at 7 and 28 days

Shrinkage

The concrete specimens were 100 x 100 mm in cross – section and 550 mm long. The experimental elements were kept in constant conditions (65 % humidity and 20°C temperature).The free shrinkage strain curve in time are presented in Figure 4.After 300 days the shrinkage strain is about 0.35 mm/m. The stabilizing tendency of shrinkage is obvious after 90 days.

Creep

The magnitude of long – term compressive load being 35 % of the compressive strength. The load is applied at the age 180 days. The experimental elements were kept in an air – conditioned room within constant conditions of humidity and temperature : U = (65 ± 5) % ;

t = (20 ± 2)° C. The overall experimental deformations ε_{cr+sh} (creep and shrinkage) as a function of time and represented in Figure 5. At time 0 is reprezented the instantaneous strain ε_e at loading.

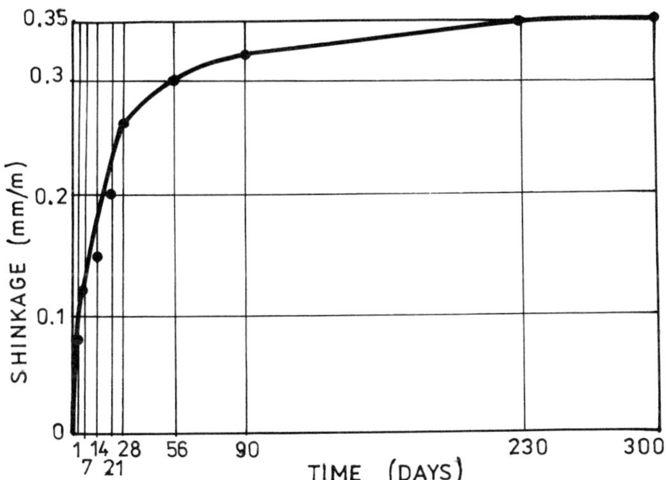

Figure 4 Shrinkage developement for high – strength concrete (65 MPa)

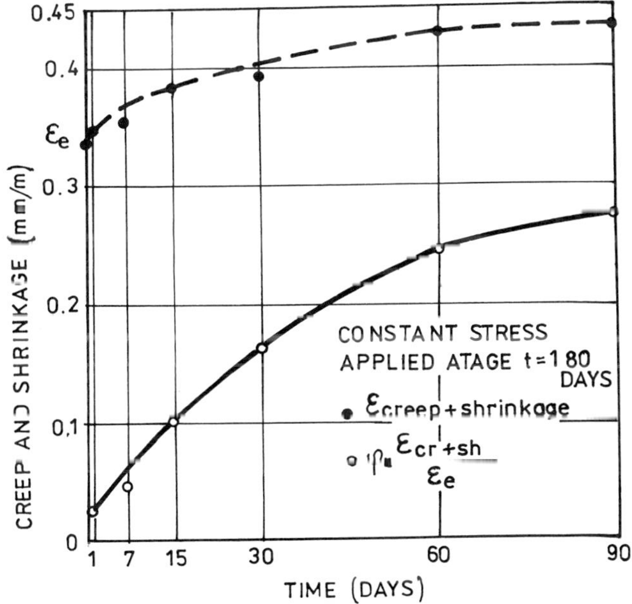

Figure 5 Long term strain development

The creep and shrinkage factor φ, experimentally determined is also represented.
The time depending deformations factor φ are determined with relation :

$$\varphi = \frac{\varepsilon_{cr+sh}}{\varepsilon_e}$$

Where ε_e is elastic, instantaneous strain.

Experimental results indicate, that the drying creep and shrinkage of high – strength concrete is reduced compared to the of normal – strength concrete, and stabilising tendency is obvious after 90 days.

CONCLUSIONS

The present paper has focused attention on high – strength concrete through the use of additives such as silica fume and superplasticizing admixtures. It has been demonstrated that significant improvements of strength and durability can be achieved with admixtures such as silica fume and superplasticiziers in concrete compositions. The research to be continued.

REFERENCES

1. CHEW, M Y L. Utilization if high strength concrete in Singapore, High Strength Concrete 1993, 20 – 24 June, Lillehammer Norway, Vol. 2, pp. 678 - 690

2. MORENO, J. The State of the Art of High Strength Concrete in Chicago – 225W. Wacker Drive. Concrete International : Design and Construction, Vol. 12, No. 1, January 1990 pp. 35 – 39.

3. RONNEBERG, H., Sandvik M., High Strength Concrete for North Sea Platforms. Concrete International Design and Construction, Vol. 12, No. 1, January 1990 pp. 29 – 34

4. AITCIN, P C. High Strength Concrete using Silica Fume, Proc. of the International Workshop on the use of Fly ash, slay, Silica Fume and other Siliceous Materials in Concrete. Ed. By W.G.Ryan, Sydney, Australia, 4 – 6 July 1988 pp. 2522 – 2267.

5. MALHOTRA, V M, CARETTE, G G, SIVASUNDARAM, V. Role of Silica Fume in Concrete : A review. Advances in concrete tehnology, Second Edition CANMET, Editor V.M. Malhotra 1994, pp. 915 - 990

6. DILGER, W H,.KRISHNA RAO, S V. High Performance Concrete Mixtures for Spun – Cast Concrete Poles, PCI Journal, July – Aug. 1997, pp. 82 – 95.

STRESS-RIBBON BRIDGES STIFFENED BY EXTERNAL CABLES OR ARCHES

J Strasky T Kulhavy

Technical University of Brno

Czech Republic

G Rayor

OBEC

United States of America

ABSTRACT. Stress-ribbon pedestrian bridges designed in the Czech Republic and in the United States are described in terms of their architecture and structural solution, the process of construction and the influence of the structural arrangement on the static and dynamic analysis. The stiffness of these bridges is mainly given by the geometric stiffness of the deck. The stress ribbon structures can be either cast in-situ or formed of precast units. In the case of precast structures the deck is assembled of precast segments that are suspended on bearing cables and shifted along to their final position. Prestressing is applied after casting the joints between the segments to ensure sufficient rigidity of the structures. The main advantage of the stress ribbon bridges is that they have a minimal environmental impact because they use very little material and can be erected independently from the existing terrain. Since they do not need bearings or expansion joints they need only minimal long-term maintenance.

Keywords: Pedestrian bridges, Stressed bibbon, Prestressed concrete, Precast segments, Post-tensioning, Composite structure, Arch, Suspension cable, Geometric stiffness.

Professor Jiri Strasky is Head of the Department of Concrete and Masonry Structures of the Technical University of Brno, Czech Republic and Technical Director of Strasky Husty and Partners, Consulting Engineer, Brno & Mill Valley, California. He specialises in the development and design of the light and transparent steel and concrete structures. Professor Strasky has published widely and serves on many Technical Committees. He is a member of the presidia of the ***fib*** - International federation for structural concrete.

Mr Gary Rayor is a Professional Engineer with OBEC, Consulting Engineers, Eugene, Oregon. He specialises in the design of prestressed concrete structures.

Mr Tomas Kulhavy is a Ph.D. student at the Department of Concrete and Masonry Structures of the Technical University of Brno and Design Engineer with Strasky Husty and Partners, Consulting Engineer, Brno, Czech Republic. He specialises in a design of stressed ribbon structures.

INTRODUCTION

Stress ribbon bridges is the term that has been coined to describe structures formed by a very slender concrete deck with the shape of a catenary. They can be designed with one or more spans and are characterised by successive and complementary smooth curves. The curves blend into the rural environment and their forms, the most simple and basic of structural solutions, clearly articulate the flow of internal forces. Their fine dimensions also correspond to a human scale [1],[2].

A characteristic feature of stress ribbon structures, in addition to their very slender concrete decks, is that their stiffness and stability comes from their geometric stiffness. They are able to resist not only uniformly distributed load but also large concentrated load created by the wheels of heavy trucks. Heavy flooding in the Czech Republic in 1997 also showed that they can resist large ultimate load. Although stress ribbon structures have low natural frequencies our experience confirmed that the speed of motion and acceleration of the deck caused by walking is within acceptable limits. Also our detailed dynamic test confirmed that vandals cannot damage these structures.

Wind tunnel tests and performance of the completed structures have shown that the stress ribbon structures have excellent behaviour to the wind load. From the dynamic analysis of the stress ribbon structures it is evident, that although the cross section of the deck is weak in torsion, the pure torsion frequency is far from the bending frequency. It is caused by the fact that the decks of the stress ribbon structures are always in a vertical curve and are fixed at the abutments.

However, the classical stress ribbon type structures have one main disadvantage - the need to resist very large horizontal forces at the abutments. In some cases, depending on ground conditions, this can outweigh the cost savings of the deck. Therefore we were looking for solutions where the classical stress ribbon structures are combined with another structural members – cables or arches. We tried to reduce or totally eliminate horizontal loading of the foundations while maintaining the excellent behaviour the stress ribbon structures. The possibility of the above combination of the structural members is demonstrated on two examples of the recently designed pedestrian bridges.

STRESS RIBBON STRUCTURE STIFFEN BY SUSPENSION CABLES

Willamette River Pedestrian Bridge

Recently, a new suspension pedestrian bridge across the Willamette River in Eugene, Oregon has been opened Figure 1. The structure is formed by very slender concrete deck flexibly anchored at abutments and suspended on suspension cables formed by monostrands grouted in steel tubes. The structural solution of the bridge was derived from the design of the Vranov Lake Pedestrian Bridge [3].

Two parts form the bridge of an overall length 178.8 metres: precast suspended spans and cast-in-place approach spans, a platform and stairs – see Figure 2a. The suspended spans are 17 metres and 103 metres for side spans and main span respectively.

The entire bridge consists of 5 spans including a curve ramp into a park on the east end. The deck width is 4.2 metres [4]. The typical deck section consists of a series of precast concrete deck segments 3.0 metres long longitudinally post-tensioned together after erection –see Figure 2b. Two edge girders and the deck slab form the cross section of the segments. At joints transverse diaphragms stiffen the segments. Design of the bridge called for an observation platform situated in the middle of the main span. Therefore the main cables have to be "threaded" through the central wider deck panels of the main span.

Figure 1 Willamette River Pedestrian Bridge

The maximum deck panel thickness is 0.42 metres resulting in a span-depth ratio of the main span of 245. This remarkable span-depth ratio is achieved by post-tensioning and damping the deck in curvature opposite to that of the main cables. The resulting deck/cable system is self-stiffened and extremely rigid minimising lateral movement and aerodynamic excitation. Both elastic and inelastic finite element modelling and aeroelastic model testing in a wind tunnel confirmed excellent stability of the bridge under transient pedestrian and aerodynamic excitation.

Bents 1 and 4 are fixed at structural steel anchorages that resist longitudinal main cable forces. At the bent 4 where the steel anchorage consists of tripod shaped support that anchors the main cables and supports a saucer shaped viewing deck above. The tripod is formed by two tension ties connected by a top diaphragm and by a compression strut founded in one common footing. The tension from the main suspension cables is transferred via ties into the ranked steel piles. The pylons have a shape of the letter A. The legs, that are slightly curved, are connected by a top diaphragm and by a crossbeam situated in the deck level. During construction the pylons are pin connected, after the erection of the deck the pylons are fixed into the foundation.

Complex bearing details typical in suspension bridges were completely eliminated by placing deck expansion joints at tower bents 2 and 3 and post-tensioning each deck series created individually. Deck joints allowed thermal movement and elastic shortening due to deck post-tensioning to occur without inducing additional longitudinal loads into the abutments, eliminated the possibility of the deck being in tension, and allowed fixing of the ramp spans 4 and 5 and spiral staircase structure. In order to eliminate sudden longitudinal movement of span 2, shock transmission devices or dampers are utilised at the deck's joints. The dampers allow slow temperature and creep and shrinkage movements to occur while damping live load and tractive forces. In addition, the dampers resist longitudinal earthquake in the main span along with the fixity of the main cables at the mid-span of span 2. In all earthquake analyses performed, the dampers were considered functioning and failing and design was for the worst case.

Although the results of finite element modelling confirm good behaviour of the bridge under aerodynamic excitation, the authors felt that it was necessary to check the response of the structure in the wind tunnel. An aeroelastic model was constructed at 1:68.7 scale and tested by Professor Miros Pirner at the Institute of Theoretical and Applied Mechanics, Academy of Sciences of the Czech Republic

The erection of the structure was developed from the erection of the classical stress ribbon structures – see Figure 2. At first central segment that form the mid-span observation platform were erected as stress ribbon structures. Temporary cables from strands anchored at the bents 1 and 4 formed the bearing cables. After the erection of these segments the main suspension cables were installed and remaining part of the segments was erected. The segments were shifted along the erection cables and suspended on the main suspension cables. After erection of all segments the joints between the segments were cast and post-tensioned.

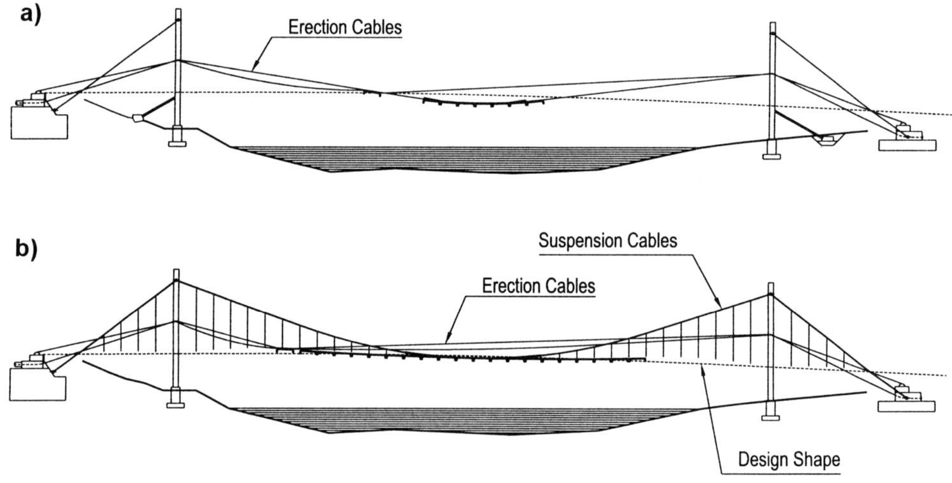

Figure 2 Willamette River Pedestrian Bridge: Erection of the deck

STRESS RIBBON STRUCTURE STIFFENED BY ARCH

Pedestrian Bridge across the Radbuza River in Plzen

A new structural system was developed for a proposed pedestrian bridge across the Radbuza River in Plzen. An 80m long structure was required, with a relatively low elevation due to the approaches on both sides. There are utilities in the banks on both sides of the river, and the rock is located about 10m below ground level. Thus the classical stress-ribbon structure would be far too expensive.

Instead a structure was designed combining a steel tube arch with a span length of 62.4m and 'boldness' of 973m (ratio of square of length L divided by rise f) with a modified stress ribbon type deck – see Figure 3. Two steel tubes filled with concrete form the arch and are supported on concrete foundations on each bank of the river. The tubes have a diameter of 45cm and a thickness of 1cm. The steel arches support the deck, which is formed by a two-span stress ribbon. The deck is a composite of precast concrete elements with a cast-in-place concrete portion. The precast concrete deck elements consist of waffle-slabs, 3.6m wide by 3m long, and the width between railings on the deck is 4.2m. The precast concrete elements are designed so that they are suspended on bearing cables situated in the areas to be filled by the in-situ concrete

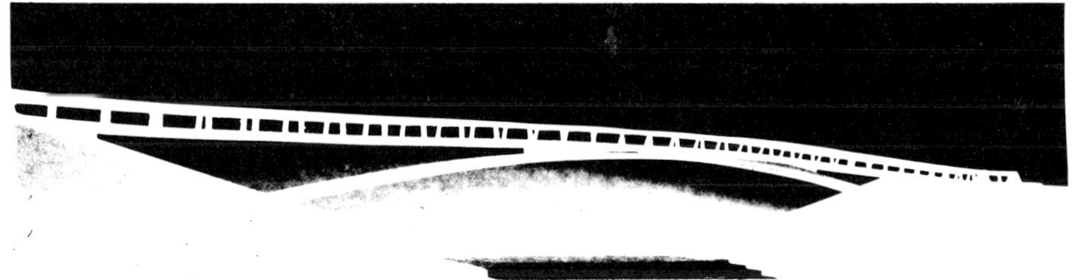

Figure 3 Pedestrian Bridge across the Radbuza River in Plzen

The deck has a fixed connection with the arch at mid-span. Steel plates supporting the deck extend for a short distance from the mid-span outward toward the sides to help keep the maximum variability in the slope of 8%. At both ends, the stress ribbon is fixed to a diaphragm supported by two inclined concrete struts fixed in the foundations of the arch. Two tension pin piles support the diaphragm and nine compression micro-piles support the foundation. The structure thus forms a self-anchoring system, where the horizontal forces from the stress ribbon are transferred by the inclined concrete struts to the foundation where they are balanced against the horizontal component of the arch.

Process of construction is as follows – see Figure 4. Once the piles have been installed, the concrete foundations and diaphragms will be cast in place. Cables anchored at the diaphragms will suspend the steel arch tubes. By increasing the lengths of the cables, the arches will be rotated to the design position. When the two sides of the arch meet, they will be connected and the tubes filled with concrete.

The next stage will be to draw the bearing cables across the river. At the mid-span of the arches, the cables will be supported on hydraulic jacks. The precast elements will be shifted from both side-spans to the middle of the bridge. When the segments have been erected, the jacks will be lowered until the segments are supported directly on the arch. Prestressing cables will then be erected, reinforcing steel placed, and the in-situ portions of the deck will be cast. Finally the structure will be post-tensioned.

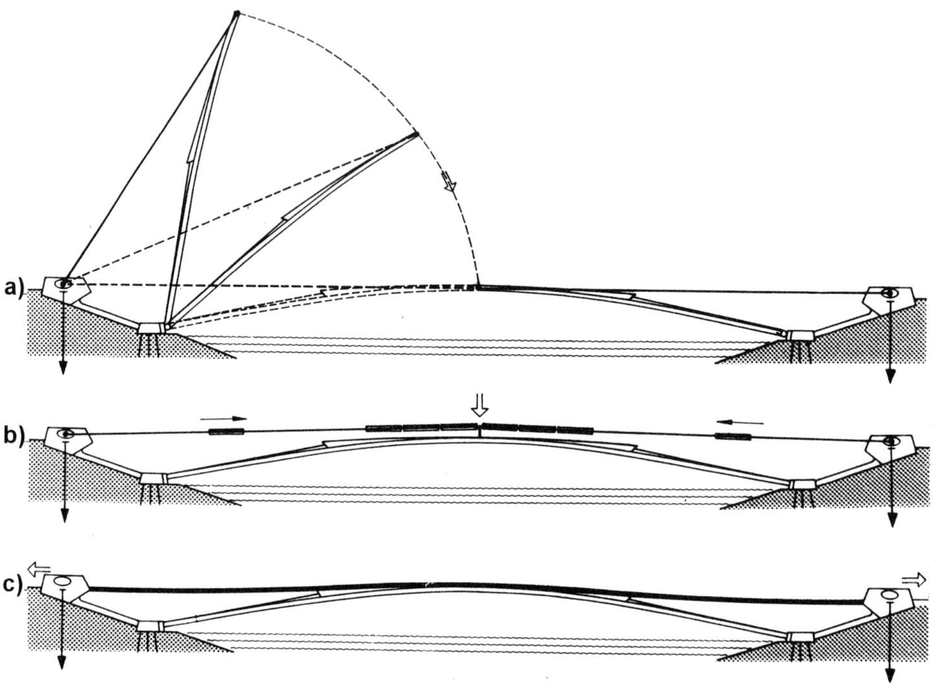

Figure 4 Pedestrian Bridge across the Radbuza River in Plzen: Erection of the structure

The function of the structure has been confirmed by detailed static and dynamic analysis, and also verified on a full aeroelastic model by both dynamic and wind tunnel tests. A static modelling of the whole structure on the scale 1:10 was also performed at the Technical University of Brno, Czech Republic. Results of the test confirmed the static assumptions and results of the analysis. At present long-term deformations of the structure are being studied.

A similar approach can be used for a structure where the stress ribbon deck is suspended from the arch and also fixed at the end abutments connected with the foundation of the arch by a compression strut.

CONCLUSIONS

The structures described in this paper present recent development of the stress ribbon structures and show further development of this exciting structural type. They have their own architectural value and have a minimum impact on the environment. Their economic form expresses recent development of science and technology. Therefore the author believe that will find further applications too.

ACKNOWLEDGEMENTS

The bridge in the Czech Republic was designed by Strasky, Husty and Partners, Czech Republic, The bridge designed in the United States was deigned by OBEC, Consulting Engineer, Eugene, Oregon with the collaboration of Jiri Strasky. Theoretical investigations of these structures and model tests were done under support of the Grant 103/96/1635 GACR of the Grant Agency of the Czech Republic.

REFERENCES

1. STRASKY,J.: Ribbon Footbridges. Concrete Quarterly 156. British Cement Association. January-March 1988.

2. REDFIELD,C.- STRASKY,J.: Sacramento ribbon. Concrete Quarterly. British Cement Association. Autumn 1992.

3. STRASKY,J.: Pedestrian bridge at Lake Vranov, Czech Republic. Proceedings of the Institution of Civil Engineers, Civil Engineering, London August 1995.

3. STRASKY, J.- RAYOR, G. Positively not Pedestrian. ASCE Civil Engineering, March 1998.

THE EXPERIMENTAL RESEARCH AND ANALYSIS OF THE FRICTION FORCES OF THE EARTH ACTING ON THE ANCHOR BAR OF ANCHOR BRIDGE DAIS

Z Li X-Y Wang W-M Liang

Taiyuan University of Technology

Y Li J-Y Zhang

Twelfth Engineering Bureau of Railway Ministry

China

ABSTRACT. The friction forces of earth acting on the anchor bar of bridge dais have been measured for 20h insitu. It has been found that the anchorage tension varies with time and temperature, as well as with the changes and distribution of the friction forces of earth. The results of this research can be used to improve design and quality of engineering.

Keywords: Earth friction force, Upper cane anchor bar.

Professor Zhu Li works in the Taiyuan University of Technology of China. His main research interests include the elastic-plastic dynamics and the experimental mechanics.

Mr Yao Li is a senior engineer of Twelfth Engineering Bureau of Railway Ministry of China. His main research interests include the elastic-plastic dynamics and the anti-seismic structure of building.

Mr Xian-Yao Wang is a senior engineer of Taiyuan University of Technology of China. His main research interests include the elastic-plastic dynamics and the experimental mechanics.

Mr Jian-Yuan Zhang is an engineer of Twelfth Engineering Bureau of Railway Ministry of China. His main research interests include the elastic-plastic dynamics and the anti-seismic structure of building.

Mr Wei-Min Liang is a lecturer of Taiyuan University of Technology, he is interested in the elastic-plastic dynamics.

INTRODUCTION

The anchor bar is of most important function in the system of anchor bridge dais, which can stabilize and reinforce the bridge dais and strengthen the function of shock proof and resistance. The forces that resist the tension of the anchor bar include three parts, which the first σ is the resistance stress of earth acting on anchor block, the second f is the friction forces of earth acting on around the anchor block, and the third F is the friction forces of earth acting on the anchor bar. The resistance σ stress of earth acting on anchor block and the friction forces f of earth acting on around the anchor block have been considered fully in the previous design. The research about the friction forces of earth acting on the anchor bar is more difficult by comparison. The friction forces of earth acting on the anchor bar are taken as an advantageous factor without considering in the present design[1-3]. This kind of design method is safe out of question, on the other hand, it is not economic also. According to the results of testing on the anchor bar in this paper, the length of anchor bar, the magnitude of the tension of anchor bar and the friction forces of earth take an important role in stabilizing and reinforcing the bridge dais . It should be considered in the future design.

EXPERIMENTAL CONDITIONS AND METHOD

Four anchor bars of full scale are selected as the test pieces. The length of each test piece is 36m. Each test piece is consisted of four strong steel bar welded together whose length is 9m and diameter is 32mm. The weld joints of two test pieces are protected against the friction forces of earth with tubing brick and the other are directly embedded in earth for researching the effect of weld joints on the tension. The two test pieces protected can be regarded as the seamless bar and the others can be regarded as the seamy bar. One end of the test piece is fixed in the bridge dais and the other end of the test piece can be fixed with nut through the duck of the anchor block. The anchor block is a rectangle$(17.2\times2\times1)\text{m}^3$. It weights 800kN. The testing pieces are all tensed with the tensioner. Before tensing, the position of the anchor bar must be firstly deployed well, then the filler that are consisted of sand, rock and earth should be filled layer by layer. The thickness of one layer is 100mm and the deepness of the filled earth is total 1m. Each the filled layer must be planished by road roller and the dry density of the filled earth must attained 2.1kN/m^3.

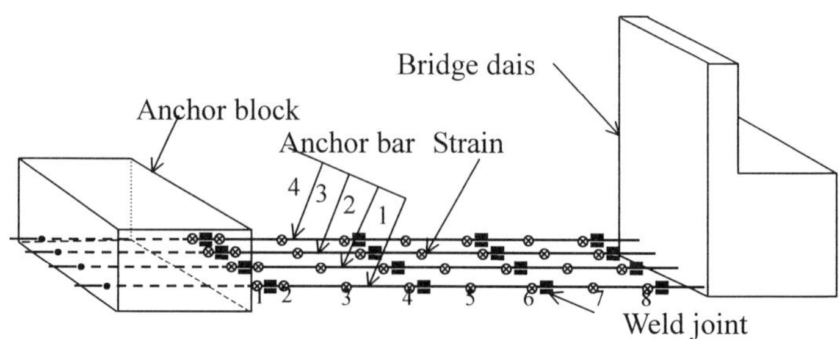

Figure 1 The sketch map of measurement points arrangement

Eight measurement points are arranged along each test piece, as shown in Figure 1. In order to eliminate the effect of bending stress, the high precision strain gauge are symmetrically pasted to measurement points. Because the strain gauges are embedded in earth for a long time, each of strain gauges is processed to protect against invalidation. The data recording system is Fluke2280b are adopted to automatically record date. The testing time is 20 hours and the changes of temperature are too recorded.

ANALYSIS AND DISCUSSION OF EXPERIMENTAL RESULTS

The maximum value of the tension of anchor bar can be gained as

$$P_{max} = P_i + F_i \tag{1}$$

Where, $P_{\max}$ is the maximum value of tension of the anchor bar, P_i is the tension of each measurement points, F_i is the friction force of earth acting on the anchor bar measured from 1 point to i point.

The variation rule of the tension with time at all measurement points including the seamless and seamy anchor bar are shown in Figure 2 and 3, which are measured as long as 20 hours on the spot. The changes of the tensions at all measurement points are basic same about two kind of anchor bar. The tensions at all measurement points become stable after beginning to tensing two hours. The testing on the spot began at night 22:00 hours and stopped at 18:30 hours the day after. Because of the changing of temperature□the filled earth will expand or compress so that it should cause the changes of the tensile stress of anchor bar. Now we can see that the tension of anchor bar change more intensively in a certainly phase. From Figures 2 and 3 we find out that the tensions of the anchor bar at all measurement points are changed slightly from 22:00 to 7:00 the day after for the average difference in temperature being only $\pm 1.5°C$, it is changed intensively from 7:00 to 11:00 for the average difference in temperature gaining $\pm 5°C$, it is also changed slightly from 11:00 to 18:30 in that the average difference in temperature is $\pm 2.1°C$. By the above analysis we can know that the tension of the anchor bar get stable after 2 hours, but it must be considered that the additional tension caused by the changes of temperature if the anchor bar embedded in shallow fill. According to the testing results in this paper, the additional tension caused by $\pm 5°C$ of average difference in temperature can achieve 20kN.

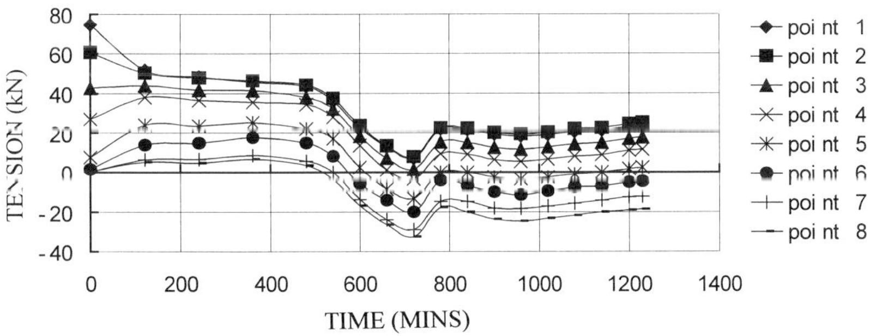

Figure 2 The curve of tension with time of the seamless bar

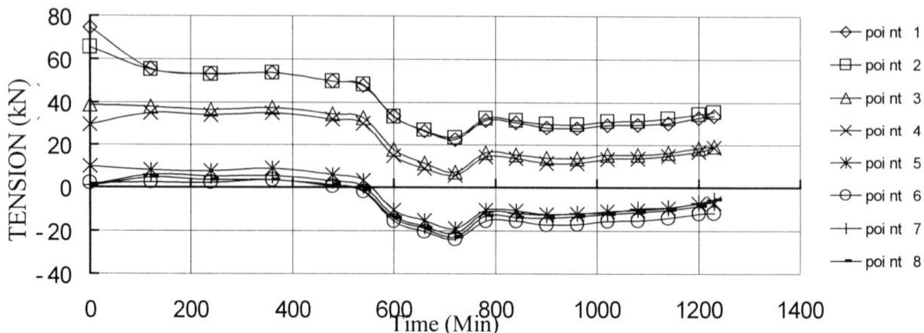

Figure 3 The curve of tension with time of the seamy bar

The distribution of the tension along the anchor bar is shown in Figures 4 and 5. The distribution of the tension is very different between the seamless and seamy bar. Because yielding very large friction force of earth at weld joints, the tensions of anchor bar change intensively to the seamy bar at weld joints. The tensions with the time along the seamy bar show ladder shape distribution, whereas, the tensions of seamless bar show linearly

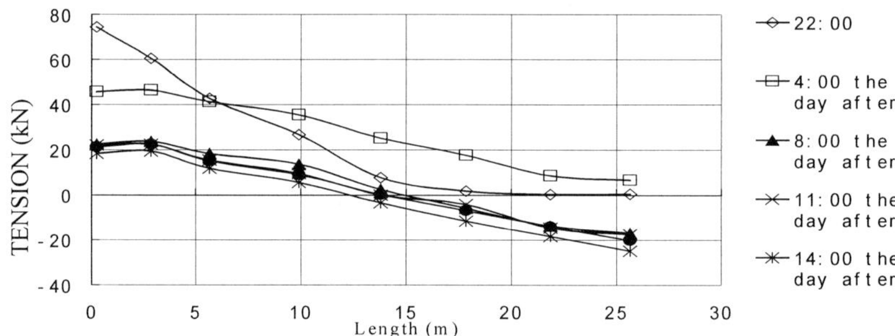

Figure 4 The curve of the tension along of the seamless bar

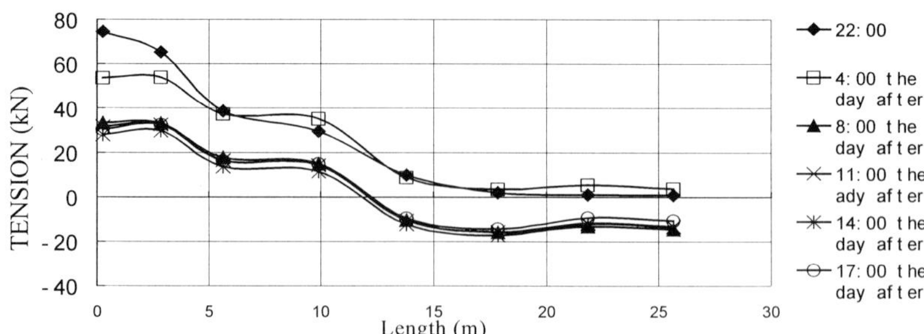

Figure 5 The curve of the tension along of the seamy bar

distribution. From the distribution regularity of the tension of anchor bar as shown in Figures 4 and 5 we can conclude that the average transferable length is about 4.2kN/m, which has an important reference value about the optimization design of the length of anchor bar. The regularity of the friction force of earth with the time is shown in Figures 6 and 7, it is the basic same about the seamless and seamy bar. The friction forces of earth acting on the anchor bar at all measurement points are very largely changed in tensing phase, but it becomes stable after 2 hours. Because the function of wave of earthquake similar to suddenly tense, which both yield very large friction forces of earth acting on the anchor bar, the destroying forces to the bridge dais can be effectively amortized by the friction of earth acting on the anchor bar.

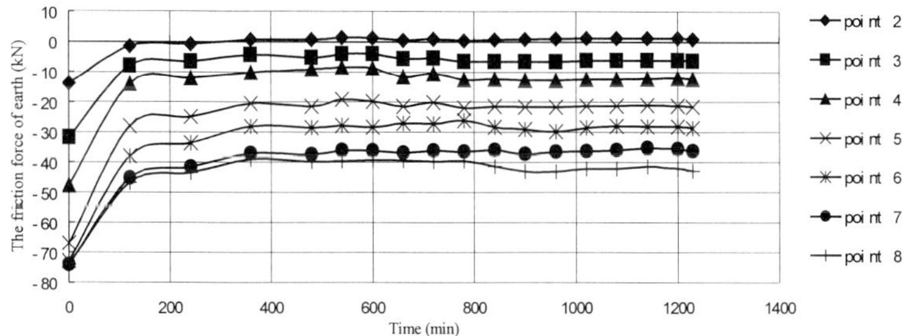

Figure 6 The curve of the friction forces with time of earth acting on the seamless bar

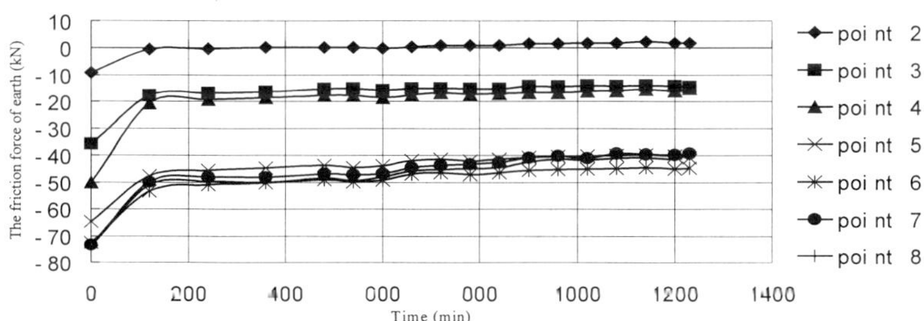

Figure 7 The curve of the friction forces with time of earth acting on the seamy bar

The regularity of the friction force of earth along the anchor bar is shown in Figures 8 and 9. The friction forces of earth along the anchor bar at all measurement points are almost the same in the process of tensing, whose distribution is shown as the exponential variation regularity. But, as time gone on the distribution of the friction forces of earth along the anchor bar at all measurement points become the same as the distribution of the tension along the anchor bar. From the distribution of the friction forces of earth along the anchor bar is the exponential variation regularity in the process of tensing, we can conclude that the friction forces of earth acting on the anchor bar can resist the dynamic load acting on the bridge dais.

CONCLUSIONS

1. The effect of temperature must be considered in the design of the tension of anchor bar.

2. The average transferable length should be considered in the design of anchor bar length.

3. If the weld joints are directly embedded in earth, the friction forces of earth acting on the anchor bar should be largely increased.

4. The friction forces of earth acting on anchor bar take an important role in the bridge protected against earthquake.

5. The friction forces of earth acting on the anchor bar is a main component of the forces that resist tensing, it should be fully considered in the design of anchor bridge dais.

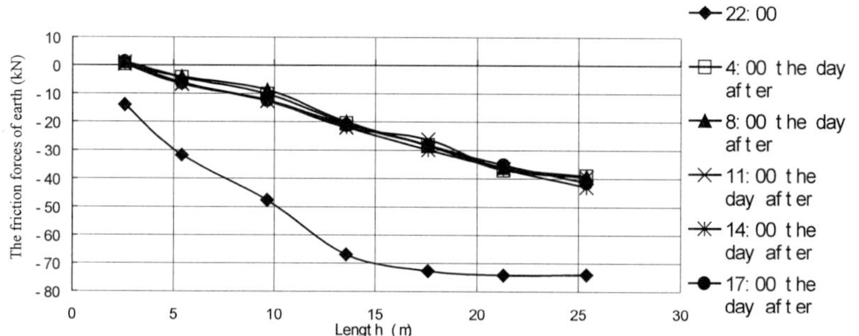

Figure 8 The curve of the friction forces of earth along the seamless bar

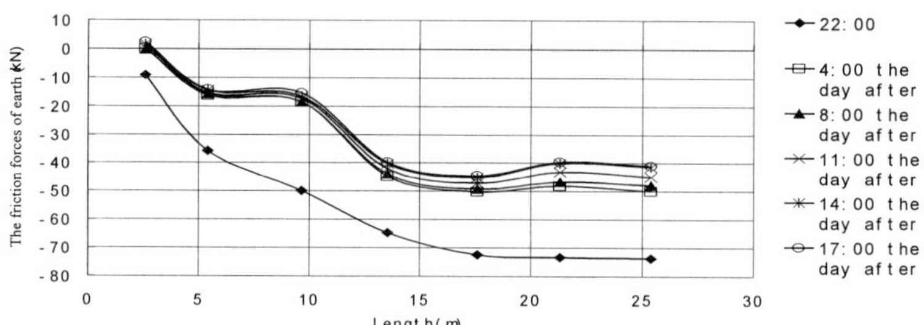

Figure 9 The curve of the friction forces of earth along the seamy bar

REFERENCES

1. ZONG-FANG YANG, The experimental research of tension technology about curve prestress, Construction Technology, 1986, Vol. 8, 16-18

2. ZONG-FANG YANG, The calculation of fixing loss of curve prestress, Building Structure , 1988, Vol. 8, 14-16

3. BING-JUN LU, The fixing plate structure about resisting pressure of earth Railway Ministry Pressing House of China Beijing, 1988, 1-105

A CONTRIBUTION TO THE STRESS-STRAIN RELATIONSHIP OF CONCRETE

H Schneeberger

University of Erlangen

Germany

ABSTRACT. A hypothesis of the stress-strain relationship of concrete is given and it is shown with some series of experiments, that the agreement of the data with the hypothesis is very good.

A1. Prisms of similar compressive strength were centrically loaded (centric means, that the load is uniformly distributed on the pressed side)with different rates of strain ε.

A2. The hypothesis σ (ε) for the stress is also valid for the most compressed fibre of the compression zone: The results of experiments with short-term-loading are given for different compressive strengths.

B. As supplement it is shown, that the same hypothesis is also valid in tension tests of steel for the region of strain hardening.

Keywords: Hooke's law, Generalisation, Exponential formulation.

Professor Dr Hans Schneeberger is a mathematician and statistician at the University of Erlangen, Germany, as former Associate Professor at the University of Technology, Munich he was acquainted with problems of analysing technical experiments.

INTRODUCTION

The experiments, which are analysed here, were made at the University of Technology in Munich and are published in [1] and [2]. The one tension test with steel is from the University of Technology in Braunschweig [3]. As in these papers no numerical data are given, but curves as result of a graphical analysis with these data, the author used points (ϵ, σ) from these curves for his numerical analysis. These points are given in the tables and plotted in the figures as stars or similar symbols.

THE THEORY

It is a well-known fact that Hooke's law

$$\sigma = E \, . \, \varepsilon \tag{1}$$

i.e. the stress σ is a linear function of the strain ε, is an approximation of the stress-strain relation $\sigma = \sigma(\varepsilon)$ only for smaller values of ε.

It will be shown in the following, that the hypothesis approximates the experimental points

$$\sigma(\varepsilon) = \acute{\alpha} \, . \, (e^{\beta \epsilon} - 1); \; \acute{\alpha}, \beta < 0 \tag{2}$$

(ε, σ) very well within well-defined much greater regions, e.g. until fractures appear in [2] or up to 90% of the utmost loading in [1].

Expanding exp.($\beta \, \varepsilon$) in a Taylor's series, one has from formula (2):

$$\sigma(\epsilon) = \acute{\alpha}. \left(\left(1 + \frac{\beta\epsilon}{1!} + \frac{(\beta\epsilon)^2}{2!} + \ldots \right) - 1 \right) = \acute{\alpha}\beta\varepsilon + \tfrac{1}{2} \acute{\alpha}\beta^2 \varepsilon^2 + \ldots \tag{3}$$

The first term on the right side is Hooke's approximation $\sigma(\varepsilon) = \acute{\alpha}\beta.\varepsilon$, $\acute{\alpha}\beta > 0$ is the coefficient of elasticity E. Formula (2) is an exponentially growing process, beginning from $(\varepsilon, \sigma) = (0,0)$ to the horizontal asymptote with $\sigma = = \acute{\alpha} > 0$. It is known from numerous experiments, that $d\sigma / d\epsilon$ is diminishing with growing ε. But in what way? We have this

$$d\sigma / d\varepsilon = \beta \acute{\alpha} e^{\beta \epsilon} = \beta(\sigma + \acute{\alpha}) = \beta\sigma + \beta\acute{\alpha} \tag{4}$$

differential equation of formula (2) is a generalisation of the famous differential equation of radioactive decreasing (more general exponential decreasing in techniques and economics), namely

$$dI/dt = \beta.I \; (\beta < 0) \tag{5}$$

with t = time and I = intensity. Formula (5) means, that the decreasing of I with time t is linearly proportional to I. The factor of proportionality is $\beta < 0$. Solving (5) gives

$$I = I_{oe}{}^{\beta t} \tag{6}$$

with intensity I_o at time t=0.

In formula (4) – compared with (5) – we have on the right side the additional factor $\acute{\alpha}\beta>0$. This is necessary, as otherwise $d\sigma/d\varepsilon$ for $(\varepsilon, \sigma) = (0,0)$ would be zero. So the stress-strain curves begin in $(\varepsilon, \sigma) = (0,0)$ with an angle γ, where tg $\gamma = \acute{\alpha}\beta$. Formula (4) means, that $d\sigma/d\varepsilon$ decreases from $\acute{\alpha}\beta$ linearly proportional to σ. The factor of proportionality is $\beta<0$. When $\sigma = -\acute{\alpha}$, $d\sigma/d\varepsilon$ becomes zero.

The Calculation

The mathematical problem is the estimation of the parameters $\acute{\alpha}$ and β in formula (2).

We assume, that the points ($x_i = \varepsilon_i$, $y_i = \sigma_i$), i-1,..., n of a stress-strain experiment are given. Then we estimate the parameters $\acute{\alpha}$ and β with the method of Least Squares from Gauss: we minimise

$$Q = \Sigma\ (\sigma_i - \sigma_i)^2 \qquad (7)$$

where σ_i are the data, $\sigma_i = \acute{\alpha}\ (e^{\beta\varepsilon i} - 1)$.

As method of calculation of the optimum values of $\acute{\alpha}$ and β I used the robust Simplex Method for Non-linear Programming of Nelder and Mead [4], (cf. [5]). Q is a measure of exactness of the minimisation process.

APPLICATIONS

A1 Centrically Loaded Prisms with Constant Rate of Strain

(Centric means, that the load is uniformly distributed on the pressed side). In 1962 in [2] the data of many such experiments at the University of Technology of Munich were published. The results are given as (ε, λ) – curves with $\lambda=\sigma/W$; W is the strength of cubes of 200.200.200 mm^3. As experimental data I took points (ε, λ) from some of these graphs, where ε=0, 0.2, ...n.0.2 up to approximately an ε-value, where first fractures were registered. N is the number used in formula (7). This n-th point also is approximately the maximum of the stress-strain curve. Instead of formula (7) I minimised $\Sigma\ (\lambda_i - \lambda_i)^2 = q$; when $(\acute{\alpha}, \beta)$ in $\lambda = \acute{\alpha}$ (exp $(\beta\varepsilon)$-1) minimise q, then a=w $\acute{\alpha}$ and b=β in σ=a (exp (bε)-1) minimise Q in formula (7). $Q=q.W^2$.

The data ε, λ, λ for these different ε-values are given in Table 1a, the corresponding plots in Figure 1. Data points (ε, λ) are given by stars or similar symbols, the curves are $\lambda=\acute{\alpha}(e^{\beta\varepsilon}-1)$ with $\acute{\alpha}$ and β according to Table 1b. In the last column of Table 1b q-values are given as a measure of correspondence of λ- and λ-values.

From these curves in Figure 1 once can see, that in the last points of the curves the correspondence can be smaller. This comes from the way, in which we choose the n points for our analysis according to formula (7). Obviously the last, n-th point is already beyond the region of validity of hypothesis (2). So for example for the curve with ε=1%/30 min the last point is clearly below the curve. Calculation without this point gives $q=.99.10^{-4}$ (Table 1b: $q=3.616.10^{-4}$).

Table 1a Data (ε, λ) and stress-strain relation $\lambda=\acute{\alpha}(e^{\beta\varepsilon}-1)$ for centrically loaded prisms with constant rate of strain

W[1]	t[2]											
24.3	7.5	ε[3]	2	.4	.6	.8	1	1.2	1.4	1.6	1.8	2
		λ[4]	.235	.422	.565	.675	750	.809	.845	.869	.886	.892
		λ	.248	.432	.567	.668	.742	.797	.837	.867	.889	.906
24.3	30	ε	.2	.4.	.6	.8	1	1.2	1.4	1.6	1.8	2
		λ	.218	.394	.517	.612	.680	.733	.770	.802	.820	.835
		λ	.226	.394	.519	.611	.679	.730	.767	.795	.816	.831
24.4	120	ε	.2	.4	.6	.8	1	1.2	1.4	1.6	1.8	2
		λ	.214	.375	.494	.583	.642	.690	.732	.755	.773	.790
		λ	.217	.377	.495	.581	.645	.691	.725	.750	.769	.782

Table 1a continued

t					
30	ε	2.2	2.4		
	λ	0.842	0.837		
	λ	0.842	0.850		
	ε	2.2	2.4	2.6	2.8
	λ	0.800	0.800	0.800	0.795
	λ	0.792	0.800	0.805	0.809

Table 1b Parameters $\acute{\alpha}$, β; q

t	$\acute{\alpha}$	β	q[5]
7.5	-0.953	-1.511	8.055
30	-0.875	-1.500	3.616
120	-0.820	-1.544	4.481

[1] W (N/mm^2) = cube strength of cubes 200.200.200 (mm^3); [2] t is time in ε = const. (1‰ /t min); [3] ε = strain (‰); [4] $\lambda=\sigma/W$; [5] q in 10^{-4}

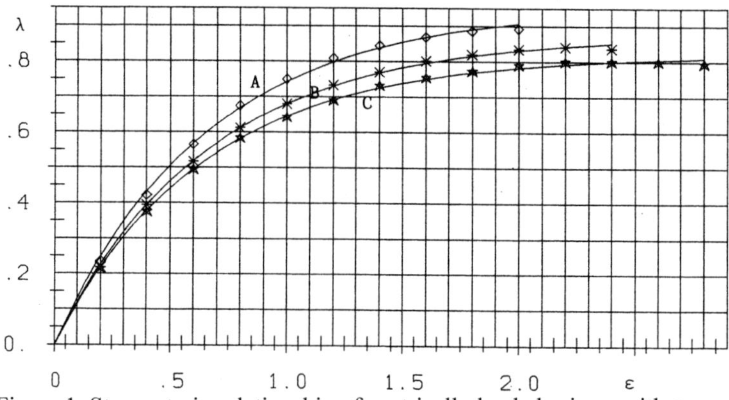

Figure 1 Stress-strain relationship of centrically loaded prisms with ε=const.: A) ε=1‰/7.5 min; B) ε=1‰ / 30 min; C) ε=1‰ / 120 min

The fact stated here for concrete is quite the same for steel (a reason to cite this material). But the steel-workers have marked the region, where formula (2) is valid, by an own name, namely region of strain hardening. Behind that the region called reduction in area begins. I shall show in Section B, that in the region of strain hardening hypothesis (2) is also valid for steel.

A2 Non-Centrically Loaded Prisms Under Short-Term Loading

The basis for this investigation was a great program of experiments at the University of Technology in Munich [1], with which at last a theory of Galilei's problem of the distribution of the stress in beams should be evolved (cf. [6]). The results of this work are published in [1]. For the sake of shortness the techniques of the loading process are given by repeating the photograph from [1] and a sketch (Figure 2).

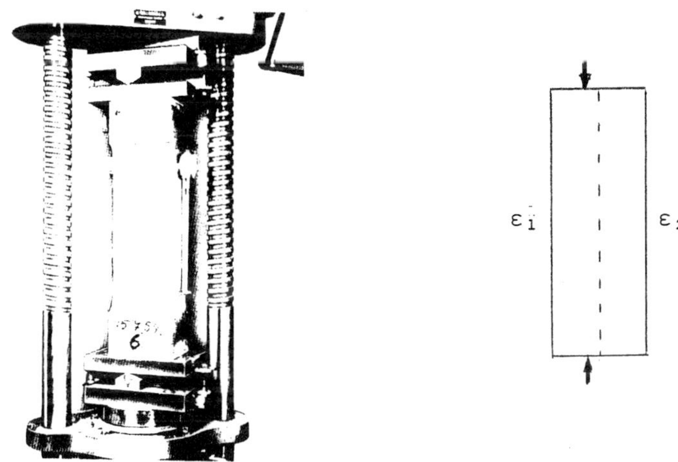

Figure 2 Experimental arrangement

Prisms of the same cube strength W were pressed with different eccentricities up to fracture. Then by intelligent graphical interpolation the (ε_1, σ)-values were found, where the strain $\varepsilon_2=0$; this ε_1 (called ε_o in [1] and here) is the strain at the most compressed fibre of the compression zone.

The results of this very extensive and skilful work, i.e. the values of ε_o and $\kappa=\sigma/\sigma_\beta$ (σ=mean stress; B for fracture=Bruch in German) are given in Table 2a together with the values of the analysis with formula.

$$\kappa=\acute{\alpha}(e^{\beta\varepsilon}-1) \tag{9}$$

In Table 2b the parameters $\acute{\alpha}$ and β are given. See also Figure 3.

Table 2a Data (ε κ) and stress-strain relationship κ for the most compressed fibre of the compression zone

W[1]										
8	ε_o[2]	.06	.13	.20	.30	.41	.55	.76	1.06	1.61
	κ[3]	.1	.2	.3	.4	.5	.6	.7	.8	.9
	κ	.098	.200	.290	.399	.499	.599	.708	.807	.892
22.5	ε_o	.11	.22	.33	.47	.64	.82	1.05	1.35	1.83
	κ	.1	.2	.3	.4	.5	.6	.7	.8	.9
	κ	.110	.208	.297	.398	.503	.597	.695	.795	.905
60	ε_o	.20	.41	.62	.82	1.03	1.26	1.50	1.76	2.07
	κ	.1	.2	.3	.4	.5	.6	.7	.8	.9
	κ	.103	.208	.309	.402	.495	.594	.692	.794	.909

Table 2b Parameters ά, β; q

W[1]	ά	β	q[4]
8	-0.940	-1.844	2.957
22.5	-1.093	-0.964	2.748
60	-2.887	-0.183	4.148

[1] W(N/mm^2)=cube strength of cubes 200^3 (mm^3); [2] ε_o (‰)= strain of the most compressed fibre of the compression zone; [3] $\kappa=\sigma/\sigma_B$; σ = mean of the stress, B for fracture; [4] in 10^{-4}

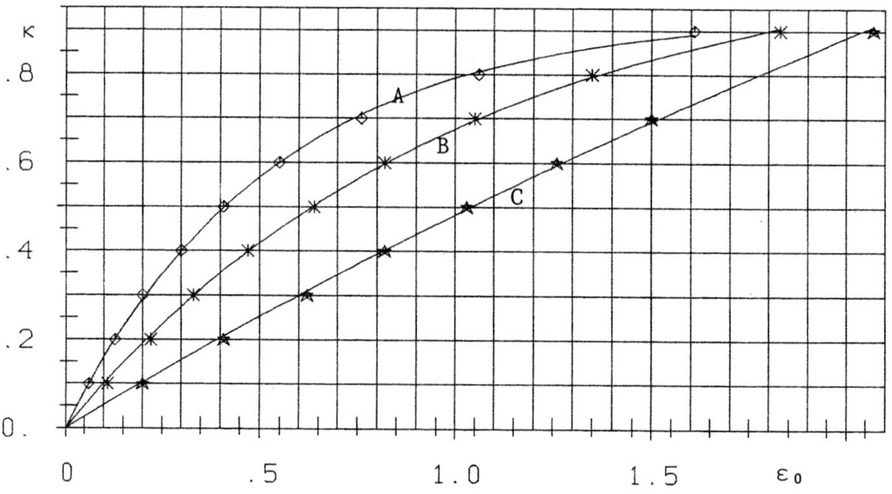

Figure 3 Stress-strain relationship for the most compressed fibre of the compression zone. A) W=8; B) W=22.5; C) W=60 N/mm^2

As experimental data those (ε, κ) with κ=0.1, 0.2, ...0.9 were used, not those with κ=1, that is, not those with the values (ε, κ) of fracture. This was done in correspondence with the comment of Professor Rusch in [1], pp 24, that it is "very doubtful, to give the values (of ε), measured between 90% and 100% of the utmost loading, great significance". It is a well-known fact that the increase of ε in this interval is unproportionally great.

One can see from Figure 3, that the correspondence of the experimental data, marked with different symbols for different values of W, with the curves according to hypothesis (2) is again very good).

B The Stress-Strain Relationship of Steel in the Region of Strain Hardening

I have to thank Professor Scheer from the University of Technology in Braunschweig for kindly sending me some of his papers, as I gave my opinion, that hypothesis (2) for concrete possible might be valid also for the stress-strain relationship of steel, or part of this curve. Fortunately in [3] the stress=strain curve in a tension test with $\varepsilon \simeq 1‰$ / sec for a steelsort (st 37) in the region of strain hardening is given. It is up to the maximum, i.e. up to the beginning of the reduction area, of the same form as the stress-strain curves of concrete in centric pressure tests, as it is dealt with in section A1.

The validity of hypothesis (2) for the stress-strain relationship of steel in the region of strain hardening is also a justification of my procedure in section A1, to take as limitation for the validity of formula (2) those points (ε, σ), where the stress-strain curves reach the maximum; in general then also fractures arise.

In table 3 and figure 4 the data (ε, σ) and the corresponding $\sigma=\acute{\alpha}.$ (exp. $(\beta\varepsilon)$-1); $\acute{\alpha}$=-47.53; β=-0.02129 are given. Q=3.076. The last, 14-th point, used in the present analysis of formula (2) differs more clearly from the curve than the other points, a sign, that the region of strain hardening is passed. By carrying out the analysis with only the first 13 points the adaptation of the curve is considerably better; Q reduces to 2.096, i.e. by circa one third.

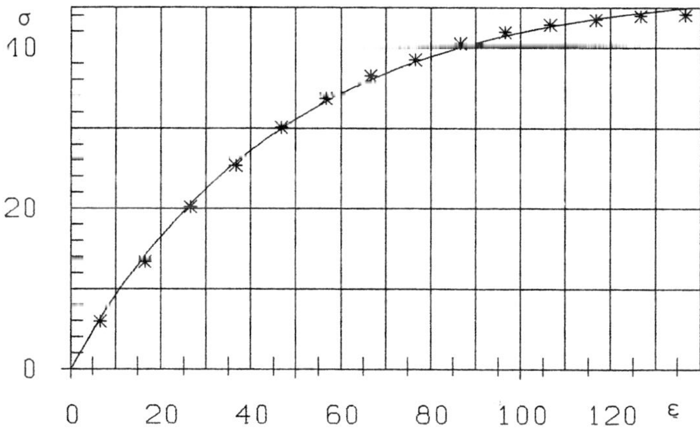

Figure 4 Stress-strain relation of steel st 37 in the region of strain hardening

Table 3 Data (ε, σ) and hypothesis σ of steel in the region of strain hardening

ε (‰)	6.66	16.66	26.66	36.66	46.66	56.66	66.66
$\sigma(N / mm^2)$	5.90	13.35	20.18	25.31	30.12	33.69	36.49
σ	6.28	14.19	20.58	25.75	29.92	33.30	36.03
cont.	76.66	86.66	96.66	106.66	116.66	126.66	136.66
	38.50	40.52	41.92	42.85	43.47	43.94	44.09
	38.23	40.01	41.45	42.62	43.56	44.32	44.93

REFERENCES

1. RUSCH, H. Versuche zur Festigkeit der Biegedruckzone. Deutscher Ausschuβ fur Stahlbeton, Heft 120, 1955.

2. RASCH, C. Spannugs-Dehnungs Linien des Betons und Spannungsverteilung in der Biegedruckzone bei konstanter Dehngeschwindigkeit. Deutscher Ausschuβ fur Stahlbeton, Heft 154, 1962.

3. SCHEER, J. Zur Qualitatssicherung mechanischer Eigenschaften von Baustahl, Institut fur Stahlbau der Technischen Universitat Braunschweig, 1987.

4. NELDER, J, A, AND MEAD, R. A Simplex Method for Function Minimisation. The Computer Journal 7, 1965, pp 308-313.

5. THE NAG FORTRAN LIBRARY MANUAL – MARK 13 (The Numerical Algorithmus Group Limited. Oxford, UK, 1960.

6. SCHOLZ, G. Festigkeit der Biegedruckzone, Theoretische Auswertung, Deutscher Ausschuβ fur Stahlbeton, Heft 139, 1960.

BEHAVIOUR OF REINFORCED CONCRETE CONTINUOUS DEEP BEAMS

A Arabzadeh

University of Tarbiat Modarres

Iran

ABSTRACT. A reinforced concrete deep beam is a beam having a depth comparable to its span, and usually they are specified by their span to depth ratio. For instance in America this ratio is less than 5, whereas in Europe this ratio is less than 2 for simply supported deep beams and 2.5 for continuous deep beams. They have useful application in buildings, offshore structures, bridges, silos, dams and caisson foundation, etc. Deep beams can be divided to: 1- Simply Supports, 2- Fixed ends and 3- Continuous deep beams. In this paper results of 12 reinforced concrete continuous deep beams, loaded from the top center are reported. The results indicate two modes of failure, Shear, and Bearing, but usually these beams failed in shear. The experimental ultimate loads and failure modes are compared with results obtained from theoretical proposed method. The results are in good agreement with test results.

Keywords: Reinforced Concrete, Deep beam, Continuous beam, Shear failure

Dr A Arabzadeh is a lecturer in Civil Engineering Department, University of Tarbiat Modarres, Tehran, Islamic Republic of Iran. He is especialised in design of reinforced concrete structures and stability of structures.

INTRODUCTION

Simply supported reinforced concrete (RC) beams with span to depth ratio less than 2 or continuous beam with span to depth ratio less than 2.5 is considered as deep beam. Since this ratio is small, they have non-linear behaviour. Therefore their behaviour is different compared with normal shallow beams. Also more than one neutral axis, high inertia in the section and high shear resistance, made these beams economical. They are used in multi-story buildings, foundation walls, offshore structures, etc as in Figure 1.

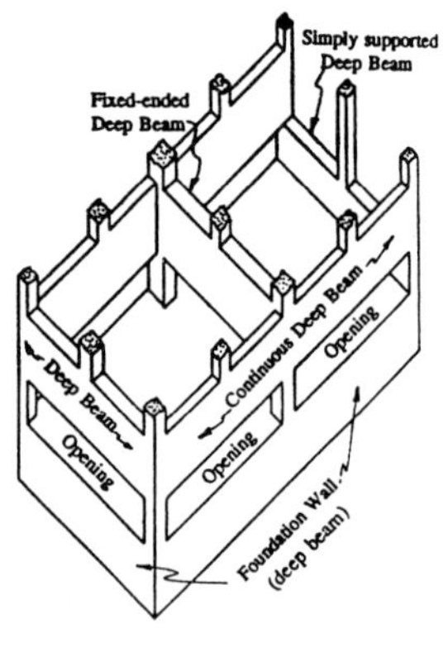

(a) Section of Multi-story building

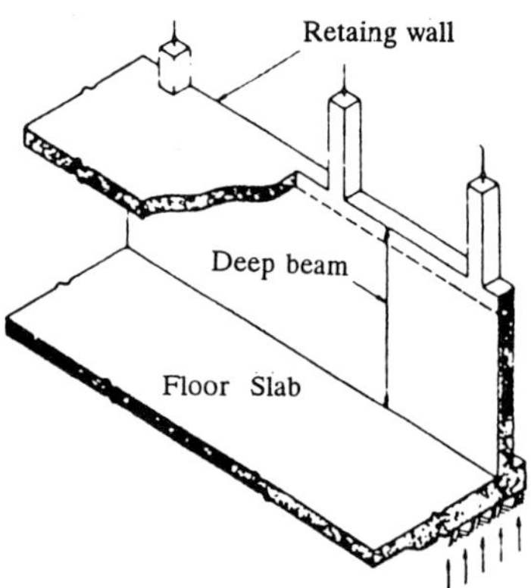

(b) Foundation wall

Figure 1 Typical deep beams in a structure

Furthermore, shear deformation of these beams compared with flexural deformation is considerable and can not be ignored. So ultimate strength and non-linear analysis for these beams should be used. In this paper a method based on equilibrium of the forces along the failure plane is proposed. The validity of the method is controlled with experimental results of 12 continuous deep beams. All the beams where loaded from the top mid-span. The results indicate that the method is simple, safe and accurate.

BEHAVIOUR

Behaviour of deep beams is different compared with ordinary shallow beams, this is because of the small span to depth ratio of deep beams. Also the geometry of these beams is such that section of the beams subjected to a state of two-dimensional stresses and distribution of stress along the sections of the beam is not linear. Transfer of load to the supports takes place by arch action and therefore shear resistance of these beams is 2 to 3 times of the ordinary shallow beams. The modes of failure of these beams are mostly shear. However, when span to depth ratio in less than 2.5, the behaviour of the beam is not linear and linear analysis of ordinary beams can not be used for these beams.

Both the shear and flexural strength influence the overall deformations of the beam. The structural behaviour of RC deep beams top point load is influenced by concrete strength, amounts and proportions of top and bottom main reinforcement. From experimental results two modes of failure, shear and bearing failure have been identified, Figure 2. It is clear from the Figure 2a, crashing of a small area of concrete near the edge of loading and lower corner take place in a number of beams. In both modes of failure, at about 20% to 30% of ultimate load major tension cracks appeared on both mid-spans. After appearance of major diagonal cracks (MDC) these cracks start to close.

The final failure was characterised by the sudden widening of one of the MDCs and crushing of the small areas of concrete under and over the loading plates Figure 2a. In the beam with high amount of top reinforcement bearing failure occurred. The formation of cracks were as for the shear failure, but at higher loads, the MDC did not show any sign of deterioration. Near the failure under load bearing spalling was noticed and failure occurred rather suddenly, Figure 2b.

(a) Beam 1CB1

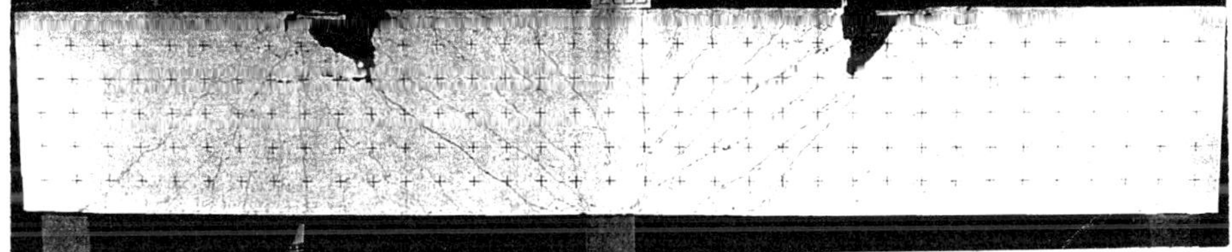

(b) Beam 2CB3

Figure 2 Crack pattern, loading condition, load bearing and failure modes of test beams

EXPERIMENTAL DETAILS

The test programme consisted of four series. The first series of beams (1CB1 and 1CB2), had span to depth ratio of 1.25 and 2.5 respectively. The dimensions of the first beam were 500 mm long (each span), 400 mm depth, 50 mm thick and second were 1000 mm long, 400 mm depth and 50 mm thick. The dimensions of the second series were 1680 mm long (each span), 600 mm depth and 75 mm thick giving span to depth ratio 2.8 (clear span to depth ratio of 2.5) and slenderness ratio of 8. The third series beam (3CB5 to 3CB9) had span to depth ratio of 2 and slenderness ratio of 8. The dimensions of third series were 800 mm long (each span) 400 mm deep and 50 mm thick.

The dimensions of fourth series (4CB10, 4CB11 and 4CB12) were 1260 mm long (each span), 450 mm deep, 50 mm thick, giving a span to depth ratio of 2.8 and slenderness of 9.

The top and bottom reinforcement represent the main reinforcement and their properties are listed in Table 1. In the first, third and fourth series a 100 mm square mesh of 6 mm diameter mild steel with yield strength of 340 N/mm^2 and 350 N/mm^2 were used respectively as web reinforcement.

In the second series, 2CB3 and 2CB4, two layers of 100 mm square mesh of rounded bar with yield strength of 403 N/mm^2 was used, as web reinforcement. Four concrete mixes were used for the beams, and their characteristics are listed in Table 2.

All the beams were loaded from the top center, Figure 2a. An upper girder was positioned to transfer equal amount of load from the loading machine to the specimen at center of each span. The load was directly recorded from the machine and Data logger.

In Figure 2a the detail of bottom and top bearing clearly is shown. The rollers at outer supports allow the rotation and horizontal movement. The middle support act as fix support, with only rotation permitted. The top bearing allows rotation and vertical displacement at the loading points.

ANALYSIS

The part of the beam separated by the potential critical diagonal cracks in the Figure 3b is shown as the idealised free body together with the associated forces. The prediction of ultimate load based on equilibrium of forces along the failure plane. The following assumptions are made:

1. The failure occurs along a plane between the load and mid-support, identified as MDC.

2. The failure plane is considered to stop at the center of the top and bottom main bars.

3. The contribution of the force along the diagonal plane of failure, depends on whether concrete or web bars controls the web strength, Table 1.

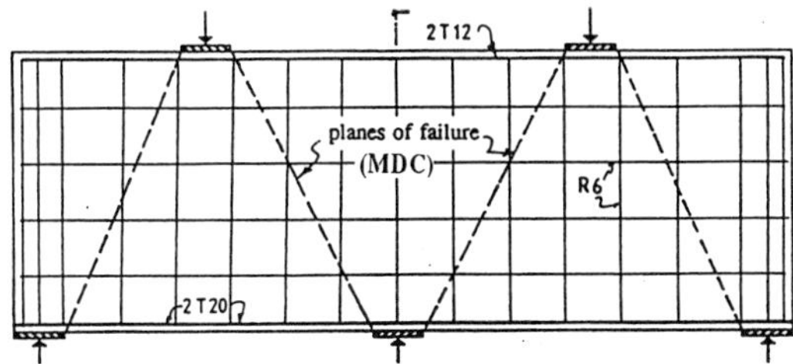

(a) Potential Planes of failure

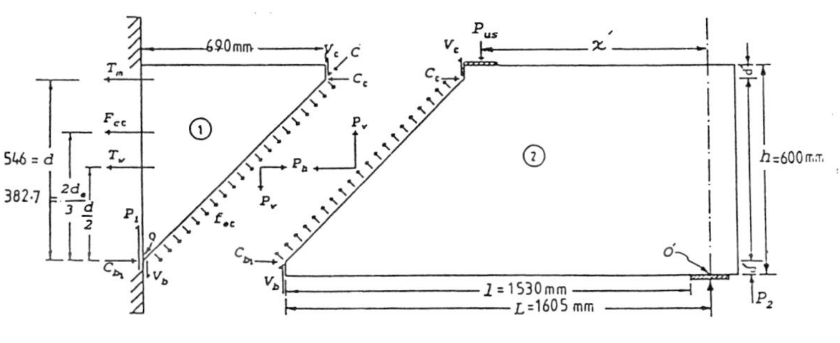

(b) Force Diagrams

Figure 4 Idealised force equilibrium body of continuous deep beams

Table 1 Criteria tests for supports and web strength

TEST	F_s	$f_{et}bd + A_hf_s$	$f_{et}bx + A_vf_s$	SUPPORTS AND WEB STRENGTH CONTROLLED BY
1	$<T_m + T_w$	$<A_hf_{yh}$	$<A_vf_{yv}$	Reinforcement
2	$>T_m + T_w$	$>A_hf_{yh}$	$>A_vf_{yv}$	Concrete
3	$>T_m + T_w$	$<A_hf_{yh}$	$<A_vf_{yv}$	Concrete
4	$>T_m + T_w$	$>A_hf_{yh}$	$>A_vf_{yv}$	Concrete

For the equilibrium of the idealised free body diagram, Figure 3 it is required that sum of the forces in horizontal direction, vertical direction and sum of the moment about any point in the diagram to be zero. The equations are; in horizontal direction:

$$C_b + C_t - f_{et}bd - P_h = 0 \tag{1}$$

in vertical direction:

$$P_2 + V_b - f_{et}bx - V_h - P_{us} = 0 \tag{2}$$

sum of the moments about O':

$$V_b L + (f_{et}bx + P_v)(L - \frac{x}{2}) - (f_{et}bd + P_h)\frac{d}{2} + C_t(L - x) - P_{us}x' = 0 \tag{3}$$

The term in Equation 1 to 3 are defined in Figure 3, and clearly explained in early papers [3,5]. The ultimate shear strength, P_{us}, is given by rearranging the Equation 4:

$$P_{us} = \frac{V_b L + (f_{et}bx + P_v)(L - \frac{x}{2}) - (f_{et}bd + P_h)\frac{d}{2} + C_t(L - x)}{x'} \tag{4}$$

Where:

$$C = 0.67 f_{cu}bd' ; \quad C_t = C\cos\alpha ; \quad V_b = \frac{d''}{d'}V_t ; \quad V_t = X\sin\alpha ; \quad f_{et} = \left[\frac{f_{cu}}{\frac{f_{cu}}{f_{it}} + (\frac{d}{x})^2}\right]$$

$$f_s = \frac{E_s}{E_c} f_{et} ; \quad P_v = A_v f_e ; \quad P_h = A_h f_s \text{ and } f_{it} = 0.3 f_c'^{2/3}$$

The results are summerised and listed in Table 2. As indicated above the proposed method is simple and based on the observed behaviour of the beams. It is believed that the method is safe and capable to predict the ultimate load and modes of failure for continuous deep beams.

RESULTS AND DISCUSSION

By comparing the crack patterns of tested beams, Figure 2, it can be understood that continuous deep beams with smaller span/depth ratio show less spalling and initial major diagonal cracks form at higher load. In fact, the first crack to form was a diagonal crack. A possible reason for the difference between the failure modes may be due to the difference in the amount of top bars. In other words, a larger amount of top reinforcement may be responsible for a stiffer beam and the load bearing type of failure.

The major diagonal cracks appeared at about 25% to 45% of ultimate load in all test beams irrespective of their geometry, reinforcement, ultimate load or mode of failure. These cracks were wider even at formation (0.1 mm to 0.35 mm). In all the beams with shear failure, one of the planes of failure passed through these cracks.

The ultimate strength mainly depends on the beam properties, such as geometry, concrete strength, amount of reinforcement, loading and support conditions. From the results it can be seen that, beams with smaller span/depth ratio can withstand higher load than beams with the same properties but larger span/depth ratio. Furthermore, excessive amount of top reinforcement could be responsible for the beam to fail in bearing [2CB3]. The overall performance of the beam did not improve beyond a certain limit by increasing the amount of the top reinforcement in compared with other beams.

Table 2 Summerised test and analysis results of tested beams

BEAM	MAIN REINFORCEMENT			TEST RESULTS				ANALYSIS		
	Top/Bottom			Failure				Failure		
	No, Size and Type	f_{yt}/f_{yb} N/mm²	E_{st}/E_{sb} kN/mm²	Mode	P_t kN	f_{cu}	F_{et} N/mm²	Mode	P_{US} kN	$\frac{P_{us}}{P_t}$
1CB1	1T16/1T16	493/493	205/205	S	332	56.5	2.3	S	245	0.75
1CB2	1T16/1T16	493/493	205/205	S	187	56.5	4.14	S	208	1.11
2CB3	2T20/2T20	515/515	203/203	B	420	44.7	3.12	S	374	0.89
2CB4	2T12/2T20	527/515	188/203	S	420	44.7	3.12	S	348	0.87
3CB5	1T12/1T12	378/378	204/204	S	135	39.6	2.69	S	140	1.04
3CB6	1T12/1T16	378/354	204/200	S	140	34.1	2.69	S	127	0.91
3CB7	1T16/1T16	354/354	200/200	S	155	34.2	2.12	S	127	6.87
3CB8	1T20/1T10	346/346	206/206	S	155	39.6	2.69	S	140	0.90
3CB9	1R6/1T16	340/340	208/206	S	155	39.6	2.69	S	140	0.90
4CB10	1T16/1T10	354/346	200/200	**	**	**	**	**	**	**
4CB11	1T10/1T20	377/346	208/200	S	160	39.6	2.69	S	169	1.05
4CB12	1T12/1T20	378/346	204/200	**	**	**	**	**	**	**

The prediction of ultimate strength was based on determining the load for the shear strength of the beam. The results indicate that the proposed method can predict the ultimate strength of continuous deep beams and the results are safe, accurate and consistent. Flexural and bearing strength of the test beams can not be analysed on the basis of the test results. They have to be designed such that, these modes of failure do not occur during useful life of the beams. Different standards such as [1,4] have given few guide lines for the design of flexural and bearing strength of RC deep beams.

CONCLUSIONS

From the experimental and theoretical results the following conclusions are made:

1. Failure mode strongly depends on the amount of top and bottom main reinforcement, and clear shear-span/depth ratio.

2. The change of mode of failure from shear to bearing was accompanied by a significant reduction in the ultimate load.

3. The diagonal cracks dominated the behaviour right from the outset. They were the major cracks at all stages of test. At failure the widths of these cracks were in excess of 0.3 mm in all the test beams except, those which experienced bearing failure.

4. The ultimate strengths of test beams were strongly influenced by the beam geometry, amount of main reinforcement and concrete strength.

5. The proposed method capable of dealing with any combination of beam parameters. In majority of cases the experimental and predicted results were found to be in good agreement.

REFERENCESS

1. ACI (REVISED Building code requirements for reinforced concrete (ACI 318-83). American concrete Institute, Detroit, P. 111. (1992).

2. ARABZADEH, A. "Truss Analogy for the Analysis of Reinforced Concrete Fixed - Ended Deep Beams". The Second International Conference in Civil Engineering on Computer Application, Bahrain 6-3 April 9 pp. (1996)

3. ARABZADEH, A, AND ERSHADI, S. "Ultimate load and Failure modes of Simply Supported Deep Beams with span depth ratio 2 to 3". fourth International Conference on Civil Engineering, Tehran, Iran. (1997)

4. CIRIA GUIDE 2. The design of deep beams in reinforced concrete. Ove Arup and Partners, Construction Industry Research and Information Association, London, P. 131. (1977)

5. SUBEDI, N K. Reinforced concrete deep beams: a method of analysis. Proceedings of the institution of Civil Engineers, Part 2, Vol. 58, March, 30 pp. (1988).

6. SUBEDI, N K, AND ARABZADEH, A. Some experimental results for reinforced concrete deep beams with fixed-end supports. Structural engineering review. Vol. 6, No. 2, (1994).

BEHAVIOUR OF PILE SYSTEM FLOATING BREAKWATER

M I Balah

E R Tolba

Suez Canal University

Egypt

H J Kaldenhoff

University of Wuppertal

Germany

ABSTRACT. The present work gives a solution to overcome the disadvantages of the Floating breakwater (F.B.W) moored with chains or cables to the bottom of the sea. The basic idea of the studied solution is to replace the mooring system using vertical piles instead of the chains. Such mooring system prevents the sway motion of the floating body and in addition, the roll motion is limited due to the existence of the vertical piles. The pontoon is allowed to move freely upward and downward with the tides at the same designed draft. The mentioned solution also helps to solve the problems of failure appear at the connections between the pontoon and the mooring lines. The behaviour of the floating pontoon supported with vertical piles was studied experimentally by measuring the heave motion of the body and its effect on wave attenuation. The efficiency of the F.B.W (consists of pontoons) moves in heave motion was determined and expressed as a function of transmission and reflection coefficients. The effect of the pile stiffens on the body motion was investigated. The characteristics of the heave motion such as, the height of the heave and its phase angle with the standing waves was investigated. The horizontal hydrodynamic wave force on the F.B.W due to wave attack was also measured. The stability of the floating pontoon moored with the new system using piles, its attenuation to the incident waves showed good results compared with the previous results of the F.B.W moored with chains and cables to the bottom of the sea.

Keywords: Floating breakwater, Mooring lines, Piles, Reflection coefficient, Transmission coefficient, Pontoons, Heave motion

Prof Mohamed Balah is professor of harbour engineering and head of Civil Engineering Department - Faculty of Engineering - Suez Canal University - Egypt.

Dr Ehab Tolba is lecturer working in Civil Engineering Department, Faculty of Engineering Suez Canal University, Egypt. His main research interests related to shore protection end coastal engineering. In 1998, he got his Ph. D. under the title of "Behaviour of Floating Breakwaters Under Wave Action"

Prof Hans Kaldenhoff is the director of Igawa University of Wuppertal, Germany. His main research interests related to hydraulics, waterways and coastal engineering.

INTRODUCTION

The increasing prosperity of the world population has resulted in an increased popularity and need for recreational facilities such as small craft marinas. Many small craft marinas have developed in the last decade, and it is obvious that many more will be built in the future. The rubble mound breakwater is limited in its potential application for marine use. Due to the mass of the rubble mound breakwater a firm subsurface soil capable of providing adequate foundations is necessary. As the breakwater extends to the full depth of water, it acts as a pervious vertical barrier to shoreline processes. This may interrupt littoral transport including local silt or scour problems or may disrupt circulation and cause water quality deterioration within a marina. It is for these reasons that alternative breakwater designs are of interest for small boat marina application.

An alternative wave attenuator is the floating breakwater. The structures are typically of finite draft and rely on the interaction of structure and wave in only the upper portion of the water column. The floating breakwater consists usually of a floating pontoon anchored by chains or cables to the bed of the sea. When this F.B.W is subjected to the attack of incident waves it will experience the rectilinear motions of heave, sway and surge, and the angular motions of yaw, pitch and roll. An attractive benefit of floating breakwaters is that their cost is relatively insensitive to water depth at a site. As the structures are buoyant, the breakwater is mobile that may facilitate realignment or removal if desired. Of importance for marina application is the dual use potential of a floating breakwater. This may permit the structure to act as both pier and breakwater. As the breakwater does not extend to the full depth of the wafer, interruption of littoral processes and local circulation would not be anticipated. On the other hand, these floating breakwaters anchored with chains or cables have some disadvantages which are as follows:-

1. Large roll motion may affect the performance of the F.B.W to be used as a pier.
2. The horizontal motion of the pontoon (sway) generates waves that are not recommended on the leeward side of the structure. In addition, the sway motions ofthe pontoon allow the body to impact with the boats when it is used as a pier.
3. The pontoon moves upward and downward with the water surface according to the tide. When it moves upward with the rising tide, the mooring lines may reach its maximum length, in the case the designed draft may changed and over topping may occur. On the other hand, when the body moves downward with the falling tide, the mooring lines will be deflected. This deflection increases the sway motion and may affect the performance of the F.B.W to be used as a pier
4. If the F.B.W is exposed to strong waves, cracks will appear at the connection between the chains and the pontoon and failure may occur at this zone.

The present work gives a solution to overcome the above-mentioned disadvantages of the F.B.W moored with chains or cables shown in Figure 1. The basic idea of the suggested solution is to replace the mooring system using piles instead of chains. Details of the suggested solution are given in Figure 2. Such a system prevents the sway motion of the body. The roll motion is limited due to the existence of the piles end the pontoon is allowed to move freely (up and down) with the tides. This system also has the same advantages of the floating breakwaters moored with chains, which are mentioned before. The problem posed for study concerns the motion of the body through this suggested system of floating breakwater.

As the body is prevented to move in sway motion, the only motion possible for the pontoon is the free heave motion. The study also determines the effect of each individual motion on the efficiency of the F.B.W as a function of transmission and reflection coefficients under different regular wave conditions and the effect of the hydrodynamic wave forces on the body which transmit to the supporting piles.

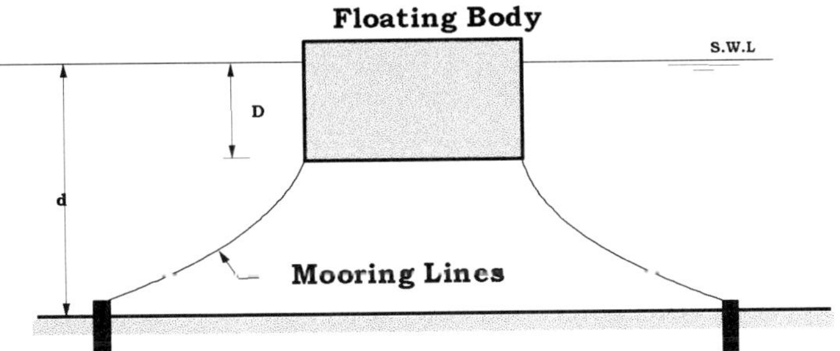

Figure 1 Floating breakwater moored with chains

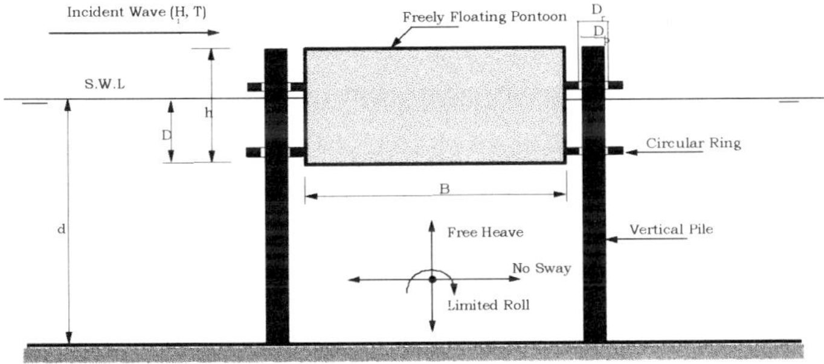

Figure 2 Definition sketch of the suggested floating breakwater

LABORATORY EQUIPMENT AND MEASUREMENTS

Experiments were conducted in a wave flume 24.0m long, 0.30m wide and 0.50m high with transparent glass side walls mounted in steel frame. Figure 3 shows the general arrangements in the wave flume. For the purpose of the study, a wave flume will be considered as a wave study facility in which the length of the flume s many times greater than the width while the depth is nearly twice the width.

The regular waves were generated using a piston type wave generator while the reflection of the propogated waves reflected from the far end of the flume was absorbed using a plastic absorber.

Measurements of Water Surface Profile

Four wave probes were used to measure the water surface profile. Three probes were installed in front of the model while the fourth one was installed behind it as shown in Figure 3. The Mansard and Funke method (1980) was used to measure the incident end reflected waves. In fact, this is a least square method to separate the incident and reflected waves from the co-existing waves in front of the model. This method requires simultaneous measurement of the waves at three positions in the flume which are in reasonable proximity to each other and are on a line parallel to the direction of wave propagation A software program was developed for this method using MATLAB.

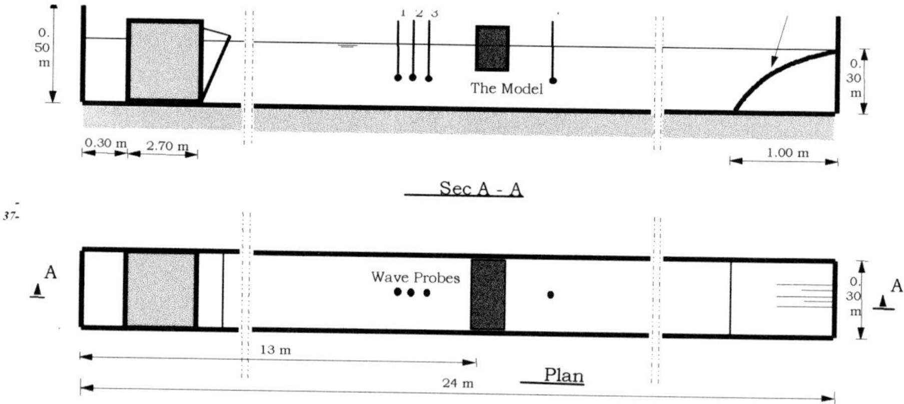

Figure 3 Definition sketch of the wave flume

The wave probes were fixed so that the distance between the front face of the model and the third probe (No. 3), was equal to the tested incident wave length L which is given by linear wave theory as:

$$L = \frac{gT^2}{2\pi} \tanh \frac{2\pi d}{L} \tag{1}$$

The distance L was chosen to minimize the effect of wall friction and the decay on the measured co-existed wave which appear significantly in the case of short crested waves. The distance between the probes itself was chosen following the recommendation given by Mansard and Funke (1 980) for accurate results. To measure the transmitted wave height, the fourth wave probe was installed (No. 4) at the lee side of the model at a distance L from the rear face of the body. By measuring the incident wave height Hj, the reflected wave height Hr using Mansard and Funke method (1980) and measuring the transmitted wave height H' directly using the fourth probe, the reflection and transmission coefficients can be calculated respectively as follow:

$$C_r = \frac{H_r}{H_i} \tag{2}$$

$$C_t = \frac{H_t}{H_i} \quad (3)$$

HEAVE MOTION TESTS

As the main purpose of this study is to investigate the effect of the motion of the floating body moving through the new mooring system (using piles) on wave attenuation, the study was divided to two groups of experiments.

The first group of experiments was to study the heave motion of the pile system F.B.W. The investigations were conducted to study the effect of heave motion on the transmission and reflection coefficients. The motion of the body in heave and its phase angle with the standing wave were measured. The second group of experiments was to study the hydrodynamic wave forces on the floating body.

The model used for this investigation is shown in Figure 1. The model was constructed from four external hollow pipes with an internal diameter of 1.9 cm fixed to the horizontal plate. Four internal pipes with a diameter of 1.8 cm were fixed to the body and extended inside the external hollow pipes, allowing the body to move only in heave motion when exposed to the attack of the incident regular waves. The horizontal plate was supported on top of the flume using *four roller supports* and connected freely to the force cell, allowing the horizontal force to transmit to the force cell without any moment.

Six pressure transducers were fixed to the faces of the body, (four on the bottom and two on the side walls), to measure the pressure affecting the wetted faces of the body. It is very important to know the effect of the different wave conditions on the amplitude of the heave motion of the F.B.W, its phase angle with the standing wave formed in front of the body. Studying these parameters was done by recording the movement of the body using a video camera.

The motion of the body in time was calculated by watching the movement of a clear point on it moving beside a vertical linear scale mounted on the transparent side walls of the flume. The motion of the body in time and its phase angle with the standing wave were observed using the slow motion video option which shows 25 frames per second. The body was tested under different wave conditions and body parameters. Adding uniform weights inside the body increased the draft D.

The model and the wave probes used for running the heave motion tests were installed in the position shown in Figure 3. The period of the tested waves ranged between 0.5 and 2.0 sec and the corresponding wave lengths from 0.389 to 3.257 m. The wave lengths were calculated using (1), which represents thelinear wave theory The dimensions ofthe tested body were B = 15, 30 cm and b = 29.5 cm.

The tested body drafts were 5, 6, 7.5 and 10 cm for the fixed water depth in the flume d=30 cm, which gives D/d=1/6, 1/5, 1/ 4, and 1/ 2 respectively.

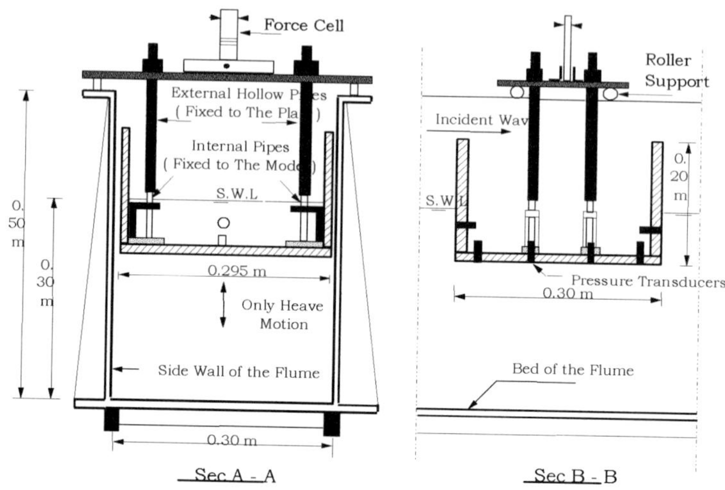

Figure 4 Details of model used for heave motion tests

RESULTS AND DISCUSSION

Figure 5 shows the effect of B/L and d/L on Cr and C,. It is clear from the figure that C, decreases with the increase of B/L and d/L. for example, in Figure 8 where D/d = 1/6, C, decreases from 0.904 at B/L = 0.046 and d/L = 0.092 (very near from shallow water zone) to 0.052 at B/L = 0.394 and d/L = 0.789 (deep water zone). On the other hand, the figure also shows that Cr increases as B/L and d/L increases. The increase of Cr in the deep water zone as unsteady due to the existence of local minimum and local maximum values of Cr. For the same values of B/L and d/L mentioned above it was found that C, increases from 0.074 to 0.721, which means that the reflection coefficient of a certain body is also affected significantly by the incident wave length end the type of the zone (shallow, transitional or deep) at which the incident wave interacts with the structure

Figures 6 and 7 give the effect of the draft D on Cr and C, in the form of the dimensionless ratio D/d. It is clear in the figures that? at the same values of B/L and d/L, Ct decreases and Cr increases with the increase of the value of D/d. The figures also show that maximum reflection and minimum transmission happens in the deep water zone (d/L >= 0.5), while the maximum transmission and the minimum reflection are found near the shallow wafer zone (d/L = 0.05). This means that C, and Cr are strongly affected by d/L.

Figures 8 and 9 show the values of Cr and Ct for both the restrained case and pure heave motion plotted versus B/L and d/L for different values of D/d. The figures show that the values of Cr and Ct of the F.B.W moving only in heave motion and fixed in position using vertical piles are always lower than the results for the restrained structure for all the values of D/d. The explanation of these results is that in the case of the restrained structure the total energy of the incident wave will be dividedintothreeparts,the energy ofthe transmitted wave, the energy of the reflected wave and the energy losses. While in the case of the F.B.W there is an additional source of energy losses, which is the energy lost in inducing the heave motion of the body.

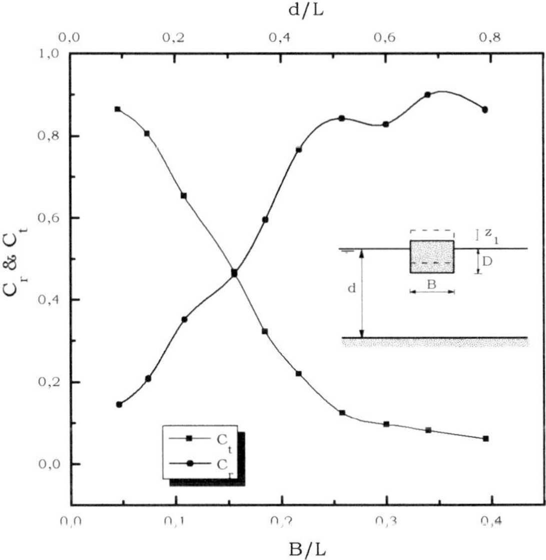

Figure 5 Variation of Cr and C, with B/L and d/L
(Only Heave Motion, D/d = 1/ 4, HJL = 0.014 - 0.048, B/d = 1/ 2)

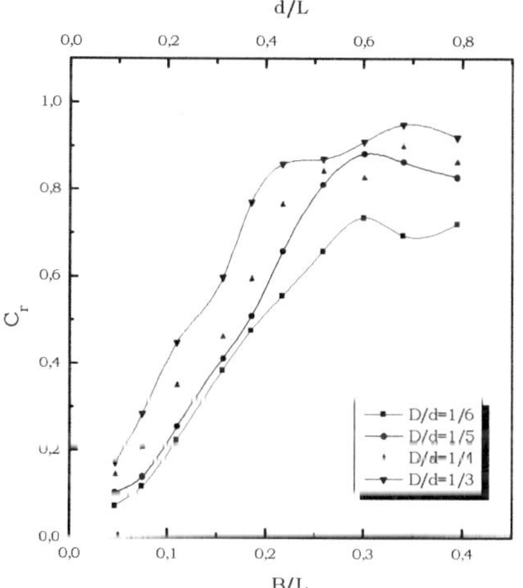

Figure 6 Effect of D/d on the Reflection Coefficient C_r
(only heave Motion, H_i/L=0.014-0.048, B/d=1/2)

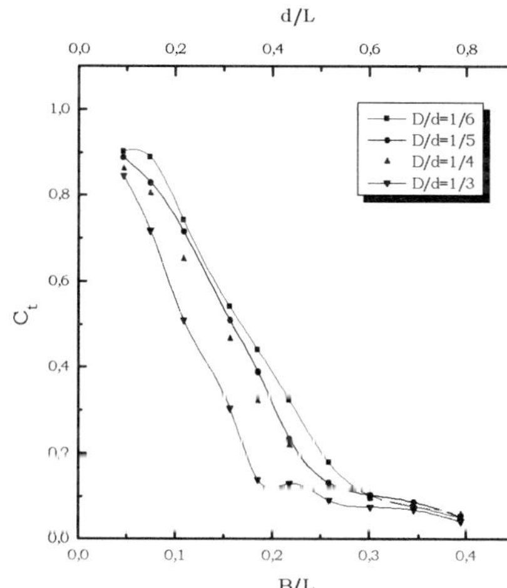

Figure 7 Effect of D/d on the Reflection Coefficient C_t
(Only heave Motion, H_i/L = 0.014-0.048, B/d = ½)

Figure 10 shows a comparison between the results of the present work for D/d = 0.25 and B/d = 1/ 2, with the work of other authors for different types of floating breakwaters. The figure shows that the suggested F.B.W is efficient compared with the other results and this effciency appears mainly in the deep water zone. The figure also shows the effect of B/L on C,, where it is clear for all the results that C, decreases it the increase of B/L.

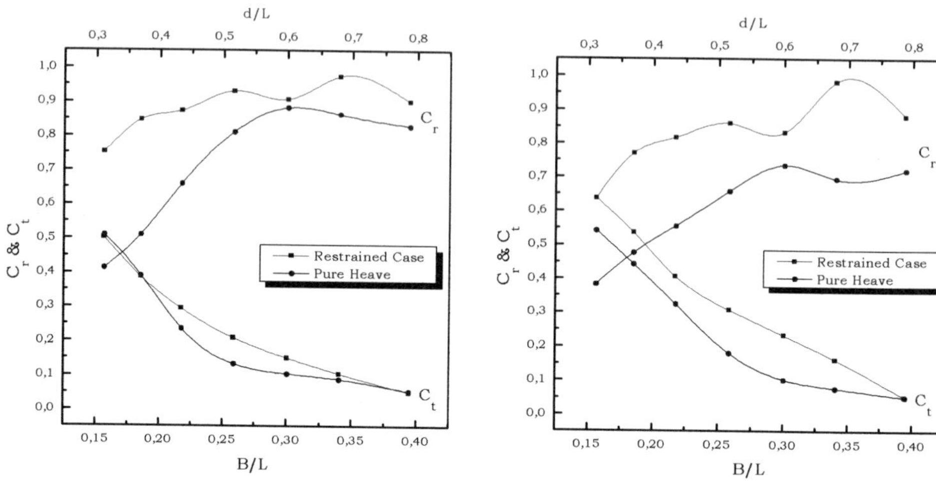

Figures 8 Effect of the Heave Motion on C_r and C_t (D/d = 1/6, B/d = ½, H_i/L = 0.0485

Figure 9 Effect of the Heave Motion on C_r and C_t (D/d = 1/5, B/d = 1/ 2, H_i/L = 0.0485)

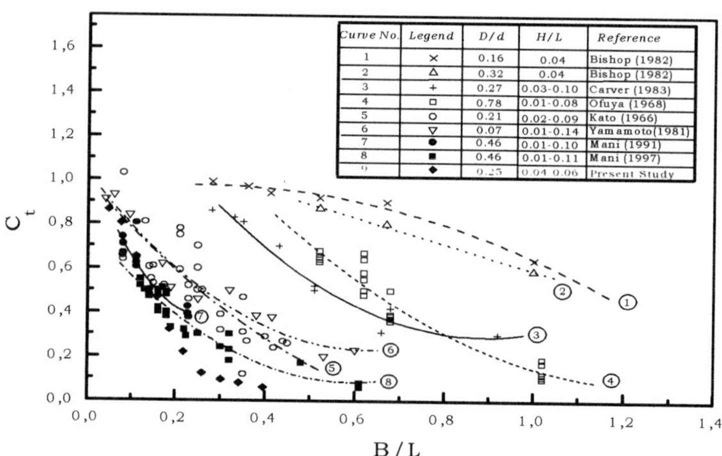

Curve No.	Legend	D/d	H/L	Reference
1	×	0.16	0.04	Bishop (1982)
2	△	0.32	0.04	Bishop (1982)
3	+	0.27	0.03-0.10	Carver (1983)
4	□	0.78	0.01-0.08	Ofuya (1968)
5	○	0.21	0.02-0.09	Kato (1966)
6	▽	0.07	0.01-0.14	Yamamoto(1981)
7	●	0.46	0.01-0.10	Mani (1991)
8	■	0.46	0.01-0.11	Mani (1997)
9	◆	0.25	0.04 0.06	Present Study

Figure 10 Performance of Floating Breakwaters – Comparison

Figure 11 gives the measured values of Hh/Hj using an incident regular wave with constant height. Hi=4 cm, for different values of D/d. The figure shows that in the shallow water zone, the body has the same dimensionless heave motion ratio and that the ratio D/d has no effect on Hl,/Hj. The figure also shows a maximum value of Hl,/Hj which is due to the resonance. Resonance normally occurs when the wave period T is equal to the body's natural period Tn. (or when Tn/T =1), at which the waves add energy to the body, increasing the amplitude of the heave motion.

The phase angle between the body and the standing wave was measured using the video camera. The angle was calculated by plotting the observed motions of both the body and the standing wave together versus time. The phase angles measured for different wave periods and body drafts are given in Figure 12. The figure shows that for small values of B/L and d/L, (near the shallow water zone), the phase angles are very small (< 30 degrees) and the body moves nearly in phase with the standing wave. As the wave period increases (B/L and d/L increase), the phase angle between the body and the standing wave increases until it reaches the maximum value, which is nearly 180 degrees (case of deep water); in this case the body is completely out of phase with the wave. In fact, this difference in phase between the body and the wave in heave motion plays a considerable part in attenuating the incident wave and this phenomena is known as *out-of-plrn.se damping.* Furthermore, the figure also shows that for all the values of B/L, the values of B/L, the value of the phase angle decreases as the D/d ratio increases (or the weight of the body increases).

The study showed that part of the wave energy is lost due to the body wave interaction. The loss coefficient C' was calculated using the measured values of Cr and C~ as follows:

$$C_l = \sqrt{1- C_r^2 - C_t^2} \qquad (4)$$

Figure 13 shows the calculated loss coefficient,C' for ditferent conditions plotted together with the reflection and transmission coefficients. The study showed that some of the incident wave energy is dissipated due to the formation of vortices around the two sharp edges of the body. The study also showed that the area of the vortex formed in front of the body is bigger in size than the second vortex formed at the lee of the body. Furthermore, the position of the vortex changes with the movement of the free surface of the wave as shown in photo 1

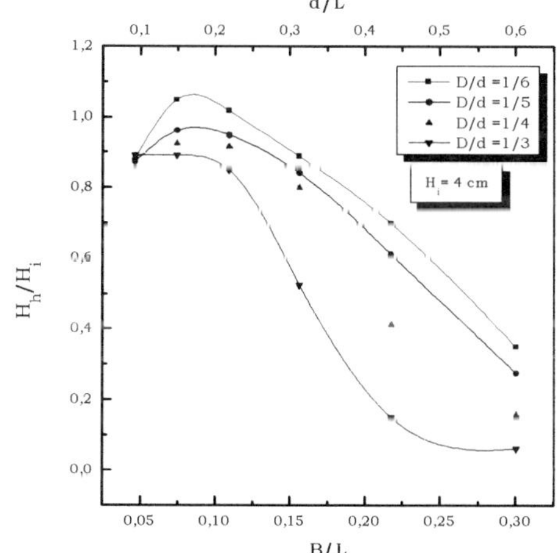

Figure 11 Dimensionless displacement of the heave motion versus B/L and d/L

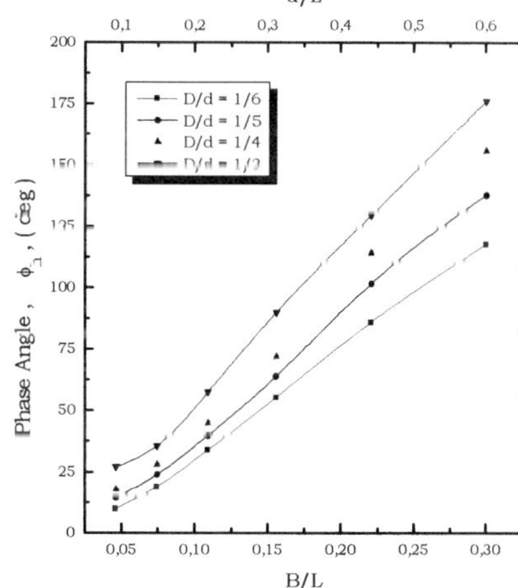

Figure 12 phase angles with the standing waves (only heave motion B/d = 1/ 2, H_i = 4 cm)

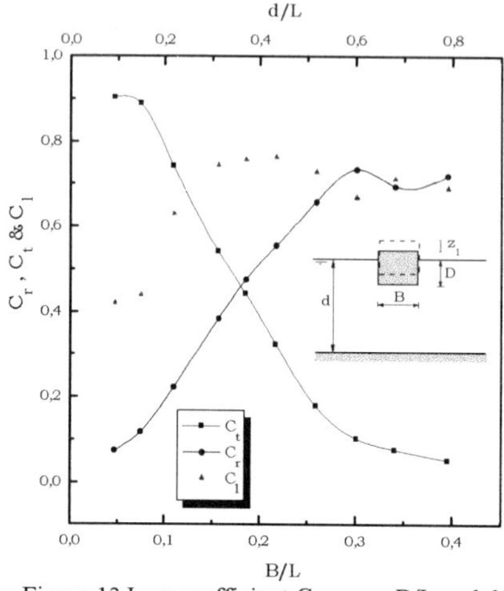

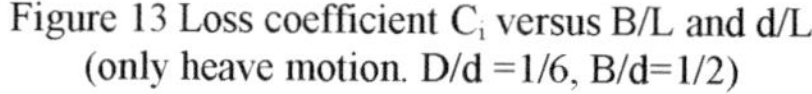

Figure 13 Loss coefficient C_l versus B/L and d/L (only heave motion. D/d =1/6, B/d=1/2)

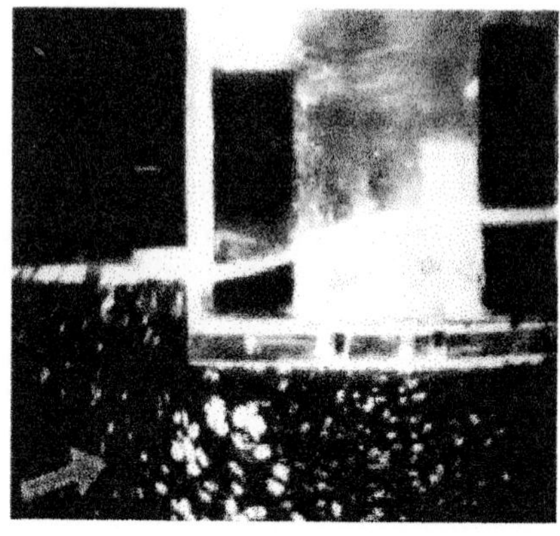

Photo 1 Vortices around the floating body in heave motion

Figure 14 shows the variation of the dimensionless absolute values of F_{rmax} with B/L and d/L. It is clear from the figures that at the low values of B/L (the case of long waves), both the forces are minimum and that the forces increase with the increase of B/L and d/L to a certain value of B/L equal to 0.25 (d/L = 0.5), where the forces become maximum.

After these maximum values, the forces begin to decrease with the increase of B/L and d/L (the case of short waves). The above mentioned distribution of the forces versus B/L and d/L was found to be similar for the three tested incident wave heights but with different force values. Furthermore, Fj,l,ax was found to be always higher than FrmaX where:

Fj = the horizontal force on the body in the direction of the incident wave (N/m).

Fr = the horizontal force on the body opposite to the direction of the incident wave (N/m)

Figure 15 shows the effect of wave steepness on the maximum measured wave pressure on the different faces of the structure for different values of d/L. The maximum pressure on the bottom of the body? PBmaX was found always to be at gauge No. 2, furthermore the maximum pressure on the body was found to be recorded by gauge No. 1 (sea side pressure Pss). It is clear from all the figures that the pressure increases with the increase of wave steepness.

As a result, this investigation showed that the zone of maximum pressure is the zone between gauge 1 and gauge 2 and near gauge 3 (harbour side pressure Plls).

One explanation for this high pressure may be due to the formation of vortices in these zones at which the wafer particle velocities increase; this may increase the impact of these particles with the structure and as a result increase the pressure at this zone of the structure.

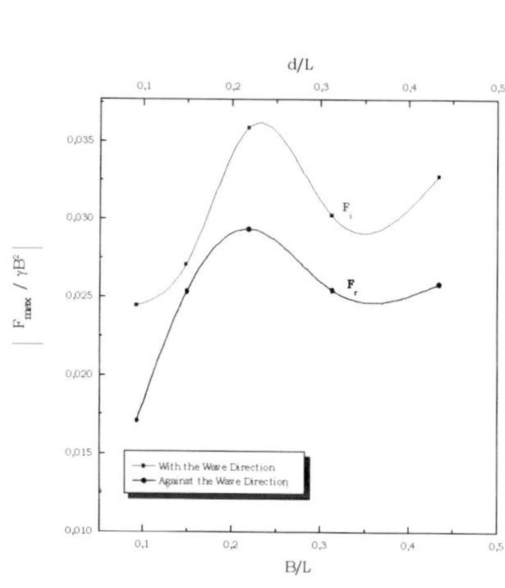

Figure 14 Max horizontal force on the floating body (D/d = 1/3.6, B/d=1, H_i/d=1/6.25)

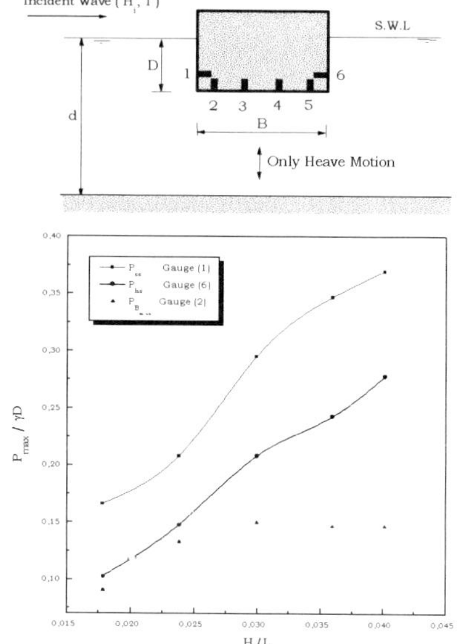

Figure 15 Wave pressure on the floating body versus (D/d = 1/3.6, B/d=1, d/L = 0.218)

CONCLUSIONS

Floating breakwaters supported by vertical piles instead of chains and/or cables to the bottom of the sea were studied. This new system can be used for marinas as F.B.W and/or piers for the small boats. The mentioned mooring system allows the floating body to move onlyin heave motion. The results showed that a floating body moving only in heave motion through this mooring system has higher wave attenuation than that of the restrained body. The results showed a good agreement with the previous work and showed higher attenuation than that of the floating breakwaters moored with chains and/or cables.

REFERENCES

1. BISHOP, C.T. (1982): "Floating tire breakwater comparison". J. Wtrwy., Port, Coast, and Engrg., ASCE, 108(3), 421-426.

2. CARVER, R.D., AND DAVIDSON, D.D. (1983): "Slopping floating breakwater model study" Proc., Coast. Struct. 83 A; Specialty Conf On Des. Constr. Maintenance and Perf. Of Coast Struct., 417-432.

3. KATO, J., HAGINO, S., AND UEKITA, Y. (1966): "Damping effect of floating breakwater to which anti-rolling system is applied". Proc.7] oti. Coast. Engrg. Conf., 1068-1078.

4. MANI, J. S. (1991); "Design of Y-frame floating breakwater". J. Wtrwy.7 Port, Coast., and Oc. Engrg., ASCE7 117(2)7 105- 119.

5. MANI, J.S. (1997): "Performance of cage floating breakwater". J Wtrwy., Port. Coast, and Oc. Engrg., ASCE, 123 (4), 172- 179.

6. MANSARD, E.P.D., AND FUNKE, E.R. (1980): "The Measurement of Incident and Reflected Spectra Using a Least Squares Method." Proceedings of the 17th Coastal Engineering Conference, Sydney, Australia, Vol. 1, pp. 154-172.

7. OFUYA, A. O., AND BREBNER, A. (1968): "Floating breakwaters" Proc. 5th Coast., Engrg. Conf., 1055-1085.

8. YAMAMOTO T. (1981): "Moored floating breakwater response to regular and irregularwaves". J. Appl. Oc. Res.7 3(1)7 114-123.

THERMAL STRESS AND TEMPERATURE CONTROL OF CONCRETE GRAVITY DAMS WITHOUT LONGITUDINAL JOINT INCLUDING RCC GRAVITY DAMS

B Zhu

P Xu

China Institute of Water Resources and Hydropower

China

ABSTRACT. There is a tendency to construct concrete gravity dams without longitudinal joint, particularly, the roller compacted concrete gravity dams. The thermal stresses and temperature control of this type of dam are studied. It is discovered that the thermal stresses in this type of dam are different from the conventional concrete gravity dams with longitudinal joints on all the following respects: the temperature drop in the dam near the foundation, the temperature variation in the vertical direction, the temperature difference between the surface and the interior of dam, transverse crack on the upstream face of dam and the extracooling of concrete around the orifice in the dam

Keywords: Thermal stress, Temperature control, Gravity dam without longitudinal joint, RCC gravity dam.

Professor Bofang Zhu is the academician of Chinese Academy of Engineering and Professor of the China Institute of Water Resources and Hydropower Research. He has been engaged in the design and research of large concrete dams, especially the shape optimization, thermal stress and temperature control and numerical analysis of concrete dams. He has published five books and more than 130 papers.

Dr Ping Xu is a senior engineer of China Institute of Water Resources and Hydropower Research. He specializes on the numerical analysis of concrete dam

INTRODUCTION

In recent years, there is a tendency to construct concrete gravity dams without longitudinal joint, particularly the roller compacted concrete (RCC) gravity dams. Much research work has been made on the thermal stress and temperature control of conventional concrete gravity dams with longitudinal joints[1], but few research work has been made on those of the concrete gravity dams without longitudinal joint. The most popular view at present about the temperature control of concrete gravity dams without longitudinal joint is that the temperature drop near the foundation must be less than that in the conventional concrete gravity dams with longitudinal joints, because the length of concrete block is longer [2].

Being entrusted by The China Three Gorge Project Authority, we have studied systematically the problem of thermal stress and temperature control of concrete gravity dams without longitudinal joint. The following problems are studied: 1) thermal stress due to temperature drop in the dam concrete near the foundation, temperature variation in the vertical direction of the dam and temperature difference between the surface and interior of dam, 2) the transverse cracks in the upstream face of the dam, 3) thermal stress due to extracooling of concrete around the orifice in the dam. It is discovered that in all these respects the concrete gravity dam without longitudinal joint is different from the dam with longitudinal joints.

When RCC (Roller Compacted Concrete) was begun to be used in dam construction, for a time it was thought that there was no problem in the temperature control of RCC because the amount of cement in RCC is much less than that in the conventional concrete. But some time later, it is discovered after research that RCC still has the problem of temperature control when it is used in dam construction [3]. We have studied systematically the thermal stress and temperature control of RCC gravity dams. The creep coefficient, i.e., the ratio of creep strain to elastic strain of RCC is less than that of conventional concrete. As there is no longitudinal joint in RCC gravity dams, the thermal stresses in this type of dam caused by temperature drop near the foundation, temperature difference between the surface and interior of dam, temperature variation in the vertical direction and the extracooling around the orifice in the dam are different from those in the conventional concrete gravity dam. Particularly, we should point out that transverse cracks on the upstream face may appear in high RCC gravity dams.

THERMAL STRESS DUE TO TEMPERATURE DROP NEAR THE FOUNDATION

For a conventional concrete gravity dam with longitudinal joints, generally the dam block is not very high and there is no water load during the artificial pipe cooling for joint grouting, so the stresses caused by weight of concrete and water load may be ignored when the temperature near the foundation drops to the final stable temperature of the dam. Only the temperature drop need be considered in the study of temperature control of the dam. The case of gravity dam without longitudinal joint is quite different because there is no artificial cooling for joint grouting. The dam is already completed when the temperature in the interior of dam drops to the final stable temperature. Thus, the temperature drop, the weight of concrete and the water load must be superimposed in the stress analysis for the study of temperature control of gravity dam without longitudinal joint.

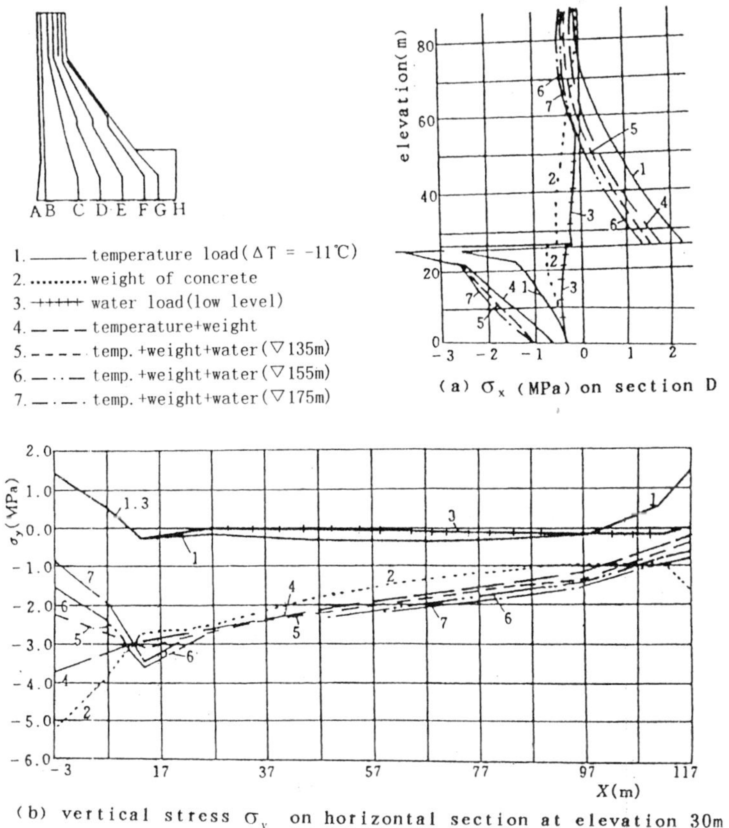

Figure 1 Stresses due to temperature drop, weight of concrete and water load in a gravity dam without longitudinal joint. (a) horizontal stress σ_x, (b) vertical stress σ_y

In order to study the influence of the weight of concrete and water load on the tensile stresses in gravity dam without longitudinal joint, the following seven loading cases are computed by FEM for the Three Gorge gravity dam:

Temperature drop alone $\Delta T = -11$□ in the dam body
Weight of concrete alone
Water load at low level (elevation 135m) alone
Temperature drop + weight
Temperature drop + weight + low water level (135m)
Temperature drop + weight + mean water level (155m)
Temperature drop + weight + high water level (175m).

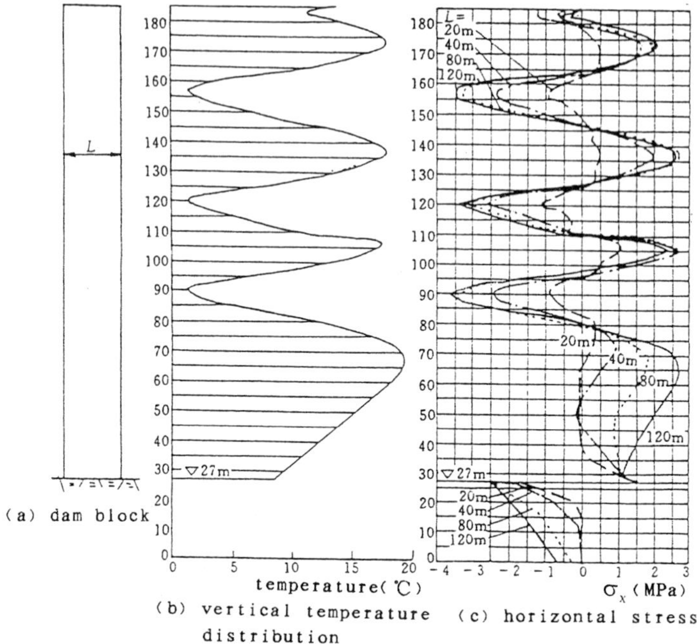

Figure 2 Relation between the length of dam block and the thermal stresses caused by vertical temperature variation

The results of computation are shown in Figure.1. It is clear that: (a) in the region of foundation restraint, the horizontal stress due to weight of concrete and water load is compressive stress which may compensate partially the tensile stress caused by temperature drop. For the maximum horizontal tensile stress in the Three Gorge gravity dam, 0.5 MPa is cut down by the weight of concrete and water load even at low water level 135m, (b) for temperature drop alone, the vertical stress σ_y near the upstream and downstream face of the dam are tensile stress, the maximum value may be 1.4 MPa. After the influence of weight of concrete and water load is considered, the vertical stresses near the upstream and downstream face of the dam become compressive (σ_y= –2.2MPa at the upstream face, σ_y = –0.4 MPa at the downstream face).

THERMAL STRESSES DUE TO VERTICAL TEMPERATURE VARIATION

The vertical temperature variation is caused by the seasonal variation of the placing temperature of the concrete and by stopping of construction due to flood or severe cold temperature in the winter. For a conventional concrete gravity dam with longitudinal joints, the thermal stresses induced by the vertical temperature variation are not high when the length of dam block is not more than 25m. Experience shows that the thermal stresses caused by vertical temperature variation may be large in gravity dams without longitudinal joint. In order to obtain the relation between the tensile stress due to vertical temperature variation and the length of dam block, concrete blocks with different length are analyzed.

According to the schedule of construction, the temperature distribution and thermal stresses are computed by FEM. The results are shown in Figure.2. For the block 20m long, there is no high tensile stress except in the region near the foundation, but for blocks with greater lengths, the thermal stresses may be high.

SUPERFICIAL THERMAL STRESSES

The variation of superficial thermal stresses caused by the temperature difference between the surface and interior of dam are quite different for the gravity dam with and without longitudinal joint, as shown in Figure 3.

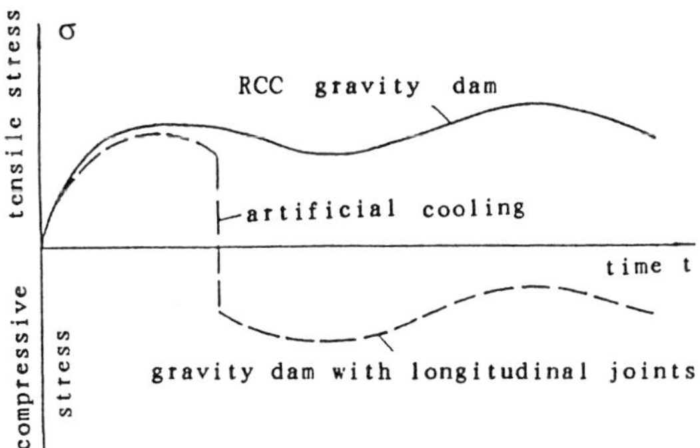

Figure 3 Superficial thermal stress due to temperature difference between surface and interior

Due to the heat of hydration, the temperature in the interior of the dam rises after casting of concrete, while the surface temperature is low because heat is dissipated into the air. As a result, there are tensile stresses in the surface of concrete dam. For a conventional concrete gravity dam with longitudinal joints, the internal temperature is reduced to the final stable temperature by pipe cooling before joint grouting, so the superficial stress will become compressive after joint grouting. It is due to this reason that the superficial cracks in the conventional concrete gravity dam with longitudinal joints appear generally in the early age of concrete, except for the dams in cold regions.

There is no artificial pipe cooling for joint grouting in dams without longitudinal joint, so the internal temperature drops very slowly, yet the surface temperature will be low in the winter, thus, high tensile stresses will occur which may cause horizontal or vertical cracks.

TRANSVERSE CRACK ON UPSTREAM FACE OF DAM

Many severe transverse cracks appeared on the upstream face of gravity dams without longitudinal joint, the depth of crack may be 20~30 meters, such as those appeared in Dworshak dam in USA and Revelstoke dam in Canada [4]. No such serious crack has appeared on the upstream face of dams with longitudinal joints.

Now we shall explain why the severe transverse cracks appeared on the upstream face of concrete dams without longitudinal joint but not on the upstream face of dams with longitudinal joints. Some superficial hair cracks may appear on the surface of concrete blocks of gravity dam with longitudinal joints in the early age, but the internal temperature is reduced to the final stable temperature by pipe cooling before joint grouting and the superficial stress will become compressive after joint grouting as shown in Figure.3, thus the superficial cracks will not extend. For dams without longitudinal joint, there is no artificial pipe cooling for joint grouting, the internal temperature drops very slowly. When the reservoir is filled with water in the early stage after completing of the dam, the temperature on the upstream face is low while the temperature in the interior is very high. The tensile stress resulting from this temperature difference between the surface and interior of dam and the splitting effect of the water pressure in the crack lead to the extending of the crack.

EXTRA COOLING AROUND ORIFICE IN DAMS

The minimum temperature in the dam without orifice is the final stable temperature. The case is different when there is orifice in the dam. The internal surface is in contact with the air or water in the orifice. The temperature of air or water in the winter is less than the final stable temperature of the dam. This is the reason why more cracks appeared on the surface of orifice in dams. The problem of extracooling around orifice is more serious for concrete gravity dams without longitudinal joint than those with longitudinal joints. The reasons for this are: **a)** The region of foundation restraint is larger. For example, if there are two gravity dams both with height 200m and length of base 150m, the distance between the foundation and the bottom of the orifice is 10m. For the dam with longitudinal joints, the width of the dam block is 15m, the orifice is practically free from the restraint of foundation, but for the dam without longitudinal joint, the orifice locates in the region of severe restraint of foundation. **b)** For the dam without longitudinal joint, the superficial cracks appeared in the course of construction are liable to develop to large cracks later, this is due to the fact that there is no artificial pipe cooling for joint grouting, the temperature of the water in the orifice is lower than that in the interior and the orifice locates in the region of foundation restraint. For the dam with longitudinal joints, the temperature difference between the surface of orifice and the interior of dam is small because there is artificial cooling for joint grouting. Furthermore, the orifice is free from the restraint of foundation, so the superficial cracks rarely extend to large cracks. Of course, if the orifice is close to the foundation, the superficial cracks may also extend to large cracks. It is clear that on account of extracooling of orifice, rather rigorous measures of temperature control must be taken for the prevention of cracking in concrete gravity dams without longitudinal joint.

THERMAL STRESS AND TEMPERATURE CONTROL OF RCC GRAVITY DAMS

In the early stage after the appearance of RCC gravity dams, it was thought for a time that there was no problem for temperature control because the cement content is very low. Some time later it is discovered that less heat is dissipated through the horizontal lift surface due to the rapid rise of dam and the extensibility and unit creep of RCC are somewhat lower than those of conventional concrete and some cracks have appeared in RCC gravity dams, so it is necessary to pay attention to the thermal stress and temperature control of RCC gravity dams.

There is no longitudinal joint in RCC gravity dams, so the conclusions in the preceding paragraphs for temperature drop near the foundation, the vertical temperature variation, the temperature difference between the surface and interior and extracooling around the orifice in the conventional concrete gravity dams without longitudinal joint are also applicable to RCC gravity dams. Generally a layer of conventional concrete with thickness 3~6m is cast on the surface of foundation as a cushion of RCC gravity dam and followed by a long period of stop for foundation grouting. It is discovered that there are remarkable thermal stresses in the cushion layer. It seems that no severe transverse crack has appeared on the upstream face of RCC gravity dams. But this may be due to the fact that the RCC gravity dams are not very high until present. There is no longitudinal joint and no pipe cooling in RCC gravity dams, the temperature of the upstream face is lower than that in the interior when the reservoir is filled after completing of construction. If superficial cracks had appeared on the surface, they are liable to develop to large cracks under the action of water pressure in the crack and the temperature difference between the surface and the interior of dam. Thus, we must pay attention to preventing the severe transverse crack on the upstream face of high RCC gravity dams.

CONCLUSIONS

1. The tensile thermal stresses due to temperature drop in the interior of a RCC or conventional concrete gravity dam without longitudinal joint are partially compensated by the action of weight of concrete and water pressure.

2. The temperature in the interior of a RCC or conventional concrete gravity dam without longitudinal joint drops very slowly, severe transverse crack may appear on the upstream face of dam. Some measures must be taken to prevent these cracks. The most effective measure is to insulate the upstream face with foamed plastics.

3. The tensile stresses caused by vertical temperature variation in a RCC or conventional concrete gravity dam without longitudinal joint are greater than those in the concrete gravity dam with longitudinal joints.

4. The tensile thermal stresses caused by extracooling around orifice in a RCC or conventional concrete gravity dam without longitudinal joint are greater than those in dam with longitudinal joints.

5. There are remarkable tensile thermal stresses in the conventional concrete cushion on the foundation of RCC gravity dams due to long time stop for foundation grouting. Measures must be taken to prevent cracking.

6. Due to low content of cement, the adiabatic temperature rise of RCC is less than that of conventional concrete, but less heat is dissipated through the horizontal lift surface because of rapid rise of dam. The creep coefficient of RCC is lower than that of conventional concrete. Attention must be paid to prevention of cracks in RCC gravity dams.

REFERENCES

1. ZHU BF, Thermal Stresses and Temperature Control of Mass Concrete, China Electricity Publishing House, 1998.

2. MINISTRY OF WATER RESOURCES AND ELECTRICITY OF CHINA, Design Specifications for Concrete Gravity Dams, Water Resources and Electricity Press, 1979.

3. DONG FP and ZHU BF, Thermal stresses in RCC dams, Water Resources and Hydroelectricity, No.10, 1987.

4. BRUNNER WJ and WU KH, Cracking of the Revelstoke concrete gravity dam mass concrete, 15th International

AN EXPERIMENTAL STUDY ON MECHANICAL PROPERTIES OF EXTREMELY LEAN CONCRETE

I Nagayama

K Watanabe

K Nishizawa

Public Works Research Institute

Japan

ABSTRACT. Mechanical properties of extremely lean concrete were studied with compression tests, split tensile tests and tri-axial compression tests. The following results were obtained. (1) Even for extremely lean concrete, the values of the uni-axial compressive strength, split tensile strength, elastic modulus and the compressive strength under the confining pressure can be obtained by simply extrapolating those of conventional concrete, as long as the VC value is set at about 20 seconds and the concrete is compacted sufficiently. (2) When the ratio of the confining pressure to the uni-axial compressive strength is constant, the normalized strain-stress curve in the tri-axial compression test indicates the same shape that is independent of the unit cement content. (3) Whether or not concrete shows plastic properties depends on the ratio of the confining pressure to the uni-axial compressive strength of the concrete.

Keywords: Extremely lean concrete, Unit cement content, Uni-axial compressive strength, Tri-axial compression test, Stress-strain curve, VC value.

Mr Isao Nagayama is Head of Dam Structure Division, Dam Department, Public Works Research Institute, Ministry of Construction, JAPAN. He specializes in dam engineering, including concrete engineering, rock mechanics and earthquake engineering. He serves on many technical committees.

Mr Kazuo Watanabe is Senior Researcher of Dam Structure Division, Dam Department, Public Works Research Institute, Ministry of Construction. He specializes in mechanical properties of stiff consistency concrete like Roller Compacted Dam-concrete. He is now engaged in the use of High Performance Concrete for concrete dam bodies.

Mr Kentaro Nishizawa is Researcher of Dam Structure Division, Dam Department, Public Works Research Institute, Ministry of Construction. He specializes in the effective use of low quality aggregate. He is now engaged in experimental studies on mechanical properties of extremely lean concrete.

INTRODUCTION

Concrete gravity dams depend on their own weight to keep the safety and the stress requirement of concrete of dams is not strict. Then, in order to achieve the construction economy, lean mix of concrete such as RCD concrete is effectively used in the internal concrete, where resistance to the durability is not important.

Unit cement content of RCD concrete is usually 120 kg/m^3 (30 % of cement is replaced by fly ash in general) in Japan. If the unit cement content of RCD concrete can be reduced further, the dam construction will be more economical.

From these backgrounds, the authors have carried out the comprehensive experimental study on mechanical properties of extremely lean concrete including the uni-axial compressive test, the split tensile strength test, and the tri-axial compressive test. The paper discusses these test results and the possibility of use of extremely lean concrete.

OUTLINE OF EXPERIMENTS

Materials and Mix proportion of concrete

Table 1 shows the properties of materials used for concrete and table 2 shows the mix proportions of concrete. The standard mixture (C175 in table 2) was equivalent to 40mm-wet-screened concrete of typical RCD-mixture with maximum size of aggregate of 150mm. Then, the unit cement content was gradually reduced, while the unit water content was adjusted to obtain a VC value of approximately 20 seconds.

Table 1 Properties of materials

MATERIALS	PROPERTIES
Cement; Normal portland cement	Specific gravity; 3.16 Specific surface; 3,320 cm^2/g
Fine aggregate; Crushed sandstone	Specific gravity; 2.65 Absorption; 1.0 % Fineness modulus; 3.04
Coarse aggregate; Crushed sandstone	Specific gravity; 2.67 Absorption; 0.45 % Fineness modulus; 7.26

Table 2 Mix proportion of concrete

MIXTURE	MAX SIZE OF AGG.	TARGET OF VC VALUE	TARGET OF AIR CONTENT	WATER CEMENT RATIO	SAND	UNIT CONTENT (kg/m^3)			
	(mm)	(sec)	(%)	(%)	(%)	Water	Cement	Fine Agg.	Coarse Agg.
C175	40	20±10	2.2±1.4	67	44	118	175	938	1203
C140	40	20±10	2.2±1.4	88	44	123	140	945	1212
C105	40	20±10	2.2±1.4	120	44	126	105	955	1225
C70	40	20±10	2.2±1.4	186	44	130	70	963	1235

Specimen preparation

Cylinder specimens with a diameter of 15cm and a height of 30cm were used in the tests. RCD-concrete Standard Specimen Compaction Device [1] was used to make the specimens. This device is widely used in compaction tests on RCD-concrete in Japan. The specimens were placed in three layers and the compaction time for each layer was 40s. The specimens were cured in water at temperature of 20°C until testing age of 91 days.

Measurement

Three types of test were carried out – a uni-axial compression test, a split tensile strength test and a tri-axial compression test. The uni-axial compression test and the split tensile strength test were as specified in JIS A 1108 and JIS A 1113 respectively [2][3]. In the uni-axial compression test, a strain gauge was attached to the surface of the specimens, and the elastic modulus was measured. In the tri-axial compression test, four values of the confining pressure were adopted: 0.5, 1.0, 1.5 and 2.0 N/mm^2. The axial loads were applied so as to keep the axial strain rate of 0.05%/min. The confining pressure was applied by hydraulic pressure via a rubber sleeve.

TEST RESULTS

The mix proportions of concrete that gives a VC value of 20 seconds

Figure 1 shows the relationships between the unit cement content and the volume of each material that gives a VC value of 20 seconds. When the unit cement content decreases, the paste volume becomes smaller and the absolute volume of the aggregate becomes bigger. The reason is believed to be that when the water-cement ratio becomes bigger, the viscosity of the cement paste decreases and it reduces the frictional resistance between aggregates.

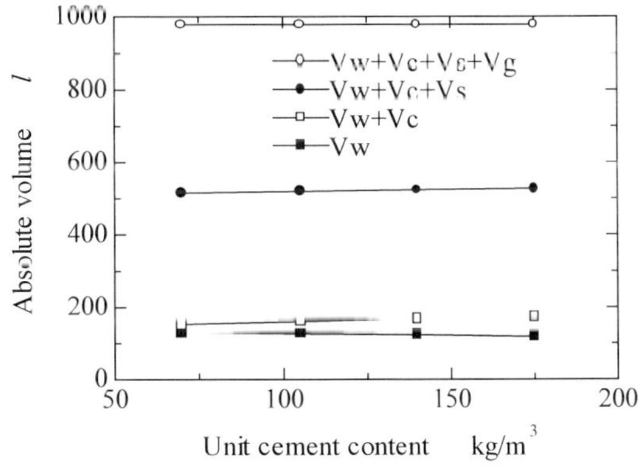

Vg: volume of coarse aggregate
Vs: volume of fine aggregate
Vc: volume of cement
Vw: volume of water

Figure 1 Relationships between the unit cement content and the volume of materials

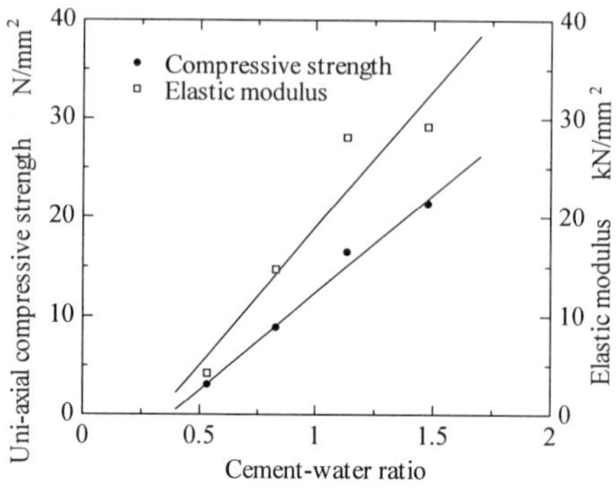

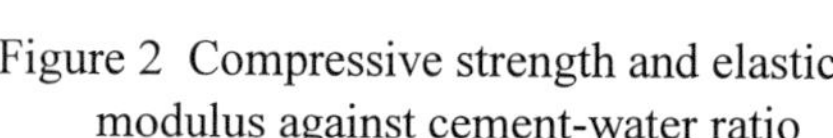
Figure 2 Compressive strength and elastic modulus against cement-water ratio

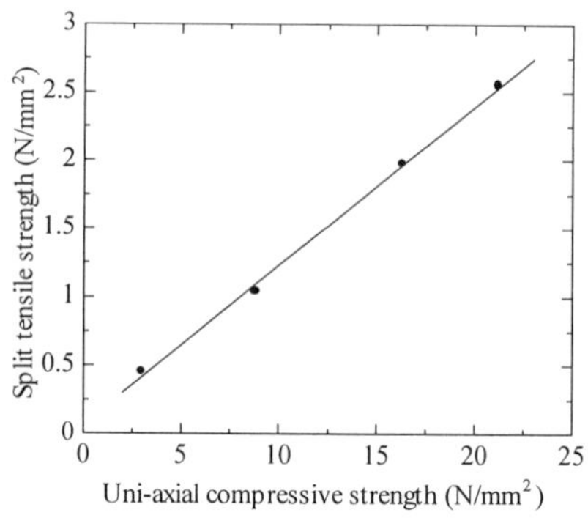

Figure 3 Relationship between split tensile strength and compressive strength

Uni-axial compression test and split tensile strength test

Figure 2 shows the relationships between the cement-water ratio and the uni-axial compressive strength, and between the cement-water ratio and the elastic modulus. This figure shows that, even for extremely lean concrete where the unit cement content is as low as 70kg/m^3, there are linear relationships between the cement-water ratio and the uni-axial compressive strength, and between the cement-water ratio and the elastic modulus. The uni-axial compressive strength is given by the law of cement-water ratio. From Figure 2, it was estimated that the uni-axial compressive strength would reach zero at a cement-water ratio of about 35%. Extrapolating the concrete mixture proportion (in Figure 1) the unit cement content would be about 50kg/m^3 at this point. Accordingly, when the maximum size of aggregate is 40mm, it is predicted that the lowest unit cement content in the concrete where the cement is expected to be effective as a binding material should be about 50kg/m^3.

Figure 3 shows the relationship between the uni-axial compressive strength and the split tensile strength. The ratio of the former to the latter is a constant value of about 10:1.

It can be concluded from these results that, even for extremely lean concrete, the values of the uni-axial compressive strength, split tensile strength and elastic modulus can be obtained by simply extrapolating those of conventional concrete, as long as the VC value is set at about 20 seconds and the concrete is compacted sufficiently. In other words, this extremely lean concrete can be treated in the same way as conventional concrete.

Tri-axial compression test

Test results

Tri-axial compression tests were carried out on each of the mixtures (unit cement content of 175, 140, 105 and 70 kg/m^3) for four different values of the confining pressure (0.5, 1.0, 1.5 and 2.0 N/mm^2). The plots of principal stress difference $\sigma_1 - \sigma_3$ (difference between the axial stress σ_1 and the confining pressure σ_3) against axial strain, and volumetric strain against axial strain are shown in Figure 4. These figures show that all of the mixtures behave as plastic materials under the confining pressure.

Strength characteristics

Figure 5 shows the relationship between the confining pressure σ_3 and the maximum principal stress difference $f_T - \sigma_3$ (the maximum value of $\sigma_1 - \sigma_3$; this represents the compressive strength under the confining pressure) for each of the four mixtures. The results of the uni-axial compression test (i.e. the case when $\sigma_3 = 0$) are also included. This figure shows that there is a linear relationship between the maximum principal stress difference and the confining pressure for each of all the mixtures.

Figure 6 shows the relationship between cement-water ratio and the maximum principal stress difference for each value of the confining pressure. There is a linear relationship between them. In other words, the compressive strength under a confining pressure is still given by the law of cement-water ratio.

Combining the results shown in Figures 5 and 6, the compressive strength of the concrete under a confining pressure can be expressed as in the following equation.

$$f_T - \sigma_3 = (a\sigma_3 + b)\ (cC/W + d) \qquad (1)$$

$$= 0.0112\ \sigma_3\ C/W + 3.64\ \sigma_3 + 0.195\ \ C/W - 6.52 \qquad (2)$$

where a, b, c and d are coefficients of two linear equations.

The coefficients in the equation (2) were calculated using multiple regression analysis, where $f_T - \sigma_3$ and C/W are expressed in units of N/mm^2 and % respectively. This equation is applicable to all types of concrete. The uni-axial compressive strength f_C is given by equation (2) with $\sigma_3 = 0$, that is,

$$(3)\quad f_C = 0.195\ \ C/W - 6.52$$

(4)

$$f_T - \sigma_3 = 0.0573\ \sigma_3\ f_C + f_C + 4.02\sigma_3$$

Eliminating C/W from equations (2) and (3),

Figure 7 shows the relationship between uni-axial compressive strength and maximum principal stress difference where symbols express test results and solid lines express the equation (4). For each value of the confining pressure, there is a linear relationship between the maximum principal stress difference and the uni-axial compressive strength

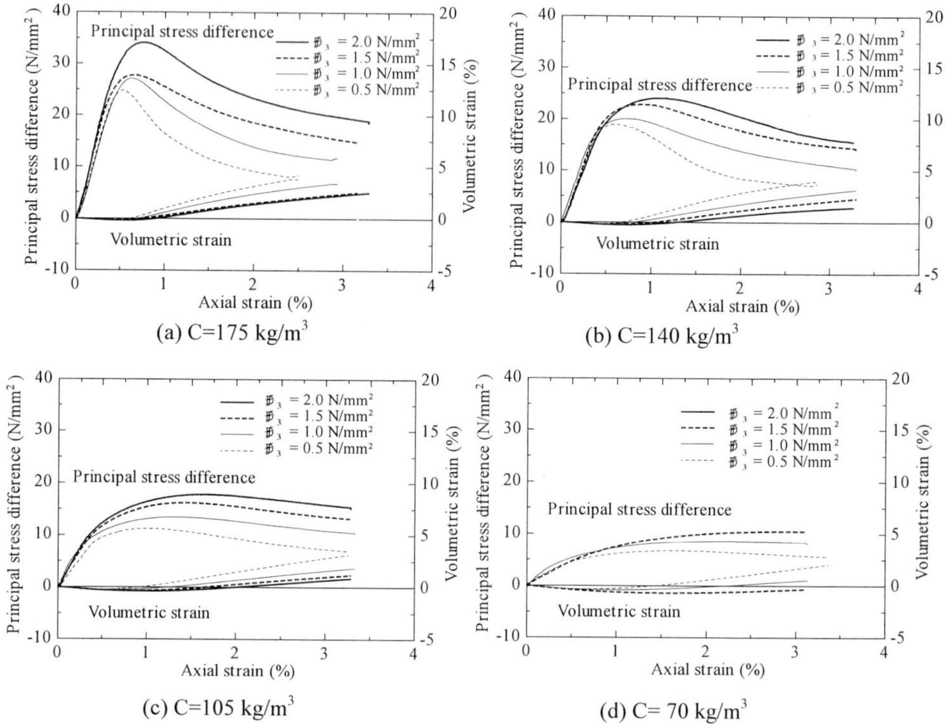

(a) C=175 kg/m^3

(b) C=140 kg/m^3

(c) C=105 kg/m^3

(d) C= 70 kg/m^3

Figure 4 Relationship between principal stress difference and axial strain and relationship between volumetric strain and axial strain

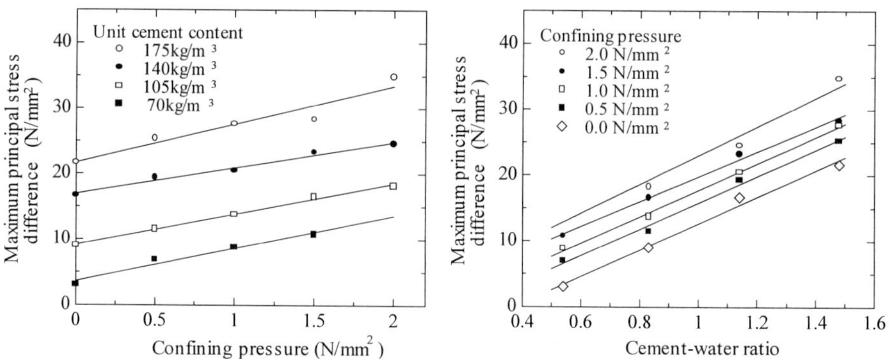

Figure 5 Relationship between confining pressure and maximum principal stress difference

Figure 6 Relationship between cement-water ratio and maximum principal stress difference

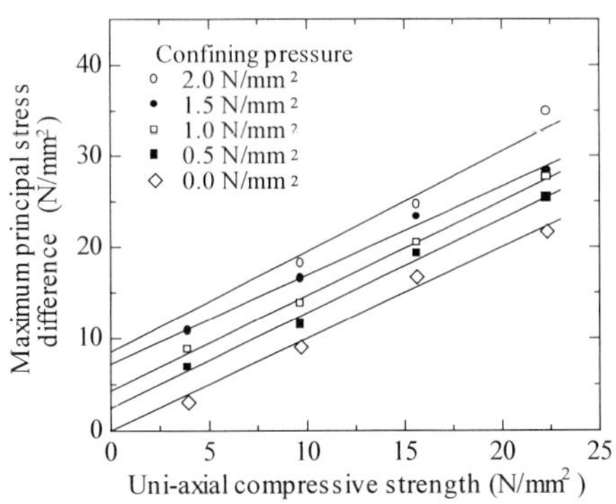

Figure 7 Relationship between uni-axial compressive strength and maximum principal stress difference

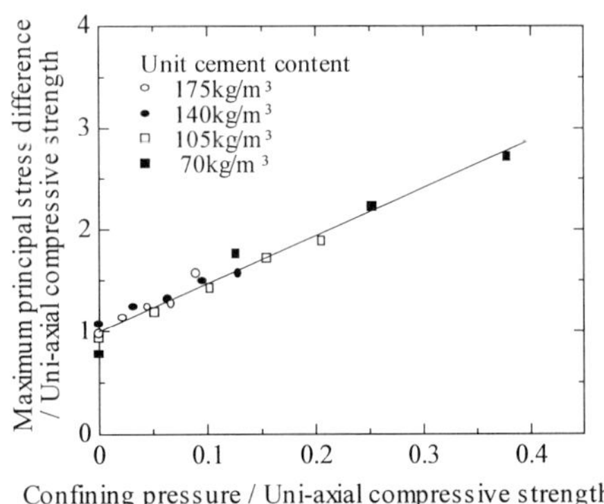

Figure 8 Relationship between normalized confining pressure and normalized maximum principal stress difference

Figure 8 shows a different representation of Figure 7; where, the maximum principal stress difference and the confining pressure have both been normalized by dividing by the uni-axial compressive strength. This figure shows that, regardless of the unit cement content, the relationship between the normalized confining pressure σ_3/f_C and the normalized maximum principal stress difference $(f_T-\sigma_3)/f_C$ are almost on the same single straight line. This shows that, even for extremely lean concrete, the strength characteristics remain the same as for conventional concrete. It is based on the fact that the coefficient of $\sigma_3 f_C$ in equation (4) is sufficiently small (0.0573).

The relationship between the cement content and stress-strain curve

Figure 9 shows the stress-strain curve when the confining pressure is 1.5 N/mm^2. This figure shows that when the unit cement content increases, the maximum principal stress difference also increases but the residual principal stress difference seems to converge to one specific value that is independent of the unit cement content. The residual principal stress difference depends only on the confining pressure.

Figure 10 show the normalized stress and strain curves for the typical two test results whose plots are almost at the same point on the regression line in Figure 8. Normalization is made in Figure 10 so that the maximum principal stress difference is 1 and the strain at the time is 1, too. The figure shows that when the ratio of the confining pressure to the uni-axial compressive strength is the same for two test results, the normalized stress-strain curves for these two test have almost the same shape. It means that the stress-strain curve of concrete is determined by the rate of the confining pressure to the uni-axial compressive strength of the concrete. In other words, whether or not concrete indicates plastic properties is determined not by the unit cement content, but rather by the rate of the confining pressure to the uni-axial compressive strength.

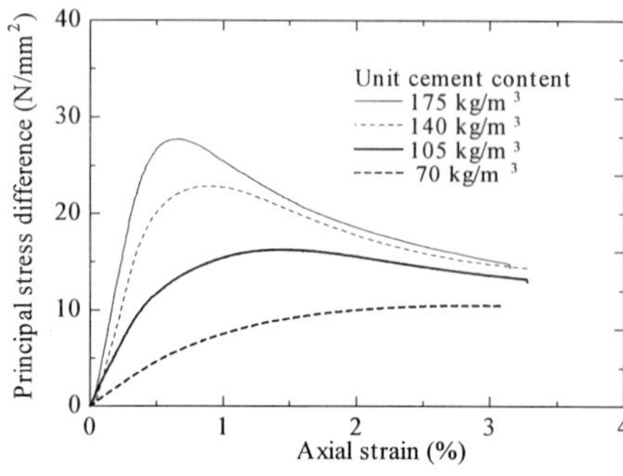

Figure 9 Stress-strain curves (confining pressure is 1.5 N/mm^2)

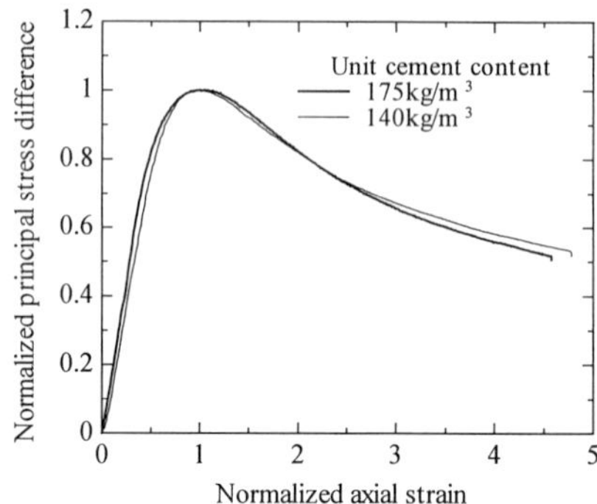

Figure 10 Normalized stress-strain curves

CONCLUSIONS

1. Even for extremely lean concrete, if the VC value is set at 20 seconds and sufficient compaction is carried out, the uni-axial compressive strength follows the law of cement-water ratio. In this research, it is predicted that the limit of the unit cement content for which the cement can be expected to be effective as a binding material should be about 50 kg/m^3.

2. There are linear relationships between the elastic modulus and the cement-water ratio, and between the split tensile strength and the cement-water ratio. The ratio of the split tensile strength to the uni-axial compressive strength is about 1/10.

3. The law of cement-water ratio can apply in the case of the tri-axial compression test under the constant confining pressure. In other words, there is a linear relationship between the maximum principal stress difference and the cement-water ratio.

4. When the confining pressure and maximum principal stress difference were normalized by dividing by the compressive strength, there is a fixed relationship between the resulting normalized confining pressure and normalized maximum principal stress difference, which is independent of the unit cement content.

5. When the ratio of the confining pressure to the uni-axial compressive strength is constant, the normalized strain-stress curve in the tri-axial compression test indicates the same shape that is independent of the unit cement content.

6. Whether or not concrete shows plastic properties depends on the ratio of the confining pressure to the uni-axial compressive strength of the concrete.

REFERENCES

1. T. FUJISAWA, I. NAGAYAMA AND K. WATANABE. A Study on Mixture Design of RCD Concrete using Standard Specimen Compaction Device. 19th International Congress, ICOLD, Florence, 1997, pp.393-412.

2. JIS A 1108-1993 (Japanese Industrial Standard) Method of test for compressive strength of concrete. Japanese Standards Association, 1993.

3. JIS A 1113-1993 (Japanese Industrial Standard) Method of test for splitting tensile strength of concrete. Japanese Standards Association, 1993.

RESISTANCE OF ALAG CONCRETE TO HYDRAULIC ABRASION AND CAVITATION IN DAM STRUCTURES

J L Cabiron

Lafarge Aluminate

S Lavigne

CERG-ACB-GEC Alsthom

France

ABSTRACT. The performance of a special concrete made of calcium aluminate cement and calcium aluminate reactive synthetic aggregate (ALAG ®) is evaluated for its resistance to abrasion and cavitation by high water velocities commonly found in hydraulic structures. ALAG concrete is compared to Portland cement concretes and other protection materials using traditional hydraulic abrasion test methods and a new cavitation test designed for measuring erosion effects on metallic materials. Other mechanical properties are also reported such as impact resistance, freeze / thaw and bond strength. Abrasion performances have been cross checked using different testing methodologies from the major hydraulic laboratories specialised in concrete dam evaluation. In almost all cases, ALAG concrete performed among the best materials tested. New structures of greater life expectancy can be designed with ALAG concrete in lieu of the more conventional materials such as steel plates or granite blocks for example.

Keywords: ALAG, Abrasion, Cavitation, Concrete, Calcium aluminate cement, Calcium aluminate aggregate, Dam, Spillways, Rehabilitation, Durability, Rapid hardening, Freeze-thaw, Impact, Corrosion, Adherence.

Jean Louis Cabiron is responsible for hydraulic civil engineering applications at Lafarge Aluminate, Paris, France. He has been in contact with the hydraulic industry for 10 years including 5 years in the USA. He coordinates the internal and external laboratory test programs on abrasion and cavitation in order to understand and improve the concrete applications in the dam structures. His technical background and worldwide expertise in this field allowed him to be involved in several international dam projects and other structure protection. He is a member of the French committee on large dams.

Dr Sylvain Lavigne is Doctor in Fluid Mechanics. He is responsible for hydrodynamic studies in relation to turbo machinery, cavitation and erosion associated with cavitation at CERG-ACB-GEC Alsthom, Le Pont de Claix, France.

INTRODUCTION

The criteria for choosing abrasion resistant materials can be summarised as cost-effectiveness and durability. Cost effectiveness may be broken down into, ease of installation and compatibility with the rest of the surrounding structure, especially adhesion to the substrate. Durability and adhesion can be evaluated by a series of tests. Table 1 summarises various abrasion test methods implemented on ALAG concrete. They have been selected for their water velocity from 3 to 50 m/sec, and their design in terms of sand load, rotating balls, water jet angle and for their correlation with real hydraulic conditions in order to estimate a life expectancy. A summary of each of them is detailed in the following paragraphs.

Table 1 Various abrasion test conditions

	ABRASION TESTS			
	C.N.R. [1] (France)	A.S.T.M. C 1138 (U.S.A.) [2]	Vattenfal (Sweden)[3]	Los Angeles Index
Water velocity	10 ms.	3 ms.	50 ms.	N.A.
Type of test	Underwater water jet + sand	Underwater rotating balls	Water jet	N.A.
Duration of test	75 minutes	24 and 72 hours	7 hours	N.A.

TRADITIONAL LABORATORY TESTS FOR ABRASION AND IMPACT EVALUATION

Test Conditions: Underwater-Sand Jet, 45° Angle, 10 m/., Equivalent to 110 000 m^3 / Year and 5 000 T Sand Load / m.

The main abrasion test used is the C.N.R. test (Compagnie Nationale du Rhône - France) which measures the relative performance after 75 minutes of exposure. A summary of the results is given in Figure 1(a) and an outline of the test method in Figure 1(b).

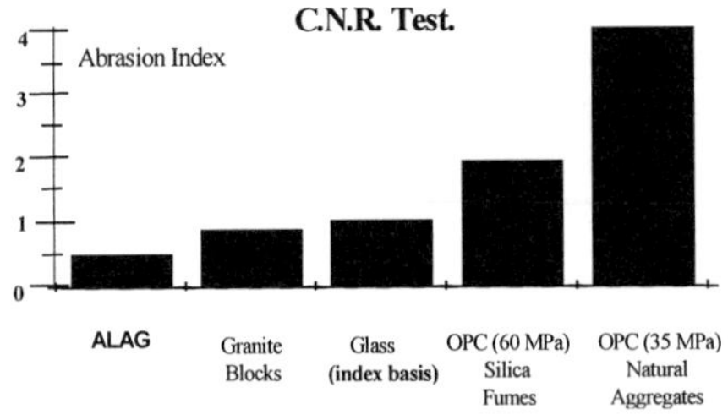

Figure 1(a) Summary of CNR test results

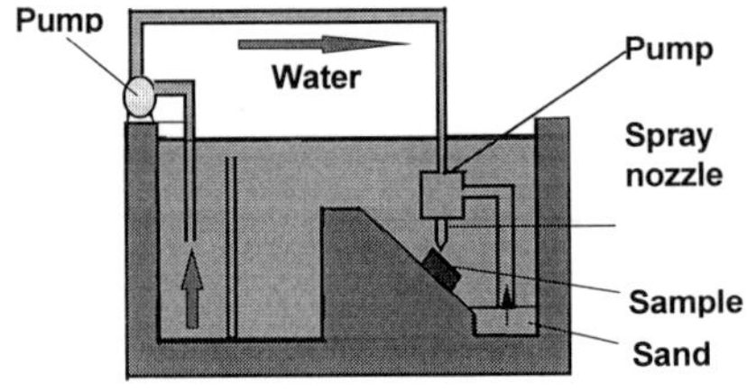

Figure 1(b) CNR test method. Water velocity = 10m/s, duration = 75 minutes

It may be seen that, based on these results ALAG concrete is more abrasion resistant than igneous rock (granite for example) and considerably more resistant than Portland cement concretes even when containing silica fume. When the economical aspect is being taken in account for each solution, the position of each material can be established as follow in the Figure 1(c) below.

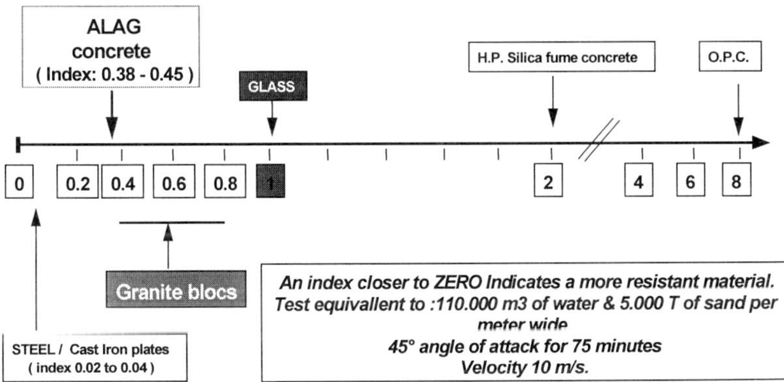

Figure 1(c) Classification scale for various materials (based on CNR test)

Test Conditions: Underwater Rotating Balls

Tests based on the ASTM C1138 procedure, or the US Corps of Engineers test C.R.D. C 63 A summary of the ASTM C1138 results is shown in Figure 2(a). An outline of method is shown in Figure 2(b).

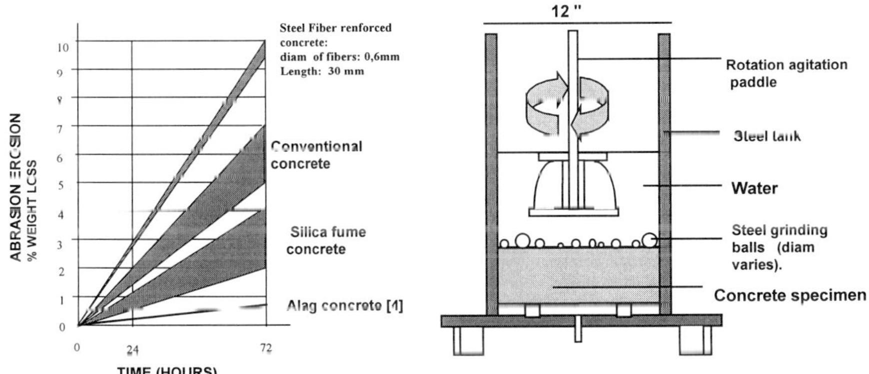

Figure 2(a) Summary of ASTM C 1138 test results at 24 and 72 hours

Figure 2(b) ASTM C 1138 test apparatus. Water velocity 3m/s

Although this test uses a lower water velocity (3 m/s - 6 ft/s), it has been argued to be at least as severe as the C.N.R. test, because there may be more turbulence generated at low water speeds and the duration is longer; 24 and 72 hours versus 75 minutes.

2 major comments can be made by observing the tests results summarised in Figure 2(a) :

1. Steel fibres do not enhance the abrasion performance of P.C. mixes.
2. Silica fume P.C. although more abrasion resistant than traditional O.P.C. are still 4 - 8 times less resistant than ALAG concrete. The abrasion resistance is not correlated with mechanical strength only. Other parameters are also of importance.

Test Conditions : 90° Angle Water-Jet at 50 m/s Velocity

The Vattenfall test (Sweden) is designed around a water jet at high speed : 50 m/s during 7 hours. No abrasion is reported on ALAG concrete compared with 8 mm depth of attack for Portland concrete.

Test Conditions : Los Angeles Index

Some countries refer to the hardness of the aggregates for designing abrasion resistant concrete.Tests run on ALAG show an index of 11-14 %, which is considered very hard for an aggregate. Usually an index below 40 % is requested.

Test Conditions : Impact Resistance

Two laboratories have developed equipment to monitor impact resistance : the C.N.R. (France) and TIWAG (Austria).Figure 3 below represents the apparatus used by the C.N.R. laboratory. The ball falls 2700 times for 3 hours on the concrete specimen.This basic test simulates impact and measures the volume loss, which is indicated in Table 3.

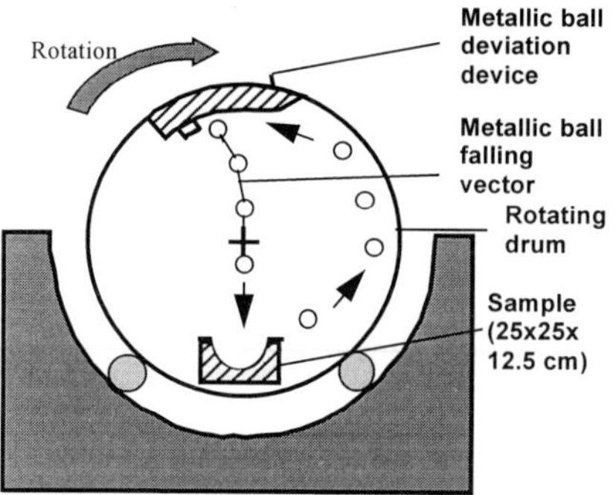

Figure 3 CNR impact test method

Table 3 Comparison of CNR impact resistance

MATERIAL	LOSS OF MATERIAL
ALAG Concrete	<120 cm³
Ordinary Portland Concrete	>300 cm³
Granite	<100 cm³

Service Life Expectancy Estimation

From the various tests above, one can correlate the testing conditions and cross check it with the hydraulic parameters found in each project to estimated life expectancy of the materials tested. The most interesting are : flow quantity (m^3/year), water velocity (m/sec), sand load quantity (ton/m^3) and sand particle sizes.

The angle of attack, the local pressure conditions and the cavitation risks as well as the nature of the water (soft, pure , corrosive...) and the weather conditions (freeze - thaw variations) are also part of the equation.Another indicator, if it exists, is the life span of the previous material used in such area (usually granite blocks, Portland concrete or steel plates).

The ALAG concrete life expectancy can be estimated in the worst conditions with the following parameters : constant angle of attack of the stream of 45°, a water velocity of 10 m / sec, fine particle (< 2 mm), sand load of 50 %. In these conditions, 1 cm of ALAG could be expected to be lost. In real scale conditions, for a 54 m wide river for example, these laboratory conditions are close to an equivalent of a 110 000 m^3 flow and 5 000 tons of sand particles carried per meter wide.

Example of material comparison with granite :

If the C.N.R. abrasion index for the local granite is twice the ALAG one (meaning half as resistant), in theory, if the granite block thickness specified is 30 cm, 15 cm of ALAG concrete should be equivalent.

In addition due to the joint weakness of the granite due to accelerated flow between the blocks, 2/3 of the granite thickness can be considered to be really efficient. It means in this example that only 20 cm of granite is really to be considered as the protection thickness.

The wear of ALAG concrete being linear, without joint problems, allows in this case to compare in fact 10 cm of ALAG against 30 cm of granite.

EXPERIMENTAL CAVITATION TESTS

The Experimental Set-up

Recently, consulting engineers have been interested in studying possible new applications of ALAG concrete for the replacement of steel plate structures. Research has therefore been undertaken at the Centre d'Etude et de Recherche de Grenoble (CERG), the hydraulic laboratory of G.E.C- ALSTHOM - ACB in France. The key objectives were :

1. To determine how this concrete would perform in cavitation conditions and high flow rates of water speed of 110 m/sec. at 54 bars pressure.
2. Evaluate the theoretical life expectancy of this material under non accelerated erosion conditions.

The special mini tunnel simulating extreme accelerated cavitation conditions was used. It can creates significant erosion losses on aluminium within a few minutes or on stainless steel within a few hours.

Previous studies have quantified precisely the influence of geometrical and hydrodynamic conditions simulating and monitoring cavitation conditions (see Figures 4 & 5).

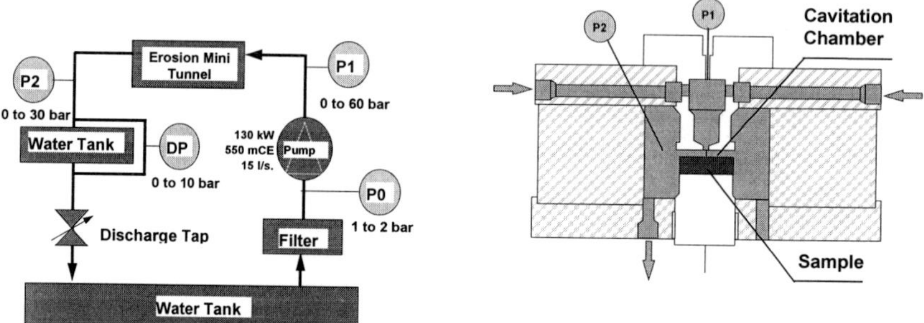

Figure 4 Details of CERG mini tunnel

Figure 5 Layout of CERG mini tunnel

Accordingly, the equipment and parameters used in the recent tests were as follow :

A high pressure pump (rating W. = 130 kW ; water load ΔH = 550 mCE ; flow rate Q = 15 l/sec.), an erosion chamber with two pressure sensors controlling upstream and downstream pressure (P1 upstream = 0-60 bar; P2 downstream = 0-30 bar), a 14 mm diaphragm equipped with a differential pressure sensor to measure the volume of water passing, a discharge by-pass vane to the main water tank.

A number of comparative tests were run at high flow speed (110 m/s) and high pressure (54 bars) for 2 minutes on four samples. Two of the samples were based on a high strength Portland concrete and mortar formulations. The other two were ALAG concrete and mortar compositions based on Ciment FONDU Lafarge and synthetic Aluminous Aggregates (ALAG).

Test Conditions

The concrete samples (see Figures 6 & 7) are fixed in the chamber and systematic tests were carried out three times on each of the four sample. The upstream and downstream pressure conditions (P1 & P2 respectively) were monitored precisely with:

$$P2 /(P1 - P2) = \sigma = 0.8$$

This σ value gives an indication of the condition of cavitation. If the σ remains constant during the two- minutes duration of the test, the erosion conditions are at the maximum. If it varies, (for example, P2 can increase due to surface aspect perturbations), the erosion conditions are less severe and cease to be aggressive when P2 tends to be closer to P1.

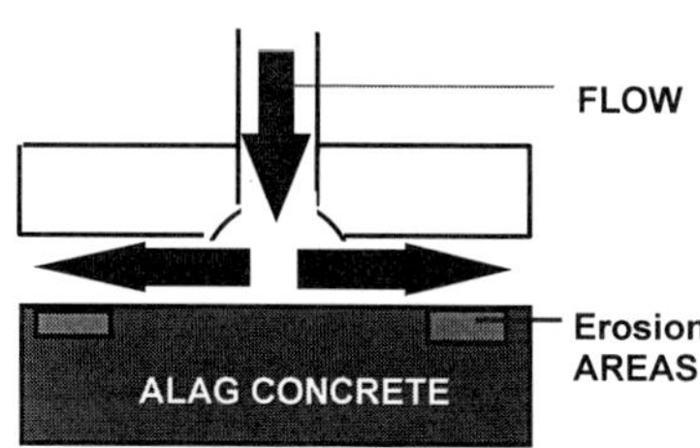

Figure 6 Principle of CERG chamber

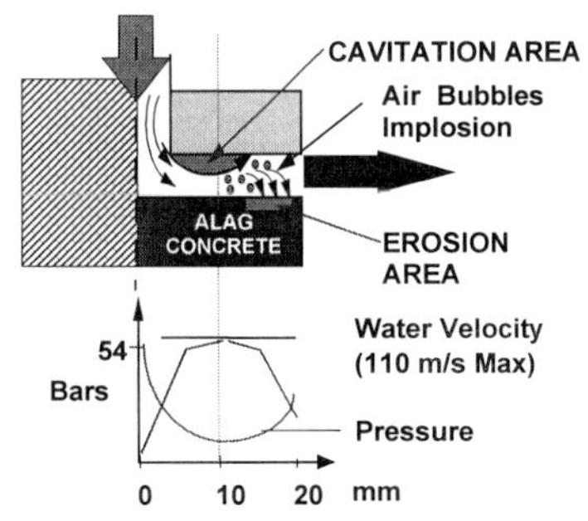

Figure 7 Details of cavitation and erosion areas

TEST RESULTS [5, 6]

Montoring the σ Pressure

With the Portland mixes, the σ value varied slightly, probably because of a variation in the perturbation of the flow on the surface of the specimen generated by the erosion. However, the σ value remained stable on the ALAG specimen. This indicates that the cavitation conditions were more severe for the ALAG mixes than for the Portland ones. Although the tests conditions were slightly different, all the concrete sample surfaces darkened forming a 25 mm-diameter ring, 5 mm wide, identical to that formed on metallic samples. This indicates that the cavitation conditions were present, and in these areas most of the damage occurred as a result of vapor bubble implosion (also referred as erosion pitting (5 - 8 mm deep) erosion due to cavitation).

Comparison of the Surface Aspects

The ALAG samples behaved very well with no sign of erosion or weight loss. The specimen also remained homogeneous, like the metallic one. No cavitation pitting, or loosening of the aggregates occurred. However, the Portland samples experienced and also aggregate loosening.

It appears that the weak points on Portland mixes are :

1. The interface between the aggregates and the cement paste. Loosening of the aggregates was the primary cause of the poor performance of the Portland mixes, including erosion of the aggregates themselves

2. The attack of the Portland cementitious matrix. The good behaviour of ALAG concrete probably results from the physical and chemical links of the elements in the matrix which are of the same mineralogical and chemical nature. In fact the synthetic ALAG aggregate is reactive. It hydrates at the interface with the cement (which has the same chemistry) to

produce a homogeneous bound between the cement and the fine and coarse aggregates. Another features is that both the cement and the aggregates are equally hard and therefore no weak zone is created.

Full Scale Transition

This accelerated test combines the three major parameters for the fast cavitation simulation of cavitation: maximum water speed, aggressive structural geometry and optimised cavitation σ (or Thomas's number of similitude). The Mean Depth Erosion Rate (MDER) for a water speed of less than 57 m/s depends of the velocity (MDER = k. V^6). From the two-minute test at 110 m/s, it was determinated that at 30 m/s which is close to the maximum authorised turbine water velocity, it is necessary to wait at least four days to observe the same cavitation damage. At 10 m/s it is necessary to wait seven years. In addition, with an optimised geometric profile, as exists in the real turbine draft tube, and with lower water speed, this test indicated that it could be 10-15 years before any cavitation damage would be observed. These results are therefore favourable as regards applications of ALAG concrete in areas subject to risk of cavitation. Also, the theoretical life expectancy can be predicted (see Table 4 & Figure 8) for formulation details and ALAG mechanical performance).

Table 4 Typical ALAG concrete mix design

ALAG CONCRETE MIX DESIGN (PER CUBIC METER)		
Cement	Ciment Fondu Lafarge	515 kg
Aggregates	ALAG FINE (0 – 2.5 mm)	1030 kg
	ALAG COARSE(2.5 – 12.5)	1030 kg
Water	W/C ratio 0.4	206 liters

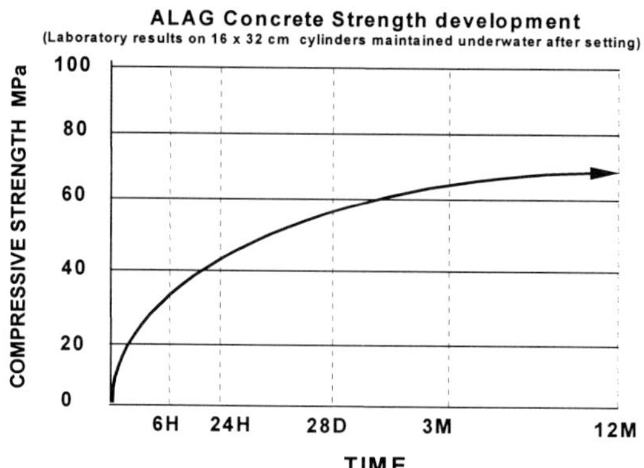

Figure 8 Strength evolution of ALAG concrete

Real tests in areas subject to cavitation now have to be undertaken to confirm these preliminary positive results. In fact some rehabilitation jobs have already been done using the same concept. However, as mentioned earlier, the test conditions are in fact more aggressive that the reality, and with this MDER it is reasonable to expect that 15 years without erosion is a good probable time range.

The three combined factors mainly responsible for damages in hydraulic structures are then documented such as discussed above :

1. Erosion by cavitation,

2. Erosion by abrasion,

3. Mechanical impact. additional parameters such as the corrosion aspects are also of interest and covered by tests.

Other Parameters to Consider

For complete material evaluation, other parameters are also important to consider such as bond strength, freeze - thaw and installation method. Each of those parameter is reviewed hereafter.

Good adhesion to support

Adhesion to the substrate is particularly important where relatively thin layers of material are applied, even more so when they are subjected to high shear forces due to large volumes of water at high velocity. Various methods are commonly used to prepare concrete surfaces for specialised surfaces. Generally the best preparation is achieved with a high pressure (450 bars) hydro demolition of the prior concrete surface to obtain an exposed aggregate finish. Then, pre-soaking the Portland cement concrete surface to saturation for 24 hours minimum is necessary prior to casting ALAG concrete.This method allows for good adhesion, up to 4 MPa, which is considered good for eliminating the risk of debonding.The failure occurs generally in the Portland slabs, and reflects the limit of tensile strength of the Portland itself. The interface bond strength is generally superior to the Portland tensile strength. This is why failure occurs in the Portland slab. The adhesion strength varies therefore according to the nature and dosage of the Portland. On a low heat mix, bond is close to 1 MPa, for example.

Freeze - thaw cycling

Many hydro-electric sites are at altitudes or in climates where low temperatures are experienced, therefore, freeze / thaw resistance may be an important requirement. Tests were carried out using ASTM C-666 and SS Vattenfall tests n° SS 13 72 44 1B and SS 13 72 44 1 A. All tests show excellent freeze-thaw resistance of the ALAG concrete.

Corrosion resistance

The durability of calcium aluminate cements when in contact with soft water or other corrosive environments is in general superior to that of Portland cements.

This is explained by the absence of portlandite (CaOH) and by the formation of special hydrates which show good when in contact with aggressive agents.

ALAG concrete behaves well in these aggressive or very aggressive conditions.

Joint system between the slabs

To minimise shrinkage cracks and to produce thin construction joints ALAG slabs are applied at a maximum 2 m x 2 m panels, cast according to a checkerboard method (see Figure 1). The in-fill panels are placed 7 to 10 days later ideally, or at least 24 hours later (calcium aluminate cements achieve 80 % of their hydration within 24 hours of mixing).

Curing for 24 hours either by water spray or approved curing membrane is also essential for the surface aspect Figure 9 : and the shrinkage control.

Figure 9 Example of checker board casting method

GEBIDEM, SWITZERLAND (E.O.S.) CASE STUDY

To correlate test conditions to real experience, job site investigations are of interest. The Gebidem dam (see Figures 10 & 11) is one example with the downstream threshold flushing gates slabs cast with ALAG concrete. This application could represent up to 15-20 years of normal classical application in rivers. The abrasion reported is minimal and far less than expected by the laboratory test.

Three flushing operations were realised between 1996 and 1998, each flushing representing in a few days 5 to 6 million m^3 of water loaded with 800 000 m^3 of gravel (440 000 m^3 of fine material 0<10 mm and 360 000 m^3 of coarse material 10<50 mm), at a flow between 12 and 50 m^3/s. The water velocity was reported between 7 and 20 m/s.

After those 3 operations, the ALAG concrete slabs of 40 m2 each were in good conditions which represents a good improvement compared to all the other traditional materials used previously.

Figure 11 The Gebidem dam flushing gates

Figure 12 Detail of ALAG slab

Table 5 Example of operating conditions follow up

LOCALISATION	AREQUIPA (PERU)	GEBIDEM (SWITZERLAND)	EMOSSON (FRANCE)
Type of structure	Spillways and peers	Desilting flushing gates	Glacier water intakes
Flow quantity (m^3 s.)	600	12 to 50	Up to 20
Total flow (m^3 per year)	N.A.	5 to 6 millions	60 millions
Water velocity (m s.)	15	7 to 20	20
Loads particals	0 to 500 mm bolders	800 000 m^3	60 micron silt
Load mineralogy	Basalt, 7 Moth	Granit, silice 7 Moth	Granit, silice 7 Moth

GENERAL CONCLUSIONS

1. In traditional abrasive conditions 3 major rules can be established :

   a) ALAG concrete can be considered as almost 8 to 10 times more resistant than Ordinary Portland concrete and approximately twice as resistant as some granite, according to the C.N.R. test.
   b) It is also 4 to 8 times more resistant than modified sophisticated P.C. with silica fume according to ASTM C 1138.
   c) Steel fibres do not enhance abrasion resistance (ASTM C 1138).

2. The range of erosion test data available (from 3 to 50 m/s flow rates in traditional abrasion conditions, and up to 110 m/s in cavitation conditions, with and without particles) together with the 10-15 years of site references, indicates that ALAG concrete technology is becoming increasingly well documented and recognised for its durability under severe conditions.

3. Specifications on areas closer to the turbines, such as in the dissipation downstream tunnels, in the galleries or in high velocity spillways is now becoming easier to design based on these new experimental data.

4. The reason for the good performance of such concrete is due to the fact that both the cement and the aggregates have the same physical and mineralogical characteristics, creating a double bound between the cement paste and the aggregates: a physical link such as with all cementicious materials, and a chemical bound at the interface around the aggregates which reacts with the cement.

5. Mechanical strength is not the only parameter to consider in designing anti-abrasion concrete. Traditional methods to increase the performance of Portland concretes such as steel fibers, are not ideal for underwater abrasion conditions.

6. This type of concrete is generally applied as a superficial slab coating at a minimum thickness of 7 to 20 cm depending on the environmental and impact conditions. It can easily be applied using conventional techniques including dry gunning.

7. Based on these test results, ALAG concrete can now be specified as the original anti-abrasion protection lining on new jobs or for rehabilitation of damaged structures. Special admixtures are also under development to allow underwater casting in order to extend even further more its application range in dam structures.

REFERENCES

1. C.N.R. TESTS. Tests reports abrasion and schocks, France 1996-97

2. US CRD C 63-80 tests (ASTM C1138).

3. VATTENFALL, Test Report, Sweden, 1997

4. FURNAS, Test Report, Brazil, 1997

5. LAFARGE ALUMINATES, Technical Tests Results, France 1997

6. GEC - ALSTHOM - ACB France Test Report R 21727, 09-97, 1997

DEVELOPMENTS IN HIGH STRENGTH CONCRETE BEAM-COLUMN CONNECTION DESIGN

S J Hamil

R H Scott

University of Durham

United Kingdom

ABSTRACT. The presence (or absence) of column ties within the connection zone of a reinforced concrete beam-column assemblage was investigated using test results from sixteen external connection specimens. Eight specimens were made using high strength concrete and the results compared with those from a similar set of normal strength specimens. Compressive strengths as high as 132 MPa and tensile strengths of up to 5.5 MPa were obtained. Two methods of anchoring the beam steel rebars within the connection zone were used - bending the tension bar down into the column and using a U-bar. The technique of internally strain gauging the reinforcement was used for measuring strains in the main beam and column reinforcement in addition to tie strains within the connection zone. Results are presented to compare the performance of the high strength specimens with those using normal strength concrete.

Keywords: Reinforced concrete, High strength concrete, Beam-column connections, Strain gauged reinforcement

Stephen J Hamil is a Research Student at the University of Durham in the U.K. He is currently working towards his PhD by investigating the monotonic and seismic loading of reinforced concrete beam-column connections. He is in his final year of study

Dr Richard H Scott is a Reader in Structural Engineering at the University of Durham. He spent ten years in industry, designing a wide range of structures in reinforced concrete, structural steelwork, load bearing brickwork and timber before joining the University of Durham in 1978. His research interests include the behaviour of reinforced concrete structural elements, particularly the measurement of reinforcement strain and bond stress distributions.

INTRODUCTION

Uncertainties exist regarding reinforced concrete beam-column connection behaviour due to the fact that it is governed by a number of variables including shear, bond and confinement which are not fully understood in themselves. Extensive world-wide research has been undertaken to investigate the behaviour of beam-column connections both under monotonic loading and simulated seismic loading. Taylor [1] addressed the problem of monotonic loading in 1974 and carried out 26 tests which resulted in BS8110's recommendations [2] for avoidance of shear cracking and connection zone steel congestion. At around the same time at the Polytechnic of Central London, Ryan [3] conducted a series of 12 tests. Both Taylor and Ryan identified the problem that, unless the beam steel percentage was limited, premature shear failure could occur within the connection zone. Sarsam and Phipps [4] followed up this work and linked it with the results from Meinheit and Jirsa's [5] cyclic work to present refined equations for initial shear cracking and subsequent failure. Scott [6,7] continued the study of monotonic specimens by measuring reinforcement strains using internally strain gauged reinforcement. As a result BS8110's equation 6 was split into equations 6(a) and 6(b) [8]. Recently there have been three PhD's presented within the U.K. proposing strut and tie modelling techniques for beam-column connections. Reys de Ortiz [9], Parker [10] and Vollum [11] have all presented strut-tie models using different approaches.

Although there has been much research world-wide regarding the monotonic loading of reinforced concrete beam-column connections there is, however, an absence of data concerning the strains developed within the column ties. There is also an absence of data concerning the performance of high strength concrete under monotonic loading. Consequently, Scott's work [6,7] has been continued by conducting further tests in which the connection zone ties were strain gauged, in addition to the main beam and column reinforcement. Once again the technique used to measure reinforcement strains was that of internally strain gauging the reinforcement. Specimens containing up to three strain gauged ties were successfully manufactured and tested to give a total of around 240 strain gauges in each specimen.

SPECIMEN DETAILS

Sixteen reinforced concrete specimens, having the geometry indicated in Figure 1, have been tested as part of a larger programme of 70 specimens. Each had a column 1700 mm high and 150 x 150 mm square into which framed, at mid-height, a beam 840 mm long, 210 mm deep and 110 mm wide. All main reinforcement was high-yield steel, with four T16 rebars being used for the column reinforcement and a pair of T16 rebars for the main beam tension steel. Column ties and beam links were all mild steel and 6 mm in diameter. Parameters investigated were concrete strength, beam tension steel detail and presence (or absence) of connection zone ties. These details are summarised in Table 1 which indicates that specimens were divided into four groups, determined by reinforcement detail and concrete strength. C6L** specimens had U-bars for the main beam steel and C4AL** specimens had the main beam steel bent down into the column (the notation used in previous tests [6,7] has been retained for consistency). The high strength concrete had compressive strengths in the range of 120-130 MPa and tensile strengths, by the cylinder splitting (Brazilian) test, in the range of 4-5 MPa. The mix incorporated micro-silica and a superplasticiser. Typical cube strengths for the normal strength specimens were 50-60 MPa.

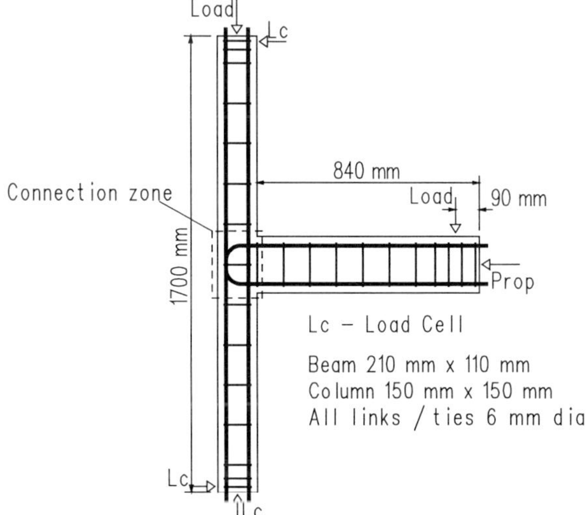

Figure 1 Geometry of a typical specimen

Table 1 Specimen details and behaviour

SPECIMEN	COMP STRENGTH (MPa)	TENSILE STRENGTH (MPa)	BEAM STEEL DETAIL	NO. OF JOINT TIES	P_{crack} (kN)	P_{fail} (kN)	FAILURE TYPE
C6LN0	64	3.4	U	0	19	24	J
C6LN1	64	3.3	U	1*	18	25	J
C6LN3	61	3.2	U	3*	18	29	J
C6LN5	46	2.7	U	5	15	34	J
C6LH0	126	4.7	U	0	25	36	J
C6LH1	127	4.9	U	1*	22	37	J
C6LH3	121	3.7	U	3*	23	41	J
C6LH5	125	5.2	U	5	26	51	B
C4ALN0	53	2.7	D	0	13	27	J
C4ALN1	57	3.4	D	1*	19	34	J
C4ALN3	52	2.9	D	3*	13	35	J
C4ALN5	63	3.2	D	5	14	39	J
C4ALH0	130	4.9	D	0	21	43	J
C4ALH1	119	3.7	D	1*	20	43	B
C4ALH3	132	5.5	D	3*	25	46	B
C4ALH5	123	5.0	D	5	27	49	B

* denotes strain gauged connection zone column ties
U: U-bars, **D**: Beam reinforcement bent down into column
J: Joint failure, **B**: Beam failure

BAR GAUGING TECHNIQUE

The technique used to measure the reinforcement strains was that of internally strain gauging the reinforcement whereby electric resistance strain gauges were mounted in a central duct running longitudinally through the centre of the reinforcing bars, thus avoiding disruption of the bond between the bars and the surrounding concrete. For the purpose of the research programme 12 and 16 mm diameter steel rebars were strain gauged with up to 70 gauges within each bar. It is believed that this is the first test programme to use ties that have been strain gauged in this fashion and specimens containing one and three gauged ties were successfully manufactured and tested. This technique has received extensive development at the University of Durham [12].

TEST PROCEDURE

The test procedure was to load the column first and then maintain the column load while the beam was loaded incrementally, vertically downwards, at a point 90 mm in from its free end. Column loads used were 50 kN for normal strength concrete and 100 kN for high strength concrete specimens, both of which produced strains of around 100 microstrain in the column bars.

Beam loading caused a diagonal crack to form in the connection zone of each specimen, followed, in due course, by failure of the specimen itself either by a plastic hinge forming in the beam at the face of the column or by further extensive cracking of the connection zone. A full set of strain gauge readings were recorded at each load increment and typically there were around 50 load increments per test generating a total of around 12 000 strain gauge readings.

RESULTS AND DISCUSSION

Specimen Behaviour

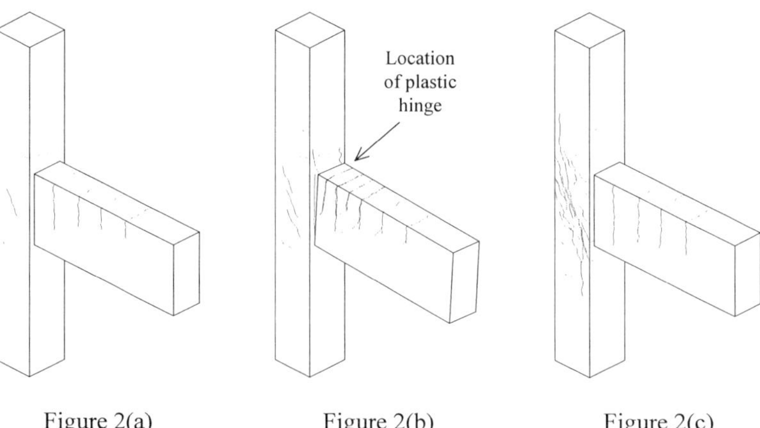

Figure 2(a) Initial joint cracking

Figure 2(b) Beam plastic hinge

Figure 2(c) Extensive joint cracking

All specimens exhibited flexural cracking in the beam and the column regions followed by diagonal cracking in the connection itself, as shown in Figure 2(a). Further shear was then carried by the concrete struts between the cracks assisted by the confinement provided by the connection zone ties.

Failure occurred when a specimen could not withstand further increase in load. There were two different failure mechanisms. If the ultimate moment of resistance of the beam was reached then a plastic hinge formed in the beam at the face of the column, as shown in Figure 2(b). If excessive shear cracking developed in the connection zone, before the beam reached its ultimate moment, then a joint failure occurred, as shown in Figure 2(c). Failure mechanisms are listed in Table 1.

Initial Joint Cracking

The load for initial joint cracking (P_{crack} in Table 1) was dependent mainly on concrete strength although the C6 specimens appeared to perform better than the C4A specimens (possibly because the U-bar reinforcement detail had a greater volume of steel in the connection zone and also provided a more rigid core than was achieved by bending the steel down into the column). Initial joint cracking was unaffected by the number of ties within the connection zone.

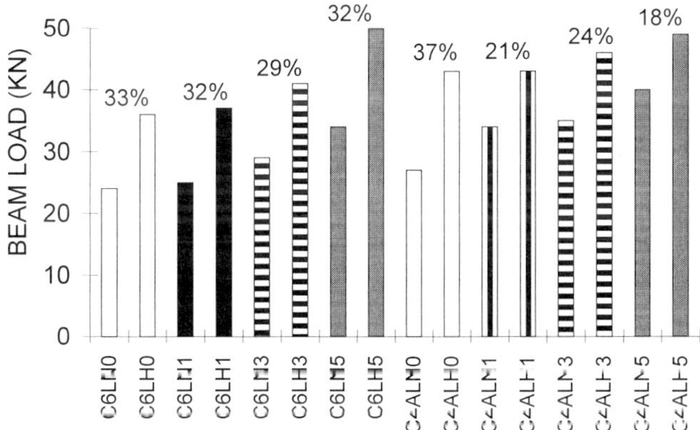

Figure 3 Comparisons of performance between normal and high strength concrete

Specimen Failure

Figure 3 shows a comparison of performance between similar specimens made from normal and high strength concrete. The enhancement, given as a percentage, is shown on the chart.

All the U-bar specimens achieved an increased performance of around 30% when using high strength concrete. The bent-down specimens all achieved a similar increased performance or changed a joint failure into a beam failure.

Throughout the testing it was clear that specimens with the main beam steel bent down into the column performed better than the specimens using the U-bar detail. It is believed that the bent down steel detail had better anchorage due its anchor leg transmitting the beam's force into the column away from the joint. All of the beam's force within the U-bar specimen was transmitted into the column within the connection zone, accelerating joint failure.

Tie Strains

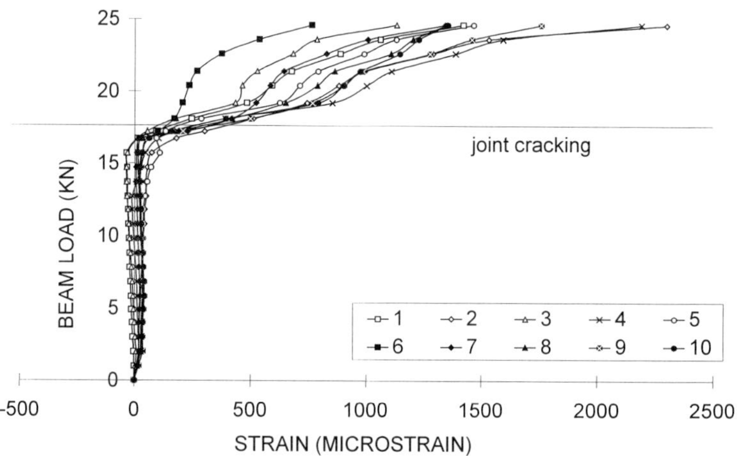

Figure 4 Tie strains within the specimen C6LN1

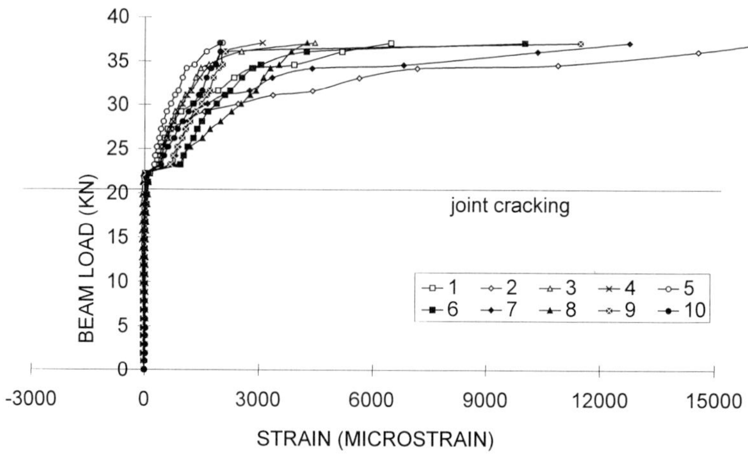

Figure 5 Tie strains within the specimen C6LH1

High strength concrete's ability to sustain a greater shear force within the joint before failure was the reason why the high strength concrete specimens performed so much better than the normal strength specimens. Figure 4 shows the individual strain values within the connection zone ties for specimens C6LN1 and Figure 5 shows the equivalent strains for C6LH1.

Figures 4 and 5 show that strain in the connection zone tie for both normal strength and high strength specimens was very low until joint cracking first occurred. After joint cracking the strain in the ties steadily increased until the connection zone failed due to shear cracking. Whereas the tie within the normal strength specimen, C6LN1, displayed strains of approximately 1 500 microstrain at failure the tie within the high strength specimen, C6LH1, yielded and strains of over 16 000 microstrain were recorded.

Figure 6 shows average strains for the three connection zone ties in specimen C6LH3, the high strength U-bar specimen. For comparison, the bold line shows the average strains for the single tie in the similar specimen C6LH1.

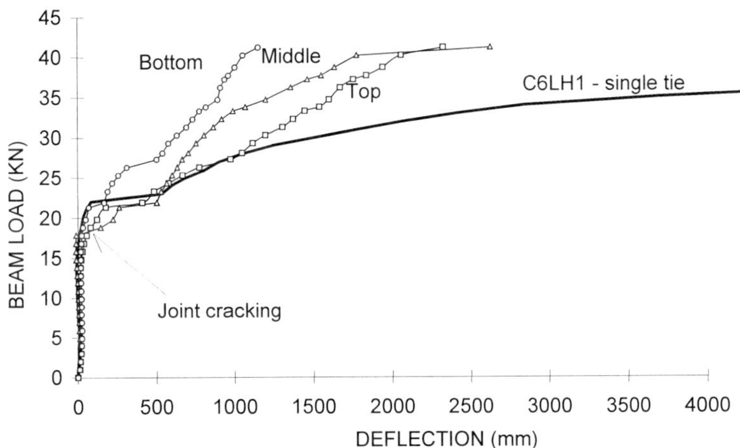

Figure 6 Strain distribution for the ties in specimens C6LH1 and C6LH3

It can be seen that load was not shared equally between the three ties within C6LH3. After joint cracking strains in the top and middle were significantly higher than those in the lower tie, probably due to the beam tension steel causing the top of the joint to be more strained than the bottom. Specimen C6LH1, with only a single tie, went through joint cracking at approximately the same load as the three tie specimen.

After joint cracking the single tie in specimen C6LH1 is seen to develop strain at a higher rate than the three ties in C6LH3. As stated previously, the single tie within C6LH1 yielded and strains as large as 16 000 microstrain were observed.

Results from the three tie specimens suggest that when only a single tie is provided in the connection zone it would perform more effectively if it were positioned between the mid height of the joint and the level of the main beam tension steel. It is recommended that if, in the design of reinforced concrete connections, shear resistance is of greatest concern then joint ties should be positioned around the mid-height level of the connection zone. If an enhancement of the beam steel anchorage is sought then the ties should be positioned around the level of the main beam steel.

Average Bond Stresses

The beam rebars contained four strain gauges within their top bend, one at the start, one at the end and two equally spaced gauges within the bend. The strain gauges within the top bend allowed detailed values of bond stress to be calculated. A comparison of the average bond stresses around this top bend is given in Figure 7. The average bond stress values were calculated by taking the average of the stress gradients between the four strain gauges within the top bend.

Both specimens C6LN1 and C6LH1 failed due to excessive joint cracking and the high strength specimen achieved a failure load over 30 % greater than the normal strength specimen. The ability of the high strength specimen to attain a greater bond stress around the top bend allowed the joint to resist a greater shear force before failure.

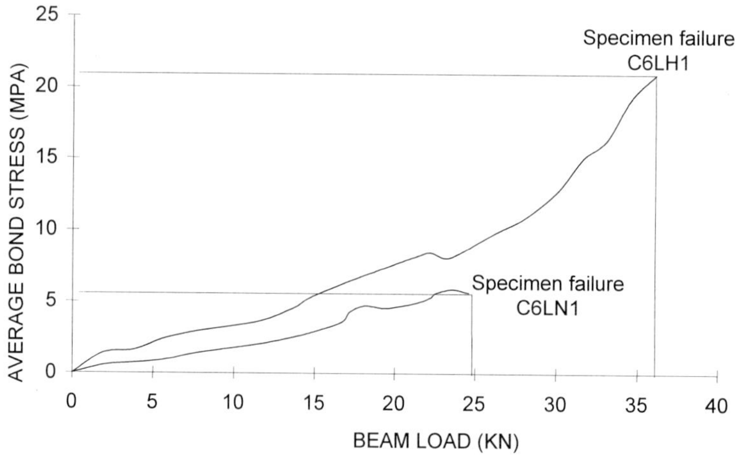

Figure 7 Average bond stress values for high and normal strength specimens

Load-Deflection Response

Figure 8 shows the load-deflection response of four specimens over their initial loading. All four specimens failed at a beam load of approximately 40 kN, thus a direct comparison of stiffness over the initial load history may be made.

Although all of the specimens failed at a similar load it can be seen that there was a clear difference in behaviour between those made with high and normal strength concrete. Both of the high strength specimens were around twice as stiff as the normal strength specimens.

It is suggested that a larger safety factor may be more appropriate for high strength structural design as a shear failure would occur with less warning.

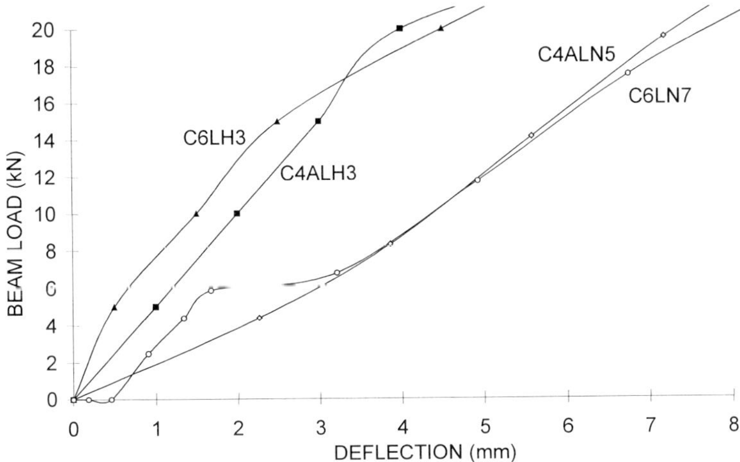

Figure 8 Load-deflection curves for normal and high strength specimens

CONCLUSIONS

1. Initial joint cracking was strongly influenced by the tensile strength of the concrete within the connection zone. Specimens made with high strength concrete had a greater resistance to initial joint cracking due to their greater tensile strength.

2. The number of connection zone ties did not effect the performance of the joint until after joint cracking had occurred.

3. Use of high strength concrete increased the ultimate shear strength of the joint and allowed higher strains to be developed within the joint zone ties.

4. Bending the main beam steel down into the column, rather than using a U-bar, increased the ultimate load at which a specimen failed.

5. Increasing the number of shear ties within the connection zone increased the ultimate shear strength of the joint, due to the enhanced confinement of the joint core.

6. The load-deflection response of specimens made using high strength concrete was seen to be twice as stiff compared with specimens made using normal strength concrete.

ACKNOWLEDGEMENTS

The financial support of the Engineering and Physical Sciences Research Council is gratefully acknowledged as are the contributions made by Tarmac Topmix Ltd, Mr Paul Brayshaw and the technical staff of the School of Engineering.

REFERENCES

1. TAYLOR, H.P.J. : 'The behaviour of in situ beam-column joints', Technical Report 42.492, Cement & Concrete Association, May 1974

2. BS 8110 STRUCTURAL USE OF CONCRETE : Part 1 : Code of practice for design and construction, London, British Standard Institution, 1985

3. RYAN, J. : 'Reinforced concrete beam-column connections', Concrete Magazine, pages 37-39, March 1977

4. SARSAM K.F. AND PHIPPS M.E. : 'The shear design of in situ reinforced concrete beam-column joints subjected to monotonic loading', Magazine of Concrete Research, Vol 37, No 130, March 1985

5. MEINHEIT D.F. AND JIRSA J.O. : 'The shear strength of reinforced concrete beam-column joints', Austin, University of Texas at Austin, Dept. of Civ Eng., 1977, Report 77-1

6. SCOTT R.H. : 'The effects of detailing on RC beam/column connection behaviour', The Structural Engineer, Vol 70, No 18, 15 Sep1992, pages 318-324

7. SCOTT R.H. *ET AL* : 'Reinforced concrete beam-column connections and BS 8110', The Structural Engineer, Vol 72, No 4, 15 Feb 1994, pages 55-60

8. BS 8110 STRUCTURAL USE OF CONCRETE : Part 1 : Code of practice for design and construction, London, British Standard Institution, 1993

9. REYS DE ORTIZ I. : 'Strut and tie modelling of reinforced concrete short beams and beam-column joints', PhD thesis, University of Westminster, 1993

10. PARKER D.E. : 'Shear strength within reinforced concrete beam-column joints', PhD thesis, Bolton Institute of Higher Education, 1997

11. VOLLUM R.L. : 'Design and analysis of reinforced concrete beam-column joints', PhD thesis, Imperial College, London, 1998

12. SCOTT R.H. AND GILL P.A.T. : 'Short-term distributions of strain and bond stress along tension reinforcement', The Structural Engineer, 65B, No. 2, June 1987, pages 39-43, 48

THEME TWO:
SUB-STRUCTURES

Keynote Paper

DELIVERING EGAN BELOW GROUND

P G Goring

John Doyle Construction

United Kingdom

ABSTRACT: The Egan Report "Rethinking Construction" concluded that the industry did not need to review what it does and do it better but to do it entirely differently. This study reviews the need for change and looks into how a specialist concrete sub-structure contractor can contribute to the improvements and benefit from achievements. On certain projects sub-structure construction is already moving forward with the introduction of an integrated approach, skills training, minimisation of waste, appropriate use of information technology, supply chain management and innovation in processes. The initiatives taken are a small but positive step and further areas of development are considered.

Keywords: Rethinking construction, Key drivers, Change of emphasis, Specialist contractors, Construction skills certification scheme, Minimising waste, Integrated approach, Information technology, Innovation, Research, Branded products.

Peter G Goring is a graduate of Civil Engineering at Imperial College, London and subsequently achieved an MSc in Construction Law and Arbitration at Kings College, London. He has worked in a management and technical capacity with contractors and is currently Technical Director with John Doyle Construction specialist concrete trade contractors. He is involved in all technical and planning aspects of the construction process from pre-qualification and scheme development to tendering and construction.

INTRODUCTION

The efficiency of the construction industry has been subject to a considerable amount of scrutiny during the past five years initially with the Latham Report "Constructing the Team" [1] and more recently the Egan Report [2].

The Government commissioned the Report of the Construction Task Force "Rethinking Construction" [2], known as the Egan Report, against a perception that the industry was generally under achieving. One of the key elements for review was the scope for improving construction efficiency by reviewing current practice and improving it by innovation in products and processes, it is this aspect of the report that will be developed in this paper.

Construction projects have been considered to be multi-user prototypes built by sub-contractors. Every project is different and all clients have their own needs and requirements. In reality each project contains thousands of different components. Standardisation of these components would minimise the inevitable learning curve every project appears to be blighted by.

This paper reviews current practice in sub-structure construction and initiatives being undertaken either unilaterally or collectively to improve efficiency.

NEED FOR CHANGE

To address the recommendations of the Egan Report [2] it was recognised that there were five key drivers to achieve the changes required. These are committed leadership, focus on the customer, integrating the process, quality driven agenda and commitment to people.

Most major office construction projects with deep basements involve a lengthy contractual chain typically the ultimate client will be a financial institution purchasing or funding the property for another party. There may be a developer, who has employed a project manager to oversee the works and procurement process. The method of procurement for the construction will depend on the client's preference but may involve a traditional main contractor or a construction management form of contract. The concrete and sub-structure works would generally involve the appointment of specialist trade contractors. The trade contractors supply the necessary labour, plant and materials to complete their works.

The contractual link between the supplier of say concrete admixture to the ultimate client involves eight different parties. As a consequence, an agenda for improvement and innovation requires the full commitment and belief of the management of all parties otherwise innovation will be stifled.

The general policy within construction is for companies to concentrate on satisfying the needs of the next employer in the chain rather than the ultimate customer. The customers rarely communicate their particular needs down the contractual chain and there is little research or audit of the industry's requirements. Developing this particular aspect is likely to be a long term goal and will involve a firm commitment and close association between a number of parties clients, project managers, designers and specialist contractors, resulting in a branded product. The adversarial nature of the industry would require several existing barriers to be broken before a fully open book approach could be achieved.

The third driver identified in the report is the requirement to integrate the construction process. The report typifies the industry as following a series of sequential and largely separate operations undertaken by individuals, who have neither any commitment to nor stake in the overall success of the project. This was a key element of the Reading Construction Forum report "Unlocking Specialist Potential" [3,4,5], which looked into the role that specialists could offer to rationalise the construction process given that they represent approximately 80% of the construction related costs. This particular aspect will be reviewed in more detail in the following section.

The emphasis in procurement particularly of trade packages is currently driven around cost. With the other essential facets of a construction project; time, safety and quality considered to a lesser degree. For the construction process to be improved each of these aspects needs to be given equal status. Figure 1 is a development of the work carried out by the Reading Construction Forum [3] to illustrate how the emphasis on the process could change. This takes a more integrated approach and requires all parties to consider these elements to improve performance and reduce costs.

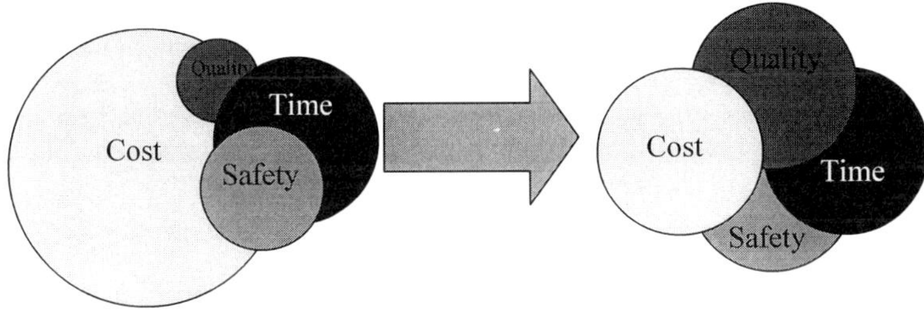

Figure 1 Change of emphasis

Quality management to ISO 9000 [6] is a basic framework and the parameters against which the certified company are assessed are set by that company. Too often certification is viewed more as a marketing asset for the client's benefit than a management tool that will achieve greater efficiency. The system needs to be developed to achieve the ultimate goal of zero defects and minimise waste. The service provided to the customer is part of this quality package.

The Construction, Design and Management Regulations (1994) [7] have focused the attentions of Clients and Designers and introduced more structure into the construction process. The obligations of well-managed contractors have not been significantly changed except for the introduction of a more integrated process requiring the format of the contractors health and safety plan to follow a standard form and consider the hazards and risks identified by the planning supervisor on behalf of the client. The introduction of the health and safety file and the requirement for the contractor to provide relevant details of products used, as built drawings and any special requirements is an important first step in requiring all parties to have a consideration for the product from concept and beyond completion.

The final driver for change is a commitment to people. This starts with providing a safe and healthy working environment. To maintain this site staff need to be trained to an appropriate level and have project specific inductions. Particular hazards and risks identified in the health and safety plan should be clearly addressed in method statements and brought to the attention of all who are exposed to the risk. All members of staff, supervisors and management should receive both vocational and developmental training to fulfil potential. The commitment is two way, from management to recognise contributions that can be made and from individuals to achieve a sustained improvement through learning. As well as commitment within organisations other parties in the supply chain can offer training by making others aware of the innovations being carried out within their own organisations.

SPECIALIST CONTRACTORS

As stated above the work of specialist contractors at site can account for as much as 80% of the contract expenditure. Giving the specialist a more pro-active role in the construction process was identified in the Latham Report [2] as a way of unlocking potential and contributing to the stated aim of reducing construction costs by 30% by year 2000.

Traditionally, particularly in the main contract form of contracting there has been an adversarial culture leading to disputes and conflict. This resulted in poor relationships and lack of trust between main contractors and sub-contractors. The Construction Act (1996) [8] introduced payment provisions aimed at assisting sub-contractors cash flow by prompt and satisfactory settlement of accounts.

The role of the sub-contractor need not also be sub-ordinate or sub-missive. Several constraints are imposed on specialist contractors in particular

- Unrealistic contract programmes

- Undue emphasis on cost rather than the other basic elements of construction contracts as outlined above, typified by the works being offered to the lowest bidder following a Dutch auction.

- Inappropriate allocation of construction risks through the contract conditions.

- Poor definition of the contract scope in particular the client's needs.

- Lack of clear communication between the parties.

- Delays to release of information resulting in disruption of site. The recovery of costs for delays and disruption through the contract are inevitably emotive, can often be intangible and employ a significant resource to justify claims.

Typically margins for specialists are low and achieving a reasonable return on capital employed is based on a regular cash flow with payments for work completed to suit expenditure. Many specialists are small firms and are vulnerable to larger more cash rich firms withholding or delaying payment to suit their own cash flow forecasts. The specialist is often caught in the trap between a strict specification requirement that precludes the use of an alternative to the specified supplier who is therefore able to command their price.

Most specialist contractors in contrast to the majority of traditional main contractors have a higher proportion of directly employed workforce. They tend to be trained in their particular trade and many already hold Construction Skills Certification Scheme (CSCS) cards. The management are experienced within their particular specialty, more so than general contractors and designers, and therefore are able to offer a particular insight into the construction process.

Often designs and specified elements are based on the material supplier's representative visiting the specifier with details of product ranges. The product is then included within the specification based only on the technical literature provided with no prior knowledge of the product. Technical literature between essentially similar products often contains test information measures against differing parameters making the assessment of an exact equivalent alternative to the specified impossible to source.

Specialists have experience of using the products and would be able to advise on appropriate applications. There may also be particular relationships between the supplier of the specified or similar product that may prove to be commercially advantageous to the project.

The general complexity and diversity of the construction process lends itself to an integrated approach utilising the combined talents of architects, consulting engineers, contractors, specialist contractors, suppliers and quantity surveyors. This involves crossing traditional boundaries to reduce the fragmented nature of the design process. Early involvement of specialist contractors will improve the release and quality of design information minimising the risk of significant costly design changes during the construction phase. Critical lead times can be identified and the design process scheduled around these dates. Construction time savings can be achieved by integrating the design and construction process rather than attempting to condense the construction period following completion of the full scheme design.

The method of procurement is largely beyond the scope of this paper, however establishing relationships with specialists may lead to procurement of a range of projects rather than individual schemes which could result in economies of scale to benefit all parties. Similarly, formal and informal partnering arrangements could be established developing interpersonal relationships and exchange of information between parties.

SPECIALIST SUB-STRUCTURE CONTRACTORS

The main specialist contractors involved in sub-structure works are piling, ground strengthening, earthworks and concrete contractors. The piling contractors and associated trades of diaphragm walling, ground investigations and anchors have long had a strong trade association in the Federation of Piling Specialists. They have detailed conditions of contract and specification for piling works, which is generally adopted as the standard for the industry and used as the basis for most scheme specifications.

The concrete frame contractors formed a trade association known as Construct in 1993 to develop the best practice, achieve common goals and standards of workmanship. It has been established as a forum for contractors, clients, designers and suppliers. There have recently been a number of initiatives aimed at raising the overall consistency of standards within the industry and improve client satisfaction.

The National Concrete Frame Specification [9] issued in 1996 has had a favourable review and is currently undergoing a thorough analysis to extend the scope to basement and sub-structure construction and toppings to steel frames. The comments of users and potential specifiers are being addressed to provide a comprehensive alternative to the National Building Specification most regularly used for commercial concrete works. The Specification promotes the involvement of the specialist within the design phase, provides an industry standard for materials, workmanship and construction tolerances. The revision of the Specification as with the original has been partially funded with a DETR Partners in Industry grant.

Complimentary to the Specification is the Construct "Guide to Contractor Detailing of Reinforcement" [10]. This initiative draws on the experience and expertise of the trade contractors at a critical stage in the design development. The main benefits enable the contractor to schedule the reinforcement to suit works at site and introduce improvements in safety, efficiency, buildability and quality. The contractor receives control of the information, there is less opportunity for delay and it produces a less adversarial approach between consulting engineer and specialist contractors.

The success of the system is based on clear information being provided by the consulting engineer for loadings or densities of reinforcement. It also provides the trade contractor with opportunity to review the reinforcement requirements to adjust the design to suit site operations. Economic use of additional reinforcement may be scheduled to reduce back propping, increase pour sizes, allow early striking of forms and prefabrication.

The Construction Skills Certification Scheme [11] has been introduced by the CITB as a certificate of competence for operatives and tradesmen. The initial transition scheme allowing automatic qualification by grandfather rights together with a basic level of safety training has now passed for most concrete related trades. The route to acquire certification is through NVQ level 2 or via a modern apprenticeship. It is recommendation of The Egan Report [2] that major clients take the lead and give preference to contractors that employ a trained workforce as evidenced by valid CSCS cards.

The commitment of companies to train individuals to achieve this status is considerable given the historic transient nature of the industry. To counter balance these concerns Construct has established a training centre where the cost of training is shared and there is general access to courses organised by member companies. A scheme has also been developed with the CITB [12] to minimise the impact of sending operatives to be externally assessed for competence, and therefore incurring costs of course fees, paying the employee during the time away from site and losing production. Company assessors can be trained by the CITB who are able to make an on the job assessment of the operatives competence. The scheme is administered by the local TEC and receives grant aid from CITB therefore minimising both the cost and time implications.

As previously stated quality management to ISO 9000 provides only a basic system and is not a general measure of a companies overall competence. Construct with partners DETR, BRE, CARES and QSRMC is developing a registration scheme for contractors. The scheme will have assessment parameters in training, quality, technical capability, financial stability, health, safety and environmental issues. This scheme also has funding through the DETR PIT scheme. It is aimed at providing a clear indication to clients that in engaging a Construct Registered company they will be dealing with a secure specialist.

MINIMISING WASTE

One of the key elements of the Egan Report [2] is the minimisation of waste. Waste can be both physical and intellectual. As well as rationalising the construction process to reduce the amount of waste materials generated all parties should strive towards the goal of zero defects. The industry as a whole uses 300 million tonnes of materials each year of this 13 million tonnes is waste.

As a general policy all operations are reviewed to assess the requirements. The days of the first operation on a concrete site involving the delivery of a standard of timber and a pack of plywood direct from the nearest merchants are numbered. Wherever possible proprietary formwork systems are used. The development in proprietary systems has been dramatic in recent years and there are few applications that cannot be tackled. Regrettably the majority of systems are of European or American development not British.

All suppliers of formwork systems have a number of schemes for rental, lease purchase, sale and maintenance which make the procurement suit most site needs. The speed of erection enhances production times and places less pressure on employing a skilled work force. The suppliers are also able to provide computer aided design packages for their products. As a result contractors are able to assess formwork requirements at tender stage and produce preliminary drawings. By involving the suppliers at an early stage contractors are benefiting from specialist advice and managing the supply chain.

Similarly, temporary works supports particularly for new works to existing basements often involves the use of a considerable amount of fabricated steelwork. This is manufactured specifically for the project and then scrapped at the end of a project. It is unusual for fabricated steel to suit other application without extensive modification. Scrap values of steel have collapsed in recent years and as a consequence the use of bespoke fabricated steel solutions is both wasteful and expensive.

Proprietary systems have become more sophisticated and design packages are available from suppliers to develop solutions to problems in-house. By developing and detailing a small amount of fabrication, proprietary systems can be used for most applications.

Intellectual waste is the duplication of work or unnecessary use of designers' time. Transfer of information between parties can improve this. Electronic transfer of design information between parties and the ability to view drawings on screen prior to printing considerably reduces the amount of paper generated. Involvement of the trade contractor assists in this process. By receiving design data electronically in CAD format a contractor can use original drawing and reduce the duplication in drafting. The concern with placing an over reliance on information technology is that too much information can be transferred and as a result pertinent details may be lost in the volume, parties need to retain not abrogate responsibility.

Cost reporting software that were previously either developed in-house or adapted from standard packages required specific computing knowledge and a dedicated department to operate effectively. These have now been rationalised to allow site data to be directly linked to head office. This can include the input of timesheets and deliveries for costing analysis and checking of invoices. Estimating breakdowns can be directly incorporated into cost plans, which are used for reporting, forecasting and preparation of submissions for interim payments.

The Egan Report [2] refers to the enormous benefits that can be gained from using modern CAD technology for the elimination of waste and rework by carrying out redesigns on the computer rather than at site.

A complex earthworks and retaining wall support operation occurred at the site covered by the case study later in this report. The retaining wall supported a busy urban road, the excavation was the subject of archaeological interest and the platforms of a Central Line underground station are within 10 m of the excavation to formation. The actual ground conditions and location of existing structures were not known until demolition works were complete at site. Well before the demolition works were completed a number of brainstorming meetings were held involving the consulting engineers, construction manager, piling trade contractor and sub-structure trade contractor.

A sequence of works was developed as a series of 2 dimensional cross sections and when the method was agreed 3 dimensional rendered images were generated to indicate how the works would be constructed. Proprietary props normally used for horizontal propping applications were adapted by fabricating waling connections for a raking solution. The hire costs associated with proprietary props gave a considerable saving over fabricated members.

There were several surprises in the existing structure that required detailed development however, Figure 2a and 2b demonstrates how accurately the situation at site was modelled and waste eliminated using the proprietary propping system.

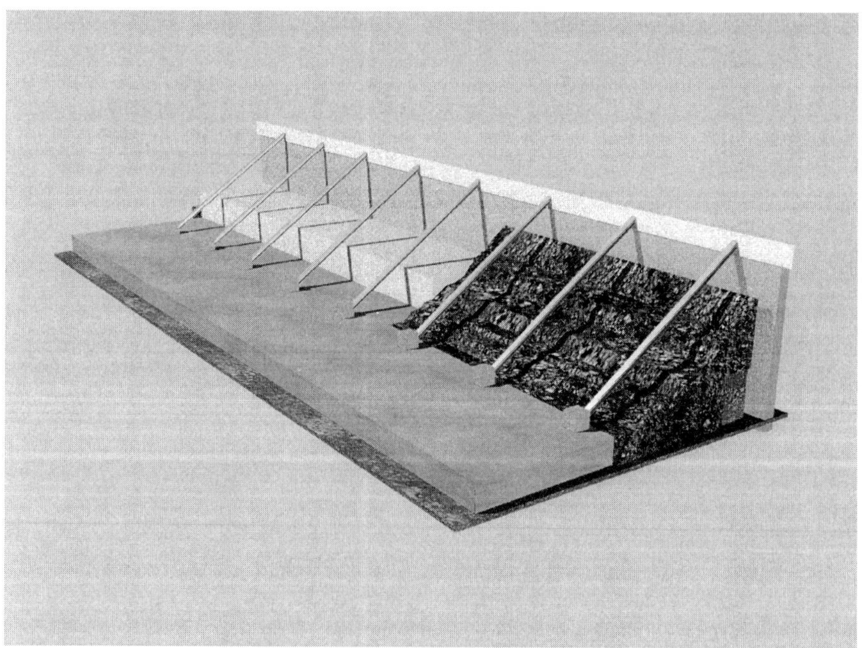

Figure 2a Computer model from September 1998

Figure 2b Progress photograph April 1999

INNOVATIONS

The Egan Report [2] made general observations about the low level of research and development expenditure. Specialist contractors operate off squeezed margins and have therefore limited resources to invest. Concrete is ultimately flexible in its application and mix designs can be varied to suit site needs. It is therefore impossible to patent concrete and as ideas are quickly adopted there is little commercial advantage to be gained by a large investment. Specialists can contribute to improve and introduce new systems by working closely with other associated organisations.

By devoting time to professional bodies and learned societies specialists can gain a commercial advantage by introducing new ideas which will give a head start. Indeed the DETR literature relating to the Construction Best Practice Programme is appropriately titled "Construction firms who embrace change make better profits" [13].

During the past year without disruption to general work John Doyle staff have been able to make a significant contribution to assist with a number of working parties, committees and other schemes. Many are reviewed and not considered to be an effective use of time or would require an excessive amount of commitment for the benefits to be derived.

As well as the close ties with the Construct group, members of staff serve on the Concrete Society Construction Committee, and working parties for reinforcement rationalisation and self-compacting concrete. Support is given to companies developing new proprietary products and construction methods.

Contributions have also been made to a number of academic bodies in terms of their research and course content. Advice from industry to balance and guide academic views is useful and where it is possible to assist in the direction of research there may be benefits to be gained by adopting innovations at site prior to general publication of research. It goes without saying that the institutions are always grateful for financial support for their departments. Ties with academia also provide access to future graduates and potential employees. Commitment to research and development in time also provides a relevant latent continuing professional development for staff involved.

An example of product development is in the use of structurally integral waterproofing additives in basement construction. The British Standards make no specific reference to the application of admixtures to produce watertight concrete. The use is attractive as it reduces the number of operations required when compared with the application of a tanked membrane.

There are presently four products available on the market that will reduce the permeability of concrete and are able to produce the necessary grade of environment. Clients insist on a performance warranty for the product although it effectively limits the supplier's liability. The result of an unusual product being incorporated into a concrete mix by a ready mix firm for supply with a guarantee is a considerable price premium over unmodified concrete. The premium is often significantly greater than the net cost of the additive due to the perceived risk. Similarly, suppliers are aware of the cost of alternative tanking systems and levy their price accordingly. As the direct cost saving to the client is not necessarily significant consulting engineers can be reluctant to risk the professional indemnity insurance on a product that is not directly covered by a British Standard.

By reviewing the supply chain for available products, the warranty liabilities and the perceived risks significant savings have been offered to clients. The admixture is purchased direct from the manufacturer, a representative of the supplier applies it at the batching plant or site and an insurance backed collateral warranty is offered against the product. As a result, the client's risks are satisfied, the consulting engineer is able to verify the product and the ready mix supplier's exposure to product liability is removed. Although the manufacturers are not able to charge their premium rate their risks are reduced, the cost advantage is improved over a tanked alternative and market share increases.

CASE STUDY

A recent project undertaken in the City of London reflects many of the initiatives outlined above. The site is a commercial office development funded by an insurance company and pre-let to a multi-national bank. The developer appointed a project manager who has a partnering arrangement with a construction manager at the start of the detailed design phase. It was agreed that the key specialist trade contractors would be appointed by negotiation and involved in the design development. The project started at site in late November 1998 and negotiations began in July.

Throughout this period there were regular meetings between the architect, consulting engineer, services engineer, construction manager and trade contractors. Regular workshops were held to discuss key areas, interfaces and programming constraints on the project. Briefing sessions and presentations were provided to the client, the trade contractors led many of these sessions to demonstrate the interaction between the parties. A typical example of the design and construction development is provided earlier in this paper.

During the detailed design phase, weekly meetings were held with the structural consulting engineer to discuss schedules of information to be provided, pour sizes, reinforcement detailing, temporary works provisions and other construction details. The reinforcement detailing was actually carried out by the consulting engineer. However as dialogue with the consultant was both close and open the resulting detailing was probably more effective in terms of site works and design than if it had been contractor detailed. Temporary works designs were discussed in detail and the construction manager and consulting engineer held controlled copies of the trade contractor's design schedule and file.

Project planning was fundamental to incorporate the various trade packages. Steel erection was due to start just 11 weeks into the project. Work zones and lay down areas were agreed between the trade contractors. The structure was designed such that the basement slab would be constructed independent of the pile caps to allow steel erection to proceed. The site was particularly congested as shown in Figure 3. Handovers and return dates between the various trades were agreed at planning meetings. The dates were shown on marked up drawings, which were duly signed by site managers.

The trade contractors site office was open plan to encourage interaction between all parties, with enclosed meeting rooms made available for confidential discussions. It was a site requirement that 60% of labour should be CSCS cardholders and training was made available through the unions to fulfill basic needs.

A draft specification for the works was tabled early in the proceedings with all areas open for discussion prior to a price for the works being finalised. The basement slab was designed to be a watertight structure with most areas providing access and car parking together with a lesser area of plant rooms that required a drier environment. From their experience of basement construction the consulting engineers allowed a relaxation of their normal specification for the size of base slab concrete pours from 40 to 100 m^2. Following a detailed review of the programme requirements, areas of waterproof concrete and site restraints a pour layout for the basement slab was developed. It was considered that there were particular areas of the basement where pours of between 200 – 300 m^2 could be accommodated without detriment to the watertight basement.

To determine the optimum pour size and methods to be adopted at site, an afternoon seminar was convened involving site and technical staff from the consulting engineer, construction manager and trade contractor. A local representative from the Concrete Advisory Service was invited to contribute to the discussion and all facets of basement slab construction and perceived alternatives discussed. This included the basement design parameters, mix designs, the use of waterproof concrete, superplasticisers, fibre reinforced concrete, curing agents, joint details, waterbars and surface finishes. The meeting concluded that slabs could be poured around 200 m^2 provided the aspect ratio of the pour was approximately square and edges restraints would not have a significant effect.

Figure 3 Trade contractor interfaces on Congested Site

Where temporary support to the existing structure was required a detailed analysis was carried out to investigate the effects on the permanent structure to minimise use of temporary materials. The close relationships between the parties and developed trust allowed various patented systems and new developments to be introduced and readily adopted.

A patented method for breaking down pile heads, which received a 1999 Innovation in Construction Award, was used to improve cycle times for pile cap construction. Proprietary temporary works supports and formwork systems were used for most situations again improving cycle times.

All excavators used on site were fitted with quick hitch adaptors allowing rapid change of attachments for bucket sizes or breakers. Carbon Fibre Reinforced Plastics were considered as an alternative to steel for waling beam construction to allow retaining wall linings to be poured to full height. Similarly, self-compacting concrete was put forward in certain locations where cost of placing column encasement concrete was less than the premium paid for the concrete itself.

Digital progress photographs of the site were taken and posted on web sites to inform staff of how works were proceeding. This technique was also used to allow head office based staff to view and respond to technical queries without the need for a site visit. The site has been nominated as a Movement for Innovation (M4I) demonstration project.

THE WAY FORWARD

This paper reviews the current state of the industry from the point of view of a specialist concrete trade contractor. Initiatives already being adopted are a small step along the path towards fulfilling the requirements of the Egan Report [2]. They are however by no means universally adopted and many hardened attitudes need to be addressed.

The Egan Report [2] and its predecessor Constructing the Team [1] sets objectives for year on year improvements for cost, time, quality, safety and profitability. Some of these parameters are relatively easy to quantify, however the unique nature of the industry makes others more difficult to baseline and monitor.

Key elements for improvement revolve around improving relationships, processes, customer focus and knowledge transfer. Many of these require the client to take the initiative by adopting an open book approach, identify and allocate the risks and benefits appropriately and develop then this trust in all members of the team.

The key processes need to be identified and wherever possible standardised. As relationships develop less reliance will be placed on bespoke designs and more emphasis can be placed on pre-determined designs. This reduces waste both intellectual and physical whilst neither stifling creativity nor preventing the client to include specific options or requirements. It is also a step along the route towards developing a branded product, which exceeds the customers' expectations and gives confidence in the reliability and integrity of the industry.

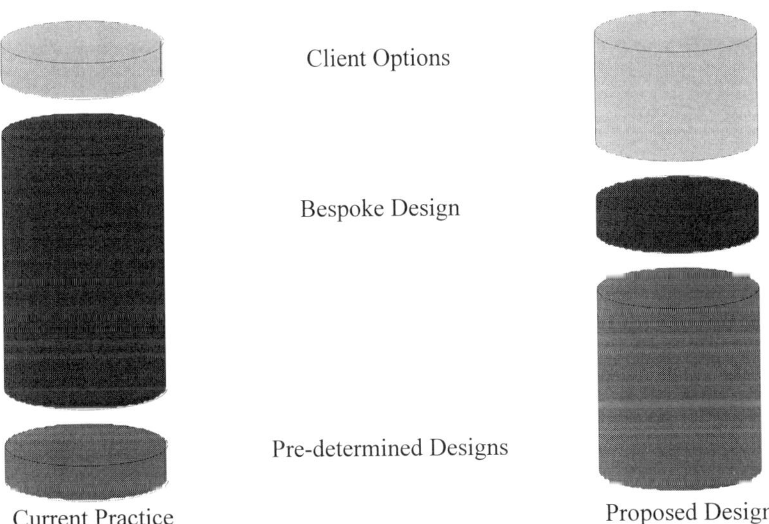

Figure 4 Towards a Branded Product (After work by Mace)[14]

The industry needs to respond to customer needs, which are constantly changing, the only way to stay in touch is by developing close relationships.

Performance and priorities should be reviewed in a no blame environment to identify improvements. Where necessary incentives should be provided to promote initiative and lateral thinking.

A culture of continuous improvement should be instilled into all parties, and suppliers can be educated into customer needs to ensure that knowledge is transferred throughout the contract chain. Parties sharing learning and undertaking joint research and development assist this process and consequently take a more proactive role in the design and construction by raising capabilities.

REFERENCES

1. M, LATHAM. Constructing the Team, HMSO, (1994).

2. THE CONSTRUCTION TASK FORCE. Rethinking Construction, [Egan Report], HMSO, (1998).

3. M, SAAD AND M, JONES. Unlocking Specialist Potential, Reading Construction Forum, (1998).

4. R, SLAVID. Making the Most of Specialists, Article in Architects' Journal, 14 May 1998.

5. G, RIDOUT. Specialists Get in on the Action. Article in Contract Journal, 13 May 1998.

6. BSI, BS EN ISO 9000. Quality Management and Quality Assurance Standards, British Standards Institute, (1994).

7. CONSTRUCTION (DESIGN AND MANAGEMENT) REGULATIONS, (1994).

8. CONSTRUCTION ACT, (1996).

9. CONSTRUCT. National Concrete Frame Specification for Building Construction, Construction Research Publications, (1998).

10. CONSTRUCT. A Guide to Contractor Detailing of Reinforcement in Concrete, British Cement Association, (1997).

11. CSCS. Construction Skills Certification Scheme, Scheme Booklet, Construction Industry Training Board, (1997).

12. CITB. Training in the Workplace, A Practical Guide for Employers, Advisers, Coaches and Learners, Construction Industry Training Board, (1998).

13. DETR. Construction Firms Who Embrace Change Make Better Profits, Construction Best Practice Programme, (1999).

14. D, CHEVIN. Mace Moves into the Risk Game, Article in Building, 27 November 1998.

DESIGN METHODS FOR CONCRETE PILES IN WEAK ROCK

G G T Masterton

Babtie Group

United Kingdom

ABSTRACT. This paper draws on the findings of a research project carried out by Babtie Group as part of CIRIA's ground engineering programme. The paper describes and compares four design methods suitable for concrete piles in weak rock and makes recommendations on improving the design and procurement process.

Keywords: Concrete, Foundations, Piling, Geotechnics, Weak rock.

Gordon G T Masterton is a Director of Babtie Group, Glasgow, Scotland, UK. He is responsible for the design and project management of the Group's bridges, industrial and building projects. He has published papers on bridge design, concrete technology, structural repairs and reinforced and anchored walls. Mr Masterton is Vice Chairman of the Glasgow and West of Scotland Association of the Institution of Civil Engineers and Visiting Professor at the University of Paisley.

INTRODUCTION

The design of piles in weak rock has traditionally been compromised by a poor understanding of the real behaviour of piles in weak rock, by the complexity of modelling the interaction between end-bearing and shaft-load transfer mechanisms, and by the modification of rock properties through the very action of piling.

The solution to this is to improve our understanding of the behaviour of piles in weak rock, to use improved models of pile/ rock socket behaviour, to use models which consider the pile performance under serviceability conditions, and to adopt flexible design procedures which permit the design to be modified in response to "as-found" conditions during construction.

This paper describes and compares four methodologies for the design of concrete piles in weak rock and makes recommendations for an improved design and procurement process.

SHAFT LOAD TRANSFER

For piles in cohesive soils, shaft resistance, τ_{su} is typically related to undrained shear strength s_u through an adhesion factor α:

$$\tau_{su} = \alpha s_u$$

where α is a function of pile type, soil type and soil history.

For pile in weak rock, the ultimate shaft resistance is more usually related to uniaxial unconfined compression strength σ_c. It is usual to assume that $\sigma_c = 2s_u$ so that the equation for shaft resistance becomes:

$$\tau_{su} = \alpha\sigma_c/2$$

Figure 1 shows the relationship between adhesion factor and unconfined compressive strength as derived by Williams et al. [2]. A further modifying factor, β, is suggested to allow for rock quality and is expressed as a function of the mass factor:

$$j = E_m/E_i$$

where E_m is the mass rock stiffness and E_i is the intact rock stiffness. This relationship for the Melbourne mudstones studied by Williams et al. is shown in Figure 2. This acknowledges a variation in behaviour dependent on roughness at the rock/concrete interface and this, together with socket cleanliness, is a key influence on shaft resistance. Kodikara et al. [3] show the adhesion factor, also for Melbourne mudstones, as mostly dependent on socket roughness and ratio of rock stiffness to rock strength, E_m/σ_c.

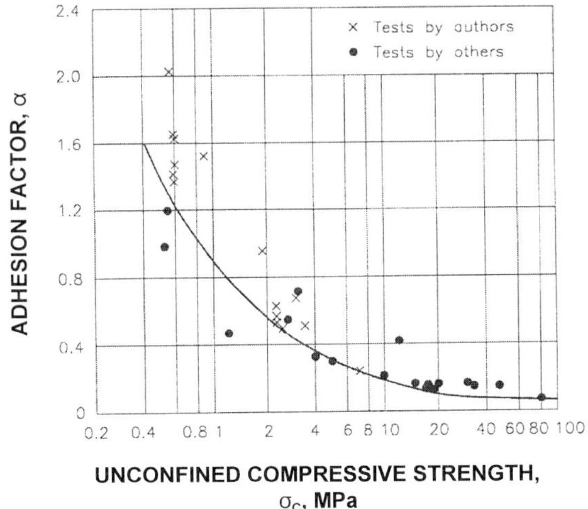

Figure 1 Adhesion factor as a function of rock strength (after Williams et al., 1980) [2]

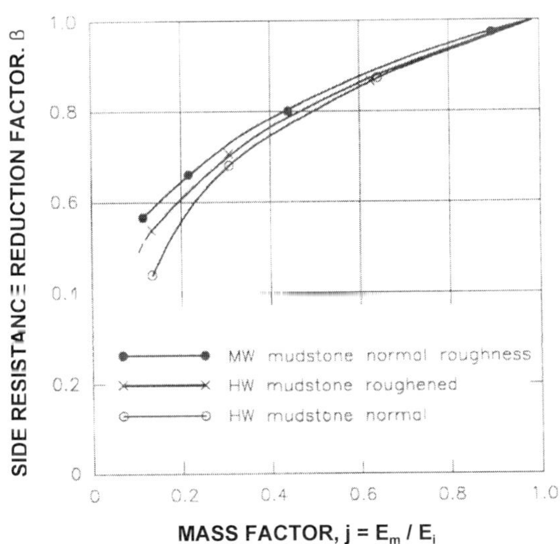

Figure 2 Side resistance reduction factor reflecting ratio of stiffness of rock mass and intact rock (after Williams et al., 1980) [2]

BASE LOAD TRANSFER

Although a pile base may not behave truly elastically except for initial low levels of load, it is unusual for pile bases ever to reach a limiting load. Williams et al. Suggest that $5\sigma_C$ is a reasonably conservative estimate of the "minimum peak base resistance for all piles in intact rock" Bearing stresses are therefore more likely to be related to a settlement criterion.

The proportion of load transferred to the base of a pile socket will depend on socket geometry as well as relative stiffnesses of pile and rock. The cleanliness of the socket base will also have an impact. The design methodologies all attempt to assess the relative contributions of shaft and base resistance using combinations of elastic analyses, modifications for nonlinearity, and various empirical adjustments to allow for field observations.

DESIGN METHODS

Williams, Johnston and Donald (1980)

Shaft capacity is estimated from the charts in Figures 1 and 2. Overall socket stiffness (allowing for both shaft and base), I_ρ, is defined as:

$$I_\rho = \rho D E_m / Q$$

where ρ is settlement, D is the pile diameter and Q is the total load. Figures 3 and 4 are used to derive I_ρ and to estimate the proportion of load carried at base level. Empirically based correction charts are used to allow for nonlinearity.

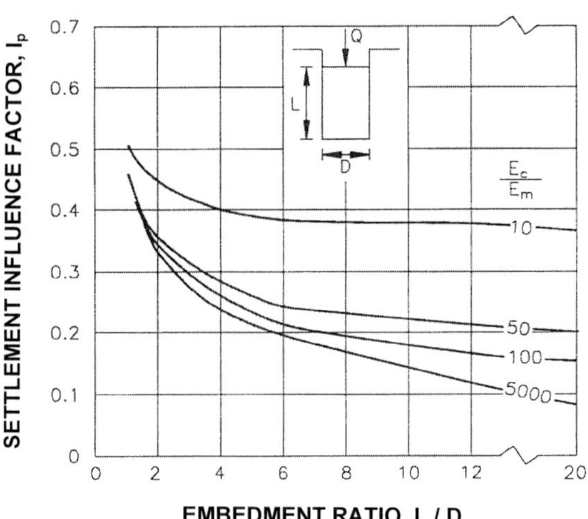

Figure 3 Elastic settlement influence factor as a function of embedment ratio and modular ratio (after Williams et al., 1980) [2]

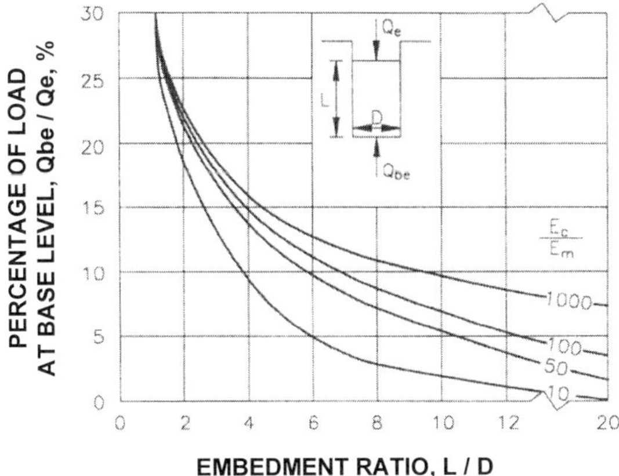

Figure 4 Elastic Load distribution as a function of embedment ratio (after Williams et al., 1980) [2]

The required settlement is the starting point, given an initial assumption of socket geometry, and the sum of the corrected shaft and base loads is compared with the design load to check that adequate capacity is available. An overall factor of safety for ultimate load of 3 is suggested with ultimate base stress taken as $5\sigma_c$.

It must be borne in mind that the empirical elements of the design are based on data from one particular type of weak rock, Melbourne mudstone.

Rowe and Armitage (1987)

Rowe and Armitage [4,5] developed their design method around finite element analyses of pile/rock response which make explicit allowance for gradual development of slip along the shaft of the socket. The analyses are summarised in a series of charts, such as Figure 5, in which the proportion of load taken to the base of the socket is shown as a function of socket geometry, and a dimensionless settlement factor, I_d:

$$I_d = \rho D E_{md} / Q$$

where E_{md} is the design value of rock mass stiffness. Different charts are given for different ratios of pile and rock stiffness. Since efficient pile design will usually imply slip along the shaft of the pile, the required socket geometry must be in the straight dashed line in Figure 5. The intersection of this line with the curve for the selected value of I_d then produces a design socket geometry.

Where a socket can support the whole load without slip occurring, a similar method to Williams et al. is proposed.

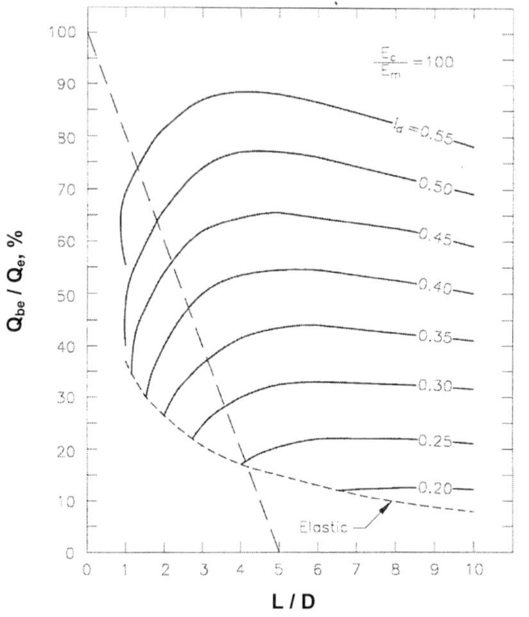

Figure 5 Design chart for a complete socketed pier
(after Rowe and Armitage, 1987a) [4]

Kulhawy and Carter (1992)

The methodology proposed by Kulhawy and Carter [6] provides an output of a complete relationship between load and settlement for a given socket geometry. It is based on an approximate closed-form analysis in which the behaviour of the pile in its rock socket is simplified into three zones (Figure 6).

In its linear elastic zone, the load:settlement relationship is based on the theoretical analysis presented by Randolph and Wroth [7]. The next zone (ignoring a gradual transition of progressive slip) represents full slip at the interface and is modelled as a Mohr-Coulomb failure with a dilatancy effect. Eventually ultimate load capacity is assumed to be reached and failure occurs at constant load.

Kulhawy and Carter present their model as equations, the complexity of which is greatly reduced if the pile can be considered rigid. A definition of rigidity is provided.

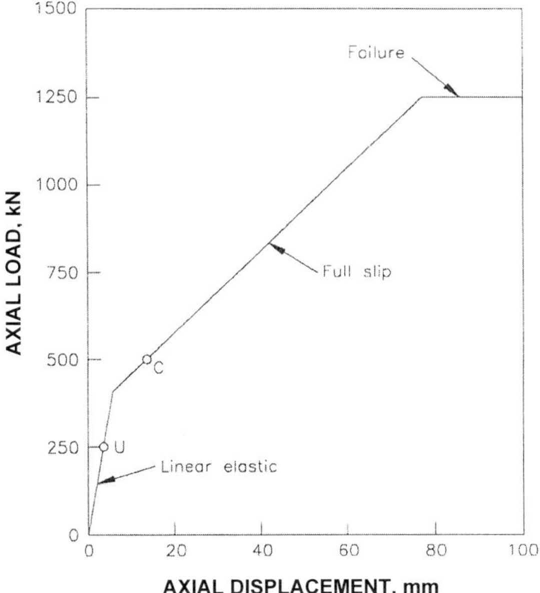

Figure 6 Predicted axial behaviour of a socketed shaft foundation in rock (L = 1.8m, D = 0.5m, E_m = 100 Mpa, σ_c = 5Mpa, Poisson's ratio = 0.25 for rock) (after Kulhawy and Carter, 1992) [6]

Fleming (1992)

The approach proposed by Fleming [8] can be regarded as an extension of that of Kulhawy and Carter, but instead of using analytical methods to describe pile response during slip, particular forms of load:settlement relationships are separately assumed for shaft and base and then superimposed to give a total response.

The relationship between pile settlement ρ and shaft load Q_S is:

$$Q_S = Q_{SU}\rho/M_SD + \rho$$

where Q_{SU} is the ultimate shaft load and Ms controls the initial slope of the shaft response.

The value of Ms can be deduced from Randolph and Wroth [7]. The base load:settlement relationship is:

$$Q_b = Q_{bu}DE_m\rho/(0.6Q_{bu} + DE_m\rho)$$

where Q_{bu} is the "ultimate" base load.

The total load is then:

$$Q_t = Q_s + Q_b$$

Further modification is made for elastic shortening.

COMPARISON OF DESIGN METHODS

The four design methodologies discussed all relate to short term axial static loading of single piles. None consider creep or cyclic loading. The design inputs are the stiffness and strength of the rock, the stiffness of the concrete pile, the required pile load and allowable settlement.

Key attributes of the four methods are summarised in Table 1 and Table 2.

The design output is either a pile socket geometry or a pile load:settlement relationship. Repeated use of the methods of Williams et al. And Rowe and Armitage can be used to generate load:settlement relationships for a given pile socket.

A comparison of these for a chosen pile is given in Figure 7.

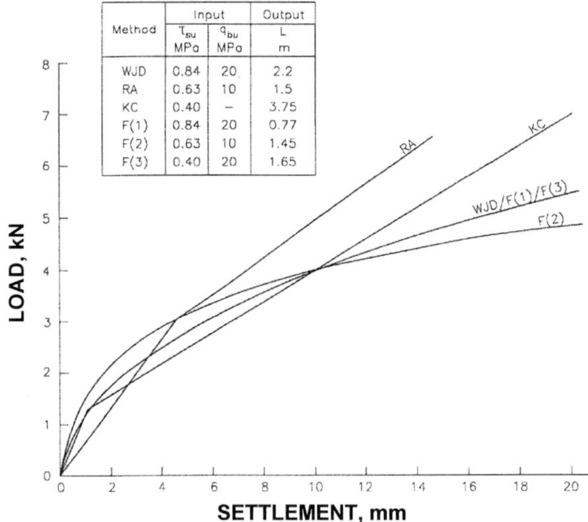

Method	Input		Output
	τ_{su} MPa	q_{bu} MPa	L m
WJD	0.84	20	2.2
RA	0.63	10	1.5
KC	0.40	–	3.75
F(1)	0.84	20	0.77
F(2)	0.63	10	1.45
F(3)	0.40	20	1.65

Figure 7 Performance of pile sockets, 0.75m diameter, designed for maximum settlement 10mm at working load 5MN (WJD: Williams, Johnston and Donald; RA: Rowe and Armitage; KC: Kulhawy and Carter; F Fleming)

Table 1 Key attributes of pile design methods

	Williams, Johnston and Donald, 1980	Rowe and Armitage, 1987
Pile types considered	Replacement	Replacement
Rock types considered	Soft rock: verified on Melbourne Mudstone	Soft rock: verified on shale
Input data required	Q, E_c, E_m σ_c, ρ_d D	ρ_d, D, Q, E_c, E_m τ_u, f_o, f_E
Elastic pile compression within socket included?	Yes	Yes
Expression for τ_u	α,β,σ_c	$1.423\ p\ (\sigma_c/p)^{0.5}$ $1.897\ p\ (\sigma_c/p)^{0.5}$
Expression for q_u used	$5\sigma_c$	$2.5\sigma_c$
Output provided	L, factor of safety	L, factor of safety
Other attributes	Assumes no cake or smear on socket walls, no base debris or artificial roughening	Not for porous crushable rocks or rocks liable to creep. Accounts for socket roughness but not for smear or base debris. Procedure for soft seams available

Table 2 Key attributes of pile design methods

	Kulhawy and Carter, 1992	Fleming, 1993a
Pile types considered	Replacement	All
Rock types considered	Not specified	All
Input data required	Q, E_m σ_c, ρ_d D, v, L	ρ_d, D, B, L, Q_{max}, Q_{bm}, E_c, E_m
Elastic pile compression within socket included?	Not for rigid shaft calculation presented but equations provided if needed for compressible pile	Yes
Expression for τ_u	$0.63\ p\ (\sigma_c/p)^{0.5}$	Not specified
Expression for q_u used	Not required	Not specified
Output provided	Load displacement curve	Load displacement curve
Other attributes		All materials, soil or rock surrounding pile are considered

The results indicate a reasonable degree of consistency for load:settlement responses but quite a wide range of recommended rock socket lengths (0.77m to 3.75m). The Kulhawy and Carter equations are based on correlations with measured rock properties which provide a lower bound to the available data, and therefore tend to suggest pile shaft capacities much lower than with other methods. With more site-specific data to draw on, it may be possible to use less conservative values.

In view of the uncertainty involved in the prediction of rock properties, it is always recommended that load tests on piles of comparable geometry, founded in similar strata and constructed using the intended techniques, should be performed as part of the design process. With test data available, Fleming's method provides a convenient way of analysing the results and extrapolating to other pile geometries.

IMPROVED PROCUREMENT PROCEDURES

Even with the benefit of good data and sound theoretical principles, pile design is not an exact science. There are many variables, particularly with weak rock, which may not be identified by boreholes and may require re-assessment during the pile installation. An improved communication process between structure designer, pile designer and constructor is required to allow design and installation techniques to be continuously appraised and reviewed during construction.

The structure designer should provide:

1. Pile design standard used (eg EC7)

2. Specified loadings at ultimate and, if required by the design code used, at serviceability limit states

3. Any limitations on displacement, or load:displacement behaviour

4. Any constraints on pile type, size or installation method

5. Factual and interpretative ground investigation reports.

The pile designer should provide to the structure designer and pile constructor:

1. Pile layout drawings, including selected pile type, size and tolerance

2. Geotechnical model (i.e. a statement of the assumed layers of soils and rock, groundwater conditions, ground properties, and design parameters which, if encountered in the field, will lead to an acceptable design)

3. Specified acceptability criteria (by testing, inspection or other method)

Actual conditions on site rarely correspond exactly to those assumed in design and it is therefore important that conditions encountered are accurately recorded, interpreted and responded to in an appropriate manner. Entering the construction stage with a geotechnical model which encompasses an appropriate range of conditions, with associated designs already in place, would be a valuable tool to enable decisions in response to changing conditions to be made quickly.

Where time and cost permits, preliminary piles are recommended, ideally loaded up to, or near, notional failure so that values of ultimate shaft adhesion, base resistance and the load:displacement response can all be determined. Scaled down model piles may be required for achievable failure loads and these should use the same socket length to pile diameter ratio as the contract piles.

Proof load testing of contract piles, although cumbersome and time consuming, remains a key part of the verification process, except perhaps for small contracts with well proven pile:ground combinations and conservative design. Non-destructive tests can supplement load tests and assess integrity of installed piles.

In the case of concrete piles, high strength concrete grades are not always practical because of increased heat of hydration. Integrity testing assists in the detection of discontinuities, voids, necking etc. Such tests do not provide direct data on pile capacity, but enhance confidence in the installation process.

Figure 8 summarises the recommended improved procurement process. The process is intended to be flexible enough to react to changing conditions as a project progresses, with tuning of the pile design being carried out as the quality of information from the field is enhanced. This process should continue through the trial piling stage into contract piling.

During pile installation, there has traditionally been little incentive for recording more than the minimal information required by most specifications. If, however, there was acknowledgement by all parties that the design would be continuously re-appraised on the basis of good quality feedback during construction, there would be an economic benefit in improving the extent and quality of monitoring and testing during pile installation.

Examples might include:

1. Description of rock arisings

2. Measurable properties of rock excavated from bored piles - eg point load tests

3. Description of rock from proving cores

4. Penetration rate for driven piles

5. Dynamic pile analysis during driving

6. Rig rotation speed, torque, thrust, concrete flow and pressure etc.

The nature of piling rigs and the piling process suggests that, given a financial incentive, manufacturers may be encouraged to develop more reliable monitoring systems which might permit data to be gathered automatically during operations such as pile driving, sinking of bore casings, augering, and any other construction activity interacting with the ground.

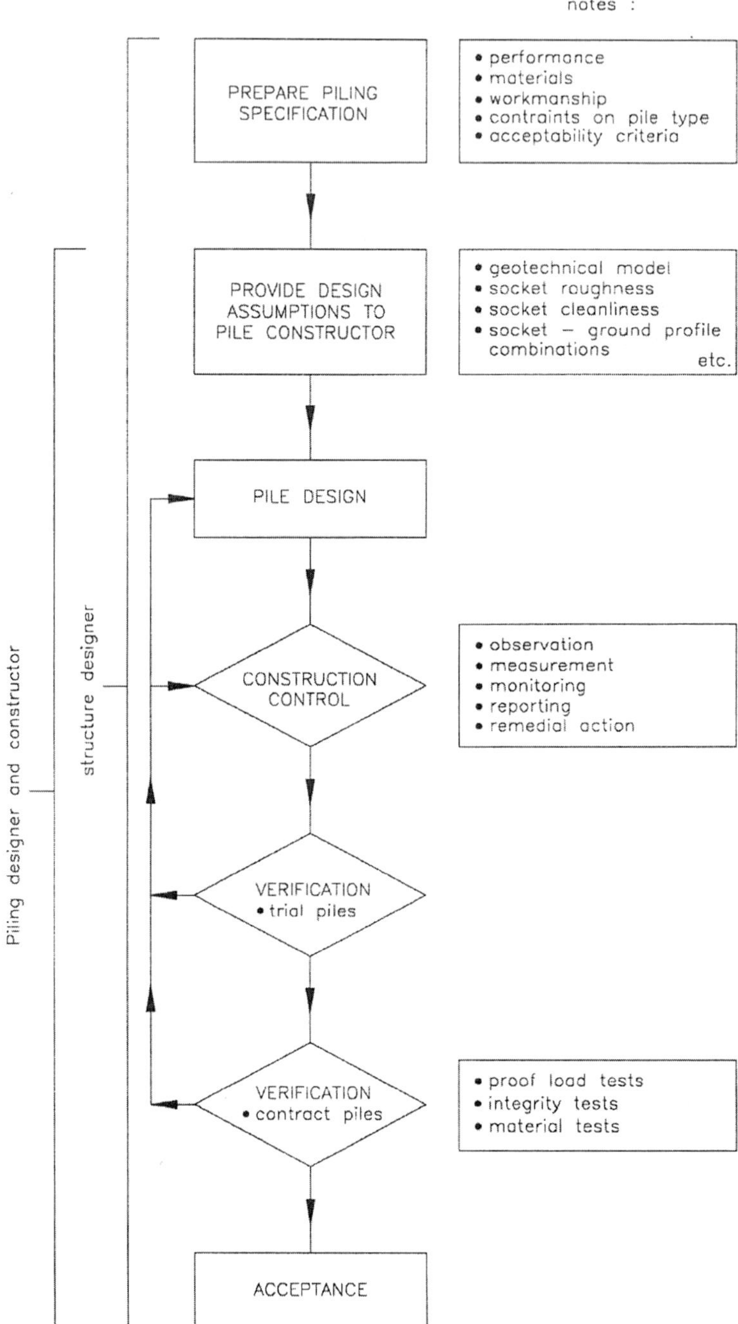

Figure 8 Flowchart for the process of procurement

With such improved feedback during construction, the development of an observational approach to the design and construction of piles could become viable.

The developing of such improved monitoring equipment might also help to ease pile design and construction gently along the route from an art to a little bit more of a science.

REFERENCES

1. GANNON, J A, MASTERTON, G G T, WALLACE, W A, AND MUIR WOOD, D. Piled Foundations in weak rock. CIRIA Research Project 509. Unpublished.

2. WILLIAMS, A F, JOHNSTON, I W, AND DONALD, I B. The design of socketed piles in weak rock. International Conference. Structural Foundations on Rock, Sydney, 1980, pp327-347

3. KODIKARA, J K, JOHNSTON, I W, AND HABERFIELD, C M. Analytical predictions for side resistance of piles in rock, 6th Australia-New Zealand Conference on Geomechanics, Christchurch, 1992, pp 157-162

4. ROWE, R K, AND ARMITAGE, H H. Theoretical solutions for axial deformation in drilled shafts in rock, Canadian Geotechnical Journal, 24,1987a, pp 114-125

5. ROWE, R K, AND ARMITAGE, H H. A design method for drilled piers in soft rock, Canadian Geotechnical Journal, 24, 1987b, pp 126-142

6. KULHAWY, F AND CARTER, J P. Socketed foundations in rock masses, Engineering in rock masses. Bell, F G, (ed), Butterworth/Heinemann, London, 1992, pp 509-529

7. RANDOLPH, M F, AND WROTH, C P. Analysis of deformation of vertically loaded piles. J Geotechnical Engineering Division, ASCE, 104, No GT12, 1978, pp 1465-1488

8. FLEMING, W G K. A new method for single pile settlement prediction and analysis. Géotechnique 42, No 3, September, 1992a, pp 411425

FOUNDATION MATERIALS – CEMENT – STABILISED QUARRY SAND

A Jasienski

FEBELCEM

R Debroux

Walloon Ministry of Equipment and Transport

C Ployaert

Belgian Cement Research Centre

Belgium

ABSTRACT. When crushing stones of different calibres, quarries produce a large amount of sand, referred to as crushed sand. This material is often considered as a by-product and its use is reserved for less noble applications than road foundation. However, when stabilised with cement, this sand has very interesting properties and can be used or many purposes. It has already been used experimentally for the foundations of various sections of the A8 motorway. Before envisaging a more extensive use, it has been decided to carry out an in-depth study specifically designed to determine the durability of such materials. In addition, a proposal has been put forward to compare sand foundations with conventional lean concrete foundations. The study showed that the use of crushed sand in road foudations is possible if the elements which are specific to the type of foundation are taken into account.

Keywords: Crushed sand, Foundation, Stabilised with cement, Durability, Lean concrete, Water content (optimum proctor).

Ir André Jasienski is Head Engineer – Infrastructure at the Federation of the Belgian Cement Industry (FEBELCEM). He is a civil and geotechnical engineer and has a great experience in road construction, mainly in concrete roads.

Ir Raymond Debroux is a civil engineer. He is Chief-Engineer Director of the Road Division of Mons (Walloon Ministry of Equipment and Transport) and is he in charge of the construction, maintenance and supervision of an important road and highway network.

Ir Claude Ployaert is an agricultural engineer. He is the technical responsible of the physical mechanical laboratory of the Belgian cement industry collective research centre. He is in charge of a lot of tests on concrete and the application of concrete in the roads.

INTRODUCTION

This paper is about a study carried out on foundation materials. When crushing stones to different sizes, every quarry produces large amounts of sand, called crushed sand. This type of sand is often considered as a by-product and is used in applications that are not as valuable as road foundations. However, when stabilised with cement, such sand has many very interesting characteristics and many outlets could be found for it. Such sand was already used, on an experimental basis, in the foundation of some sections of the A8 highway between Tournai and Lille [1].

Before considering using them on a wider basis, it was decided, in agreement with the Walloon Ministry of Equipment and Transports, to carry out a comprehensive study aimed, among other things, at determining the durability of these materials. Moreover, it was proposed to compare these sand foundations with foundations in lean concrete: on the one hand with traditional lean concrete, having a moist earth consistency, and on the other hand with lean concrete containing fly ashes, which can be placed by means of the slipform machine.

The study is not only of interest for highway foundations, but, taking the competitive price into account, this type of foundations may also be considered for a large amount of local or rural road foundations.

SAND CHARACTERISTICS

Four different types of sand were analysed:

- Two sands obtained by crushing limestone,
- Two sands obtained by crushing porphyry.

The main characteristics of this crushed sand are given below, in Table 1. An essential characteristic is their filler content, when the fillers are not washed.

Table 1 Main characteristics of the crushed sands

	FILLERS, %	SAND EQUIVALENT AT 10 % OF FINES, %	VALUE WITH METHYLENE BLUE, g/100 g OF FINES
Limestone 0/2.5	16.9	71	0.60
Limestone 0/4	14.1	71	0.60
Porphyry 0/4	8.4	59	0.31
Porphyry 0/7	18.7	59	0.42

STUDY OF CRUSHED SAND

In order to make it easy to compare them and taking their specificity into account, three different cement contents type CEM III/A 42,5 LA were chosen for the study: 60, 120 and 180 kg/m^3 of mix in place.

The main tests carried out on this sand are as follows:

1. Determination of the optimal water content on Proctor Standard specimens on the base of the volumic mass and the compressive strength at 7 days, and manufacturing of the specimens by means of optimum Proctor,
2. Compressive strength at 7 and 56 days
3. Splitting tensile strength at 56 days,
4. Durability in water at 56 days,
5. Durability in frost at 90 days,
6. Static modulus of elasticity at 56 days.

The results, presented as a synthesis, are given in the following tables:

Table 2 Optimal water contents

CEMENT CONTENT	60 kg/m^3	120 kg/m^3	180 kg/m^3
Limestone 0/2.5	8.5	8.5	8.5
Limestone 0/4	8.0	8.0	8.0
Porphyry 0/4	5.0	5.8	6.5
Porphyry 0/7	8.0	8.2	8.4

Table 3 Compressive strengths at 7 days and 56 days

CEMENT CONTENT	R'$_c$, (N/mm²)	60 kg/m^3	120 kg/m^3	180 kg/m^3
Limestone 0/2.5	at 7d	4.9	9.8	16.4
	at 56d	8.3	17.4	27.8
Limestone 0/4	at 7d	4.7	10.4	17.6
	at 56d	8.5	18.5	27.9
Porphyry 0/4	at 7d	3.3	7.2	12.8
	at 56d	5.0	10.9	16.3
Porphyry 0/7	at 7d	4.4	10.0	13.6
	at 56d	9.0	16.4	20.8

The compressive strengths were determined on Proctor Standard moulds (diameter 101.5 mm; height 117 mm) with the optimal water content. The values given below represent the averages of three samples.It was noted that for a cement content of 120 kg/m^3, all the strengths obtained at 56 days exceeded 15 N/mm^2, except for porphyry 0/4, whose filler content was inferior to 10 %. This test was carried out in accordance with ASTM D 559-89. It involved 12 cycles of moisturising-drying (5 hours under water, 42 hours at 71°C, brushing). The specimen weight losses were expressed as % of the specimen's initial dry mass.

Table 4 Durability in water at 56 days (% of loss)

CEMENT CONTENT	60 kg/m^3	120 kg/m^3	180 kg/m^3
Limestone 0/2.5	5.7	0.9	0.5
Limestone 0/4	6.7	1.0	0.5
Porphyry 0/4	16.6	2.0	0.8
Porphyry 0/7	12.1	2.6	1.1

This test was carried out in accordance with ASTM D 560-89. It involved 12 freeze-thaw cycles (24 hours at -18°C, 23 h at +20°C and more than 95 % relative humidity, brushing). The specimen's weight losses were expressed in % of the specimen's initial dry mass.

Table 5 Durability in frost at 90 days (% of loss)

CEMENT CONTENT	60 kg/m^3	120 kg/m^3	180 kg/m^3
Limestone 0/2.5	8.3	1.1	0.7
Limestone 0/4	14.4	1.1	0.7
Porphyry 0/4	16.2	1.9	1.1
Porphyry 0/7	8.4	1.9	0.9

For the durability in water as well as for the durability in frost, it was noted that the specimens behaved satisfactorily when the cement content is 120 kg/m^3. Indeed, the total loss observed was less than 3 %.

STUDY OF LEAN CONCRETE

As regards lean concrete, two different types were manufactured: one traditional, i.e. with a moist earth consistency, which was placed by means of the grader and roll, the other being placed by means of the slipform machine. In this latter case, a certain quantity of fly ashes was added. The three cement contents type CEM III/A 42,5 LA held were: 80, 100 and 120 kg/m^3. The 100 kg/m^3 content corresponds to the minimum laid down at present in the type specifications W 10 dealing with works on highways and regional roads in the Walloon Region.

In addition to the consistency measurements, the same tests as those carried out on crushed sand were carried out on lean concrete. The results are given in the following tables:

Table 6 Compressive strength at 7 and 56 days on cubes (20x20x20 cm)

CEMENT CONTENT	R'_c, (N/mm²)	80 kg/m³	100 kg/m³	120 kg/m³
Traditional lean concrete	at 7d at 56d	5.4 10.1	8.2 15.0	10.4 17.3
Slipform-Fly ashes	at 7d at 56d	3.9 11.4	5.7 15.0	7.5 19.1

Table 7 Compressive strength at 56 days on cylindrical specimens (100 cm²)

CEMENT CONTENT	R'_c, (N/mm²)	80 kg/m³	100 kg/m³	120 kg/m³
Traditional lean concrete	at 7d at 56d	10.3	13.6	17.3
Slipform-Fly ashes	at 7d at 56d	15.7	18.9	22.9

Table 8 Durability in water at 56 days (% of loss)

CEMENT CONTENT	80 kg/m³	100 kg/m³	120 kg/m³
Traditional lean concrete	6.9	7.4	1.1
Slipform-Fly ashes	0.6	0.3	0.1

Table 9 Durability in frost at 90 days (% of loss)

CEMENT CONTENT	80 kg/m³	100 kg/m³	120 kg/m³
Traditional lean concrete	1.2	1.0	0.4
Slipform-Fly ashes	7.1	1.3	0.4

A compressive strength of 15 N/mm² at 56 days of age was reached for lean concrete with a cement content of 100 or 120 kg/m³. As regards durability in water, traditional lean concrete containing 100 kg cement by m³ exceeded 3 % of loss. As regards durability in frost, concrete containing fly ashes had the worse behaviour.

With a view to all the results obtained, a rather good correlation can be made between stabilised sands at 120 kg cement and lean concrete at 100 kg cement, respectively, for the compressive strengths, durability in water and durability in frost. It is therefore noted that these materials are comparable.

As regards the moduli, it can be held that, on average, stabilised crushed sands at 120 kg cement have an average static modulus of elasticity of 20,000 N/mm² whereas lean concrete at 100 kg has average moduli of 30,000 N/mm². A sizing design of the road structures should therefore be made so as to take these differences in values into account.

CONCLUSIONS

The laboratory study shows that stabilised crushed sand at 120 kg/m³ cement reaches an average resistance greater than 15 N/mm² at 56 days of age in the laboratory for a water content of approx. 8 %.

Obtaining such results on site depends naturally on a number of execution parameters which have to be respected carefully. The requirements imposed on site shall therefore be less strict in order to take these various elements into account. Among other things, there is:

1. Quality and homogeneity of the mixing. Incorporating 120 kg cement into sand requires a priori a longer mixing than incorporating this very quantity of cement into lean concrete.

2. Respect of the chosen compositions, and above all, of the water content, of the mixes. Orientation tests on site showed that if the optimum Proctor is not respected, decreases in resistances occur rapidly. Table 10 gives values obtained on test sections on sites from a foundation made of crushed sand porphyry 0/7, for various water content values.

Table 10 Compressive strengths at 7 days, 28 days and 90 days of crushed sand porphyry 0/7 for various water content values

WATER CONTENT OF STABILISED SAND (%)	COMPRESSIVE STRENGTH (N/mm²) ON PROCTOR STANDARD SPECIMENS		
	7 days	28 days	90 days
7.23	5.5	7.8	8.9
7.83	8.5	12.0	11.8(*)
9.08	6.4	8.5	10.3
10.14	4.0	6.5	6.6

(*)doubtful result

3. Rapidity of placing and compacting after manufacturing the mixes. Tests on site showed that when compaction is done 2 hours after mixing, the decrease in the specimens' resistances can reach 25 % in relation to the same mix compacted immediately after mixing.

4. Quality of compaction. It was already showed to a large extent that the quality of compaction, and therefore the density of the specimens is directly proportional to the resistance obtained, with an equivalent water content.

5. Effective protection of the foundation against desiccation. As for every foundation stabilised with cement, this protection is essential. Generally, it is applied in two phases: the first phase is sprinkling the material's surface with water immediately after the last time the compaction machine(s) passes. The second phase takes place at the end of the day and consists in applying a bitumen emulsion at the rate of at least 0.7 l/m^2 followed by sand spreading at the rate of 3 kg/m^2.

As a conclusion, the use of crushed sand in road foundations is possible if the elements which are specific to this type of foundation are taken into account. For important applications, it is advisable to make test sections on site with the materials intended to be used in order to determine precisely, among other things, the water content to be respected and the importance of the compaction tools required.

REFERENCES

1. DEBROUX, R. and DRUEZ, M. Valorisation du sable de concassage 0/4 porphyre par stabilisation au ciment. 7ème Symposium International des Routes en Béton - CEMBUREAU - Vienne, 3-5 octobre 1994.

2. JASIENSKI, A. Les recherches récentes de FEBELCEM dans le domaine des routes en béton. 4ème demi-journée provinciale d'information Hainaut - Brabant Wallon - Chièvres, 12 mars 1997.

LEAKAGE THROUGH CRACKS IN REINFORCED CONCRETE ELEMENTS SUBJECTED TO COLD-SPOT LOADING

K van Breugel

T Wermann

Delft University of Technology

Netherlands

ABSTRACT. The research deals with the response of structures to a localised extreme temperature load, a so-called cold-spot load. Two different types of experiments were performed to evaluate the impact of the parameters w/c-ratio, crack width, depth of the element, reinforcement ratio and cold-spot size on structural and leakage behaviour. In a series of 60 experiments on 150mm cubes the parameters mix composition (w/c ratio: 0.3, 0.4, 0.55) and crack width (0.1, 0.2, 0.3mm) were investigated. Nine full-scale beams were used to investigate the influence of depth (20, 60cm), reinforcement-ratio (0.85/0.5% , 1.2/1.51%) and cold-spot size (20,60/100, 200cm).

Keywords: Leakage, Cryogenic, Liquids, Structures, Thermal cracking, Cold-spot

Dr K van Breugel is associate professor at Delft University of Technology. His main interests are: risk assessment, environmental matters, development of concrete at very early age.

Dipl-Ing T Wermann has worked as an assistant researcher at Delft University of Technology between 1994 and 1998 on the topic "Transport of (cryogenic) liquids through cracked concrete structures". Since 1998 he is an employee of Philipp Holzmann AG, Contractors, Düsseldorf, Germany.

INTRODUCTION

As a result of increased environmental consciousness the potential of concrete retaining structures for environmental protection has received a lot of attention during the past years. In case of storage of hazardous products storage vessels or retaining structures have to remain liquid tight in order to prevent contamination of soil and ground water in case of leakage of the stored products. This also holds for vessels for storage of cryogenic liquids. Many of these storage vessels consist of a steel inner tank (or a steel membrane) and a concrete outer tank. In case of local leakage of the inner tank the outer tank will be subjected to a "cold spot loading". This cold spot loading is considered one of the most severe loadings. Only little experimental data on cold spot loads is available. This data, however, is partly contradictory and can, therefore, only serve as a very rough guidance for design against this type of loading.

To gain more insight into this problem, which might enable us to formulate more adequate design concepts for tanks for refrigerated gases, the formation of temperature-induced cracks and leakage through these cracks has been investigated in an experimental study. Some of the data available from earlier studies [1,3] suggests that even cracks with a width of up to 0.3mm exhibited no noticeable leakage. Some sources suggest that freezing of pore-water might be a reason for this behaviour. Therefore two different types of experiments were set up to investigate first the material dependent aspects and secondly the more structural aspects of crack formation and leakage.

Section `PROBLEM ANALYSIS` will describe general aspects of cold-spot loads on a structure and basic material properties of concrete at low temperatures. Section `EXPERIMENTS` describes the experimental approach and the experimental setup. Section `TEST RESULTS` reports the main results of the experiments. In section `EVALUATION` an evaluation of the test results is presented and an attempt is made to assess the impact of the findings on the design approach. The final section `CONCLUSIONS` will list the conclusions that can be drawn from the research so far.

PROBLEM ANALYSIS

Due to a cold-spot load a concrete shell structure is prone to cracking. Cracking appears at the interface between the cooled and the warm part of the structure.

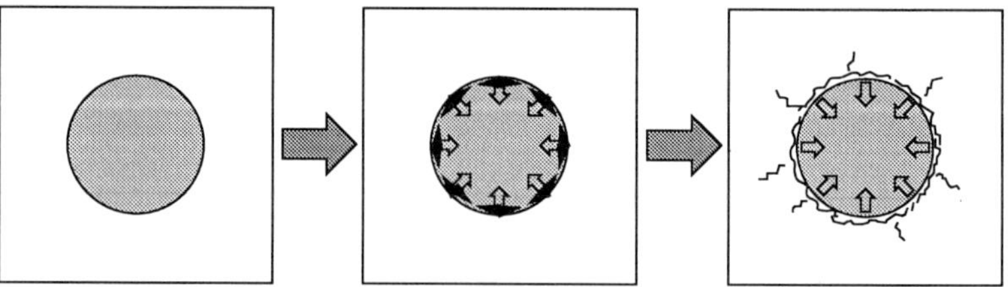

Figure 1 Crack pattern in slab with cold-spot

In the few experiments performed so far it has been found that the cooled section itself remains uncracked. This phenomenon was related to the rapidly increasing strength of concrete at dropping temperatures [4].

Tests on reinforced concrete slabs [1] in Germany revealed further, that cracks with crack widths of up to 0.4mm formed. The general mechanism of cold-spot structural damage is depicted in Figure 1. The dominant crack type for the reinforced slabs used by [1] was the circular type. Up to four circular cracks formed during the experiments, with maximum crack width of up to 0.4mm. No leakage through the slab was observed however. These observations lead to the assumption that some re-tightening mechanisms might exist, which prevent leakage through those cracks. These mechanisms were assumed to be related to the free pore water in concrete. This water would diffuse towards the crack due to the temperature gradients, and form ice in the crack. No experimental evidence on this re-tightening has been collected, however, to support these theories. Therefore experiments were designed at Delft University of Technology to investigate the phenomenon. Of big relevance for the actual design of structures regarding load definition is a possible dependency of crack width on cold-spot size. This problem was addressed by another series of experiments on large scale beams. Those experiments focused on the interrelationship between cold-spot size, crack development and leakage.

EXPERIMENTS

This section describes the experimental approach chosen to investigate the behaviour and possible re-tightening of concrete structures under a cold spot load. The approach was made in two steps. In the first step it was attempted to find indications of re tightening or freezing of cracks. For this purpose small concrete cubes were used. These experiments are described in section `Experiments on cubes`. In the second step we intended to incorporate the findings of the experiments on cubes into experiments to investigate structural behaviour which might lead to leakage. For these experiments large scale beams were used. These experiments are described in section `Experiments on beams'.

Experiments On Cubes

The experiments on cubes were used to investigate the influence of the mix-composition (pore water content) and crack width on leakage through a crack. Three different mixes were

Table 1 Characteristics of the mixes used during cube experiments

MIX	A	B	C
Cement type	CEM III/B 42.5	C EM III/B 42.5	CEM I 52.5
Cement [kg/m^3]	350	350	475
Water [kg/m^3]	192	140	8.25
Silica fume [kg/m^3]	-	-	25
Tilman OFT 3 [kg/m^3]	-	5.25	-
Tilman OFT 4 [kg/m^3]	-	-	13.6
w/(c+s)	0.55	0.4	0.3
Gravel (4mm-16mm)	809	862	1005(broken)
Sand (0mm-4mm)	737	1065	790
f_{ccm}	33.2	56.5	108.4

used, the composition of which is shown in Table 1. These were two regular mixes with w/c ratios of 0.4(mix A) and 0.55(mix B) and one HPC mix with w/b ratio 0.3(mix C). The cube dimensions were150x150x150mm with triangular notches on both top and bottom side to ensure proper crack localisation. Two rebars, diameter 4mm, were cast in to prevent the specimen from falling apart during crack generation. After casting and curing circular notches were drilled into which later the LN_2 pressure cells were glued. Cracks were introduced into the specimen by splitting in a standard testing machine.

Cube with retention frame

Cube with brass pressure cell

Figure 2 Cube specimen

To ensure a more uniform crack width distribution, a retention frame was used, by means of which the deformation of the specimen could be locally restrained during splitting. The crack widths generated were 0.1, 0.2 and 0.3mm. The pressure cells for the application of LN_2 - brass cylinders- with membrane thickness of 0.1mm were then glued into the notch on the top side of the specimen. (Figure 2). The side faces of the cracks where sealed witch the same brass foil. Then the cubes were placed in a steel outer container, equipped with 18 thermocouples for temperature profile measurement. Measuring pens for deformation measurement were attached and the annulus between specimen and steel container filled with insulating PUR-foam. The so prepared specimen was placed in the test setup, attached to a gas-meter, LN_2 reservoir, and pressure regulating mechanism. Then LN_2 was filled into the pressure cell on top of the specimen. A constant pressure of 0.5 bar was maintained in the reservoir. Leakage was measured by means of a precision wet type gas meter. Figure 3 shows a schematic representation of the cryogenic setup.

Experiments On Beams

In a second test series, beams were used to study crack formation and leakage under a cryogenic load. Parameters of the tests were depth of the beam (20cm / 60cm), size of the cooled area (0.2 / 1.0 / 2.0m) and reinforcement ratio (0.5% / 1.3%). A concrete B45 with w/c 0.39 was used, since the results from the first series of experiments suggested that re-tightening of cracks due to freezing was not likely to occur even with mixes

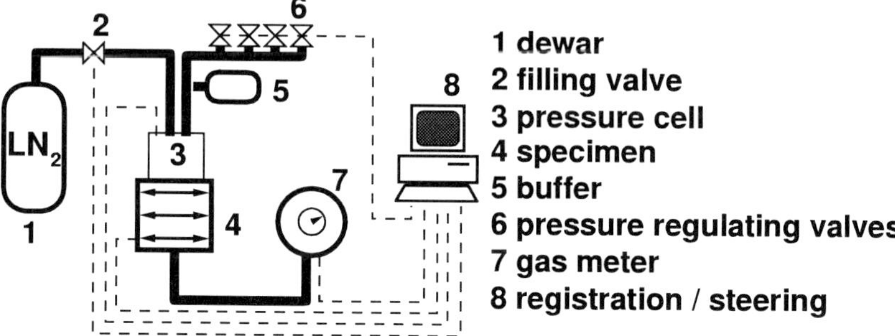

Figure 3 Diagram of test setup for leakage through cracks in cubes

Table 2 Overview of experiment parameters

EXPERIMENT	DEPTH [mm]	COLD-SPOT [cm]	CONTROL LENGTH [cm]	REINFORCEMENT [cm]
beam1	20	20	400	12
beam2	20	200	400	12
beam3	20	20	220	16
beam4	20	100	320	16
beam5	20	200	400	16
beam6	20	20	220	12
beam7	60	200	400	16
beam8	60	200	400	25
beam9	60	60	160	25

with a higher w/c-ratio. A w/c-ratio of 0.45 is regarded as the upper limit with regard to durability and tightness under severe environmental conditions in Dutch regulations. Table 2 shows the main parameters of the experiments.

The overall specimen size was (0.2/0.6 x 0.4 x 7.2m.). The total length of the cooled part plus one meter on each side were kept constant by means of a hydraulic actuator. As steering signal the mean value of two LVDT's, one on each side of the beam, were used. Figure 4 shows the general system setup. The beams were equipped with 50 thermocouples for temperature measurement, 28 LVDT's for deformation measurement and steering purposes. To the top of the specimen three brass containments were fitted, one for the application of the cooling liquid and at both sides smaller pressure cells for application LN_2 for leakage measurement. The top and the sides of the specimen were insulated with 4cm of EPS-foam plates to simulate the thermal situation in a slab. Figure 5 shows the testing rig equipped with a 60cm beam.

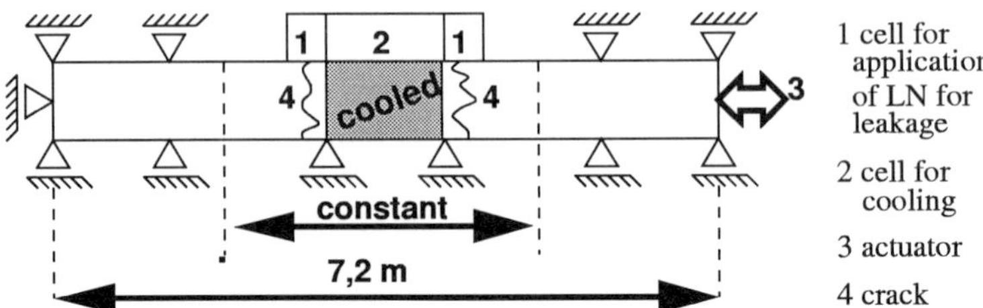

Figure 4 Schematic representation of beam-experiment setup

The bottom side was left uninsulated. During the test the middle compartment was filled with LN_2 ,and a constant layer of liquid was kept on top of the specimen. No pressure was applied. When the temperature and the associated deformation reached an almost steady state, which was for the 20cm beams at t 6h while the cooling of the 60cm beams was truncated after approximately 36h, first one and then the other of the smaller compartments was filled with LN_2 and penetration of the liquid was measured. A slight constant pressure of up to 0.5bar was kept in the leakage compartments.

Figure 5 Test rig for cryogenic beam experiments

TEST RESULTS

Cubes

Results of the cube experiments suggest, that no re-tightening takes place, or that their effect is so small that it could not be measured. In some cases with mix A (see Table 1) and small crack widths a decrease of leakage rate could be monitored. This can be attributed to the fact that concrete with a high water content expands in the temperature range between -20 and -100 [4].

Figure 6 shows the results from leakage measurements of the cube experiments. Only peak leakage is shown in this graph. The flow is almost stable at low level, and increases about two orders of magnitude when the liquid front reaches the bottom side. A likewise behaviour could be observed with all of the specimens that exhibited liquid leakage. The flow-rate of the state when liquid reached the bottom side, remained constant. The maximum duration of the experiments was 6 hours. Figure 6, shows maximum leakage versus crack width for all cube experiments. It can be seen that for small crack widths the leakage rate of the HPC specimen (mix C) is higher than for mixes A and B. This can be attributed to the relatively smooth crack surfaces in the concrete with higher strength. Figure 7 shows photographs of cracks in three specimens made from the tree different mixes that have been injected with a fluorescent epoxy resin after the cryogenic experiments. In Figure 7 the strength of the mix increases from left to right. The externally

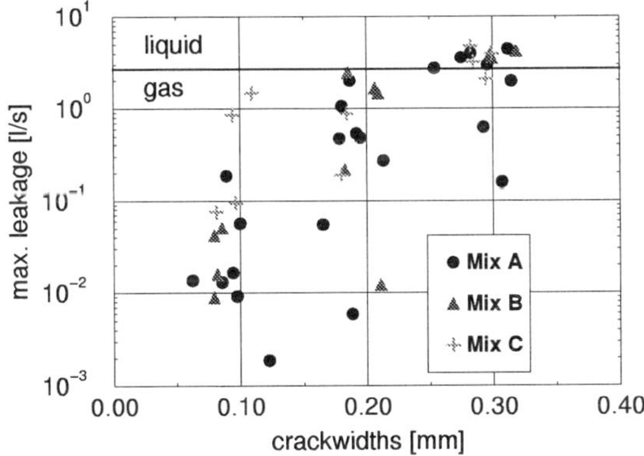

Figure 6 Maximum leakage during cube-experiments vs. initial crack widths

apparent crack width was in each case 0.2mm. Evidently the localisation of the crack increases with increasing strength of the concrete mix. Also it can be noted that the specimen made from mix A shows considerable surface damage. With larger crack widths the influence of the roughness of the crack surface is less, and consequently the values for leakage differ less. The values for leakage are of the same order of magnitude as the results published by [?]. The crack width criterion of 0.1mm to make sure that no liquid reaches the bottom side of the specimen seems sensible for normal strength concretes.

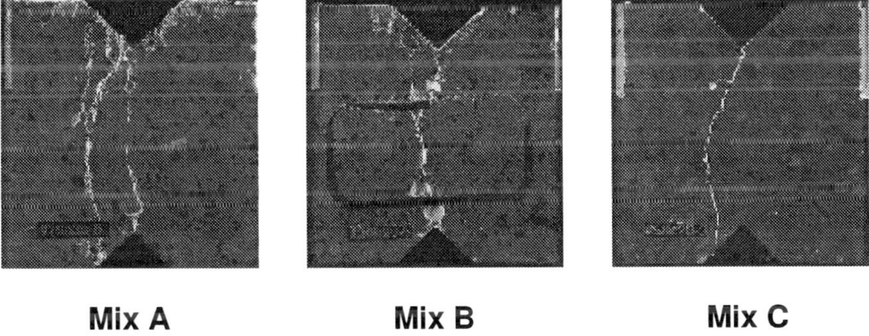

Figure 7 Cracks in cube specimen

Beams

Results from the beam-experiments suggest that the overall reaction of the beams to the cold-spot loading is according to the expectations. Crack formation is indeed depending on the size of the cooled area. With the shallow beams no crack formation could be observed in the cold-spot area. For the deep beams with the large cold-spot sizes some cracking could be observed in the cold-spot area. The cracks could be traced from the top of the specimen to approximately one third of the depth measured from the top. None of the observed cracks in the cold-spot zone was a separating crack. The contraction of the cold area was found to be around 0.3mm, 1.2mm and 2.5mm for cold spot sizes 20cm, 100cm and 200cm respectively. Leakage could only be measured in two of the six tests with the shallow beams and in none of the tests with the deep beams. Even so, the measured leakage rates were very small, only 9 x 10^{-4} l/s and 7 x 10^{-4} l/s were recorded. The corresponding crack widths were very small. This indicates that the cracks were no clean separating cracks. After warming up of the beam, the cracks were visually inspected and they indeed showed some branching. This highlights the importance of taking into account the crack morphology into leakage considerations.

The explanation for the fact that the deep beams show some cracking in the cold-spot zone can be found in temperature distribution along the depth of the beams. Figure 8 shows the development of even temperature and temperature gradient during the cryogenic beam experiments.

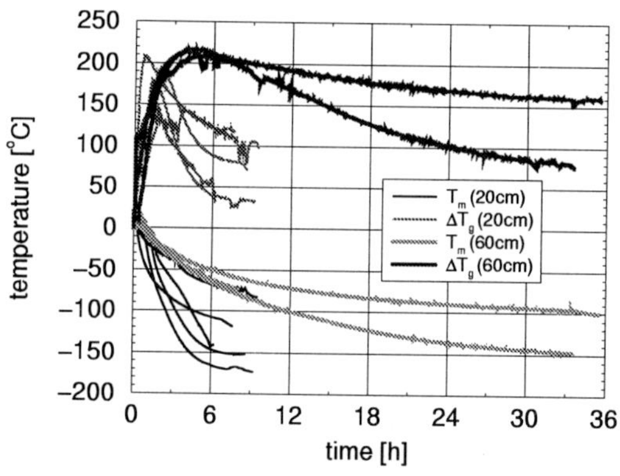

Figure 8 Development of temperature gradient and mean temperature in beams

EVALUATION

From the cube experiments it is apparent, that all non-compressed separating cracks exhibit leakage. No re-tightening could be observed with the given specimen and mixes. Looking at the performance of different mixes, it appears that the low strength / high w/c-ratio mix exhibits the least leakage especially with the experiments with the smaller crack widths. The HPC mix exhibits the poorest performance as far as leakage is concerned. The plausible reason for this behaviour is the branching of the cracks and the generally smoother crack faces with increasing strength. Especially at smaller crack widths/flow-rates, HPC has higher leakage rates than normal concrete due to smooth crack surfaces. A cause for reduced leakage rates can be irregularities of cracks, like branching or overlapping, which make the crack non-separating.

If leakage is to be limited to low values, liquid penetration must be prevented by controlling crack width to a value around 0.1mm, depending on structure thickness. More insight has to be gained into the crack formation process and crack morphology due to a cryogenic load.

CONCLUSIONS

The following conclusions can be drawn from the experiments conducted at Delft University

- Leakage of cryogenic liquids through cracks in concrete structures is governed mainly by the parameters:

  a: effective crack width

  b: length of the crack in the direction of flow

  The total rate of mass-flow depends on the position of the phase transformation front. When the liquid phase penetrates completely, the total rate of mass flow increases by a factor between 10 and 100.

- The mix composition has an impact on the leakage behaviour insofar as the crack morphology depends on the concrete strength. For very high water contents, the deformation behaviour of the concrete can lead to reduced cracking and hence reduced leakage. A high water content of the concrete, has adverse effects on the durability.

- In the experiments there was no evidence of re-tightening of cracks due to freezing. Rather it was found that the morphology of the cracks, e.g. branching dominates the performance of concrete structures with regard to leakage.

- The impact of the depth of a structure on cracking and leakage behaviour is twofold:

  - In shallow structures the assumption of a cold-spot zone remaining uncracked seems to hold.

  - In deep structures crack formation in the cold-spot zone was observed in the experiments. These cracks lead to a certain relief of the total imposed inner strain and subsequently reduced cracking in the zone adjacent to the cold-spot. In no case the cracks inside the cold-spot were separating cracks. Furthermore the chances for branching of the cracks are increased due to longer crack path.

The implications for the dimensioning of concrete structures subjected to a cold-spot load are that deep structures might perform better than expected on the basis of the assumption of the cold-spot zone remaining uncracked. A realistic definition the damage scenario with respect to size and intensity in the time domain is necessary for proper assessment of a structure's performance.

REFERENCES

1. G. IVÁNYI AND M. SCHÄPER. Kälteschockversuche an bewehrten und unbewehrten Betonplatten. Forschungsberichte aus dem Fachbereich Bauwesen 25, Universität-Gesamthochschule Essen, Essen, 1984, 87 pages

2. K. MAEKAWA ET AL. Gas-liquid two phase flow trough a cracks in concrete. In The third East Asia-Pacific conference on structural engineering & construction, Shanghai, April 1991, pages 47-52.

3. L. VAN BIERVLIET AND F. MORTELMANN. On the cryogenic behaviour of a reinforced concrete slab 400mm in depth by 5m in diameter, locally subjected to a sudden temperature drop of 214 degrees Celsius. In First International Conference on Cryogenic Concrete, Newcastle, March 1989, 6 pages. (conference proceedings not consecutively numbered)

4. C. VAN DER VEEN. Properties of concrete at very low temperatures. Technical Report 25-87-2, TU Delft, Delft, 1987, 123 pages.

IMMOBILISATION OF RADIOACTIVE WASTES IN PORTLAND CEMENT-BASED MATRICES

S Kumar

Bundelkhand Institute of Engineering and Technology

India

ABSTRACT. Ordinary Portland cement or one of its variants is attractive matrices for the immobilisation of radioactive wastes. But, in most of the cases cements may have to immobilise salt - rich materials or perform in unusual environments. The mechanism of immobilisation of individual radio-nuclides in cements observe close study in order to quantify the strength and nature of its bonding mechanism. A more rigorous approach towards the selection of composite binder components in the appropriate standards is required for their use in immobilisation of radioactive wastes. The impact of higher temperature is also an important unknown factor. Systematic research efforts are needed for developing more reliable standard tests and better models to predict performance and service life of cement matrices used for nuclear waste immobilisation.

Keywords: Blast furnace slag, Blended cement, Durability, Fly ash, Ordinary Portland cement (OPC), Pozzolans, Silica fume, Sulfate attack, Radioactive waste, Waste management.

Dr Sunil Kumar is a Reader in Civil Engineering at Bundelkhand Institute of Engineering and Technology, Jhansi, India. He received his B. Sc. Engineering in Civil Engineering from Aligarh Muslim University, Aligarh and M. Tech. from Kurukshetra University, Kurukshetra. He holds a Ph.D. degree in Civil Engineering on the topic entitled "Structural Reliability under Random Loads and Environmental Effects". Prior to joining the faculty of Civil Engineering Department at Bundelkhand Institute of Engineering and Technology, Jhansi, he was a Lecturer at Harcourt Butler Technological Institute, Kanpur. His main research interests include the durability of concrete structures in aggressive environments, structural fire engineering, structural safety and reliability analysis. His interest also includes the use of industrial wastes in brick manufacturing. Dr. Kumar has published research papers widely in International Journals and Seminars.

INTRODUCTION

The problem of environmental pollution is causing increasing public anxiety and concern all over the world and in this context safe and economic management of radioactive wastes from nuclear industry is extremely necessary for the successful implementation of nuclear power programme and for improving prospects of further growth. Radioactive wastes, though comparatively small in quantity, do pose health problems.

Unlike pollutants from conventional sources, the radioactive pollutants cannot be rendered harmless by normally available physico-chemical means. However, they decay by themselves in a natural manner and as such their hazard comes to an end after a certain period of time. These periods range from few seconds to hundred thousands of years.

The U.S. Nuclear Regulatory Commission (N.R.C.), in conjunction with National Institute of Standards and Technology (N.I.S.T.), has invested significant efforts to develop strategies for storing low-level radioactive wastes in underground concrete vaults. These vaults are expected to have service life of at least 500 years. Independent research is being conducted at several other public agencies and research Universities [2]. Durability of concrete in aggressive environments is one of the most important factors influencing performance and service life of cement concrete used in immobilisation of radioactive wastes.

Ordinary Portland cement or one of its variants is one of several materials, which may enter into the formulation of cement matrices intended for radioactive waste immobilisation. Amongst the supplementary materials used in blending are fly ash, slag, silica fume, and natural pozzolans. The advantage of these supplementary materials is that they are widely available, relatively inexpensive, and eventually react with cement to become chemically and physically part of the binder.

The main immobilising potential of cement systems comes from their high internal pH allowing precipitation of many radio-nuclides as hydroxides. The mechanisms of immobilisation of individual radio-nuclides in cement deserve close study in order to quantify the strength and nature of its bonding mechanics.

More reliable standards tests and models are needed to predict performance and service life of several hundred to several thousand years for concrete exposed to aggressive environments. It has been suggested [1,3] that some of the existing models are not reliable to predict long-term performance. The current need for reliable predicting of long-term durability performance and service life, based on short term accelerated test, is more than ever justified to convince the public and public interest groups of the safety of cement based matrices used in radioactive waste immobilisation.

Investigations of the behaviour of three important radio-nuclides I, Sr and U are described in this paper. The various factors affecting durability and durability design of cement based matrices are also discussed in this paper. Further research is needed for better understanding of sulfate attack mechanism and for developing more reliable standard tests and models to predict performance and service life of cement based matrices.

CEMENT MODIFIERS

Substances are added to modify the composition of ordinary Portland cements for various purposes. Binder type has an important role in the long-term durability. Mineral additives such as slag, fly ash and silica fume are the principal modifiers.

Fly ash is perhaps the most commonly used supplementary material. Fly ash arises as a product of coal combustion. Since fly ash is commonly employed in radioactive waste formulations, the importance of having adequate characterisation data for the fly ash component is essential.

Slags intended for cement blending are normally glassy, or nearby so and this is obtained from iron blast furnace as a by-product. Since blast furnace slag may comprise up to 90% of cementitious blends used for radioactive waste immobilisation, so characterisation data for the slag are important parameters necessary to predict the future performance of slag as a blend.

Silica fume is obtained as a by-product of ferro-silicon manufacture. Silica fume cannot be added to cement in large quantities and must be used in conjunction with a super plasticizer if low water solid ratios are to be maintained.

RADIONUCLIDE-CEMENT INTERACTIONS

It is important to establish the behaviour of radio-nuclides in a cemented waste form. The behaviour of important radio-nuclides is described below.

Iodine

Iodine-129 is one of the radio-nuclides of most concern [4,5]. This is due to its long half-life (1.7 x 10 years), its poor sorption properties on most geological materials, and its radio toxicity [5].

The immobilisation mechanism of I^- is sorption and incorporation in/on the cement hydrates phases. Sorption studies have shown [6] that calcium hydroxides and (carbonated) hydrotalcite like phases possess negligible sorption capacity for I^-. Experiments extending up to one-year shows that iodide sorption increase with increasing C/S ratio of the gel and decreasing solvate concentration [7]. The effect of chloride addition shows that the addition of chloride anions in 1000-fold excess over I causes some dissorption of I^- [6].

In order to maximise immobilisation of iodide in the solid phase; cement should be formulated to be rich in AF_m, and high C/S ratio C-S-H. It was observed [8] that the formulation for I immobilisation lies in the blending of 30-40% slag in Portland cement. Aqueous radio-iodide might be immobilised in a cemented waste by the addition of a soluble Ag salt such as Ag NO_3. However, there are several problems with this approach. The use of Ag as selective precipitant for radio-iodide is probably not a sound long-term solution [9].

Strontium

Radio-strontium will be diluted in cement matrices to some extent by the non- radioactive strontium content of ordinary Portland cement which varies in between 0.05 to 0.1% by weight. Any immobilisation of strontium will be due to incorporation or sorption in/on the major cement hydrates, or co-precipitation.

Individual cement clinker phases and PC were hydrated in strontium containing solutions. More than 99% of strontium added in the hydrating fluid is immobilised in the solid phase substituting for Ca in certain hydrate phases. Calcium silicate hydrates (C-S-H) are responsible for some of this immobilisation [6]. The hydration products of C_4AF and C_3A can also accommodate completely low levels of strontium.

The extent of strontium immobilisation appears to be proportional to C_3A content. High alumina cement displays the best sorption characteristics for strontium. Leach experiments on strontium loaded pastes, in conjunction with consideration of effect of carbonation, have indicated co-precipitation of $SrCO_3$ to be an effective solubility limiting process for radio-strontium release [7].

Uranium VI

The interaction of uranium with cement assumes that the uranium will be present as schoepite, $UO_2(0H)_2H_2O$ or under reducing conditions, UO_2. Immobilisation mechanisms for uranium VI, differ substantially from those of other cations. At a high pH and with low uranium loading, uranium is sorped in C-S-H gel. At higher uranium loading, precipitation of discrete uranium salts occurs, initially as poorly crystalline solids, but improving with age. Solubility experiments of many months duration show [7] that uranium solubilities can decrease substantially with age. Therefore, it is important to determine steady state solubilities on well-cured samples. Further experimentation is required to determine phase equilibria for systems containing sulfates and carbonate ions.

FACTORS AFFECTING DURABILITY OF CEMENT BASED MATRICES

The good performance of concrete in service, including its durability, is a major factor for its success as a radioactive waste immobilisation material. It is important that the factors capable of adversely affecting the service life of concrete be well understood. The most important issue that needs urgent attention is the durability.

Concrete structures exposed to severe environments such as temperature or humidity extremes and aggressive chemicals tend to deteriorate earlier than their intended service life. To improve the performance of concrete, numerous new technologies and materials are adopted. Sulfate attack on concrete can be one of the problems in concrete durability especially for applications such as nuclear waste management. Low permeability concrete will last longer when exposed to environment that contains low concentration of sulfate ions usually found in many natural sulfate waters.

However, waters containing more than 2000 ppm sulfate are capable of entering in to destructive chemical reactions with cementing constituent of Portland cement based matrices. It has also been observed that sulfate attack is seldom the sole cause of concrete damage.

Cracking of concrete has a great influence on the permeability and, therefore, on durability to a variety of physical and chemical attacks. We must do everything possible in selecting materials, mix proportions, and concrete construction practice to prevent cracking of structures in service. This can be achieved by developing durability oriented recommended procedures for selection of materials and mix proportions.

DURABILITY DESIGN OF CEMENT BASED MATRICES

In order to mitigate deterioration, various concept of durability design have been proposed. Prolonged durability of concrete to sulfate attack can be achieved when in addition to a low C_3A content of Portland cement, adequate steps are taken to reduce the available $Ca(OH)_2$ in the hydrated cement paste through pozzolans addition to cement. This explains why the presence of pozzolans improves the durability of Portland cement concrete [10].

A low C_3A content, such as Type V Portland cement alone does not offer adequate protection against the long time sulfate attack of the acid type. The low permeability concrete showed no evidence of deterioration, although some of the concrete mixtures contain cement having relatively high C_3A content (12%-15%).

Changes in physical and chemical properties in the mortars with different replacement of cement by fly ash and silica fume when immersed in 2% H_2SO_4, 10% Na_2SO_4 and 10% $MgSO_4$ for 3 years were investigated by Torii and Kawamura [11].

The test data showed that the replacement of Portland cement by fly ash and silica fumes effectively improve resistance of mortar to the sulphuric acid and sulfate attack. Rasheedozzafar et al [12] observed that a 20 % microsilica-blended Type I(C_3A = 14%) cement shows 1.4 times better durability performance against attack by sulfate compared to a plain Portland Type V cement (C_3A = 1.88%) after 180 days of exposure to a highly accelerated and severe sulfate environment. In addition to C_3A content, the C_3S/C_2S of the cement has also observed to be a significant factor in determining sulfate resistance.

Tikalasky et al [13] observed that fly ash with high amount of CaO and amorphous C_3A increase the susceptibility of concrete to sulfate attack. When blended cement, silica fume and slag in particular, are used in light sulfate environment, especially when the sulfates are associated with magnesium cations, additional protective measures are necessary [14].

A 20% flyash blended cement concrete was observed much more resistant than the corresponding plain PC concrete in sodium sulfate solution (18000mg $MgSO_4$/l), and equal in performance to S.R.P.C. concrete [15].

A minimum glass content of the slag (61%) was found to be necessary for sulfate resistance of the blend[15] The performance of all blended cement, particularly those made with silica fumes were generally excellent in the Na_2SO_4 environment, but their performance in the $MgSO_4$ environment was not satisfactory [14].

The type of cement did not have any significant influence on the performance of either plain or blended cement in both environments. Several other researchers [10, 12, 16, 17] have confirmed that limitations of C_3A and C_4AF contents are not the ultimate answer to the problem of sulfate attack. The use of blended cement with flyash, silica fumes, blast furnace slag is therefore recommended in sulfate environment. However, performance of blended cement is influenced by (a) alumina content of blended material and/or of cement; (b) CaO content and amorphous calcium aluminate; and (c) quantum of blended material used.

Various researchers confirmed that cement paste and plain concrete tend to loss significant residual compressive strength at higher temperatures. The concrete with replacement of PC by 10% silica fumes appeared to be even worse than the concrete containing 100% PC [18]. The inclusion of a proportion of silica fume higher than 10% is to be avoided because of the risk of explosions during heating. The specimens containing flyash and slag perform significantly better than the specimens containing 100% PC or silica fumes at higher temperature [18].

RESEARCH NEEDS

Considerable research is ongoing to develop methods for safe disposal of nuclear wastes. The current state of our knowledge on concrete durability is lacking. The long-term durability of cement based matrices is primarily affected by sulfate attack. The review of current state of knowledge in the field of sulfate attack of cement based matrices reveals that further extensive research is needed to develop concrete containment vaults, for storing radioactive wastes, which will be durable for thousand of years.

The mechanism of sulfate attack on Portland cement is not fully understood [19]. There is no universally accepted hypothesis on ettringite formation. Crystal growth pressure from ettringite formulation and the swelling pressure from water adsorption by poorly crystalline ettringite are two of the several hypotheses that have been proposed. This has limited the confidence that can be placed in and hence the reliability of existing standard tests and models to predict performance and service life of cement concrete matrices subjected to sulfate attack. A more sophisticated theory is necessary to better understand the phenomenon of sulfate attack mechanism and to developing more reliable standard tests and models to predict performance and service life of concrete containment vaults.

Blended cements are increasingly used to improve the durability performance of Portland cement concrete. Data have to be developed to study the effect of cement composition on durability performance of plain cement concrete and blended cement concrete against sulfate attack. There is a concern as to the performance of cement concrete in sulfate environment, particularly in magnesium sulfate. The type of cement, Type I, II or, V to be used in blended cements for use in sulfate environments has not been clearly established.

The research has to be conducted to understand the interaction between the chemical and mineralogical composition of flyash and slag with the sulfate resistance of concrete containing fly ash or slag. Our understanding of fundamental mechanism of physio-chemical actions of silica fume in Portland cement concrete is inadequate.

Hence, research and field data on the effect of pozzolans on durability and service life are incomplete, or at last inconclusive. In general, the effect of one environmental variable at a time is known, and little is known about the synergistic effects of two or more variables present simultaneously, the actual combinations of environmental factors exposed during its service life, are often ignored in modelling the environment.

The effect of the conjoint presence of other chemicals along with sulfate on the sulfate resistance of hydrated Portland cement is inconclusive. Therefore, investigation has to be carried out to evaluate the mechanism of sulfate attack in environments characterised by the concomitant presence of other aggressive chemicals such as chloride salts.

Temperature effect on cement based matrices is one of the most important parameter for durability study. An assessment of the durability of concrete at higher temperatures has to be made in which both material and environmental factors are to be considered. The current test methods and specifications are in need of a critical examination and a thorough review.

Most laboratory methods aimed at studying the sulfate attack mechanism are based on accelerated tests. Numerous methods have been developed to accelerate sulfate attack in concrete. There is a need to compare the various methods for accelerating the sulfate attack for establishing the best method for predicting long-term performance and service life of cement based matrices used in radioactive waste immobilisation.

An appropriate property, or indicator, which will best represent the nature and mechanism of sulfate attack, has to be established. The continuous deterioration of the matrix due to sulfate attack can be modelled better by using a continuous damage theory. There is no consensus as to what values of loss of mass, expansion, and strength loss best represent failure. Hence, there is need to establish failure criteria. In response to the need for quantitative data on projected service life of containment structures, computer generated expert systems based on mathematical models can be developed. The three key elements of a reliable method for mathematical prediction of service life are;

(a) a precise definition of the material
(b) a precise definition of environment
(c) a dependable data base from accelerated test methods on durability

The currently available laboratory methods are inadequate for the purpose of predicting the field performance of concrete in a containment structure. The effect of pozzolans in cement based matrices is to be examined according to the failure criteria and the sulfate attack mechanism. The changes in microstructure of cement with pozzolans in the presence of sulfate solution are to be studied.

CONCLUSIONS

Cements are attractive matrices for the immobilisation of radioactive wastes. The supplementary blending materials are less well specified. Since these materials may comprise the bulk of many immobilisation systems, it is important that they receive adequate characterisation. The behavour of radio-nuclides is described in this paper.

The isolation of specific mechanisms of cement radio-nuclides interactions helps to quantify the immobilisation potential of cements. Sorption on C-S-H is a significant form of iodine immobilisation. The use of silver to precipitate iodine as AgI is theoretically ineffective under reducing conditions, which lead to its breakdown, with solubilization of Iodine. In the long-term, Sr is liable to substitute for Ca in cement phases. Urarium VI can react with cement to form solubility limiting compounds with Ca and Si.

Thermodynamic modelling of cement chemistry in higher temperature regimes, even for temporary temperature exercursions, is not very well advanced. Considerable experimental input is required before those systems can be treated with reasonable confidence. The primary factor adversely affecting the service life of cement based matrices is sulfate attack. Prolonged durability of concrete to sulfate can be achieved with a low C_3A content cement, by addition of pozzolans to cement and with low permeability concrete. Several other researchers have expressed that limitation of C_3A is not the ultimate answer to sulfate attack. There is a pressing need expressed by worldwide public concern to develop concrete containment vaults that will be durable for thousand of years. Hence, extensive research is needed to study the long-term effect of sulfate salts. Further research is necessary to help in obtaining the answers to numerous fundamental questions that have remained unanswered even after decades of research. Some of these questions that need to be specifically investigated, relate to the mechanism of sulfate attack, identification of an accelerated test, role of concomitant presence of other chemicals on durability, establishing the failure criteria and the modelling of cements in higher temperatures regimes.

REFERENCES

1. CLIFTON, J R AND KNAB, L I. Service life of concrete. Report from U.S. Department of Commerce, National Institute of Standards and Technology, NISTIR 89-4086, 1989, pp 119.

2. COHEN, M D AND MATHER, B. Sulfate attack on concrete - research needs. Journal ofthe American Concrete Institute, Vol 88, No 1, 1991, pp 62-68.

3. ATKINSON, A, GOULT, D J, and HEARNE, J A. An assessment of the longterm durability of concrete in radioactive waste repositories. Proceedings of the Material Research Society Symposium - Basis for Nuclear Waste Management IX, Vol 50, Stockholm, 1985, pp 239-246.

4. NAGRA. Project gewahr, nuclear waste management in Switzerland: feasibility studies and safety analysis. NAGRA NGB 85-09, Switzerland, 1985.

5. THOMSON, B G J AND BROYD, T W. Dry run 1: an initial examination of a procedure for the post closure radiological risk assessment of an underground disposal facility for radioactive waste. DOE/RW/86.076, Atkins ES, Leatherhead Surrey, 1986, pp 76.

6. GLASSER, F P et al. Immobilisation of radwaste in cement based matrices. DOE/RW/89, Aberdeen University, Scotland, 1989, pp 133.

7. ATKINS, M AND GLASSER F P. Application of Portland cement - based materials to radioactive waste immobilisation. Journal of Waste Management, Vol 12, 1992, pp 105-131.

8. ATKINS, M, BECKLEY, A N AND GLASSER, F P. Encapsulation of radioiodine in cementitious waste forms. Proceedings of the MRS Symposium (Boston), Pittsburgh, PA, Vol 176, 1990, pp 15.

9. TAYLOR, P, LOPATA, V J, WOOD, D D AND YACSHYN, H. Solubility and stability of inorganic iodides: candidate waste forms for iodine-129. Standard Technical Publication 1033, ASTM, Philadelphia, PA, 1989, pp 287.

10. GJORY, O E. Long time durability of concrete in sea water. Journal of the American Concrete Institute, Vol 68, 1971, pp 60-67.

11. TORII, K AND KAWAMURA, M. Effects of fly ash and silica fume on the resistance of mortar to sulfuric acid and sulfate attack. Cement and Concrete Research, Vol 24, No 2, 1994, pp 361-370.

12. RASHEEDUZZAFAR, FAHD, H D, SAAD AL-GAHTANI, A, A LSAADOUN, S S AND BADER, M A. Influence of cement composition on the corrosion of reinforcement and sulfate resistance of concrete. Journal of the American Concrete Institute, Vol 87, No 2, 1990, pp 114-122.

13. TIKALSKY, P S AND CARRASQUILLO, R L. Influence of fly ash on the sulfate resistance of concrete. Journal of the American Concrete Institute, Vol 89, No 1, 1992, pp 69-75.

14. AL-AMOUNDI, O S B, RASHEEDUZZAFAR, MASLEHUDDIN, M, AND ABDULJAUWAD, S N. Influence of chloride ions on sulfate deterioration in plain and blended cements. Magazine of Concrete Research Vol 46, No 167,1994, pp 123.

15. LAWRENCE, C D. Sulfate attack on concrete. Magazine of Concrete Research, Vol 42, No 153, 1990, pp 249-269.

16. BICZOK, I. Concrete corrosion, Concrete Protection, 8th Ed., Akademiai Kiado, Budapest, 1980.

17. AL-AMOUNDI, O S B, MASLEHUDDIN, M AND SAADI, M M. Effect of magnesium sulfate and sodium sulfate on the durability performance of plain and blended cements. Journal of the American Concrete Institute, Vol 92, Nol, 1995, pp 15-24.

18. SARSHAR, R AND KHOURY, G A. Material and environmental factors influencing the compressive strength of unsealed cement paste and concrete at high temperatures. Magazine of Concrete Research, Vol 45, No 162, 1993, pp 51-61 .

19. MEHTA, P K. Concrete technology at the crossroads - problems and opportunities. Proceedings of the International Seminar on Civil Engg. Practices in the Twenty l7'rst Century, Roorkee, India, 1996, Vol 2, pp 563-586.

THEME THREE:

HIGH-RISE STRUCTURES

Keynote Paper

INVESTIGATING SHEAR FAILURE IN REINFORCED CONCRETE STRUCTURES

M P Collins

E C Bentz

University of Toronto

Canada

ABSTRACT. The 1991 one-billion dollar loss of the Norwegian Sleipner offshore gas platform and the 1995 catastrophic collapse of the piers of the Hanshin Expressway in Japan provide evidence of the vulnerability of reinforced concrete structures to shear failure. Understanding the causes of such failures and developing techniques to predict the shear strength of reinforced concrete structures have been the objectives of a long-term research project at the University of Toronto. This paper will summarize some of the major results of this project by illustrating the application to actual structures, of the analytical tools that have been developed.

Keywords: Reinforced concrete, Structures, Shear strength, Design, Analysis, Failure, Forensic analysis, Software.

Professor Michael P Collins is Bahen-Tanenbaum Professor of Civil Engineering at the University of Toronto, Canada, where he has led a long-term research project aimed at developing rational procedures for the shear design of reinforced concrete structures. He serves on numerous code committees and, in addition to writing many technical papers, has co authored two textbooks on the structural behaviour of reinforced and prestressed concrete structures.

Mr Evan C Bentz is a Ph.D. candidate at the University of Toronto, Canada. His research interests include the numerical modelling of reinforced concrete, with an emphasis on shear critical cases. As part of his thesis he has written a number of programs for the analysis of reinforced concrete, which have been placed in the public domain, including Response-2000 for beams/columns and Membrane-2000 for membrane elements.

INTRODUCTION

Reinforced concrete is the most widely used structural material of the twentieth century because it economically combines the ability of concrete to resist high compressive stresses with the ability of reinforcing steel to resist high tensile stresses. Without reinforcing bars a concrete beam would collapse at a relatively low load as soon as the tensile stresses reach the cracking stress of the concrete. Appropriate reinforcing bars will control the opening of the cracks and will make it possible for the beam to resist five, ten, or even twenty-five times more load. In the structural design of reinforced concrete structures the essential task is to decide where such reinforcement needs to be placed and what quantities of reinforcement should be used so that the structure has appropriate margins of safety against collapse.

In designing for flexure engineers have available a simple, general, rational method, called the "plane sections" theory, capable of predicting not only the flexural strength, but also the complete load-deformation response of reinforced concrete sections. Because of this, there is little disagreement between different engineers, or different design codes as to the flexural strength of a given reinforced concrete structure, or the quantities of reinforcement needed to ensure ductile flexural behaviour. There is however, substantial disagreement as to the magnitude of the shear strength of structural elements and the reinforcement requirements needed to ensure ductile shear response. Further, rather than the simple, general, behavioural theory available for flexural design, the building code procedures for shear design typically consist of a collection of complex, restricted, empirical equations for shear strength. In view of the disparity between the state-of-the-knowledge in flexure and the state-of-the-knowledge in shear, it is perhaps not surprising that, while failures of reinforced concrete structures due to deficiencies in flexural design are extremely rare, failures due to deficiencies in shear design occur much more frequently.

The shear failures of the warehouse roof beams, see Figure 1(a), at the Wilkins Air Force Depot in Shelby, Ohio, which occurred in 1955, are still of considerable importance because many of the current shear design rules are based on research that was conducted as a result of these failures. These large, lightly reinforced beams failed at a shear stress of about 0.5 MPa, which was only about one-half of the failure shear stress implied by the design procedures used at the time.

(a) Air Force warehouse roof beams

(b) Hanshin Expressway piers

Figure 1 Shear failures

More recent examples of shear failures include the August 1991 loss of the Sleipner offshore gas platform. Here, a reinforced concrete wall 550 mm thick failed in shear during deck-mating causing the platform to be reduced to rubble and resulting in a total economic loss of nearly one billion dollars. During the January 1995 Kobe earthquake several hundred catastrophic shear failures of reinforced concrete structures occurred. The most widely reported of these was the shear failures and resulting collapse of the piers of the Hanshin Expressway. See Figure 1(b).

During the last two decades, a considerable amount of research has been conducted worldwide with the aim of developing behavioural models for reinforced concrete in shear, comparable in rationality and generality to the plane sections theory for flexure [1]. One group of such models is based on a collection of assumptions about the behaviour of reinforced concrete that is known as the "modified compression field theory". This paper will summarize the key aspects of this theory and will illustrate the use of models based on this theory in investigating shear failures.

MODIFIED COMPRESSION FIELD THEORY

Perhaps one of the reasons why the "shear problem" was so difficult to understand was that the traditional type of shear experiment conducted in research laboratories, while simple to perform, was difficult to analyze. In such a test (see Figure 2) the behaviour of the member changes from section to section along the shear span and also changes over the depth of the beam. Thus, for example, if a relationship is sought between the magnitude of the shear force and the strains in the stirrups, it will be found that the strains are different for every stirrup and also differ over the height of the stirrup. In developing the modified compression field theory, experiments were conducted on elements subjected to uniform stresses (see Figure 2). While these tests were more difficult to perform, they were easier to analyze.

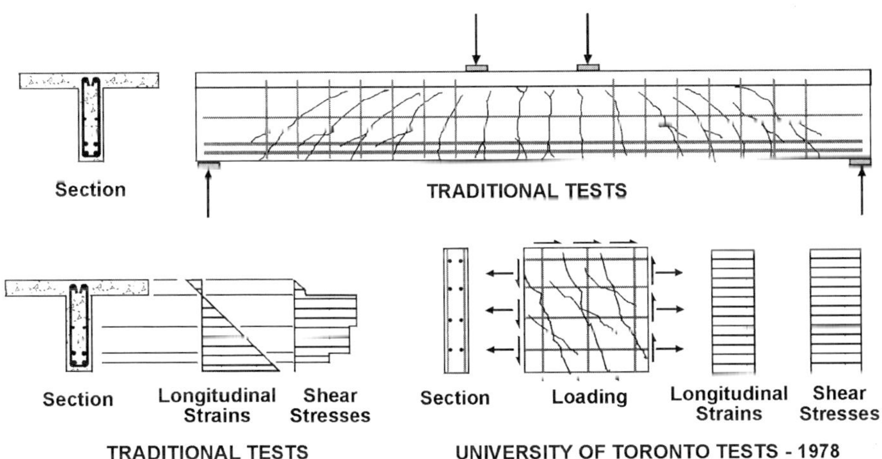

Figure 2 Testing reinforced concrete in shear

The modified compression field theory was developed by observing the load-deformation response of a large number of reinforced concrete membrane elements loaded in pure shear in the University of Toronto's membrane element tester [2,3] and shell element tester [4]. Figure 3 compares the calculated and observed response for one of these elements, which was called SE6 [4]. The problem addressed by the modified compression field theory is to predict the relationship between the shear stress applied to such an element and the resulting shear strain.

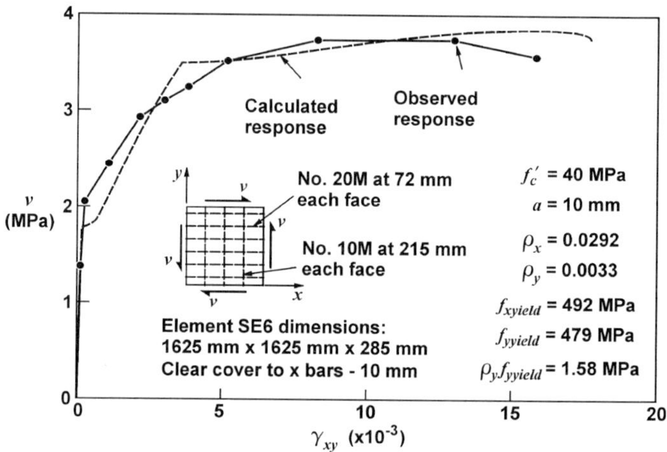

Figure 3 Comparison of calculated and observed shear response of membrane element SE6

Cracked reinforced concrete transmits shear in a relatively complex manner involving opening and closing of pre-existing cracks, formation of new cracks, interface shear transfer at rough crack surfaces, significant tensile stresses in the cracked concrete, and great variation of local stresses in both the concrete and reinforcement from point to point in the cracked concrete, with the highest reinforcement stresses and the lowest concrete tensile stresses occurring at crack locations. The modified compression field theory attempts to capture the essential features of this behaviour without considering all of the details. In lieu of following the complex stress variations in the cracked concrete, only the average values of the stresses (that is, stresses averaged over a length greater than the crack spacing) and the stresses at the crack locations are considered.

Figure 4 summarizes the equilibrium, compatibility and stress-strain relationships used by the modified compression field theory. In these relationships θ is the angle between the x axis and the direction of the principal compressive average strain. Note that these average strains are measured over base lengths that are greater than the crack spacing. The basic simplifying assumption of both the compression field theory [5] and the modified compression field theory [6] is that "the direction that is subjected to the largest average compressive stress will coincide with the direction that is subjected to the largest average compressive strain." For specified applied loads, the angle θ, the average stresses and the average strains can be determined by solving the given equilibrium equations in terms of average stresses, the given compatibility equations in terms of average strains and the given average stress-average strain relationships.

The maximum shear stress that the element can resist may be governed not by the average stresses, but rather, by the local stresses at the crack locations. In checking the conditions at a crack, the actual complex crack pattern is idealized as a series of parallel cracks, all occurring at angle θ and spaced at a distance s_θ apart. It is assumed that for crack widths greater than

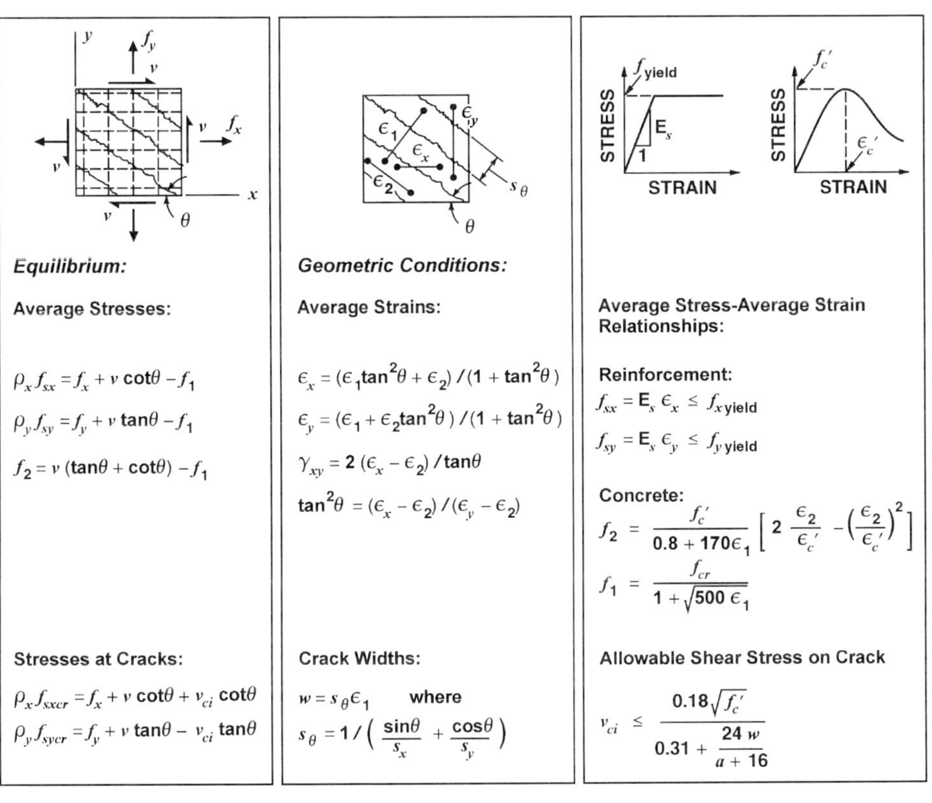

Figure 4 A summary of the relationships used in the modified compression field theory

about 0.05 mm no significant tensile stresses can be transmitted normal to the crack. However, shear stresses, v_{ci}, can be transmitted across the crack. The maximum possible value of v_{ci} is assumed to be related to the crack width, w, and to the maximum aggregate size, a.

Solving all the equations given in Figure 4 is of course very tedious if performed by hand but is quite straightforward with a programmable calculator or other small computer. A short (about 300 lines of BASIC) program called MEMBRANE, which performs these calculations is given in the textbook *Prestressed Concrete Structures* [7]. Program RESPONSE, which is also given in this textbook, is somewhat more convenient to use than program MEMBRANE and can be used to solve for membrane elements provided that there is no applied axial stress in the y direction. The results obtained from program RESPONSE for element SE6 are plotted in Figure 3.

PREDICTING THE SHEAR RESPONSE OF BEAMS

The modified compression field theory is a procedure for predicting the stress-strain response of elements of reinforced concrete. In using this procedure to predict the response of a beam a number of additional assumptions will usually be required.

Engineering beam theory assumes that plane sections remain plane and hence, can study the response of a beam section-by-section, without being concerned with the details of how the forces are introduced into the member. The modified compression field theory has been used as the basis for a number of such sectional models, the most powerful of which is called the "dual-section" model [8,9] and forms the basis for program Response-2000. In this analysis, the biaxial stresses and strains and the manner in which they vary over the height of the beam are considered. It is found that the inclination, θ, of the principal compressive stress changes continuously over the height of the beam, becoming steeper near the flexural tension face and shallower near the flexural compression face. By considering two adjacent cross sections the model can determine the distribution of shear stresses over the cross section. It is found that as failure approaches, there can be a considerable redistribution of shear stresses resulting in higher shear stresses near the stiffer flexural compression side and lower shear stresses near the more flexible flexural tension side. Response-2000 is available for downloading from the World Wide Web at the following address: http://www.ecf.utoronto.ca/~bentz/r2k.htm

It is important to note that in the modified compression field theory the shear stress which can be resisted by cracked reinforced concrete is related to the crack widths. Larger beams have wider cracks and hence, they may fail at lower shear stresses. The observation that large, lightly reinforced beams fail at lower shear stresses than geometrically similar smaller members is called the "size effect" in shear. This size effect, which is predicted by the modified compression field theory, is illustrated in Figure 5, which summarizes the results of an extensive experimental program conducted in Japan by Shioya et al. [10,11]. It can be seen that, for these lightly reinforced (0.4% of longitudinal reinforcement, no stirrups), uniformly loaded, simple span beams, the shear stress at failure decreases, both as the member depth Increases and as the maximum aggregate size decreases. For these particular beams, increasing the effective depth of the beam from 200 mm to 3 m decreased the shear stress at failure by a factor of about 3.

In the evaluation and design of reinforced concrete beams the engineer often uses expressions for shear strength that are typically based on tests of relatively small specimens. For example, the basic American Concrete Institute [12] expression for the "concrete contribution", V_c, was derived in 1962 [13] from the 194 test results shown in Figure 6, where, for MPa units,

$$V_c = 0.158\sqrt{f'_c}\, b_w d + 17.24\rho_w \frac{Vd}{M} b_w d \leq 0.291\sqrt{f'_c}\, b_w d \qquad (1)$$

The average depth of these 194 beams was 340 mm, while the average amount of flexural tension reinforcement was 2.2%. The above expression for V_c implies that the shear stress at failure will be greater than $0.158\sqrt{f'_c}$. However, some of the large, lightly reinforced beams shown in Figure 6 failed at shear stresses that were only about 50% of this value.

The 1962 ACI-ASCE Committee 326 recommendations for changes to the ACI shear design provisions were initiated because of the August 1955 collapse [14] of the Wilkins Air Force Depot warehouse in Shelby, Ohio and the 1956 collapse of a similar warehouse at the Warner-Robbins Air Force Base in Georgia. These collapses were caused by the failure of 915 mm deep beams which, at the failure locations, did not contain stirrups, and only had 0.45% of longitudinal reinforcement. See Figure 1 (a). The beams failed at a shear stress of only about 0.5 MPa. At this time, the sensitivity of the failure shear stress, of this type of member, to size was not recognized and so, in investigating the failures at the Portland Cement Association (PCA) the research engineers chose to use one-third scale models. The PCA experiments [15] indicated that the 305 mm deep beams failed at a shear stress of about 1 MPa; that is, about twice the failure shear stress of the prototype. When an axial tensile stress of about 1.4 MPa was applied to the model beams the shear stress at failure was reduced by about 50%. The PCA engineers concluded that the presence of tensile stresses, in the prototype beams, caused by the restraint of shrinkage and thermal movements was the reason why the beams had failed at such low shear stresses. It so happens that the beams tested by Shioya et al., and shown in Figure 5, had rather similar characteristics to the Air Force warehouse beams.

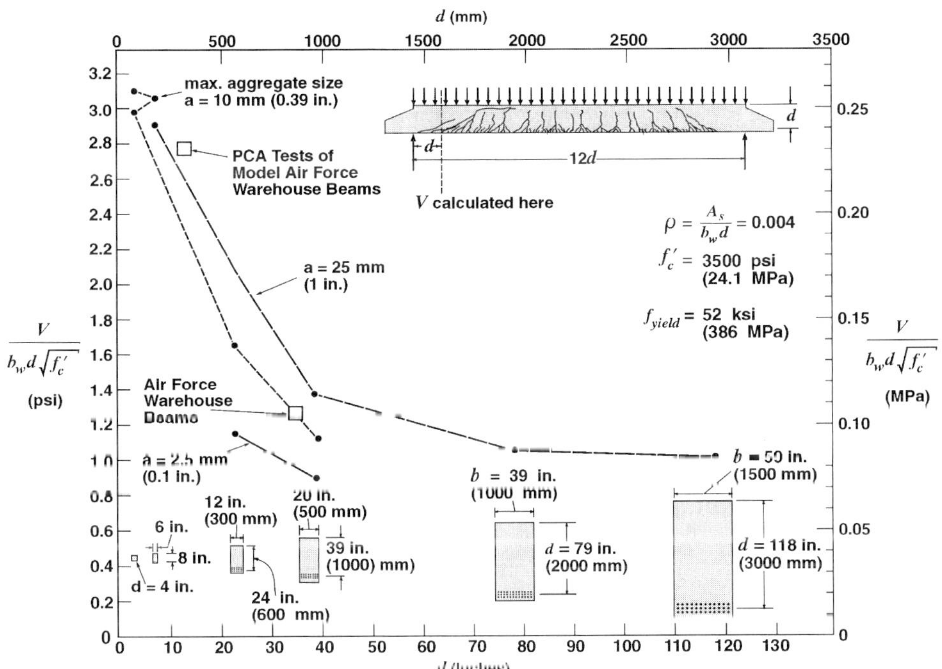

Figure 5 Influence of member depth and maximum aggregate size on shear stress at failure

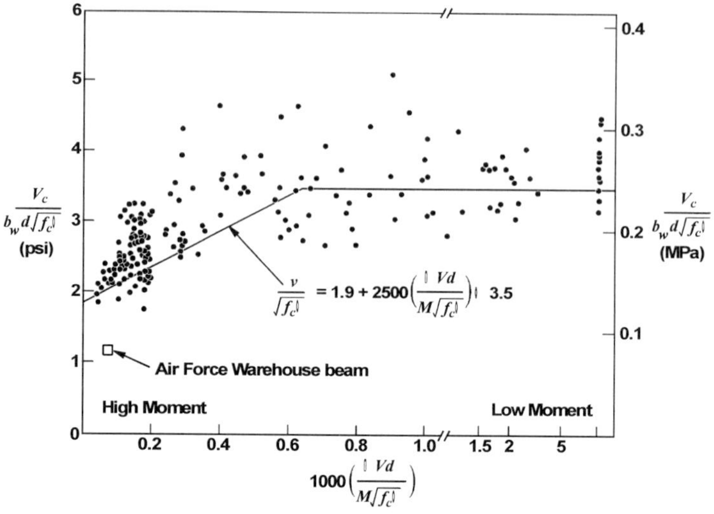

Figure 6 Derivation of ACI expression for diagonal cracking shear V_c (Adapted from [13])

The failure shear stress for the prototype warehouse beams and the PCA model beams without axial tension have also been plotted in Figure 5. From this figure it seems clear that the 50% reduction in failure shear stress between the model and the prototype for the Air Force beams was primarily due to the size effect in shear, rather than the influence of axial tensile stresses. Program Response-2000 predicts that the full size Air Force warehouse beams would fail in shear at a shear stress of about 0.6 MPa, even if there was no axial tensile stresses caused by restraint.

SHEAR FAILURE OF THE HANSHIN EXPRESSWAY PIERS

A structure can survive a severe earthquake if it is able to continue supporting its own weight while undergoing the substantial horizontal deformations caused by the ground displacements. Thus, bridge piers supporting an elevated roadway, such as the Hanshin Expressway, may need to be able to undergo lateral displacements equal to about 3% of their height while still supporting the weight of the roadway. Such large deformations are possible if the bending moment in the pier causes the flexural reinforcement to yield before the shear force causes a shear failure. In an appropriately designed member, yielding of the flexural reinforcement results in the formation of a "plastic hinge", which permits large deformations to occur while maintaining high load carrying capacity. A shear failure, on the other hand, will usually result in a bridge pier losing its load carrying capacity at quite small deformations.

The 3.1 m diameter, lightly reinforced piers of the Hanshin Expressway, which failed in the 1995 Kobe earthquake, were designed in the late 1960's. At this time, shear design used expressions such as Equation (1), which were based on testing specimens about one-tenth the size of the Hanshin piers. Because of the size effect in shear discussed above, these design expressions were unconservative for these large piers.

Figure 7 illustrates the predicted lateral load carrying capacity of the Hanshin piers. It is estimated that prior to the earthquake the axial compression, N, supported by the pier was about 14000 kN. During the earthquake, this vertical load on the pier would vary and hence, in Figure 7 a range of loads from zero to 28000 kN is considered. When the axial compression is 14000 kN it would take a lateral force, V, of 11500 kN to cause a flexural hinge to form at the weak section, B, near the base of the column. Unfortunately, before this flexural hinge can form, a shear failure is predicted to occur. The section model, Response-2000, predicts that this shear failure would occur at a lateral load of 8370 kN. A more elaborate analysis, using a non-linear finite element program called TRIX [16], which is based on the modified compression field theory, predicts shear failure at a lateral force of 9900 kN. The TRIX analysis predicts that failure will occur, with a diagonal band of cracking inclined at about 18° to the vertical axis of the piers, when the lateral displacement at the top of the piers is about 65 mm. This failure displacement is only 0.54% of the pier height indicating that the pier does not have the ability to sustain the deformations imposed by the large earthquake. See Figure 1 (b).

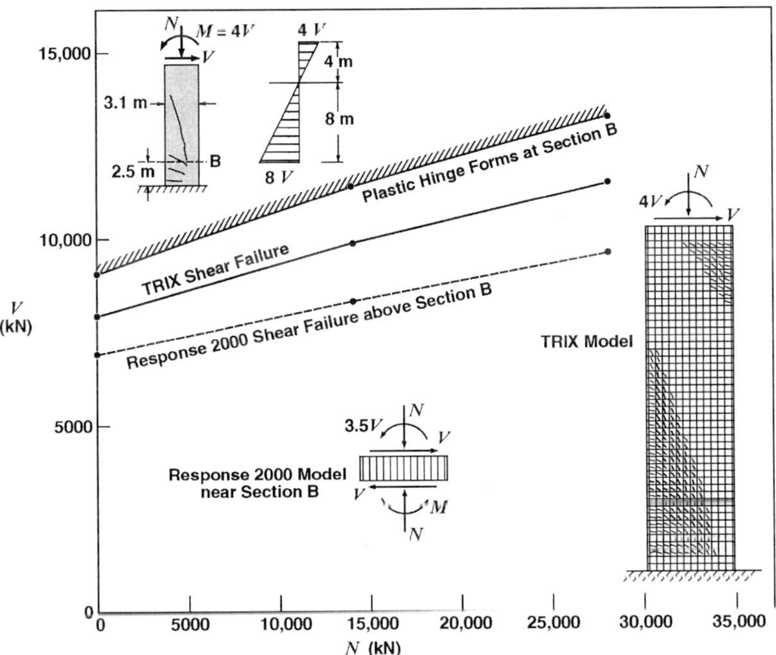

Figure 7 Predicted lateral capacities of Hanshin Piers

INVESTIGATING THE SHEAR FAILURE OF A CONCRETE OFFSHORE STRUCTURE

One significant advantage of rational design models is that they can be used to approach unusual design problems. Deciding whether the heavily loaded walls of concrete offshore structures need to contain stirrups is an example of such a problem.

Whereas, for land based structures, the sectional forces are usually determined by analyzing an equivalent frame, for offshore structures it has become the practice to find these forces using an elastic finite element analysis. The loading demand placed on a particular location of the structure can be expressed in terms of eight stress resultants, that is, three membrane forces (N_x, N_y and V_{xy}), three moments (M_x, M_y and T_{xy}), and two transverse shear forces (V_{xz} and V_{yz}).

Note that these stress resultants are obtained by integrating the stresses obtained from the finite element analysis over the thickness of the element and hence, they are expressed in terms of force or moment per unit width (e.g., $N_x = 5000$ kN/m). Even with the aid of modern super-computers, finding these stress resultants for several thousand locations in the structure, for several hundred different loading cases, is a major engineering challenge. Recent finite element models typically involve more than 500,000 degrees of freedom and formulating and checking such large models takes a substantial team of engineers many months.

After the sectional forces are calculated the required reinforcement at each location in the structure must be determined. It must be demonstrated that the section has adequate strength and also satisfactory service load performance (e.g., the crack widths must remain suitably small). Again, this is a major task with considerable difficulties being encountered in extending traditional simple beam procedures to the more complex situations encountered in shell structures. It does not help that in this problem we need to consider the triaxial stress-strain response of cracked concrete.

In the last few years a shell section analysis program called SHELL474, which can be regarded as a generalization of program RESPONSE, has been developed [17]. This program was first formulated as part of a verification study [18] of the new Canadian code for concrete offshore structures, CSA S474-M89 [19]. It is hoped that procedures such as those incorporated in SHELL474 will replace the existing "equivalent beam" shear design procedures, which are believed to be unsatisfactory.

The shear failure of the Sleipner A platform in August 1991, as it was being ballasted down prior to deckmating, dramatically demonstrated some of the deficiencies of existing design practice. A wall in one of the water-filled interior tricells (see Figure 8) failed, allowing water to rush into a drillshaft. The emergency de-ballasting procedure could not keep up with the water flow and hence, the structure sank. As it went deep into the fjord, the buoyancy cells imploded, totally destroying the structure.

The wall that failed was 550 mm thick, spanned about 4.4 m and was subjected to water pressure corresponding to about 67 m of water (i.e., 670 kN/m^2). At the elevation where it failed the wall did not contain any stirrups (see Figure 8). In addition to the high shears and high moments caused by the wall acting as a beam spanning across the opening, the wall was also subjected to high axial compression resulting from the inward push of the water pressure on the exterior walls. The wall did not contain stirrups because the finite element analysis seriously underestimated the magnitude of the applied shear, while the sectional design procedure overestimated the beneficial effects of the axial compression on the shear strength of the wall.

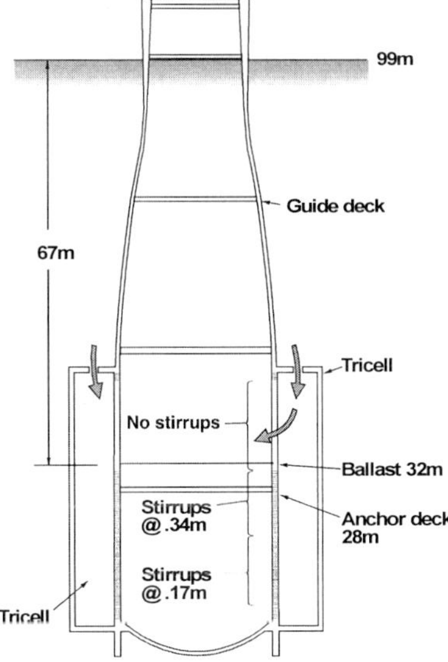

Figure 8 Cross-section of drill shaft D3 of Sleipner A platform

When a section is subjected to a loading in which both the shear and the axial compression are increased at the same time, the predicted failure load is very sensitive to the assumed interaction between axial load and shear force. This is illustrated in Figure 9, which compares the predictions of two sectional models, one based on the ACI Code and the other calculated by program SHELL474, which is based on the CSA offshore code. For the loading ratio shown, the ACI procedure results in a predicted failure load about three times higher than the CSA procedure.

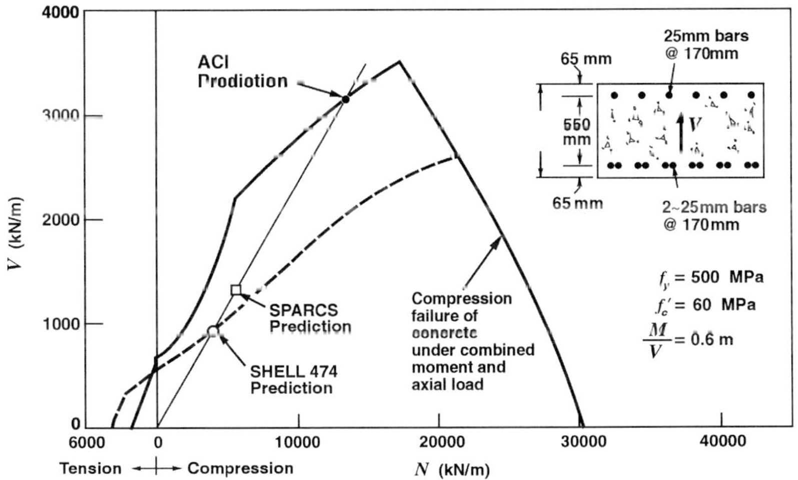

Figure 9 Shear force - axial load interaction diagram for wall section

In the weeks following the Sleipner accident, use was made of a number of the modified compression field theory models to study the cause of failure [20]. For example, Figure 10 shows the results of a number of studies using the three-dimensional, non-linear finite element program SPARCS [21]. These studies were aimed at determining how the failure pressure and failure mode changed as the length of the T-headed bar across the throat of the tricell changed. The studies indicated that failure would have been avoided if either the T-headed bar had been longer or if stirrups had been provided.

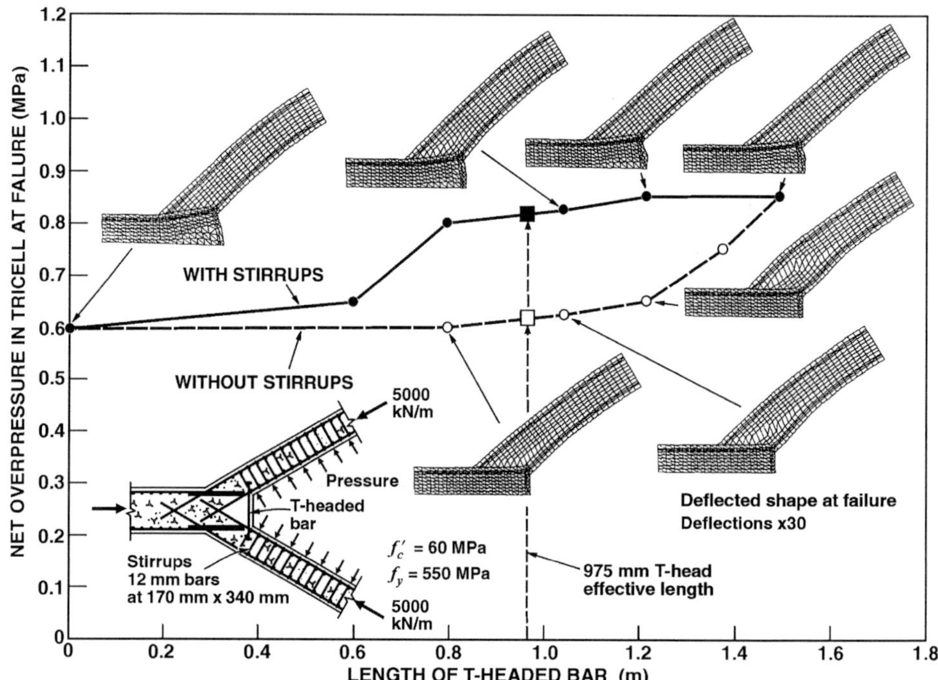

Figure 10 Influence of stirrups and length of T-headed bar on failure pressure and failure mode of Sleipner A

INVESTIGATING A POTENTIAL SHEAR FAILURE OF A HIGH RISE APARTMENT BUILDING

The final example of the use of analytical models based on the modified compression field theory as forensic tools in structural investigations concerns the high rise apartment building shown in Figure 11. The upper floors of this structure were supported by 1.2 m deep beams at the second floor level. The ends of these beams, which were visible from the outside of the building, developed wide diagonal cracks. Analysis of the structure revealed that the prime cause of these shear cracks was the differential settlement of the mat foundation. Further analysis of the interior portions of the beams predicted that a more dangerous situation existed in parts of the beams that were not visible. See Figure 12. When the ceiling was removed and these portions were examined, wide shear cracks were discovered in the locations and in the orientation predicted. A temporary support system was immediately provided.

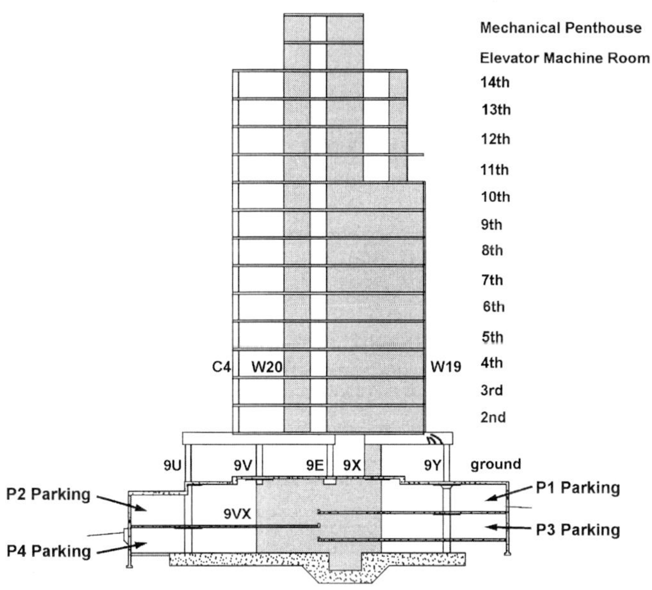

Figure 11 Visible shear cracks in exterior portion of second floor beams of high-rise apartment building

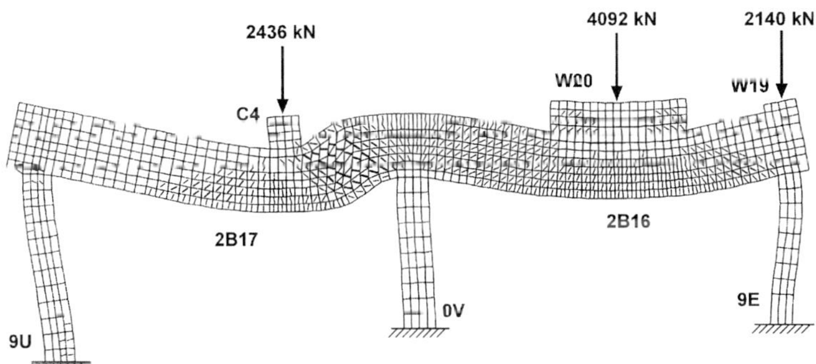

Figure 12 Predicted deformations magnified by 100 and crack patterns for interior portion of second floor beams of high-rise apartment building at load factor of 1.20

CONCLUDING REMARKS

This paper has demonstrated the use of analytical models based on the modified compression field theory in investigating potential shear failures in reinforced concrete structures.

The August 1955 Air Force warehouse failures demonstrated the deficiencies of the shear design procedures of that time, which neglected the influence of both axial load and member size. It was perhaps unfortunate that the investigations that followed concentrated on the detrimental influence of axial tension and continued to neglect the detrimental influence of large size. As a result, the shear provisions developed in the 1960's gave a large penalty to members with axial tension and a large bonus to members with axial compression. The failure of the Sleipner platform in August 1991 demonstrated that members designed by such provisions could be unsafe. The failure of the Hanshin Expressway piers resulted from shear design procedures that neglected the size effect. The models based on the modified compression field theory remove these deficiencies and bring the state-of-the-art of shear design closer to that now existing for flexural design.

REFERENCES

1. ASCE-ACI COMMITTEE 445. Recent Approaches to Shear Design of Structural Concrete. ASCE, Journal of Structural Engineering, Vol. 124, No. 12, 1998, pp. 1375-1417.

2. VECCHIO, F.J., and COLLINS, M.P. The Response of Reinforced Concrete to In-Plane Shear and Normal Stresses. Publication No. 82-03, Department of Civil Engineering, University of Toronto, 1982, 332 pp.

3. BHIDE, S.B. and COLLINS, M.P. Influence of Axial Tension on the Shear Capacity of Reinforced Concrete Members. ACI Structural Journal, Vol. 86, No. 5, 1989, pp. 570-581.

4. KIRSCHNER, U. and COLLINS, M.P. Investigating the Behaviour of Reinforced Concrete Shell Elements. Publication No. 86-9, Department of Civil Engineering, University of Toronto, Sept., 1986, 209 pp.

5. COLLINS, M.P. Towards a Rational Theory for RC Members in Shear. Journal of the Structural Division, ASCE, Vol. 104, 1978, pp. 649-666.

6. VECCHIO, F.J. and COLLINS, M.P. The Modified Compression Field Theory for Reinforced Concrete Elements Subjected to Shear. ACI Journal, Vol. 83, No. 2, 1986, pp. 219-231.

7. COLLINS, M.P. and MITCHELL, D. Prestressed Concrete Structures. Prentice Hall, Englewood Cliffs, 1991, 766 pp.

8. COLLINS, M.P. Reinforced Concrete in Combined Shear and Flexure. Proceedings, Mark Huggins Symposium, University of Toronto, 1978, pp. 154-171.

9. VECCHIO, F.J. and COLLINS, M.P. Predicting the Response of Reinforced Concrete Beams Subjected to Shear Using Modified Compression Field Theory. ACI Structural Journal, Vol. 85, No. 3, 1988, pp. 258-268.

10. SHIOYA, T., IGURO, M., NOJIRI, Y., AKIYAMA, H., and OKADA, T. Shear Strength of Large Reinforced Concrete Beams. Fracture Mechanics: Application to Concrete, SP-118, American Concrete Institute, Detroit, 1989, 309 pp.

11. SHIOYA, T. Shear Properties of Large Reinforced Concrete Member. Special Report of Institute of Technology, Shimizu Corporation, No. 25, 1989, 198 pp.

12. ACI COMMITTEE 318. Building Code Requirements for Structural Concrete (ACI 318-95) and Commentary ACI 318R-95. American Concrete Institute, Detroit, 1995, 369 pp.

13. ACI-ASCE COMMITTEE 326 Shear and Diagonal Tension. ACI Journal, Vol. 59, 1962, pp. 1-30, 277-344, and 352-396.

14. ANDERSON, B.G. Rigid Frame Failures. ACI Journal, Vol. 53, 1957, pp. 625-636.

15. ELSTNER, R.C. and HOGNESTAD, E. Laboratory Investigation of Rigid Frame Failure. ACI Journal, Vol. 53, 1957, pp. 637-668.

16. VECCHIO, F.J. Nonlinear Finite Element Analysis of Reinforced Concrete Membranes. ACI Structural Journal, Vol. 86, No. 1, 1989, pp. 26-35.

17. ADEBAR, P.E. and COLLINS, M.P. Shear Design of Concrete Offshore Structures. ACI Structural Journal, Vol. 91, No. 3, 1994, pp. 324-335.

18. COLLINS, M.P., ADEBAR, P. and KIRSCHNER, U. SHELL474 - A Computer Program to Determine the Sectional Resistance of Concrete Structures in Accordance with CSA Standard S474-M89. Canadian Standards Association, Rexdale, Canada, 1989, 32 pp. plus appendices.

19. CANADIAN STANDARDS ASSOCIATION COMMITTEE S474. Preliminary Standard S474-M1989 Concrete Structures. Part IV of the Code for the Design, Construction, and Installation of Fixed Offshore Structures. CSA, Rexdale, Canada, 1989, 39 pp.

20. COLLINS, M.P., VECCHIO, F.J., SELBY, R.G. and GUPTA, P.R. The Failure of an Offshore Platform. ACI Concrete International, Vol. 19, No. 8, 1997, pp. 29-35.

21. VECCHIO, F.J. and SELBY, R.G. Toward Compression-Field Analysis of Reinforced Concrete Solids. Journal of Structural Engineering, ASCE, Vol. 117, No. 6, 1991, pp. 1740-1758.

CONSTRUCTION OF TALL RCC CHIMNEYS IN DESERTS

S C Patodiya

Rajasthan State Electricity Board

India

ABSTRACT. A RCC multi-flue chimney of 220m height has been constructed for a thermal power plant in the Indian Thar Deserts. This desert area is characterised by highly undulated ground with unstable and shifting sand dunes having collapsible soil structure, high compressability and low load bearing capacities. Cohesion intercept 'C' value for these soils is zero and they have higher settlement values. Hard stratum and subsoil water is not met upto explored depth of 50m. This desert area faces extreme weather conditions with high air temperature going upto 50^{0}C, low relative humidity, high wind velocity and dust storms. Construction of an RCC structure of a chimney of 220m height in such conditions meeting durability and performance requirements was a challenging task which was met with innovation and responsible application of concrete construction technology with conventional materials and production methods taking adequate precautions in design, construction planning and execution. Construction of tall structures like a chimney requires high performance concrete which is high in strength, smooth, crack free, impermeable and can resist factors affecting deterioration of concrete such as hot weather, rapid fluctuations in temperature, alkali silica reactions, chloride & sulphate attack etc. Such concrete is capable of maintaining high sustained stresses for a much longer durations with high modulus of Elasticity, low creep and low shrinkage and thus meets the performance level expected.

Keywords: High performance concrete, Tall structures, Slip form, Extreme weather concreting, Practical issues.

Mr S C Patodiya is a Superintending Civil Engineer with Rajasthan State Electricity Board, India. He specialises in concrete construction with experience of more than 25 years in the construction of infra-structure projects. He has published 35 papers on concrete construction technology, many of which have been widely appreciated and awarded with merit certificates.

INTRODUCTION

The site for the Suratgarh Thermal Power Project is in the 'Thar-Deserts', the origin of which is attributed to a long, continued and extreme degree of aridity of the region, combined with the sand drifting action of the south-west monsoon winds which sweeps through Rajasthan without precipitating any part of their contained moisture. The soils have gone on accumulating year after year into the form of dunes due to continuous transportation of the surficial sands all the year round by the action of the wind. The chimney is a very heavily loaded structure and it is critical from a geotechnical as well as structural design point of view. The construction of a huge chimney structure in deserts is highly complex in nature, involving large quantity of concrete and needs several special precautions.

SALIENT FEATURES

1.	Height of chimney	220m
2.	No. of flues	2 Nos. of 4800mm dia each (Figure-2)
3.	Dia of RCC shell at bottom	28000/ 26000mm
4.	Dia of RCC shell at top	15800/ 15000mm
5.	Shell thickness	1000mm at bottom 400mm at top
6.	Shell configuration	Conical from bottom upto (+) 115m Cylindrical above (+)115m
7.	Foundation	42m dia & 4m thick raft foundation at 12m depth from G.L.
8.	Construction technology	Slip-form method

GEOTECHNICAL & FOUNDATION FEATURES

Wind blown (aeolian) sand classified as metastable or collapsible soil and generally dry with low degree of saturation is the principal deposition in the area. This sand is highly compressible and its behaviour is not governed by the normal laws of soil-water relationship. It goes through radical re-arrangement of particles upon wetting with or without load application. It has no clay content and has only about 10% silt. Through careful planning it was ensured that the chimney foundation do not bear on the unstable or loose portion of the dunes. A raft foundation was the only suitable foundation scheme in this desert area. (Figure-1 and 3).

Main parameters used for foundation design are as under :

Cohesion intercept 'C'	0
Angle of internal friction	32 degree
Bulk density of soil	1.65 gm/cc
Permissible settlement	25mm
Factor of safety	2.5
Water table correction factor R_W	0.6
Safe bearing pressure at 12m depth	550 kPa

MATERIALS & CONCRETE MIX

Good quality concrete requires stringent control on grading & uniformity of materials such as cement and aggregates so that it meets the requirement of workability, strength and durability. Grade of concrete was M-20 for foundation and M-25 for shell as per Indian standard.

Cement

Special cement having specific surface area not less than 3600cm^2/gm and initial setting time more than 50 minutes was used for concreting of the shell. For foundation work ordinary portland cement of 43 grade was used. Cement from a single source was used for the entire work.

Aggregates

The project site is located in deserts with no source of aggregates either from flowing river or from deposits of rock in a nearby area. Aggregates, see Table 1, were arranged from long distance of more than 250km. Uniformity in the quality of aggregates throughout the execution of the work was ensured. For ensuring durability to the chimney's concrete, alkali aggregate reactivity was tested by an accelerated method and a chemical method, the aggregates were found to be within permissible limits.

Table 1 Physical properties of aggregates

FINE AGGREGATES			COARSE AGGREGATES		
Fineness modulus	:	2.8	Water absorption	:	0.146%
Specific gravity	:	2.471	Specific gravity	:	2.68
Silt content	:	4.0%	Soundness	:	0.36%
			Impact value	:	6.45%

Admixtures

Superplasticisers are a great boon to preparation of high performance concrete which dramatically enhances flow properties and workability without affecting strength even with a low water-cement ratio. Lignosulphonates based water reducing admixtures (super plasticisers) in liquid form were used to reduce water demand and to increase the workability of concrete.

Concrete Mix Design

Table 2 gives details of the proportions of teh concrete mix constituents used.

Table 2 Details of concrete mix

MATERIAL	CONCRETE MIX (M-20) FOR FOUNDATION	CONCRETE MIX (M-25) FOR SHELL
Cement (in kg. per m³ of concrete)	360	400
Maximum aggregate size (mm)	20	20
Mix proportion by weight	1:1.94:3.12	1:1.65:2.65
Water-cement ratio	0.50	0.50
Admixture (ml per 50kg of cement)	120	120

CONSTRUCTION FEATURES

Equipments used for mixing, handling and placement of concrete were simple and readily available conventional type. Slip-forming technology was used for the RCC shell. High performance concrete cannot be produced merely by increased addition of cement alone. It requires greater attention on better grading of aggregates, good control on production, transportation and placement, lower water-cement ratio, improved compaction, adequate curing etc. For this, proportioning, batching and mixing of concrete was arranged close to the placement point to deliver the concrete at site and place in position in the minimum time.

The foundations for the chimney was provided at a depth of 12m in the form of 42m dia. and 4m thick raft (Figure-1). Total quantum of concrete in raft foundation was 4950m³. It was difficult to place so much concrete in single continuous operation, therefore, it was divided in 9 segments (Figure-3) of which 8 outer segments had 539m³ of concrete each and the ninth central segment had 638m³ concrete. For construction joints between each foundation segment, the surface was chipped, roughened and provided with epoxy based jointing compound. Besides this, continuity of reinforcement was ensured. The concrete surface was covered with gunny bags immediately after completion of each foundation segment to prevent evaporation of moisture from the surface of concrete and wet curing started after 8-10 hours from concreting.

Concrete was cast in the RCC shell through adopting slip-forming technology, 8 batching points with 8 concrete mixers and 8 winches were provided uniformly at the periphery of the shell. The characteristics of the concrete have a significant effect on the slip-forming operations as slip-form shutters slides on prematured concrete. Proportions of fine aggregates was kept higher in the concrete mix for the shell than that required in normal mixes to ensure adequate lubrication of the face for slip-forming and to ensure an absence of segregation and bony patches. In slip-forming concrete is left unsupported at very early stages after about 8 hours of placement. The concrete leaving the form was not subjected to much vertical load but it was subjected to large horizontal loads caused by high velocity sand laden winds which is a normal feature in this area. Finer cement having specific surface area more than 3600cm²/gm was used to ensure that the concrete emerging from the forms posses the required strength and gets sufficient setting to retain its shape. Erection of both the bottom and top central main rings and the slip-forming equipment started simultaneously with the completion of each segment of the foundation. By the time all segments were casted, the slip-

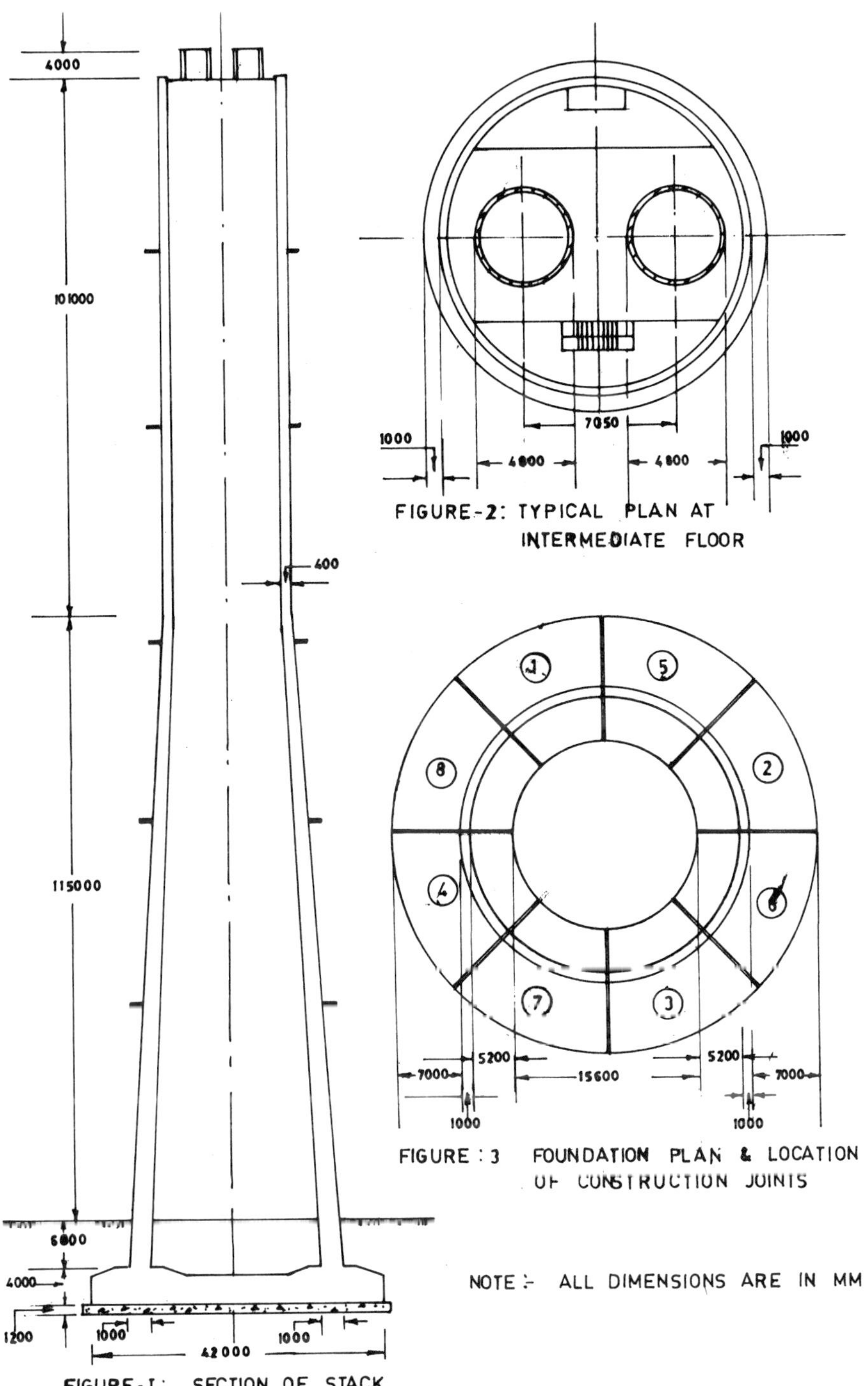

FIGURE-1: SECTION OF STACK

FIGURE-2: TYPICAL PLAN AT INTERMEDIATE FLOOR

FIGURE : 3 FOUNDATION PLAN & LOCATION OF CONSTRUCTION JOINTS

NOTE :- ALL DIMENSIONS ARE IN MM

forming equipment erection was also nearing completion. This was made possible by provision of a special supporting structure in the central segment of the raft foundation.

CONCRETING IN EXTREME WEATHER

The desert area experiences extreme weather conditions. Part of the work was executed in hot weather whereas other parts in cold weather. For most of the duration in a year, the area experiences strong dust storms with high velocity sand laden winds, higher temperature and low relative humidity. In such weather there is a rapid loss of water due to evaporation which affects workability of fresh concrete and leads to cracking of the surface due to plastic shrinkage, drying shrinkage and differential thermal stresses.

The cement content of the concrete mix was kept at the minimum which was just enough to meet the strength and durability requirements. Great emphasis was given to curing of concrete which was started immediately after the form removal. Continuous wet curing of exposed concrete surface with sprinklers was arranged to prevent plastic shrinkage. Temperature of aggregate was kept low by sprinkling water on them. Ice cool water was used in very hot weather for mixing concrete. The concrete was placed in layers not exceeding 300mm in thickness.

SPECIAL PRECAUTIONS

Since it is extremely difficult to alter, rectify or improve concrete once placed, adequate control and precautions at every stage of production, placement, compaction and curing of concrete are very important. It was always ensured that concrete mass is cohesive and uniform in colour and consistency.

The slip-form does not move continuously but in a series of small increments. At times lifting of all the jacks may not be uniform due to the existence of different frictional force exerted by the concrete on all directions of surface. This leads to the entire slip-form equipment to go out of plumb. Verticality of the slip-form equipment was checked with the help of a laser theodolite at every 3m interval and necessary adjustments were made. There was a tendency of rotation of the slip-form equipment on the periphery of the stack. This was also checked and corrected regularly. The average rate of sliding was 2m per day.

Vibration of concrete in slip-forming differ from that for static work as the vibration is allowed in the upper most concrete layers only. Penetration through the lower layer was discouraged because the extra vibration damages the partially hardend concrete in the lower layer. Specially made pocker vibrator needles of 400mm in length against the standard length 600mm were used to overcome problem of the deep and heavy vibration.

There were unplanned halts in slipping on two occasions due to unavoidable reasons. At such halts, joint preparation and treatment was done before proceeding for further concreting. There were small duration breaks in concreting during high velocity dust storms. After such breaks, the surface cleaning and joint treatment was done as usual.

CONCLUSIONS

Construction of such tall structure in desert sands in extremely unfavourable weather and site conditions required adequate precautions in design, construction planning and execution. Quality assurance was given considerable attention in this work. Under strict quality control high performance concrete meeting required properties was provided for this tall chimney structure.

EVALUATION OF HP-SFRC POST-CRACK BEHAVIOUR

J Šušteršič

A Zajc A Šajna

IRMA Ljubljana

Slovenia

V Ukrainczyk

University of Zagreb

Croatia

ABSTRACT. 4-year-old steel fibre reinforced concrete (SFRC) specimens were investigated to determine some parameters for evaluation post-crack behaviour of the HP-SFRC. Following parameters for evaluation post-crack behaviour of concrete were calculated from load –CMOD curves: absorption energy of failure zone W_{FZ} and equivalent strength at the appointed crack width f_{CW}. Three point bending configuration was used to determine load – CMOD curves. Beams with dimensions of 10×10×40 cm with notch in the middle of the span were tested. Depth of notch was 5 cm, and span was 30 cm. Two types of hooked steel fibres with length of 32 mm and aspect ratio l/d_1 = 64 and l/d_2 = 100 were used. The results of investigations show, that high-performance behaviour remain after 4 years only SFRC with steel fibres with aspect ratio of 100.

Keywords: High performance fiber reinforced concrete, Steel fibres, Toughness, Absorption energy of failure zone, Equivalent strength at the appointed crack width.

Dr Jakob Šušteršič is a Senior Research Fellow in Concrete Technology, Institute for Research in Materials and Applications Ljubljana, Slovenia. His main activities are associated with fiber reinforced concrete and other special concrete.

Professor Velimir Ukrainczyk is Director of the Department of Materials, University of Zagreb Faculty of Civil Engineering, Croatia. His activities are associated with concrete technology. His research interests include special concrete, their properties and behaviour, as well as durability. He is the member of many international and national committees.

Dr Andrej Zajc is Director of Institute for Research in Materials and Applications Ljubljana, Slovenia. He specialises in the durability of concrete, with particular reference to freeze-thaw resistance with and without de-icing salts.

Dr Aljoša Šajna is a Young Research Fellow in Concrete Technology, Institute for Research in Materials and Applications Ljubljana, Slovenia.

INTRODUCTION

High-performance steel fibre reinforced concrete (HP-SFRC) has load-deflection curve with large area under the curve [1]. In other words, absorption energy of HP-SFRC is very high, and its post-crack behaviour is very ductile. If concrete has higher compressive strength, the slope of the descending part of the curve is steeper. Because of concrete strength increases in old age, its response becomes more brittle. But, efficiency of fibres in the FRC is increased too with the age, because the bond between fibres and hardened cement paste has improved.

Some results of investigations into the mechanical and other physical properties of high-performance concrete with and without fibres using high strength artificial aggregate (carbure slag aggregate) were reported already on the last International Conference held at the University of Dundee [2]. These results were obtained from investigations of 28-day-old specimens. After, our interest was correlated to the ageing effect, for this reason 4-year-old specimens were investigated to determine some parameters for evaluation post-crack behaviour of the HP-SFRC.

EXPERIMENTAL DETAILS

Materials and Mix Proportions

All used materials are the same as they were described previously [2]: Portland cement, silica fume, a high-range water-reducing admixture, air-entraining admixture, carbure slag aggregate with a grading of 0/16 mm and two types of hooked steel fibres with length of 32 mm and with aspect ratios $l/d_1 = 64$ and $l/d_2 = 100$. Contents of both types of fibres are 0.0, 1.0 and 1.5 vol.%.

Parameters For Evaluation Post-Crack Behaviour of Concrete

Beams with dimensions of 10×10×40 cm with notch in the middle of the span were tested. Depth of notch was 5 cm, and span was 30 cm. Specimens were loaded with concentrated load in the middle of the span, where CMOD was also measured. From load - CMOD curves, which are shown in Figure 1, parameters for evaluation of post-crack behaviour of concrete were calculated: absorption energy of failure zone W_{FZ} and equivalent strength up to appointed crack width f_{CW} (CW = 0.1, 0.2, 0.3 and 0.4 mm).

Absorption Energy of Failure Zone

Because of the fibres which bridge a crack are present, it is very difficult to measure load - CMOD curve of FRC until complete material separation occurs and obtain complete area of the curve (absorption energy). This can be seen from the load - CMOD curves obtained during the investigations (Figure 1). Therefore, the load-CMOD curve was divided in four straight line segments, so that load decreases linearly to zero with increasing CMOD in the last segment.

Division of load -CMOD curves in straight line segments were carried out by the computer program. With such simplification, absorption energy of failure zone W_{FZ} can be calculated:

$$W_{FZ} = W_{II} + W_{III} + W_{IV} \qquad \text{(Nm)} \qquad (1)$$

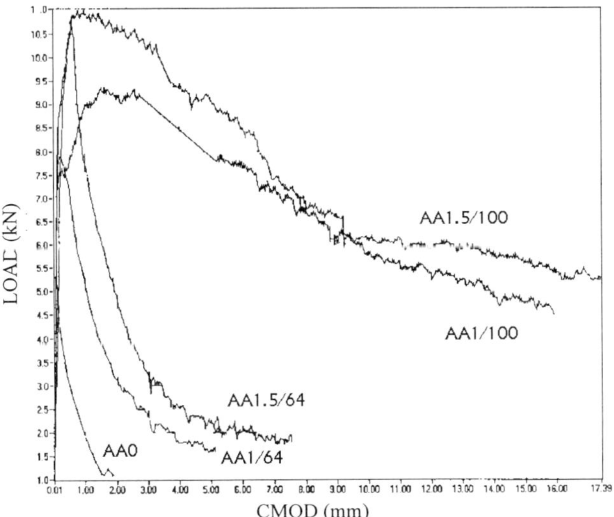

Figure 1 Typical load - CMOD curves of tested SFRC

Equivalent Strength at the Appointed Crack Width

When concrete element is loaded, small individual cracks begin to appear inside the concrete at a certain load. Those cracks bind up themselves into the continual crack, which can be seen already on the surface of concrete element. At this point of load - CMOD curve, the slope of the curve departs from linearity. Load and CMOD at this point are denoted as first crack load, first crack CMOD, respectively. Because of the end of linearity, this point is denoted also as the limit of proportionality. There are difficulties relating to precise determination of the location of the first crack. ASTM C 1018-97 defines first crack as the point on the load – deflection curve at which the form of the curve first becomes non-linear. Determination of the point of first crack has been proposed [5, 6] as the point at which the slope of the curve departs from linearity by more than 5 % and lasts for an interval of more than 0.01 mm.

The computer program has been developed in our Institute, which works in graphical form, for automatically drawing load-CMOD curves and for calculation of parameters for evaluation of concrete behaviour in the different type of loading. The point of first crack, limit of proportion (LP) respectively, is determined by the method shown schematically in Figure 2. The part of the load (F) – CMOD curve (from 0 to F_{max}) is evaluated (Figure 2a). First, ΔF – CMOD curve is drawn (Figure 2b), where ΔF is difference of load between curve and straight line passing through 0 and F_{max} (Figure 2a).

The point of limit of proportion (LP) is found out, when maximum point (ΔF_{max}) of ΔF – CMOD curve is chosen (LP in Figure 2b). It is drawn automatically on the F – CMOD curve up to F_{max} (Figure 2a). In the moment, when point of LP is reached, the crack width begins to propagate with further loading. CMOD over $CMOD_{LP}$ represents crack width (CW). Absorbed energy is expressed as product of load and crack mouth opening displacement. If this absorbed energy is divided by appointed crack width CW, average load up to the CW should be obtained.

Equivalent strength up to appointed crack width (CW = 0.1, 0.2, 0.3 and 0.4 mm), which is obtained by average load, is denoted as toughness index. It is expressed by equation:

$$f_{CW} = \frac{W_{CW} + W_I}{CMOD_{LP} + CW} \cdot \frac{3 \cdot L}{2 \cdot a \cdot b^2} \tag{2}$$

where: f_{CW} is equivalent strength up to appointed crack width, W_I is energy absorption of elastic area from 0 up to the LP, W_{CW} is energy absorption from LP up to the CW, $CMOD_{LP}$ is crack mouth opening displacement at the limit of proportion, CW is crack width, and beam dimensions are: L – span, a – width, b – high.

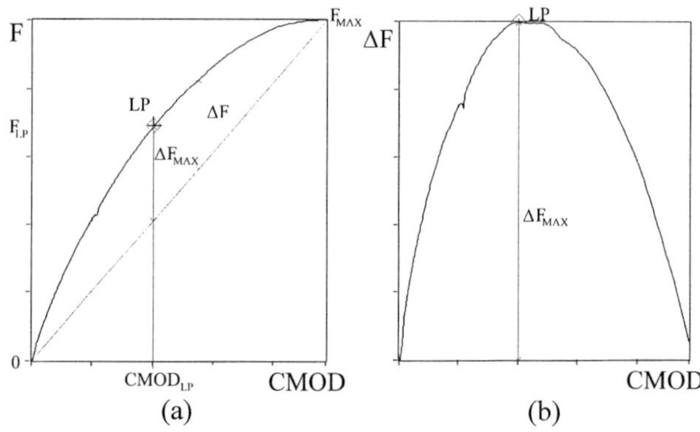

Figure 2 Graphical method for determination of point of first crack, limit of proportion (LP) respectively, of load – CMOD curve

Equivalent strength is calculated up to crack width of 0.4 mm, which is still permissive crack width for indoor structures without any aggressive attack.

RESULTS AND DISCUSSION

Absorption Energy of Failure Zone W_{FZ}

Post-crack behaviour of SFRC is improved with increase in content of steel fibres. Aspect ratio of fibres has significant effect on that behaviour. Higher aspect ratio of fibres correspond to large area of load – CMOD curve after the point of limit of proportion (Figure 1).

For that reason, absorption energy of failure zone W_{FZ}, which is calculated by Equation (1), is much higher when steel fibres with l/d_2 = 100 are used instead of fibres with l/d_1 = 64 (Figure 3).

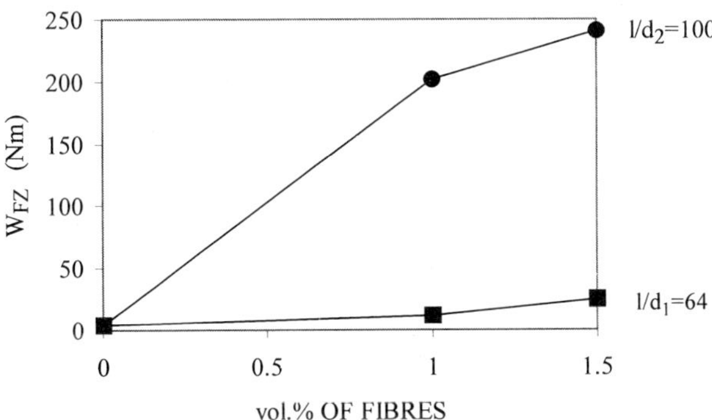

Figure 3 Influence of fibre content on increase of absorption energy of failure zone W_{FZ}

Equivalent Strength at the Appointed Crack Width

Significant effect of fibres in SFRC is evident, when post-crack behaviour is evaluated by equivalent strength which represent toughness indices up to appointed crack width of 0.1, 0.2, 0.3 and 0.4 mm (Figure 4).

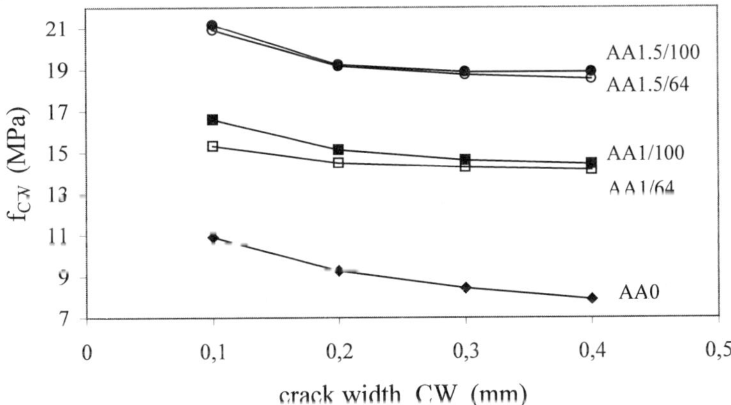

Figure 4 Equivalent strength at the crack width of 0.1, 0.2, 0.3 and 0.4 mm

Durability capacity of concrete should be measured in accordance with equivalent strength. This will prevent the crack from opening only to the appointed width. Concrete should have higher durability, if equivalent strength at the appointed crack width intends to be higher.

In more aggressive environment, concrete have to be more durable. Therefore, crack width has to be as small as possible to prevent entering of destructive substances in concrete. Influence of aspect ratio l/d of fibres on the equivalent strength at all measured crack width f_{CW} (Figure 4) is much smaller, than the influence on the absorption energy of failure zone W_{FZ} (Figure 3). Only moderate increase in f_{CW} is achieved, when fibres with $l/d_2 = 100$ is used instead of fibres with $l/d_1 = 64$.

CONCLUSIONS

The results of investigations into the post-crack behaviour of 4-year-old SFRC specimens show, that high-performance behaviour remain after 4 years only SFRC with steel fibres with higher aspect ratio (in our investigations l/d = 100). Due to crack propagation, the number of activated fibres on the fracture surface becomes higher. But, bridging effect of fibres is not so significant, when fibres with smaller aspect ratio are used (in our investigations l/d = 64).

Post-crack behaviour of SFRC was evaluated by absorption energy of failure zone W_{FZ} and equivalent strength at the appointed crack width f_{CW}. W_{FZ} can be calculated, if complete area of load – CMOD should be obtained. Complete area of the curve is obtained by dividing it in four straight line segment, so that load decreases linearly to zero with increasing CMOD in the last segment. W_{FZ} increases with increase in volume percentage of steel fibres, where aspect ratio of fibres has significant effect.

The additional fibres significant increase equivalent strength at the appointed crack width. Aspect ratio of fibres has lesser effect at the smaller crack widths (CW < 0.4 mm, approximately) than at more width cracks. Durability capacity of concrete should be measured in accordance with equivalent strength which prevent that the crack should be open up to appointed width. Further investigations are needed to find out the proper use of this parameter of post-crack behaviour for evaluation of concrete durability.

REFERENCES

1. NAAMAN, A E, REINHARDT, H W. (Eds.) High Performance Fiber Reinforced Cement Composites 2 (HPFRCC2). E&FN Spon. 1996. 502 p.

2. ŠUŠTERŠIČ, J, ZAJC, A and KOMEL, M. High Performance Concrete Using High Strength Artificial Aggregate. Concrete In The Service Of Mankind, Concrete for Environment Enhancement and Protection. Eds. Dhir,R.K. and Dyer,T.D., E&FN Spon. 1966. pp 569 - 578.

3. MARTINEZ S, NILSON, A H., SLATE, F O. Spirally-Reinforced High-Strength Concrete Columns. Research Report No. 82-10. Department of Structural Engineering, Corenell University, Ithaca. Aug. 1982.

4. ACI COMMITTEE 363 State-of-the-Art Report on High-Strength Concrete. ACI Manual of Concrete Practice 1997, Part 1. Materials and General Properties of Concrete. ACI. 1997. pp. 363R-1 - 55.

5. CHEN, L, MINDESS, S, MORGAN, D R. Toughness Evaluation of Steel Fibre reinforced Concrete. Proceedings of 3rd Canadian Symposium on Cement and Concrete, Ottawa. 1993. pp. 16 – 29.

6. MORGAN, D R, MINDESS, S, CHEN, L. Testing and Specifying Toughness for Fibre Reinforced Concrete and Shotcrete. Fiber Reinforced Concrete. Modern Developments. Eds.: N. Banthia and S. Mindess. The University of British Columbia, Vancouver. 1995. pp. 29 – 50.

LOW WATER/BINDER RATIO CONCRETES WITH LOW CRACKING RISK BY USING LIGHTWEIGHT AGGREGATES

K van Breugel
Delft University of Technology

J de Vries
Ministry of Transport

Netherlands

K Takada
Kajima Technical Research Institute
Japan

ABSTRACT. Low water/binder concretes are increasingly used for the production of concrete with a high strength and low permeability. A drawback of these mixtures is the high risk of cracking at early ages due to the high autogenous shrinkage of these concretes. Autogenous shrinkage is strongly related to the relative humidity in the pore system. It has been proposed to replace a part of the dense aggregate by saturated lightweight aggregate. Water is released from the saturated aggregate particles when the relative humidity in the paste decreases with progress of the hydration process, resulting in a smaller drop of the relative humidity. In the Stevin Laboratory of TU-Delft an experimental test program has been performed on mixtures, water/binder ratio 0.37, in which the effect of partial replacement of dense aggregates by saturated lightweight aggregate on autogenous shrinkage and strength development was investigated. Mixtures were considered with replacement percentages of 10%, 17.5% and 25%. The autogenous shrinkage of all-lightweight aggregate concretes was investigated as well. Even with low replacement percentages the autogenous shrinkage could be reduced significantly. For replace-ment percentages up to 25% only a minor reduction of the strength was observed. In case of 100% replacement no autogenous shrinkage was observed. The type of lightweight aggregate turned out to influence the volume changes and exerted stresses quite substantially. Conclusions are drawn with regard to the reduction of the risk of cracking.

Keywords: Hydration, Lightweight aggregate, Hybrid aggregate concrete, Curing, Self desiccation, Stresses development, Cracking risk, Curing

K van Breugel Dr ir PhD civil engineering, is senior researcher and lecturer at Delft University of Technology. His research topics are young concrete, design for imposed defor-mations and design of concrete structures for environmental protection.

K Takada MSc Civil Engineering, is a research engineer of Kajima Technical Research Institute, Japan, and a guest researcher of Delft University of Technology, The Netherlands. His research interests are in behaviour of fresh and early age concrete, especially Self-Compacting concrete.

J de Vries BSc Civil Engineering, is a research co-ordinator of the Ministry of Transport, Civil Engineering Division, The Netherlands. He is specialised in materials research on technology and durability of concrete and cement-bentonite.

INTRODUCTION

In recent years the use of low water/binder ratio concretes has become increasingly popular to improve strength and durability. The use of a low water/binder ratio, however, goes along with substantial shrinkage deformations of the concrete in the early stage of hardening. If these shrinkage strains, known as autogenous shrinkage, are restrained then tensile stresses can cause cracking of the concrete. Self-desiccation of the concrete is considered to be the major reason of autogenous shrinkage. Self-desiccation im-plies a decrease of the relative humidity in the pore system, which is accompanied by changes in the free surface tension of the gel particles and of the forces exerted by the capillary water. The mechanisms responsible of the autogenous shrinkage have been discussed extensively during a workshop "AUTOSHRINK'98" in Hiroshima, Japan [1]. It was concluded that although a comprehensive model for quantification of autogenous shrinkage is not available the influencing factors were known [2]. Relative humidity also affects the autogenous shrinkage. In this context it has been proposed to add saturated lightweight aggregates to the mixture. The saturated aggregate particles will act as water reservoirs, which release water when the relative humidity in the hardening paste drops [3,4,5]. The use of lightweight aggregate instead of dense aggregate, however, is expected to result in a lower strength. It is still questionable, therefore, whether the achieved reduction of the autogenous shrinkage will indeed cause a reduction of the risk of cracking. It could be that the reduction of the shrinkage-induced tensile stresses is offset by the reduction of the tensile strength. With the aim to get more information about the efficiency of the use of saturated lightweight aggregate as a means to reduce the risk of shrinkage induced cracking an experimental test program has been performed. Concrete mixtures were considered with a water/binder ratio of 0.37. The aggregates were ordinary dense aggregates, lightweight aggregates or a mixture of the two. Both the autogenous defor-mations, stresses and the strength of the concretes were determined.

EXPERIMENTAL DETERMINATION OF VOLUME CHANGES AND STRESSES

Measurement Equipment for Volume Changes and Stress Development

The volume changes in hardening concrete were determined on concrete prisms, 1000 x 150 x 150 mm^3. The concrete was cast in a mould, which was foreseen with an external insulation material. At the interior surface the mould consisted of thin steel plates (Figure 1). The steel plates can be cooled or heated by a system of tubes located between the plates and an insulating material. With this temperature control systems it is possible to impose an arbitrary temperature regime to the concrete. Temperature differences within the cross section of the specimen can be kept as low as 1.0°C to 1.5°C.

The length changes of the hardening concrete were measured over a length of 750 mm with two glass rods to which LVDT's were connected. The glass rods were connected to two steel frames, which were cast in the concrete at a distance of 750 mm. The rods could be fixed to the cast-in frames after the concrete had gained sufficient strength. After casting the top surface of the concrete was covered with a plastic foil in order to avoid moisture loss to the environment.

Stress development at early ages was determined with a Temperature Stress Testing Machine (TSTM). This testing device consists of a steel frame (see Figure 2). In this frame a mould is fixed in which the concrete is poured. The mould of the TSTM-specimen is similar to that of the shrinkage specimen (the dummy). The specimens have a prismatic shape, except at the ends where dovetailed heads are made. Two rigid steel claws hold the dovetailed specimen.

One of the claws is fixed to the frame while the other is movable. Strain measurements on the concrete are performed in the same way as in the dummy specimen. The imposed strain to which the concrete in the TSTM is subjected is deduced from the free deformations of the dummy. The deformations can be restrained in whole or in part.

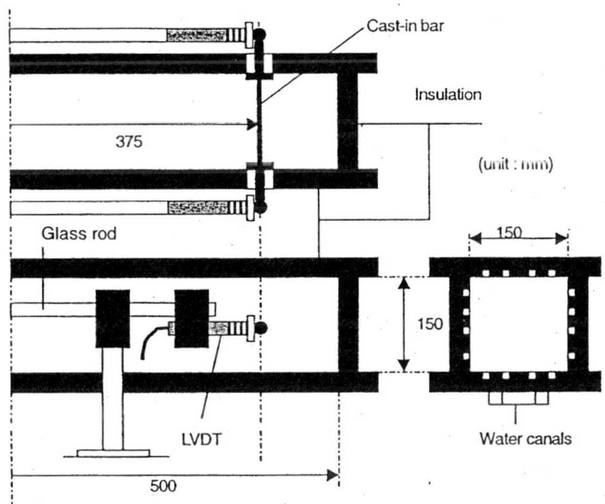

Figure 1 Sketch of (half of the) mould for concrete specimens used for measuring autogenous deformation

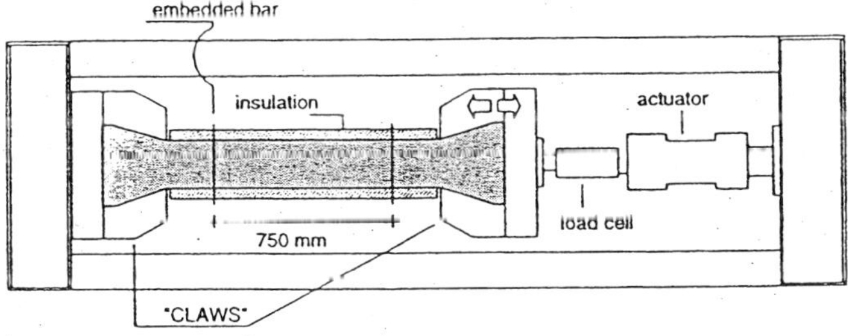

Figure 2 Top view of experimental set up for determination of stress development in hardening concrete (Temperature Stress Testing Machine)

Mixture Composition

The concrete used in the experimental program was made with a cement content of 475 kg/m^3, a blend of 50% blast furnace slag cement (CEM III/ B52.5) and 50% Portland cement (CEM I/52.5 R) and with a w/b ratio 0.37. A silica slurry, 50% water and 50% silica fume, was added in order to realise the target cube compressive strength of 85 MPa (characteristic value). In the reference mixture a siliceous aggregate was used, consisting of 45% sand (1-4 mm) and 55% coarse aggregate (4-16 mm). Two types of plasticizers were used, i.e. lignosulphonate and naphthalene sulphonate. Air content 2.5%. This reference mixture was modified by replacing a part of the dense aggregate by lightweight aggregate. The lightweight aggre-gates used were Liapor F10 and Lytag. The replacement percentages were 10%, 17.5%, 25% and 100%. Besides the volume changes and the stress development also the compressive strength was measured. Details of the mixture compositions are summarised in Table 1.

The lightweight aggregates used in the concrete were saturated with water prior to mixing. Saturation took place under water during 24 hours. The actual degree of saturation was measured afterwards and turned out to be a bit lower than the target value of 100%. The actual degree of saturation of the aggregates used in the all-lightweight aggregate concretes is indicated in the different figures shown later in this paper.

Table 1 Mix proportions of various mixtures with different replacement percentages. W/C ratio = 0.37. Target cube compressive strength at 28 days is 85 MPa

MIXTURE COMPONENT	unit	REPLACEMENT PERCENTAGE				
		0%	10%	17.5%	25%	100%
Cement CEM III/B42.5 LH HS	kg	237	237	237	237	237
cement CEM I52.5 R	kg	238	238	238	238	238
Water (incl. water in admixtures)	kg	176	176	176	176	176
Crushed gravel 4-16 mm	kg	945	710	850	780	-----
Sand 0-4 mm	kg	775	775	775	775	775
Liapor F10, 4-8 mm (100% saturated) [1]	kg	-----	60	106	150	615 [2]
Lignosulphonate (dry substance 36%)	kg	0.95	0.95	0.95	0.95	0.95
Naphthalene sulphonate (dry substance 40%)	kg	7.60	7.60	7.60	7.60	7.13
Silicafume slurry (50% water)	kg	50.0	50.0	50.0	50.0	50.0

[1] Actual degree of saturation is indicated in the legend of figures
[2] Instead of Liapor, also Lytag has been used for the all lightweight aggregated mixtures

TEST RESULTS

Development of Cube Compressive Strength

Strength values, presented in Table 2, show that partial replacement of dense aggregate by Liapor F10 lightweight aggregate by up to 25% by volume hardly affect the 28 days strength. The compressive strength of the concrete made with replacement percentages of 10% and 17.5% was even higher than that of the unblended mixtures. This might be caused by delayed hydration, which is made possible by suction of the water from the aggregates towards the hydrating paste. The all-lightweight aggregate concretes exhibited a lower strength. Also the strength at 28 days was still lower than that of the mixtures with blended aggregates and the target cube compressive strength of 85 MPa (characteristic) was not achieved.

Table 2 Compressive strength of mixtures with partial replacement of dense aggregate by lightweight aggregate, water/cement ratio 0.37, isothermal curing

CONCRETE MIXTURE	REPLACEMENT PERCENTAGE [%]	MEAN CUBE COMPRESSIVE STRENGTH [N/mm²]				
		1d	2d	3d	7d	28d
Mixture with dense crushed	0	41.9	52.0	57.5	70.4	82.5
Aggregate and blended	10	38.8	51.0	57.7	70.0	89.6
With Liapor F10	17.5	43.4	52.5	58.9	71.7	94.3
	25	37.6	47.4	53.1	67.5	87.8
	100	24.4	------	46.5	58.2*)	79.0
Mixture with Lytag	100	21.6		37.5	46.8*)	63.4

*) measured after 6 days.

Obviously the strength of the aggregates becomes the strength limiting factor. The strength of the mixtures with 100% Lytag was a bit lower than that of the 100% Liapor mixtures. This was expected, since the Liapor F10 particles are a little stronger than the Lytag particles.

Non-Thermal Length Changes of Concrete Cured Isothermally at 20°C

In Figure 3 the total deformation of the reference mixture and the mixtures with partial replacement percentages of 10%, 17.5% and 25% are presented. For the reference mixture made with dense aggregate shrinkage could be measured after about 6 hours after placing of the concrete.

After 12 hours a slight expansion was measured, whereas after 15 hours conti-nuous shrinkage is observed. For this particular mixture the shrinkage after 144 hours reached 200 10^{-6}. Replacement of the dense aggregate by 10%, 17.5% and 25% Liapor resulted in a substantial reduction of the shrinkage strains. With a replacement percentage of 25% the shrinkage after 144 hours was less than half the shrinkage of the reference mixture.

For all-lightweight aggregate mixtures made with Liapor and Lytag the isothermal deformation curves are shown in Figure 4. The mixtures exhibit expansion right from the start of the measurements.

The expansion reached a maximum of about 115 10^{-6} after 24 hours and then remained almost constant for more than 5 days. The Lytag concrete shows a tendency to ex-pand a bit more at later ages than Liapor concrete.

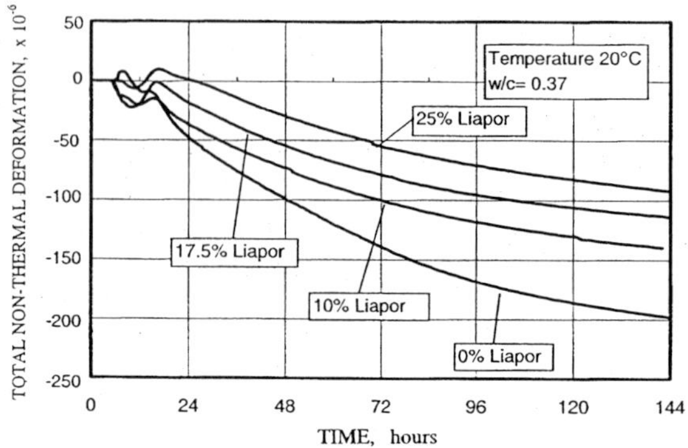

Figure 3 Non-thermal deformation of the reference mixture and mixtures with blended aggregates, replacement percentages 10%, 17.5% and 25%

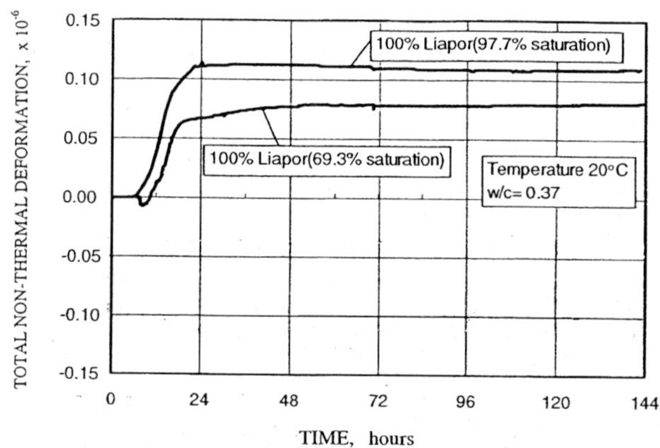

Figure 4 Non-thermal deformation of all-lightweight aggregate concrete made with saturated Liapor and Lytag

Temperature Curves and Deformations of Concrete Cured Semi-Adiabatically

In Figure 5 the semi-adiabatic temperature curve is presented that was imposed on two mixtures, viz. the reference mixture made with 100% dense aggregate and the mixture with 25% re-placement of dense aggregate by Liapor. The observed deformations are shown in Figure 6. After cooling both concretes exhibited shrinkage. This clearly demonstrates the presence of autogenous shrinkage. Shrinkage of the concrete with 25% replacement of dense aggregate by Liapor F10, however, was much less than that of concrete without Liapor.

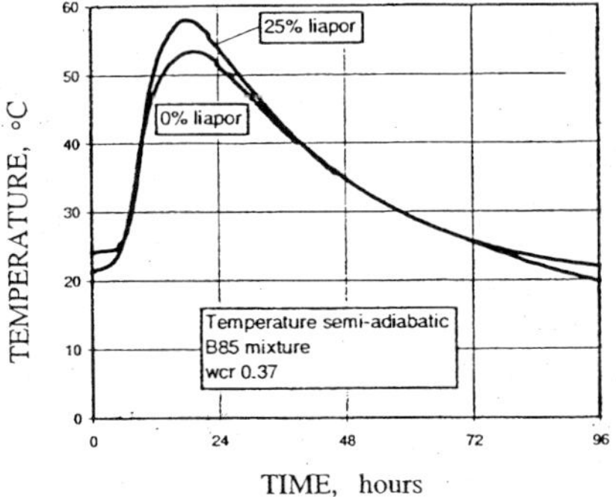

Figure 5 Imposed semi-adiabatic temperature curve in hardening concrete

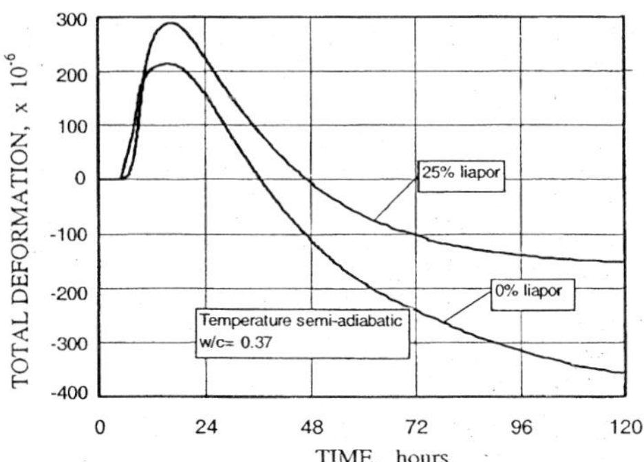

Figure 6 Total deformation of mixtures cured semi-adiabatically

For all-lightweight aggregate mixtures made with Lytag and Liapor deformation curves observed in semi-adiabatic tests are presented in Figure 7. After cooling of the concrete a net expan-sion is observed of about $130x10^{-6}$ for the mixture with Liapor and $150x10^{-6}$ for the mixture with Lytag. Note that the actual degree of saturation of the aggregates deviates a bit from the values of the isothermally cured specimen (see Figure 4). In these tests the actual values of the degree of saturation were 94% for the Lytag and 95% for the Liapor

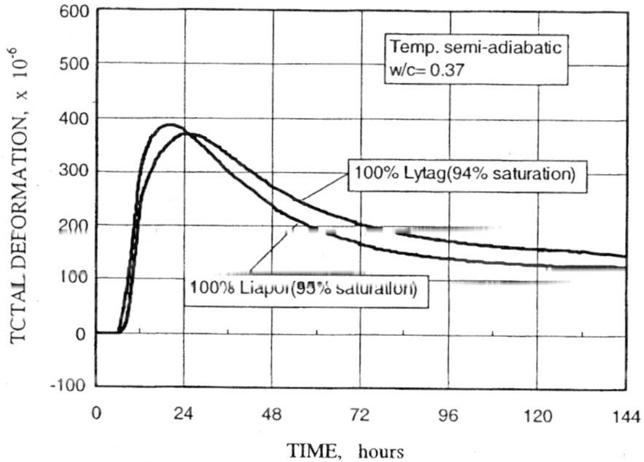

Figure 7 Total deformation of all-lightweight aggregate concrete made with saturated Liapor and Lytag, cured semi-adiabatically

Stress Development Caused by Non-Thermal Deformations – Full Restraint

The stress development in isothermally cured concretes with normal density aggregate, i.e. the reference mixture, and with partial replacement by 25% Liapor is presented in Figure 8.

The stresses in the concrete without Liapor are substantially higher than in the concrete with 25% replacement by Liapor. In the concrete without Liapor cracking occurred after 210 hours at a tensile stress of 3.5 MPa. The mixture with 25% Liapor did not crack even after 336 hours after casting. At that time the tensile stress was 2.2 MPa.

The stress curves yielded by isothermally cured all lightweight aggregate concrete are shown in Figure 9. For the Liapor and Lytag mixtures, which both exhibited swelling, compressive stresses were observed. The stresses were low, up to 0.8 MPa for the Liapor concrete and 0.5 MPa for the Lytag concrete.

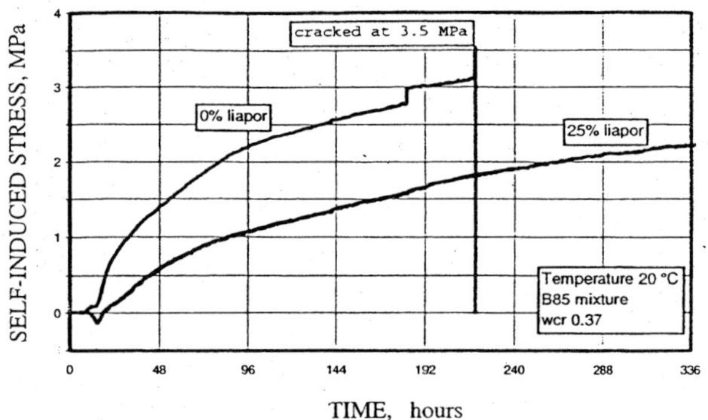

Figure 8 Stress development in mixture with dense aggregate and with 25% replacement by Liapor F10 - Isothermal curing at 20°C

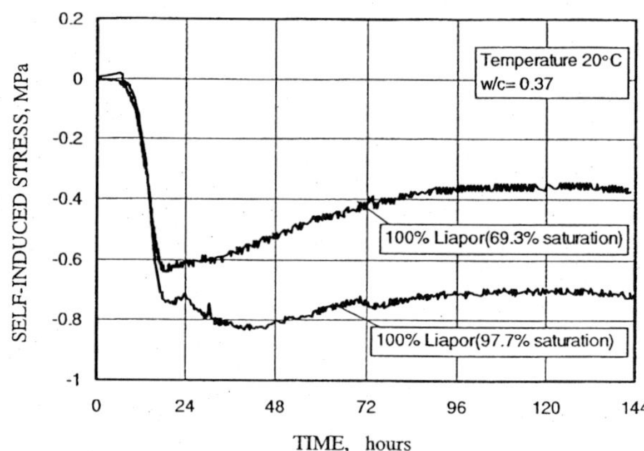

Figure 9 Stress development in all-lightweight aggregate mixture - Isothermal curing

After the peak compressive stress the Liapor mixture exhibited a small decrease of the stress. The Lytag, which continued to swell up to 144 hours after casting (Figure 4), showed a continuous increase of the compressive stress.

Stress Development in Semi-Adiabatically Cured Concrete – Full Restraint

Stress development in semi-adiabatically cured mixtures, made without and with 25% replacement of dense aggregate by Liapor, is shown in Figure 10. In both cases cracking occurred at a stress of 3.4 MPa after about 30 hours. In the mixture with partial replacement by 25% cracking occurred only a bit later. In this case there is hardly any positive effect of partial replacement as far as a reduction of the risk of cracking is concerned.

The stresses produced in a semi-adiabatically cured concretes with saturated Liapor and Lytag are presented in Figure 11. The Liapor concrete generated the highest stresses. Cracking took place after 78 hours at a tensile stress of 3.1 MPa. The stresses generated in the Lytag mixture were smaller. No cracking took place during the testing period.

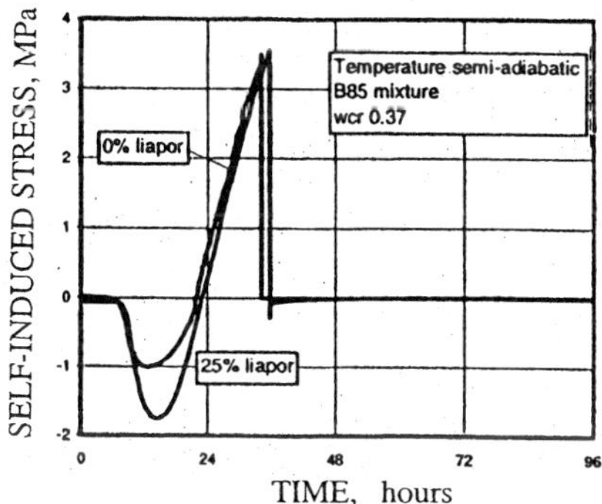

Figure 10 Stress development in mixtures with 0% and 25% replacement of dense aggregate by lightweight aggregate. Semi-adiabatic curing

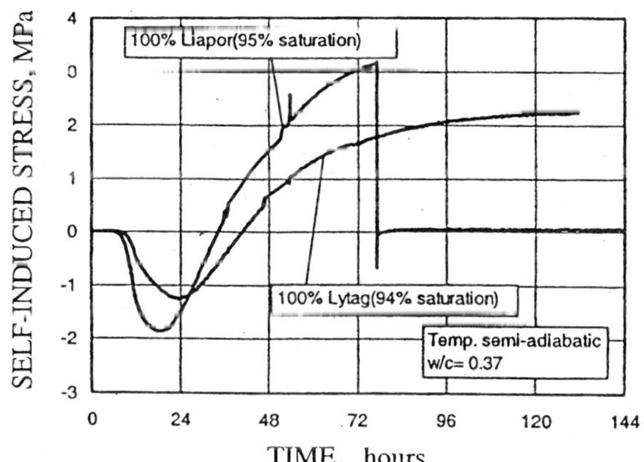

Figure 11 Stress development in all-lightweight aggregate concrete Semi-adiabatic curing

DISCUSSION AND CONCLUSIONS

Low w/b concretes have been developed with the aim to produce concretes with a high strength, a high density and hence a high durability. These concretes, however, are prone to crack in the early phase of hardening due to the high autogenous shrinkage of these concretes. Even normal density concretes made with w/c = 0.40 may exhibit autogenous shrinkage deformations large enough to cause cracking of the concrete [6].

In order to get a better per-formance of low w/b concretes as far as the risk of cracking is the effect of partial (and full) replacement of the dense aggregate by saturated lightweight aggregate has been investigated experimentally. For the concretes considered in this paper it was found that:

1. Partial replacement, up to 25%, of dense aggregate by lightweight aggregate (Liapor F10) had no negative effect on the short-term compressive strength
2. With replacement percentages up to 25%, a substantial reduction of the autogenous shrinkage and of the risk of cracking can be obtained under isothermal curing conditions.
3. In accordance with Hammer et al. [7], the autogenous deformations of all-lightweight aggregate concrete were found to differ significantly from normal density concretes. Expansion instead of shrinkage was observed for concretes made with Lytag and Liapor.
4. Reduced autogenous shrinkage as obtained by the use of saturated lightweight aggregate reduces the risk of cracking. In isothermally cured specimen full restraint of the deformations did not result in cracking if 25%, of the dense aggregate was replaced by saturated lightweight aggregate (Liapor).
5. In case of full restraint of deformations generated in semi-adiabatically cured specimen with partial replacement up to 25% the moment of cracking was slightly postponed. Cracking, however, did occur.
6. In case of full restraint of deformations generated in semi-adiabatically cured all-lightweight aggregate concrete made with Liapor F10 cracking still occurred about three days after casting. Concrete made with Lytag remained uncracked for at least 144 hours after casting, i.e. the duration of the test.

There is no doubt about the positive effect of the use of saturated lightweight aggregate in view of the reduction of the autogenous shrinkage and hence of the risk of cracking. In real structures, however, the contribution of temperature-induced tensile stresses may be so high that cracking will still occur.

The concretes considered in this paper all had the same binder content and the same w/b ratio. More research is required to generalise the results obtained for these concretes in this experimental program.

Further experimental and theoretical research is also needed concerning the observed swelling of all-lightweight aggregate concrete made with saturated aggregates and the moisture transport mechanisms that occur in the hardening paste.

ACKNOWLEDGEMENTS

This research was financially supported by the Dutch Ministry of Transport, Civil Engineering Division, of The Netherlands. Part of the experimental results were generated in the framework of the BRITE-EURAM project EuroLightcon, with participants from Norway, UK, Spain and Iceland. The assistance of Mr. E. Horeweg and Mr. F.P.J. Schilperoort in performing the experiments is greatly acknowledged.

REFERENCES

1. TAZAWA, E. (ed.) AUTOSHRINK'98. Proceedings of International Workshop on Autogenous Shrinkage. June 1998, Hiroshima, Japan. 358 p.

2. SUGIYAMA, T., OHTA, A., TANAKA, Y. Shrinkage Reduction Type of Advanced Superplasticizer. Proc. 4th Canmet/ACI/JCI Int. Symp., Sp179, Tokushima, 1998, pp. 189-200

3. WEBER, S. Nachbehandlungsunemphindlicher Hochleistungsbeton. Institut für Werkstoffe im Bauwesen. PdD, Stuttgart, 1996, 211 p.

4. KOHNO, K., OKAMOTO, T., ISIKAWA, Y., MORI, H. Effect of artificial lightweight aggregate on autogenous shrinkage of concrete. Paper submitted at the Int. Workshop AUTOSHRINK'98, Hiroshima, Japan, 1998, 4 p.

5. VAYSBURD, A.M. Durability of lightweight concrete bridges in severe environments. Concrete International, Volume 18, No. 7, 1996, pp. 33-38.

6. BREUGEL, K. VAN, DE VRIES, J, TAKADA, K. Mixture optimization of low water/binder ratio high strength concretes in view of reduction of the autogenous shrinkage. Proc. Int. Symposium on High-Performance and Reactive Powder Concretes, Sherbrooke, 1998, Vol. 1, pp. 365-382.

7. HAMMER, T.A., BJØNTEGAARD, Ø., SELLEVOLD, E.J., Cracking tendency of high strength lightweight aggregate concrete at early ages. Shrinkage Reduction Type of Advanced Superplasticizer. Proc. 4th Canmet/ACI/JCI Int. Symp., Sp179, Tokushima, 1998, pp. 53-64.

EFFECT OF THE RELATIVE HUMIDITY OF THE ENVIRONMENT AND THE LEVEL OF LOADING ON THE CREEP AND SHRINKAGE OF HIGH PERFORMANCE CONCRETES

S Setunge

B Toyne

Monash University

Australia

ABSTRACT. High Performance Concrete (HPC) with compressive strengths between 50 to 120 MPa can be produced using conventional production methods and are observed to have better durability and strength characteristics. Very little work has been reported on creep and shrinkage of HPC and most of the reported work has been limited to investigating the creep and shrinkage of standard specimens under standard conditions (50% relative humidity and 23^0 C).This paper reports the early results of an experimental study conducted to investigate the creep and shrinkage of HPC with relative humidity of the environment and the level of loading as major variables. The study is part of a major investigation in which the long-term deformations of concrete is investigated in relation to the distribution of the internal humidity of concrete.

The paper presents the creep and shrinkage results of concrete with compressive strengths between 50 and 120 MPa produced using Australian materials and silica fume. The results are compared with the predictions of concrete design standards in the world and recommendations are made for the creep coefficient and the drying shrinkage of HPC.

Keywords: High-performance concrete (HPC), Silica fume, Creep, Drying shrinkage, Relative humidity of the environment, Load level

Dr Sujeeva Setunge is a Senior Lecturer in Structural Engineering at the Gippsland School of Engineering of Monash University, Australia. She is currently conducting research in to the time dependent deformations of high-performance concrete, constitutive behaviour of laterally confined high-strength concrete, and the moment curvature behaviour of reinforced concrete beams.

Mr Bradley Toyne is a final year Civil Engineering Student at the Gippsland School of Engineering of Monash University. He has been conducting research in to the time dependent behaviour of HPC for the past year.

INTRODUCTION

High Performance Concretes (HPC) are concretes with attributes that are superior to the conventional concretes. These are produced with a low water to binder ratio, a high binder content and a lower proportion of aggregate compared to conventional concretes. Two main types of admixtures are incorporated in HPC : water reducing admixtures and pozzolanic additives. Resulting concrete has a more refined microstructure compared to ordinary concrete. As a result, the long-term properties of HPC are potentially different to those of normal strength concrete (NSC).

Shrinkage of concrete is usually divided into the three broad areas of plastic shrinkage, autogenous shrinkage and drying shrinkage. Plastic shrinkage occurs during the first few days after placing of the concrete due to the evaporation of water from the surface of fresh concrete. Due to its high paste content, High Strength Concrete is expected to be more susceptible to plastic shrinkage than ordinary concrete. Autogenous shrinkage occurs due to the reduction of water in the concrete due to hydration. The volume of hydrates formed is substantially less than the sum of the volume of water and cement. Because of the high cement or binder content used in HPC, autogenous shrinkage is expected to be higher in High-Strength Concrete. Drying shrinkage occurs due to evaporation of water from the surface of concrete after it attains the final set. Due to the high volume of paste, an HPC mix potentially has a higher drying shrinkage. On the other hand, due to the high stiffness of the rich paste, HPC may have a lower drying shrinkage than NSC. Lower porosity of HPC will prevent interior water from escaping and reduce the drying shrinkage whereas the reduction in aggregate content shall have an opposite effect.

Creep is usually determined by subtracting from the total measured strain in a loaded specimen, the sum of initial elastic strain due to applied stress, the shrinkage and any thermal strain in the identical free specimen. In addition to the parameters, which affect shrinkage of concrete, applied stress level is a major variable in any study of creep of concrete. Creep is also divided into two areas as basic creep and drying creep. Basic creep is the deformations observed when the specimens are sealed from the environment whereas the drying creep is dependent on the moisture distribution as in drying shrinkage.

The exact physical origin of creep of concrete is poorly understood and modelling creep of concrete is difficul. Bazant (1) observed that concrete without any free water does not creep. Some investigators related the creep of concrete to the volume of hydrate in the concrete paste. Gilbert (2) observed that the creep of concrete consists of two parts as delayed elasticity of coarse aggregate and viscous flow of water in the cement gel. In high-performance concrete, low w/c ratio ensures a very small amount of free water in the hardened concrete. Also, because of low water content, the volume of hydrate is observed to be less in HPC compared to NSC.

Therefore creep of HPC is expected to reduce with reduction of w/c ratio or increase in compressive strength. Literature indicates that any mix design change, which affects the compressive strength of concrete also, influences creep.

Studies directed at investigating shrinkage and creep of HPC have been restricted to measuring creep and shrinkage of standard specimens in an environment of standard humidity and temperature (23°C and 50% RH). There is no reported work in which the tests are conducted with relative humidity of the environment as a major variable. Therefore information about the creep and shrinkage behaviour of HPC under arid and coastal conditions are not currently available. This paper reports the early results of a study planned to study the effect of the environment on the creep and shrinkage of HPC.

EXPERIMENTAL PROGRAM

Figure 1 shows in schematic form, the experimental program. Three sets of High Strength Concrete Specimens were loaded to three loading levels in two different environments (Arid and Interior). A fourth set of specimens of NSC was loaded to the standard service load level (0.4f'c).

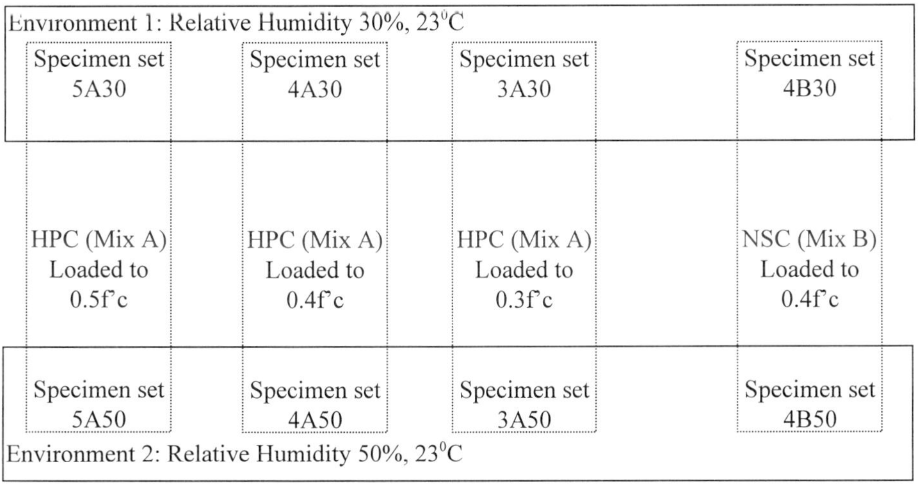

Environment 1: Relative Humidity 30%, 23°C

Specimen set 5A30	Specimen set 4A30	Specimen set 3A30	Specimen set 4B30
HPC (Mix A) Loaded to 0.5f'c	HPC (Mix A) Loaded to 0.4f'c	HPC (Mix A) Loaded to 0.3f'c	NSC (Mix B) Loaded to 0.4f'c
Specimen set 5A50	Specimen set 4A50	Specimen set 3A50	Specimen set 4B50

Environment 2: Relative Humidity 50%, 23°C

Figure 1 Experimental program

Two concrete mix proportions are shown in table 1 given below.

Table 1 Concrete Mix Proportions (/m³)

MIX	CEMENT (kg)	SILICA FUME (kg)	W/C	FINE AGG. (kg)	COARSE AGG. (kg)	SUPER-PLASTICISER (lt)	f_c MPa
A	500	50	0.25	700	1185	13	105
B	300	-	0.55	771	1202	-	53

Type A ordinary Portland cement was used in both the mixes. The HPC mix had 8% of cement content replaced by silica fume and Rheobuild 1000 by Masterbuilders used as a superplasticiser. A blended concrete sand and a crushed basalt 12 mm-14 mm graded from a local source were used as fine and coarse aggregate respectively.

Shrinkage measurements were taken on specimens from both mixes in the two environments on standard shrinkage specimens (75-mm X 75 mm X 280 mm prisms) according to the AS1012.13 (3) procedures. Specimens were water cured for seven days after which measurements and drying commenced.

Creep tests were performed on unsealed (100 mm diameter x 200 mm) cylinders under a constant load equivalent to 40% of the 28 day compressive strength of concrete for both NSC and HPC and under two other load levels (30% and 50% of compressive strengths) for HPC. Loading frame was made of four VSL high strength steel bars and the load was applied via a Flat Jack type VSL 300C. A Haskel pump MS-36C was connected to the system to maintain the load at the required level for the period of testing.

Mechanical De-mec strain gauges were used to measure the deformations of creep specimens. These results were complemented with measurements taken on SR4 strain gauges mounted on two out of four specimens tested in each creep rig.

RESULTS

Creep

Figures 2 to 4 shows the results of creep strain vs time for all the tests. Figures 2 and 3 shows that increase in loading level from 0.3fc to 0.4fc increases creep considerably, whilst increase in load from 0.4fc to 0.5fc does not appear to cause much difference in creep of HPC. In NSC three loading levels appear to correlate with creep strains well. Figure 4 shows that an increase in the compressive strength of concrete decreases the amount of creep strain as expected. Similar observations have been made by Ngab et al (4) on concrete with compressive strengths up to 60 MPa at 50% relative humidity. Most interesting observation is the effect of environment on the creep of both types of concrete. Arid environments appear to increase the creep significantly as shown in Figure 3.

Shrinkage Results

In the early days there is not much difference between shrinkage vs drying time curves of the two different concretes investigated. Curves start diverging after about 30 days. Relative humidity does not appear to have any effect on the drying shrinkage of both types of concrete. This is contradictory to the observations for NSC (8). There is a strong indication that the compressive strength of concrete has an inverse relationship with the drying shrinkage of concrete, Figure 5.

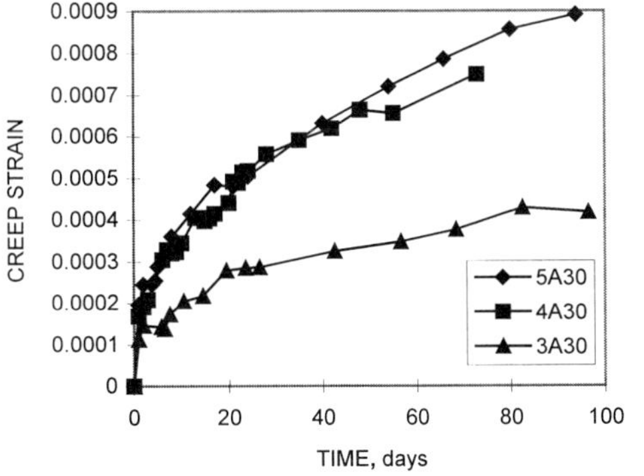

Figure 2 Creep strain vs Time- Mix A at 30% RH

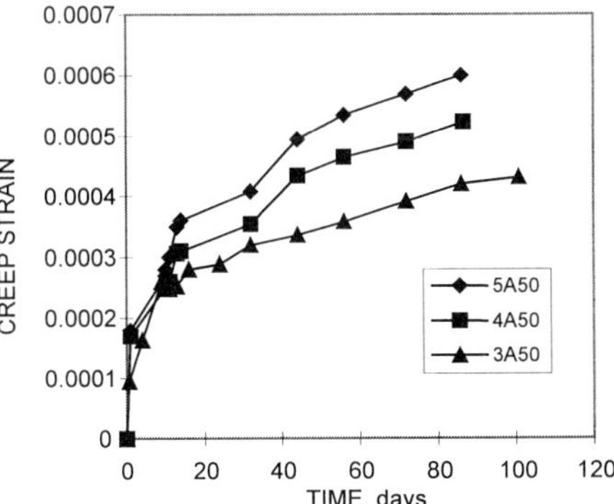

Figure 3 Creep Strain vs time - Mix A at 50% RH

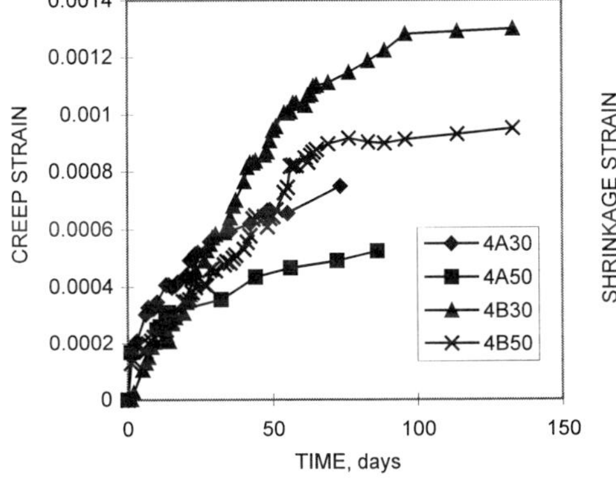

Figure 4 All the Results at Standard loading (0.4fc)

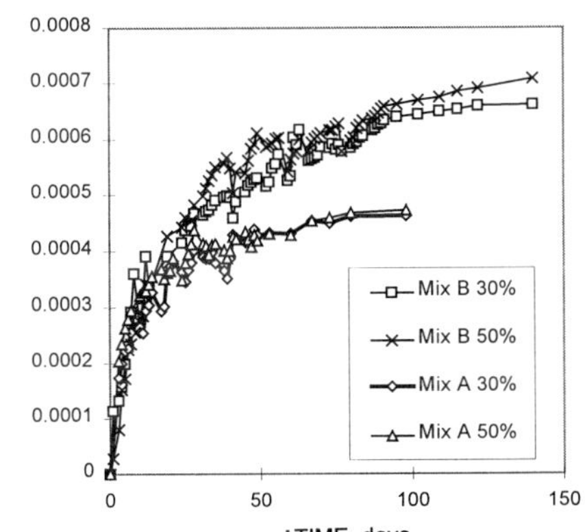

Figure 5 Shrinkage Results

COMPARISON WITH AVAILABLE RELATIONSHIPS

Figures 6 to 9 shows a comparison of experimental creep results of the specimens loaded to 0.4f'c, with the models proposed in three concrete design standards. Australian standard AS3600 (5) is applicable to concrete with compressive strengths up to 50 MPa. The Norwegian standard, NS3473 (6) is applicable for concrete with compressive strengths up to 85 MPa whereas the European code, CEB-FIP (7) is recommended for concrete with compressive strengths up to 80 MPa.

For Mix A, all three standards grossly overestimate the creep coefficient of the specimens tested.

The best prediction made by CEB-FIP (7), overestimates the coefficient by approximately 50% compared to the experimental results. All three standards provided a reasonable prediction of the creep deformations of Mix B.

Figures 10 to 13 show comparisons of shrinkage results with the models proposed in the same three standards

It is evident that NS3473 (6) and CEB-FIP (7) code predictions seriously underestimate shrinkage of HPC in arid environments. These equations should not be used in calculation of design shrinkage strains of HPC. It is interesting to note that they are the two codes, which cover concrete with compressive strengths up to 80 MPa. AS3600 (5) which only cover concrete with compressive strengths up to 50 MPa, give very conservative predictions of shrinkage of concrete.

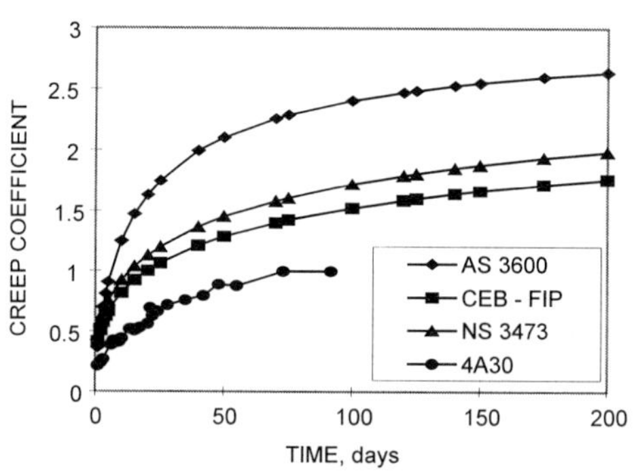

Figure 6 Comparison of 4A30 with the existing models

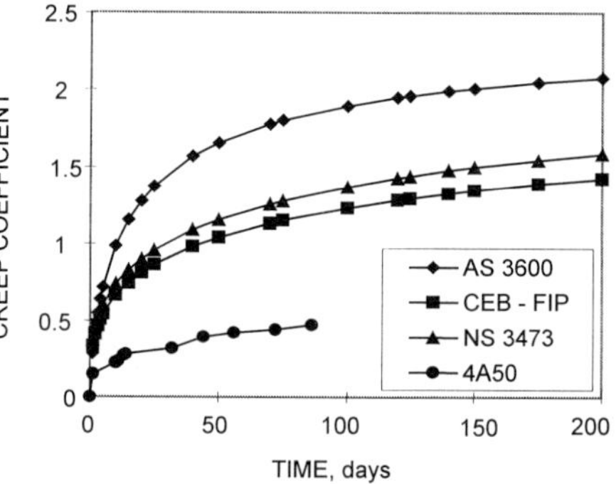

Figure 7 Comparison of 4A50 with the existing models

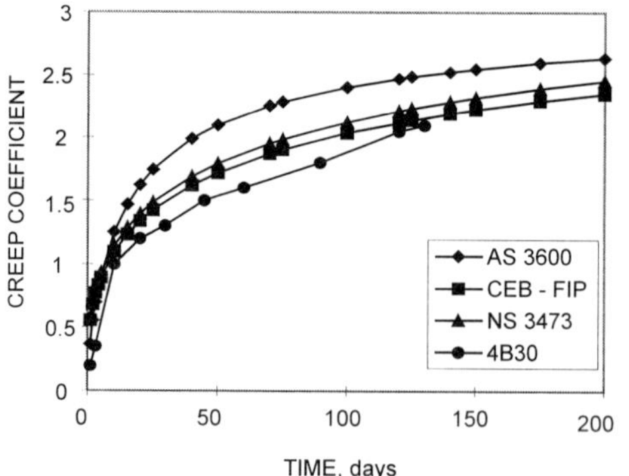

Figure 8 Comparison of 4B50 with the existing models

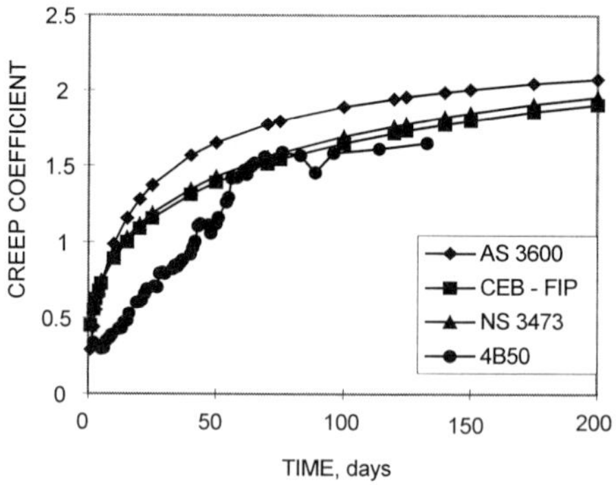

Figure 9 Comparison of 4B50 with the existing models

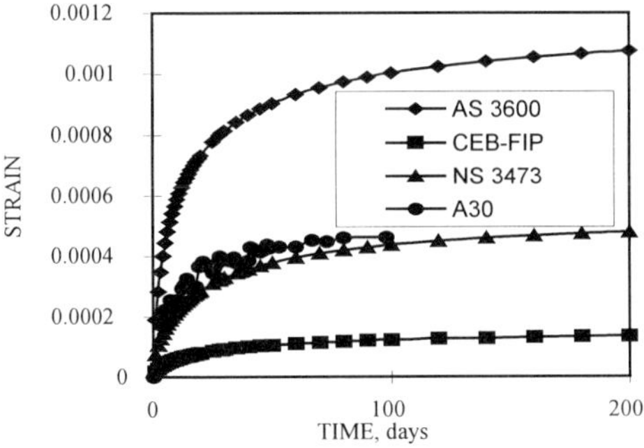

Figure 10 Comparison of shrinkage strains of Mix A at 30%RH with Existing relationships

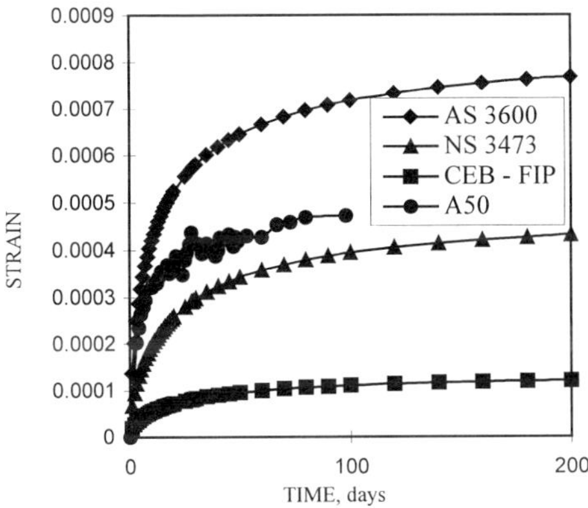

Figure 11 Comparison of shrinkage strains of Mix A at 50% RH, with existing models

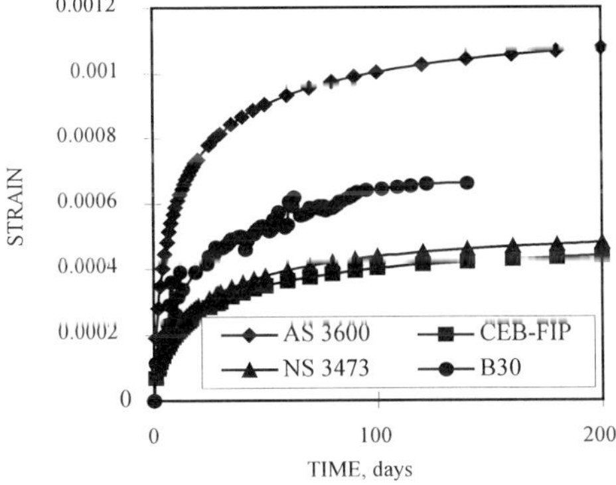

Figure 12 Comparison of shrinkage strains of Mix B at 30% RH, with existing models

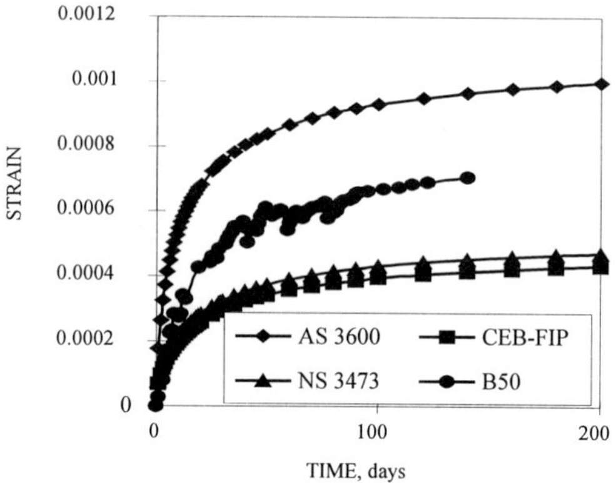

Figure 13 Comparison of shrinkage strains of Mix B at 50% RH, with existing models

DISCUSSION AND RECOMMENDATIONS

Creep deformations of the two concretes studied here shows the traditional trend of reduction of creep with increase in compressive strength of concrete. The basic creep factor, $\phi_{cc.b}$ is the element, which accounts for the compressive strength of concrete in the AS3600 method of calculation of creep deformations of NSC, which is given below.

$$\text{Creep strain} = \phi_{cc} \text{ X immediate elastic strain}$$

where, $\phi_{cc} = k_2\, k_3\, \phi_{cc.b}$ (1)

and

$$k_2 = (k_7\, k_8\, t^{0.7}) / (t^{0.7} + k_9) \quad (2)$$

where, $k_7 = 0.76 + 0.9^{\,e\,-0.008\,th}$
$k_8 = 1.37 - 0.011\ h$
$k_9 = 0.15\ t_h$

where, t_h = hypothetical thickness
h = relative humidity

the hypothetical thickness t_h is obtained from the following equation:

$$t_h = 2\,(A_g / U)$$

where, A_g = Cross sectional area of the member in mm^2
U = Perimeter of the cross section exposed to drying in mm

$\phi_{cc.b}$ is known as the basic creep factor and is defined as the ratio of ultimate creep strain to elastic strain for a specimen loaded at 28 days under a constant stress of $0.4 f'_c$. The value of $\phi_{cc.b}$ can be attained from three alternative techniques, which consist of the following:

◊ Table 6.1.8.1 of AS 3600, where a range of $\phi_{cc.b}$ for different characteristic compressive strengths between 20 - 50 MPa are given as shown below.

◊ Determined from measurements on similar local concrete

◊ Determined by tests in accordance with AS 1012.16 (3)

Basic creep factors (100 day) calculated from the experimental results reported here are plotted together with those given in table 6.1.8.1 of AS3600 in Figure 14 against compressive strength of concrete. A power relationship was found to give the best fit to the values plotted in Figure 14 with a coefficient of correlation of 0.97.

$$\phi_{cc.b} = \frac{312}{f_c^{1.3}} \qquad (3)$$

where, f_c is the compressive strength of concrete in MPa.

Figure 14 shows the above relationship.

In the experiments reported here, the shrinkage strains were measured on unsealed specimens and therefore the measurement gives the total shrinkage of the specimen, which includes both autogenous and drying shrinkage. More results are required before establishing a relationship with shrinkage and the compressive strength of concrete.

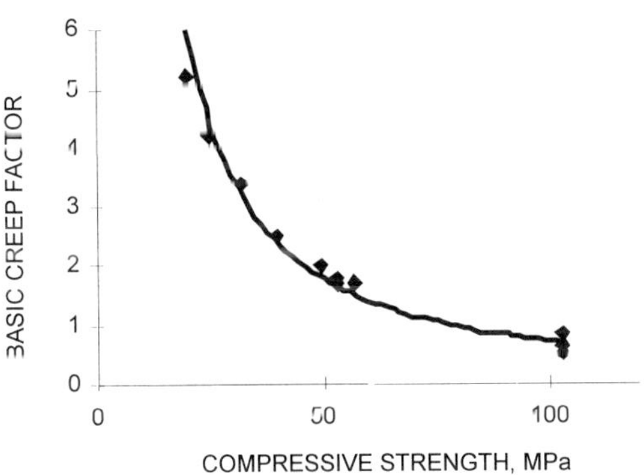

Figure 14 Basic creep factor

CONCLUSIONS

Based on the limited work reported herein following conclusions can be drawn.

1. Total shrinkage of HPC appears to have an inverse relationship with the compressive strength of concrete.
2. NS3473 (6) and CEB-FIP (7) code predictions seriously underestimate shrinkage of HPC in arid environments. These equations should not be used in calculation of design shrinkage strains of HPC.
3. Relationship between the loading level and the creep strain of concrete appears to be significantly different for HPC compared to NSC. More work is required to confirm this.
4. Equation, $\phi_{cc.b} = \dfrac{312}{f_c^{\,1.3}}$ can be used to calculate the Basic Creep Factor to be used with the AS3600 method of calculation of creep strains with reasonable accuracy.

ACKNOWLEDGMENTS

Special thanks to Mr. Brett Evans of CSR Readymix in Morwell, Australia, Mr. Peter Russel of Master Builders (Australia), and Mr. Harry Lymberatos of VSL Prestressing (Australia) for donating materials for the work reported herein. Technical support provided by Mr. Eddie Green, Mr. Ian Bowden and Mr. Ken Phelps of Monash University and the contribution of Mr. Randy Dear are gratefully acknowledged.

REFERENCES

1. BAZANT, Z.P., ASGHARI, A.A. AND SCHMIDT, J., "Experimental Study of Creep of Portland Cement Paste at Variable Water Content", Materials and Structures, RILEM, Vol.9, No.52, 1976, pp 279-290.
2. GILBERT, R.I., "Time Effects in Concrete Structures", Elsevier, 1988.
3. STANDARDS ASSOCIATION OF AUSTRALIA, AS1012, "Methods of Testing Concrete", 1986.
4. NGAB, A.S., NILSON, A.H. AND SLATE, F.O., "Shrinkage and Creep of High-Strength Concrete", ACI Journal, Vol.78, No.4, July-August, 1981, pp 255-261.
5. STANDARDS ASSOCIATION OF AUSTRALIA, AS3600, "Concrete Structures", Australian Standard, Sydney, 1994.
6. NORWEGIAN STANDARD NS 3473, Concrete Structures, Design Rules, Oslo, Norway, 1989.
7. CEB-FIP, EUROPEAN CONCRETE COMMITTEE, CEB-FIP Model Code for Concrete Structures, 1990.
8. BHAL N.S. AND MITTAL M.K., "Effect of relative humidity on creep and shrinkage of concrete, The Indian Concrete Journal, January 1996, pp 21-27.

OPTIMUM MIX PROPORTIONING OF SELF-COMPACTING CONCRETE

P L Domone J Jin

University College London

United Kingdom

H-W Chai

Taiwan Construction Research Institute

Taiwan

ABSTRACT. An extensive laboratory investigation has shown that self-compacting concrete can be produced with readily available UK materials. Limits to the proportions of materials for satisfactory self compacting properties have been defined; these are wider when using 10mm maximum aggregate size than when using 20mm aggregate. A mix design procedure, based on the simple general method proposed by Japanese workers has been developed. This makes use of a linear optimisation technique to produce an optimum mixture of water, powders and aggregate for further testing. An initial estimate of the amount of admixtures required can be obtained by tests on the mortar component of the concrete. This minimises the amount of trial mixes on concrete.

Keywords: Self-compacting concrete, Test methods, Fluidity, Segregation resistance, Passing ability, Mix design, Cement replacement materials, Superplasticizers

Dr P L Domone is a senior lecturer in concrete technology at University College London. His main research interests include high performance concrete, with an emphasis on fresh and early age properties. He is a member of the RILEM Technical Committee on Self Compacting Concrete.

Dr H-W Chai is a former lecturer in civil engineering at the Nan Ya Junior College, Taiwan, and has recently completed a PhD at University College London on self-compacting concrete. He is currently an associate researcher at the Taiwan Construction Research Institute.

Mrs J Jin is a graduate research student at University College London, working on self-compacting concrete. She is a graduate of the Harbin Architecture University, and was a research engineer with the China Academy of Railway Sciences before her studies at UCL.

INTRODUCTION

Self-compacting concrete (SCC) was first shown to be feasible just over ten years ago [1], and has subsequently had significant and increasing development and use, initially in Japan but more recently in other countries, notably France, Sweden, Thailand, China, Taiwan, Holland and Canada. The advantages of SCC include reduced demand for skilled labour, shorter construction times, improved compaction and hence durability of concrete in areas of high reinforcement density and reduced noise during construction [2].

Successful self-compacting concrete must have high fluidity (for flow under self-weight), high segregation resistance (to maintain uniformity during flow) and sufficient passing ability so that it can flow through and around reinforcement without blocking or segregating. These are three distinct properties.

There is no unique solution for a particular application, and a wide variety of materials and mixes have been used. Water/binder ratios are generally less than 0.5, and mixes have a lower coarse aggregate and a higher paste content than conventional concrete [3]. Admixtures and cement replacement materials contribute to both fluidity and segregation resistance. Admixtures invariably include a superplasticizer, often with a viscosity agent of some form, typically cellulose or polysaccharide based. Viscosity agents are reported to allow higher water contents to be used, and to make the concrete more tolerant to fluctuations in mix proportions during batching [4]; they do not obviate the need for careful mix design.

Okamura and Ozawa [5] have suggested relatively simple guidelines for mix design and proportioning which will ensure more than adequate self-compacting properties for most applications. Other have developed more detailed mix design methods [6,7,8,9], although these have necessarily resulted from the use of materials available in their respective countries, and may not be fully applicable elsewhere.

This paper summarises the outcome of a laboratory based project which has established the feasibility of producing SCC using commercially available UK materials, and has developed and extended the mix design procedure suggested by Okamura and Ozawa [5]. An extensive series of tests were carried out on concrete and the component pastes and mortars [10,11], and use was made of the increasing amount of internationally published information.

MATERIALS

A class 42.5N Portland cement with C_3S, C_2S, C_3A and C_4AF contents of 55%, 14%, 11% and 9% respectively, and a specific surface area of 355m^2/kg was used. The ground granulated blast furnace slag (ggbs) had a surface area of 420m^2/kg and the pulverized fuel ash (pfa) had 85% of its particles finer than 45μm. The limestone powder had 90% of its particles finer than 50μm. of cement.

Table 1 gives the retained water/powder ratios and the deformation coefficients of the powders; these were obtained from a flow spread test on the paste at progressively increasing water contents [2,12]. The retained water/powder ratio is a measure of the amount of water adsorbed on the powder surface together with that required to fill the voids in the powder system and to provide sufficient dispersal of the particles for flow under self-weight to be about to commence. The deformation coefficient is a measure of the sensitivity of the fluidity characteristics of the paste to increasing water content in excess of the retained water/powder ratio.

Table 1 Retained water/powder ratios and the deformation coefficients of the powders

	PORTLAND CEMENT	GGBS	PFA	LIMESTONE POWDER
Retained water/powder ratio (by vol.)	0.96	1.17	0.69	0.77
Deformation coefficient	0.044	0.043	0.03	0.037

The fine aggregate was a natural river sand with a fineness modulus of 2.6. The coarse aggregate was a river gravel with a maximum size of either 10 or 20mm.

The superplasticizer used for the majority of the study was a sulphonated vinyl copolymer. No viscosity agent or other admixtures were included in the work.

All of the materials used can be considered typical of those available in the UK, and complied with the relevant British Standard.

TEST METHODS AND CRITERIA

The tests used to evaluate the properties of the concrete were selected from those developed and used by other workers [13-16]. These were:

1. Slump flow. This is the mean diameter of spread of the concrete after a normal slump test, with the exception that concrete is not rodded after placing in the cone. This is a measure of the fluidity of the concrete.

2. V-funnel flow time. A funnel (Figure 1) with an outlet 75mm square is filled with 10 litres of concrete; the flow time is that between opening the orifice and the first daylight appearing when looking vertically down through the funnel. This is a measure of the viscosity of the concrete, and hence gives an indication of its segregation resistance.

3. A U-test (Figure 2), in which the concrete flows around a bend under self-weight through a mesh of reinforcement. One side of the apparatus is filled with concrete, the partition gate is then raised and the final height of the concrete when flow has ceased measured. This therefore measures the passing ability of the concrete. This was found to be an effective and convenient test, with the measurement of final height being readily interpreted in terms of an acceptance criterion.

A combination of experience of using the tests and recommendations from others who have designed, produced and placed SCC lead to the adoption of the following criteria for satisfactory self-compacting behaviour:

1. Slump flow between 650 and 700mm for concrete with 20mm aggregate and between 600 and 700mm with 10mm aggregate. Below the lower limit the concrete may have insufficient fluidity to pass through and around obstacles, and above the upper limit segregation is likely to occur.

2. V-funnel flow time between 4 and 10 seconds for concrete with 20mm aggregate and between 2 and 10 seconds with 10mm aggregate. If the flow time is higher than 10 seconds, then the concrete is either too viscous for satisfactory handling and placing, or so unstable that the aggregate particles bridge and block the flow. Flow times below the lower limit indicate insufficient viscosity for adequate segregation resistance.

3. For both aggregate sizes, a U-box filling height in excess of 300mm (equal final levels give a value of 360mm). This is possibly the most important indication of successful self-compacting properties.

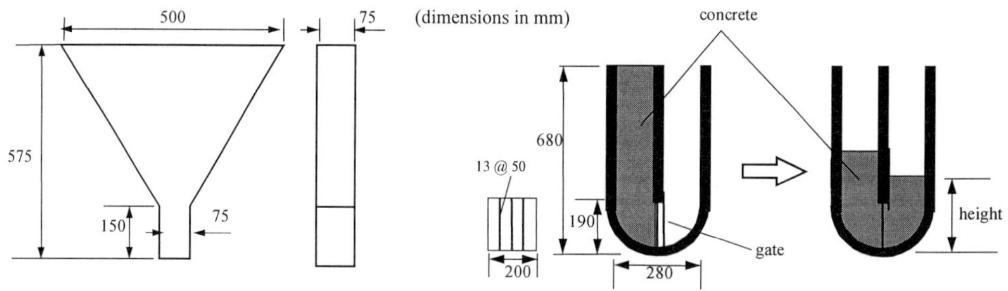

Figure 1 V-funnel

Figure 2 U-test for assessing passing ability

BINDER COMPOSITION

Cement replacement materials (CRM's) were found to have the following benefits for self-compacting properties:

- *Pulverised fuel ash (pfa)* improves the fluidity and can therefore be useful at lower water/binder ratios. It also reduces the loss of workability with time.

- *Ground granulated blast furnace slag (ggbs)* gives higher viscosities and can therefore enhance the segregation resistance

- *Limestone powder* gives similar workability advantages to fly ash, but it does not contribute to long term strength gain.

The use of CRM's will of course also effect other properties of the concrete, such as heat of hydration effects and long term strength and durability, and the overall limits for the proportion of each material should follow conventional concrete practice.

LIMITS TO THE PROPORTIONS OF THE MIX CONSTITUENTS

The test programme showed that satisfactory self-compacting properties could be achieved if the mix proportions were within the overall limits given in table 2. These are wider for concrete with 10mm aggregate than with 20mm aggregate, mainly due to the higher segregation resistance with smaller size aggregate. The use of 10mm aggregate has some of the advantages reported as resulting from the use of a viscosity agent, and therefore it could be considered as an alternative to this.

Table 2 Limiting mix proportions for successful self-compacting concrete

	MAX. AGG. SIZE 20mm		MAX. AGG. SIZE 10mm	
Coarse aggregate content (kg/m^3)	0.5 x dry rodded unit wt.		0.5 - 0.54 x dry rodded unit wt	
Max. water content (kg/m^3)	200			
Water/powder ratio by wt (w/p)	0.28 - 0.4		0.28 - 0.5	
Water/(powder + fine Aggregate) ratio by wt	0.12 - 0.14		0.12 - 0.17	
Paste volume (m^3/m^3 concrete)	0.38 - 0.42			
Volume sand/volume mortar (v_{fa}/v_m)	w/p	v_{fa}/v_m	w/p	v_{fa}/v_m
	<0.3	0.4	<0.3	0.4
	0.3 - 0.34	0.4 - 0.45	0.3 - 0.34	0.4 - 0.45
	0.34 - 0.4	0.45 - 0.47	0.34 - 0.4	0.45 - 0.47
	0.4 - 0.5	do not use	0.4 - 0.5	>0.45

PROPOSED MIX DESIGN METHOD

The above suggestions and limits have been incorporated into the following mix design process, which extends that suggested by Okamura and Ozawa [5].

Step 1. A typical value of air content is chosen (1-1.5% for a non air-entrained mix).

Step 2. The coarse aggregate content is set.

Step 3. A powder composition is chosen from general guidelines given above, and from past experience or limitations on the availability of materials.

Step 4. The maximum water/powder ratio is obtained for each of the following conditions:

a) to ensure that the paste has sufficient inherent plastic viscosity (i.e. before the addition of the superplasticizer) to provide adequate segregation resistance to the concrete. This can be estimated with the following relationship, derived from rheological and flow tests on pastes [12]

$$\mu = \frac{23.6}{E_p}\left(\frac{[w_w/w_p]sg_p}{\beta_p}\right)^{-6.38}$$

where μ is the inherent plastic viscosity, sg_p is the specific gravity of the powder, and E_p and β_p are the powder characteristics of retained water/powder ratio and deformation coefficient respectively, obtained from the values for the separate powder components given in Table 1.

The value of μ should be greater than 280mPa.s if 20mm aggregate is to be used in the concrete, or greater than 70mPa.s if 10mm aggregate is to be used.

b) to ensure that concrete meets its strength requirement. The generalization of Feret's formula proposed by de Larrard et al [8] for concrete containing cement replacement materials was found to be applicable to the materials used:

$$f_{cyl} = \frac{K_g R_c}{\left(1 + 3.1\frac{(W + A)}{C(1 + K_1 + K_2) + GGBS}\right)^2}$$

where

f_{cyl} = cylinder compressive strength, in N/mm^2;
K_g = an aggregate coefficient, found to be 4.6 for the aggregate used in the project;
R_c = the cement strength (N/mm^2, ISO mortar - sand:cement:water = 3:1:0.5);
W = free water content, in kg/m^3 (including that in the admixtures);
A = entrapped air volume, in l/m^3;
K_1 = a pozzolanic activity coefficient = 0.4PFA + 3SF/C ($K_1 \leq 0.5$);
K_2 = a limestone activity coefficient = 0.2 LP/C ($K_2 \leq 0.07$);
C, PFA, SF, LP and GGBS are the contents, in kg/m^3, of cement, fly ash, silica fume, limestone powder and blast furnace slag respectively.

To obtain cube strengths, the conversion given by CEB-FIP (1990) [17] can be used:

$$f_{cube} = 1.25\ f_{cyl} \quad \text{for } f_{cyl} \leq 40 \text{ MPa}$$
$$f_{cube} = f_{cyl} + 10 \quad \text{for } f_{cyl} \geq 40 \text{ MPa}$$

c) to comply with any durability or other requirements.

The minimum value of water/powder ratio from (a), (b) and (c) is used.

Step 5. The volume of sand in the mortar is then chosen .

Step 6. The paste content is calculated, and adjusted if outside the limits given in table 2.

Step 7. The water and powder contents are calculated. The water content should be less than the maximum given in table 2. If it is not, then this becomes a limiting factor, and it should be set before step 5.

All of the above calculations can be conveniently set up on a Microsoft Excel spreadsheet program. This allows the use of a linear optimisation tool ('Solver') to determine an optimum solution within the overall constraints of the mix design parameters and concrete properties. A target of a minimum paste content has been found be effective. This may result in a minimum cost mix, but also minimises such effects as heat of hydration and shrinkage. The Solver output also gives information about which parameters have been limited by the imposed restraints; this is useful in mix adjustment during trial mixes.

Step 8. An estimate of the superplasticizer dosage is then made from tests on the mortar component of the concrete, using the flow spread test and the funnel test suggested by Okamura et al [2]. These are similar to the slump flow and V-funnel tests used for concrete, but use smaller volumes of material. The truncated cone for the spread test has top and base diameters of 70 and 100mm respectively, and a height of 60mm. The V-funnel has an aperture 30mm square, and uses 1.134 litres of mortar. Sufficient superplasticizer is added to give a flow spread value greater than 300mm and a V-funnel flow time between 4 and 10secs when 20 mm aggregate is to be used, and between 2 and 10secs for 10mm aggregate. Both criteria must be met, and some adjustment to the water/powder ratio or powder composition may be necessary to achieve this.

Step 9. A trial mix of the resulting concrete is then made, using a similar mixing procedure to that used for the mortar, and tested for slump flow, V-funnel time and U-box height. The results are then compared to those for satisfactory self-compacting properties given above; it is important that all three criteria are met.

The use of a target plastic viscosity for the mortar, the generalized Feret formula and the Solver optimisation tool to obtain an initial estimate of the mix proportions (excluding admixtures) minimise the amount of laboratory work required. Furthermore, maximum use is made of mortar tests, which are more convenient than tests on concrete. However, as with any mix design process, testing the concrete is an integral and essential requirement.

CONCLUSIONS

1. The extensive test programme has shown that production of self-compacting concrete with readily available UK materials is feasible.

2. The slump flow, V-funnel and U-tests enable the fluidity, segregation resistance and passing ability of the concrete to be assessed. Using criteria suggested by experience in this project and successful practice elsewhere, limiting values for mix proportions have been defined for concrete made with either 10 or 20mm gravel aggregate.

3. A mix design method based on the approach of Okamura and Ozawa has been developed. This incorporates limits for mix proportions, ensures that the water/powder ratio is

sufficiently low for adequate inherent plastic viscosity of the paste and strength of the hardened concrete, and, using a linear optimisation analysis, produces mixes for laboratory testing. Experimental work on the mortar component gives a good estimate of the required admixture (superplasticizer) dosage, which can then be confirmed by tests on concrete.

4. The significant problems of maintenance of workability and the translation of the successful laboratory mixes to larger scale factory, plant or site production remain to be addressed.

ACKNOWLEDGEMENTS

The authors would like to express their thanks to the Taiwanese Ministry of Education and the UCL Graduate School for the financial support to Dr Chai and Mrs Jin respectively. We are very grateful to all the technical staff at UCL who assisted with the apparatus development and the experimental work, and in particular Mr Owen Bourne. Finally our thanks go to Fosroc International Ltd, Sika Ltd, Rugby Cement Ltd, Ash Resources Ltd, and Civil and Marine Ltd for supplies of materials.

REFERENCES

1. OZAWA K, MAEKAWA K and OKAMURA H High performance concrete with high filling capacity. Proceedings of RILEM International Symposium on Admixtures for Concrete: Improvement of Properties, ed E Vazquez, Barcelona, May 1990, pp 51-62

2. OKAMURA H, MAEKAWA K, and OZAWA K High performance concrete. First ed, (in Japanese) Gihoudo Pub Co Ltd, Tokyo, 1993, 232p

3. OKAMURA H, and OZAWA K Self-compacting high performance concrete Structural Engineering International, 4, 1996, pp 269-270

4. SAKATA N, MARUYAMA K and MINAMI M Basic properties and effects of welan gum on self-consolidating concrete. Production Methods and Workability of Concrete, ed P J M Bartos, D L Marrs and D J Cleland, E&FN Spon, Glasgow, 1996, pp 237-254

5. OKAMURA H and OZAWA K Mix design for self-compacting concrete. (Translation from Proc. of JSCE) Concrete Library of the Japanese Society of Civil Engineers, 6, June 1995, pp 107-120

6. TANGTERMSIRIKUL S and VAN B K Blocking criteria for aggregate phase of self-compacting high performance concrete. Proceeedings of Regional Symposium on Infrastructure Development in Civil Engineering, SC-4, Bangkok, Thailand, 1995, pp 58-69

7. TANGTERMSIRIKUL S, VAN B K and AOYAGI Y Optimum proportioning method for the aggregate phase of highly durable vibration-free concrete. MSc thesis, Asian Institute of Technology, Bangkok, Thailand, 1994

8. SEDRAN T, de LARRARD F, HOURST F and CONTAMINES C Mix design of self-compacting concrete (SCC). Production Methods and Workability of Concrete, ed P J M Bartos, D L Marrs and D J Cleland, E&FN Spon, Galsgow, 1996, pp 439-450

9. SEDRAN T and de LARRARD F Rene-LCPC Software to optimize the mix design of high performance concrete. BHP 96, Proc. 4th Int. Symp on Utilization of High Strength/High Performance Concrete, Paris, May 1996, pp 169-178

10. DOMONE P L and CHAI H-W Design and testing of self-compacting concrete Production Methods and Workability of Concrete, ed P J M Bartos, D L Marrs and D J Cleland, E&FN Spon, Glasgow, 1996, pp 223-236

11. CHAI H-W Design and testing of self-compacting concrete. PhD Thesis, University of London, 1998

12. DOMONE P L and CHAI H-W Testing of binders for high performance concrete. Cement and Concrete Research, V27, No 8, 1997, pp1141-1147

13. DOMONE P L and CHAI H-W The slump flow test for high workability concrete. Cement and Concrete Research, V28, No 2, 1998 pp 177-182

14. OZAWA K, SAKATA N and OKAMURA H Evaluation of self compactability of fresh concrete using the funnel test (in Japanese). Proceedings of the Japanese Society of Civil Engineers, No 490, Vol 23, 1994, pp 71-80

15. NISHIBAYASHI S, YOSHINO S, INOUE S, KURODA T and KUME T A study on the flow of highly superplasticized concrete. 4th CANMET/ACI Int. Conf. on Superplasticizers and other Chemical Admixtures in Concrete, ed V Malhotra, ACI SP-148, Montreal, 1994, pp 177-187

16. SHINDOH T, YOKOTA K and YOKOI K Effect of mix constituents on rheological properties of super workable concrete. Proc. of the Int. RILEM conf. on production methods and workability of concrete, Paisley, E&FN Spon, London, 1996 pp 263-270

17. CEB-FIP Model Code 1990, 437 pp, Thomas Telford, London, 1993

USE OF HSC IN TALL BUILDINGS

P A Mendis

A R Hira

University of Melbourne

Australia

ABSTRACT. The advancement of material technology and production has led to higher grades of concrete strengths. The application of High-strength Concrete (HSC) elements (concrete strength, f'c > 50 MPa) in civil engineering structures has increased significantly, with economy, superior strength, increased stiffness and greater durability being the principal reasons for its popularity. Concrete strengths up to 130 MPa are readily available using conventional production procedures. The enhanced strength/unit cost, strength/unit weight and stiffness/unit ratios associated with HSC has made use of reinforced concrete in highrise construction highly competitive with other materials. Other benefits in using HSC in highrise construction includes reduced creep and drying shrinkage characteristics which minimises differential shortening effects of vertical elements. This paper provides an overview on applications of HSC in tall building structures.

Keywords: High-strength concrete, Highrise structures, Design, Reinforced concrete, Optimisation, Walls, Columns.

Dr Priyan A Mendis is a Senior Lecturer in Structural Engineering at The University of Melbourne. He is a specialist in the fields of structural dynamics, earthquake engineering and design of tall buildings, as well as in the technology and properties of high-strength, high performance concretes. He has several years of recent experience designing a number of concrete structures including some tall buildings in Australia.

Mr Anil R Hira is a Senior Lecturer in Structural Engineering at The University of Melbourne. He has extensive experience in designing tall buildings structures in Australia and South-East Asia. He has conducted several training courses on structural design of highrise structures in Australia, Indonesia, Singapore and Middle East.

INTRODUCTION

In the 1990s, many tall record-setting concrete buildings were built worldwide including the world's tallest building, the 450m high twin Petronas Towers in Kuala Lumpur. The lower levels of the columns and the shear core of this building were constructed with 80MPa concrete. High-strength concrete (HSC) is also adopted for the lower levels of the core and the mega-columns of the 88 storey Jin Mao Building in Shanghai, which is the third tallest building in the world. The proposed "The Grollo Tower", planned for Melbourne, Australia, which will be the world's tallest building, will be adopting reinforced concrete for the primary structural elements. The use of HSC has been a significant factor in making these structures economically feasible.

Although HSC is often considered a relatively new material, its development and application has evolved over many years with bounds defining HSC continually changing. In the 1950s, concrete with a compressive strength of 35MPa was considered high-strength. In the 1960s, commercial use of concrete with compressive strength of 40 to 50MPa was achieved, increasing to about 65MPa a decade later. More recently, compressive strength of 100MPa has been commonly applied and strengths as high as 140MPa have been used in cast-in-place buildings.

For many applications, the cost benefits of using HSC more than compensates the additional cost of the materials due to additives such as silica fume and super-plasticisers and the extra quality control needed to produce them. The biggest advantages of HSC that makes its use attractive in highrise construction are the larger strength/unit cost, strength/unit weight and stiffness/unit cost ratios compared to normal strength concrete and other competitive building materials. Various applications of HSC in tall buildings, identified by the authors, are presented in Table 1. Some of these applications are already utilised by designers.

Two case studies are presented in this paper to highlight the advantages of HSC. However designers must be aware of the limitations and other concerns associated with the use of high-strength concrete. Some important design considerations are briefly discussed in the next section.

DESIGN CONSIDERATIONS

The current Australian concrete code, AS3600-1994[1], British Code, BS8110[2], American Code, ACI318[3] and many other concrete codes worldwide are based on research conducted on structural members made of normal strength concrete (NSC), defined as $f'c<50$ MPa. Extrapolating these design rules for HSC applications may not be appropriate. Due to the variations in fracture modes, microstructure and the differences brought about by various additives such as silica fume, fly ash, superplasticizers, the empirical design rules originally intended for NSC, needs to be re-evaluated. Some of the empirical design rules intended for NSC, when applied to HSC could either be :

a) unsafe
b) conservative (with varying degrees) or
c) satisfactory

Individual design rules codified in any of the national and international design codes must be experimentally verified and categorised with a view of taking further action. Although the application of high-strength concrete has been around since the 70's, some design issues are still not fully addressed. In recent years, many research projects have been undertaken to develop design formulae suitable for the design of high-strength concrete members. Designers are now in a position to use high-strength concrete without the uncertainties that existed previously. However, there still remains some unanswered questions regarding the design of high-strength concrete members, especially relating to ductility. There is also a need to consolidate and deliver the knowledge out to the practitioners for whom the research was intended in the first place. Key areas affected by concrete strength that requires addressing for design applications particularly for highrise structures, include ductility, stiffness, creep, shrinkage, shear capacity and long-term deflections. These are briefly discussed in the paper.

Table 1 Typical applications where high-strength concrete has advantages

TYPE OF MEMBERS	APPLICATIONS	ADVANTAGES OF HSC
General	• Columns in high rise buildings • High rise frames • Service cores • Shear walls and outriggers	• Strength • Reduction of cross-sectional dimensions • Greater rigidity of the frames • Overall economy and faster construction, most cost effective building material, Overall a more favourable life cycle cost • Structures with HSC result in a larger rentable area • Low maintenance needs; improved durability • Low creep and shrinkage • The use of HSC results in smaller cross-sections and savings in foundations and material handling costs
Precast/ Prestressed Members	• Precast beam and column elements • Tiltup panels • Prestressed beams	• Allows for a higher degree of prestress • Reduction in weight; improved handling • Higher cracking loads • Reduced stripping time for removal of formwork and reduced time for transfer of prestress, reduced time for lifting.
Beams and Slabs	• Slabs on ground • Plantroom floors • Other floor slabs and beams	• Improved punching shear resistance in slabs • Shallower floor system leading to reduced height of building • Better abrasion resistance • Overall a more life cycle cost saving • Ability to increase spans; favourable
Earthquake resistant design	• Frames • Walls	• Reduced inertial loads due to reduced dimensions • Enhanced ductility under flexure (Note:less ductility under high axial compression) • Higher-stiffness for sway control

Ductility

A primary concern of HSC application has been the reduction in ductility observed under uniaxial compression for increasing concrete strength. However it can be shown, experimentally and theoretically, that HSC flexural members possess a similar or higher ductility when compared to normal-strength concrete, with ductility demands of heavily loaded HSC compression members being satisfied by providing additional extra confining steel.

Defining the appropriate level of ductility in a concrete column may depend on the importance of the structure or the level of earthquake risk in the particular area. The design of lateral reinforcement in a column is based on following three criteria:

1. To prevent longitudinal steel from buckling. The strain at which buckling of longitudinal steel occurs, is independent of the compressive strength of concrete. Therefore the spacing requirements suggested for NSC are applicable to HSC columns.

2. To prevent fracture of lateral steel due to sudden concrete cover loss. Unconfined cover concrete of HSC columns deteriorate at a higher rate compared to NSC. This may cause a sudden transfer of energy to the confining steel, which may lead to steel fracture. In order to investigate this uncertainty, an energy balance method was adopted. It was shown that increasing compressive strengths from 50 to 80 MPa causes only a small reduction in the hoop fracture strain.

3. To provide adequate confinement for ductility. In a project completed recently[4], a formula was proposed to calculate the spacing of lateral reinforcement by modifying the present requirements in the Australian Concrete Code. In the method suggested by Mendis[4], adequate ductility is provided for a high-strength concrete column by comparing its moment-curvature characteristics with that of a normal strength concrete column. The level of ductility obtained from a 50 MPa concrete column is therefore also ensured for the same column made with high-strength concrete. Limits are also determined for the longitudinal steel buckling and lateral steel fracture.

Elastic modulus

Researchers around the world agree on the fact that the existing formulae overestimate the static modulus of elasticity of HSC. In the absence of other information the following equation suggested by Mendis [4] can be used to find the elastic modulus.

This equation is similar to the equation given in AS3600 and ACI318, for normal strength concrete, but applicable for both normal and high-strength concrete.

$$E_c = 0.043\eta\rho^{1.5}\sqrt{f'_c} \pm 20 \qquad (1)$$

where, $\eta = 1.1 - 0.002f'_c. \leq 1.0$

Creep

The creep characteristics of HSC is expected to reduce with reduction of water/cement ratio or increase in compressive strength. This property minimises the expected differential shortening between HSC columns with high stress levels and HSC walls with lower stress levels often encountered in tall building design. Table 2 [5] shows the reduction in basic creep factor of concrete with increasing compressive strengths up to and including 100 MPa.

Table 2 Basic creep factor

Characteristic Compressive Strength MPa	20	25	32	40	50	60	70	80	90	100
Basic Creep factor	5.2	4.2	3.4	2.5	2.0	1.5	1.2	0.9	0.7	0.5

Shrinkage

Shrinkage of concrete is usually divided into three broad areas: plastic shrinkage, autogenous shrinkage and drying shrinkage. Plastic shrinkage occurs during the first few days after placing of concrete due to evaporation of water from the surface of fresh concrete. Due to the high paste content, HSC is expected to be more susceptible to plastic shrinkage than NSC. Autogenous shrinkage occurs due to reduction of water in concrete due to hydration. The volume of hydrates formed is substantially less than the sum of the volume of water and cement. Autogenous shrinkage is expected to be higher in High-Strength Concrete due to the higher cement or binder content. Drying shrinkage occurs due to evaporation of water from the surface of concrete after it attains the final set. Typically HSC possesses a lower drying shrinkage component. This information can be obtained from the concrete supplier.

Shear

The shear strength of high-strength concrete does not proportionately increase in strength with f'c. The lack of aggregate interlock mechanism, which resists a substantial portion of the post- crack shear forces, is the factor that reduces the shear strength HSC. Unlike for NSC, fracture surfaces in HSC are smooth with fractures passing through aggregates. Pendyala and Mendis [6] suggested an additional reduction factor to compute the shear strength of HSC members. The shear resisted by concrete can be calculated from Equation (2).

$$V_{uc} = \beta_1 \beta_2 \beta_3 \beta_4 b_v d_o \left(\frac{A_{st} f'_c}{b_v d_o} \right)^{1/3} \qquad \text{Eqn. (2)}$$

where

$\beta_4 = 1$ for $f'_c < 50$MPa

$= 1.25 - 0.005 f'_c \geq 0.9$ for $50\text{MPa} \leq f'_c \leq 100\text{MPa}$

$\beta_1, \beta_2, \beta_3$ as in AS 3600

Deflections

Due to the lower creep characteristics of HSC the long-term deflections of HSC beams are expected to be significantly less than that of NSC. In addition, the effect of compression steel in reducing the long-term deflection in beams is less for HSC, which is not accounted for in the code formula. Mendis [4] suggested a revised formula for the long-term factor K_{cs} in AS3600 considering the aforementioned factors. This formula has been validated by comparing with experimental results. The revised formula is given below.

$$K_{cs} = \mu\left[2 - 1.2\left(\frac{A_{sc}}{A_{st}}\right)\mu\right] \geq 0.8 \qquad (3)$$

$$\mu = \frac{50}{f_c'} \qquad \mu \leq 1$$

Walls

High-strength concrete shear wall systems are now widely accepted around the world as a rational and economical part of multi-storey construction, but limited guidance is available on its design methodologies. Fragomeni and Mendis [7] showed that the strength of HSC walls do not increase proportionately with concrete strength. It was suggested that for a 60% increase in concrete strength (from 50MPa to 80MPa) a 37.5% increase in wall strength is appropriate, taking account of the test results and other research conducted on HSC concrete panels. However as the present formula for NSC given in AS3600, which is based on BS8110 formula, is conservative, the predicted values using this formula for HSC walls were below the experimental values up to f'c of 80 MPa. Therefore the present formula can be retained for $f'_c < 80$ MPa, however should be used with caution for f'c > 80 MPa.

The optimisation of core wall elements is complex. The designer is required to optimise between concrete quantity, reinforcement quantity and concrete strength. The designer requires accurate knowledge on the cost rates for each of the components. Studies by the authors based on Australian cost rates have clearly shown that increasing concrete strength is by far the most economical means of increasing the load capacity of a typical core wall. This is illustrated in Figure 1. Therefore, given a core geometry, the designer should ideally aim to minimise wall thickness and reinforcement quantity by using HSC.

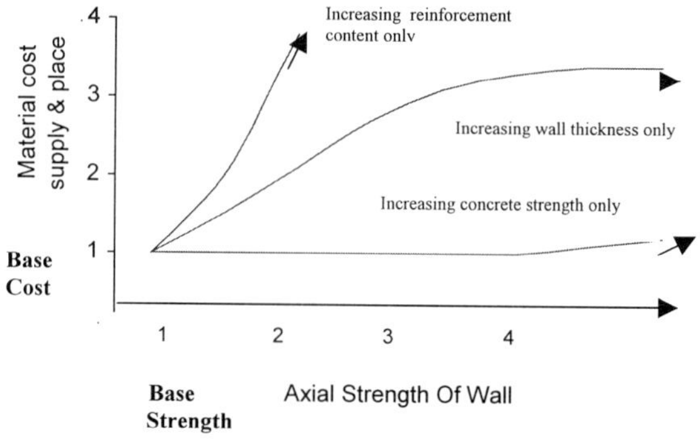

Figure 1 Cost trends to increase reinforced concrete wall strength

STRUCTURAL OPTIMISATION

Although concrete strengths of up to 140MPa are readily available the costs of production associated with HSC and the lack of awareness of the advantages of using this versatile material effectively have discouraged many designers from specifying it for tall building applications. Extensive studies have shown that the cost benefits associated with optimal use of HSC leading to increased lettable space due to reduced wall thickness are significant, with the benefits increasing with increasing rental rates.

A systematic optimisation process based on principle of virtual work has been applied by the authors [8] on tall buildings with the primary lateral load resisting structural systems consisting of cantilever structural walls. By investigating the participation of element stiffness of each wall component at each level towards the lateral displacement at the top of highrise building, often the governing design criteria, the study investigates the benefits of using HSC. The study incorporates many factors including buildability aspects (eg restricting the number of wall thickness transitions), material cost savings, reduced design lateral inertial seismic loads from mass variations due to thinner walls, increased stiffness associated with HSC and most important of all yield and rental rates which impact on the capitalised value of the increased lettable space produced by use of thinner walls through use of HSC.

The computer program "OPTIC"[8] was developed to calculate the optimum solutions for cantilever structural walls with different concrete strengths and designated number of wall thickness transitions. The study revealed the following conclusions.

1. Overall cost savings of up to 3 to 4 times the material cost associated with the cantilever walls can be made by use of HSC.
2. The benefits of using HSC increases with higher yielding buildings corresponding to prestigious buildings in central business districts.
3. The single biggest cost benefit in utilising HSC is the capitalised value of the increased lettable space generated by reduced wall thickness.

CASE STUDY 1

In most of the tall buildings, concrete shear walls function as the main lateral load resisting structural element. The walls can be either planar, open sections or closed sections typically located around elevator and stair cores, or located at the ends of the building. The shear wall in essence behaves as a cantilever in resisting lateral loads. These walls are normally subjected to axial forces and moments. It would be futile to attempt to introduce all the available systems in this paper because of the significant number of variations that are possible in highrise buildings. Instead, this paper attempts to summarise the advantages of using HSC shear walls through a case study.

To illustrate the advantage of using HSC shear walls a typical concrete core structure in a 197m. high, 47 storey office building located in Melbourne is taken as an example. The cross-section is modified for the purposes of this case study. The modified cross-section and elevations are given in Figure 2. The typical core section is given in Figure 3 with the shear walls in core box 2 analysed for this study.

The analysis involves dividing the core box into wall elements so that sectional properties are calculated. The total ultimate axial design load is determined for each core element. This design load is then compared to the ultimate capacity of the core element. The ultimate capacity is found by adding the capacities of individual wall elements. Provided the ultimate capacity exceeds the ultimate applied load then the overall ultimate conditions are satisfied. Also the ultimate stresses at the critical points (normally corner points) are determined by the following expression:

(P/A + or - M/Z)wind + (P/A)dead and live (or dead only) Eqn. (4)

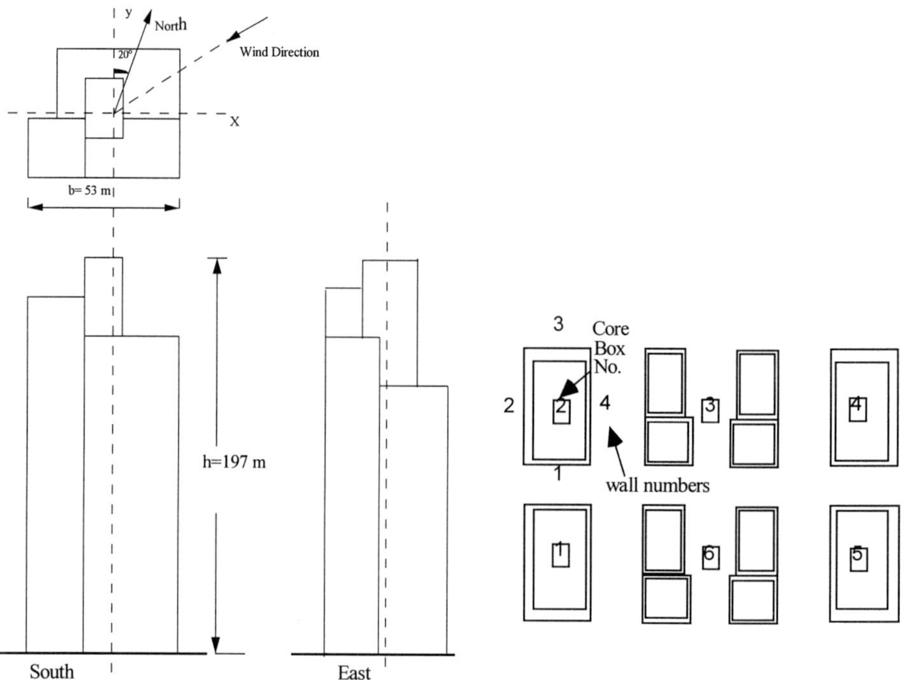

Figure 2 Typical cross-section and elevations

Figure 3 Typical core section

These stresses are compared with the capacities of individual wall elements. The wall stresses are determined for both ultimate and serviceability conditions. Serviceability capacity is found by reducing the ultimate capacity by a reduction factor (eg. 0.7). If these conditions are not satisfied, the core must be redesigned. This procedure is described more extensively by Mendis et al. [9].

The AS3600 wall equation was used with an effective length factor of 0.8 and eccentricity of t/6. The thicknesses were than optimised for various values of concrete strength. The resulting thicknesses are shown in Figure 4. Evidently, a substantial decrease in wall thickness and cost can be obtained using high-strength concrete.

To illustrate the cost savings involved in this case, concrete strengths between 40 and 80MPa were selected. The 40 to 80MPa range is predominately adopted in the construction industry today.

The capitalised net lettable area saving used is $3500/m^2 and the concrete cost used is $250/m^3 for 40MPa increasing proportionately to $306/m^3 for 80MPa. These costs represent some typical values and may change according to the place and time of construction. The results are given in Table 3. The concrete strength of 40MPa is used as the base for cost comparison. The values provided are per storey height. Therefore if all 47 levels are considered then the cost saving is enormous. The advantages of using higher concrete strengths is obvious from Table 3, with 80MPa being the most optimum solution for this case.

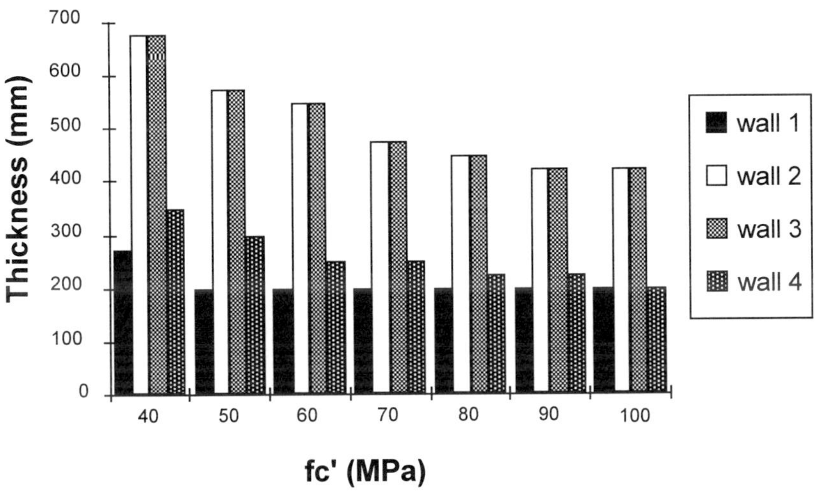

Figure 4 Change in wall thickness with varying concrete strength

These observations are equally applicable for HSC columns. Moreno and Zils [10] examined several factors associated with the optimum design of highrise buildings such as the lateral forces, building drift, foundation systems and cost of concrete material and placement, reinforcement and column formwork by using three different column sizes.

The cost-load evaluation indicated that cost per unit axial load decreased as the concrete strength increased. They showed that the most economical way to design columns was to use the highest available concrete strength corresponding with minimal reinforcing steel. Smith and Rad [11] investigated the economic advantages of using HSC in columns in low-rise (5 storey) and medium-rise buildings (15 Storey) using ACI318 provisions.

They showed that for their case study the relative reduction in the column construction cost was in the order of 26% for 55 MPa concrete and more than 40% for 83 MPa concrete when compared with 28 MPa concrete.

Table 4 Cost savings per wall for increasing concrete strength

		COST SAVINGS PER WALL / STOREY ($)			
Concrete Strength (MPa)	Wall 1	Wall 2	Wall 3	Wall 4	Total Savings Per Core Box / Storey
40	0	0	0	0	0
50	824	3481	1344	1735	7384
60	797	4181	1269	3515	9762
70	770	6840	2076	3403	13089
80	743	7606	2309	4270	14928

CASE STUDY 2

Time-dependent axial shortening in the cores and columns of tall concrete buildings is primarily due to creep and shrinkage which occurs during construction. These deformations result in both cumulative and differential shortening of the vertical building elements and are of particular interest to designers. The shear core of a tall building is constructed normally in advance of the outer frame. This will allow separate construction crews to work independently and enables the anchoring of the crane and other ancillary equipment. Accurate estimations of differential shortening between the core and the outer columns is of prime importance during the design and the construction phase. The axial shortening depends on several factors such as the construction loads, occupancy loads, creep, shrinkage and the construction schedule. It is an extremely complex task to manually calculate the axial shortening due to long-term effects. The computer program COLECS [12] is a useful tool to determine the axial shortening of cores and columns. This program was modified by the authors to include high-strength concrete. A 55 storey building constructed in Sydney, Australia was used as the case study to show the influence of HSC in reducing the long-term deformations. To simplify the problem, uniform core properties were used throughout the building. The area of the core was equal to 32.8 m^2 and the core perimeter was equal to 127.2 m. The storey height was taken as equal to 3.8 m. The shear core was modelled as a single column.

Three cases were considered.

i. Ground to Storey20 -f'c=50MPa; Storey21 to Storey55 -f'c=25MPa; Shrinkage= 550□s
ii. Ground to Storey20 -f'c=100MPa; Storey21 to Storey55 -f'c=50MPa; Shrinkage= 550□s
iii. Ground to Storey20 -f'c=100MPa; Storey21 to Storey55 - f'c=50MPa;Shrinkage = 400□s

The total long-term (30 years) axial shortening values at the top of the core were 122.5 mm, 77.5 mm and 58.2 mm for cases (i), (ii) and (iii) respectively. The axial shortening at different storey levels are illustrated in Figure 5.

As seen from Figure 5, there is a significant reduction in axial shortening of the core when high-strength concrete is used for lower-storeys. For cases (i) and (ii), a basic shrinkage value of 550 μs was used. As mentioned earlier, drying shrinkage in high-strength concrete is lower compared to normal strength concrete. A lower shrinkage value (400 μs) was used in case (iii) to demonstrate the reduction in axial shortening of the core. Different components of axial shortening for case (i) are shown in Figure 6. Creep component is the largest with a value of 48.6 mm at the top of the core. Other components were shrinkage (43.9 mm) and elastic shortening (32.2 mm).

There is a small reduction in shortening due to the stiffening effect of the reinforcement (2.2 mm). This case study shows the importance of accurate estimation of creep, shrinkage and elastic modulus in long-term deformation calculations of tall buildings.

CONCLUDING REMARKS

In 1990s, many tall record-setting concrete buildings were built worldwide including the world's tallest building, the 450m high twin Petronas Towers in Kuala Lumpur. The use of High-strength concrete (HSC) has been a significant factor in making these structures economically feasible. An overview on applications of HSC in tall building structures is provided in this paper. Key areas affected by concrete strength that requires addressing for design applications particularly for highrise structures, include ductility, stiffness, creep, shrinkage, shear capacity and long-term deflections. These aspects are briefly discussed in the paper.

To illustrate the advantage of using HSC shear walls, a typical concrete core structure in a 197m. high, 47 storey office building located in Melbourne is analysed in the first case study. It is shown that significant cost savings can be achieved by using higher concrete strengths, with 80MPa being the most optimum solution for this case. Time-dependent axial shortening in the cores and columns of tall concrete buildings due to creep and shrinkage which occurs during construction are of particular interest to designers. A 55 storey building constructed in Sydney, Australia is used as a case study to show the influence of HSC in reducing the long-term deformations of a shear core. There is a significant reduction in axial shortening of the core when HSC is used.

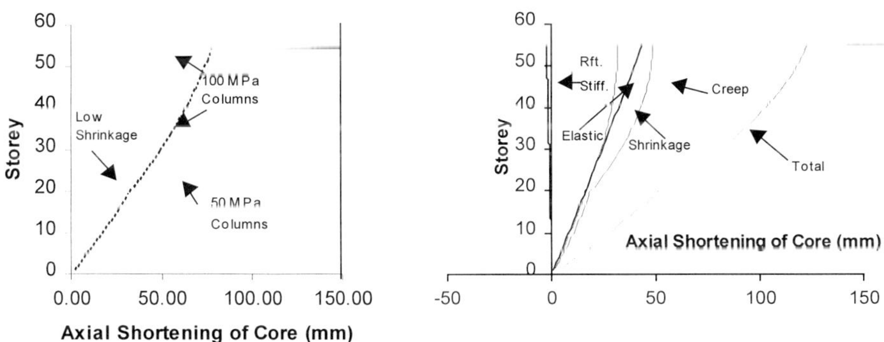

Figure 5 Shortening of core for cases (i)-(iii) Figure 6 Components of shortening for case (i)

REFERENCES

1. SAA, AS3600 Concrete Structures, Standards Association of Australia, Sydney, 1994.

2. BS8110 "Structural Use of Concrete, Part 1 - Code of Practice for Design and Construction", British Standards Institution, London, 1997.

3. ACI318 - 95 "Building Code Requirements for Reinforced Concrete", American Concrete Institute, Detroit, 1995.

4. MENDIS, P., "Design of High-strength Concrete Members: State-of-the-Art", To be published, 1999.

5. PENDYALA, R. AND MENDIS P.A., "Shear in High-strength Concrete", Civil Engineering Transactions, Institution of Engineers of Australia, Vol SE1, No. 1,1998, pp 67-74.

6. SETUNGE, S. AND PADOVAN, "Early Creep and Shrinkage in HPC in Arid Environments", CIA Biennial Conference, Adelaide, 1997, pp. 37-43.

7. FRAGOMENI, S. AND MENDIS, P.A., "Improved Axial Load Formulae for Normal and High-strength Concrete Walls", Civil Engineering Transactions, The Institute of Engineers, Australia, Vol. CE38, Nos. 2,3 & 4, 1996, pp. 71-81.

8. BONG, B.G., HIRA, A., AND MENDIS, P.A., "Optimal Application of High-strength Concrete in Tall Buildings" , Proceedings of ACMSM 15 Conference, Melbourne, 1997, pp. 243-250.

9. MENDIS, P.A., KARANTZAS, V. AND FRAGOMENI, S. "Efficient design of concrete core structures", Proceedings of the 18th Concrete and Structures Conference, Singapore, August, 1993.

10. MORENO, J. AND ZILS, J., "Optimization of highrise concrete buildings", Special Publication 97, Analysis and Design of Highrise Buildings, American Concrete Institute, pp. 25-92.

11. SMITH, G. AND RAD, F., "Economic Advantages of High-strength concrete in Columns", Concrete International, Vol. 11:4, pp. 37-43.

12. BEASELY, A., Program COLECS, Civil Engineering Dept., University of Tasmania, 1987.

HIGH PERFORMANCE CONCRETE FOR EFFICIENT CONCRETE FRAME CONSTRUCTION

S J Crompton

RMC Ready Mix Limited

United Kingdom

ABSTRACT. In early 1998 a seven-storey concrete frame building was constructed inside the BRE research facility at Cardington. High strength concrete with a characteristic strength of 85 N/mm^2 was used to construct the lower storey columns in order to maintain equal column size throughout the height of the building. This study, which is part of a major research project into the performance of concrete frame structures, deals with the mix design and performance of the high strength concrete. Mixes were designed with microsilica or metakaolin and the relative performance of each was examined. The effects of incorporating high dosages of polypropylene fibres (3%) was examined and the effects of maintaining high workability on early stripping times of formwork was investigated.

Keywords: High strength, Metakaolin, Microsilica, Superplasticiser, Fibres.

Mr Steven J Crompton is Divisional Manager – Technical Services of RMC Readymix Limited. He is Head of Technical Services in the UK and is responsible for the provision of central technical services for RMC Readymix Limited.

The role includes responsibility for research and product development along with the development of quality assurance

INTRODUCTION

The use of high strength concrete (HSC) in buildings is normally associated with the construction of high rise structures. However, an innovative research project at the BRE research facility at Cardington investigated the use of high strength concrete in the construction of medium rise concrete frame buildings.

A full-scale seven storey concrete frame was built inside the research facility to provide controlled conditions to monitor the performance of the structure both during and after construction. The building was designed on a 7.5m grid with flat slab construction and the same column sizes on each floor. The columns used C85 concrete from the ground floor to the third floor and C37 concrete on the remaining storeys. Column sizes were maintained at 400mm x 250mm at the edge and 400mm x 400mm in the centre to enable the multiple reuse of the same formwork. The performance of the concrete was monitored to establish the rate of stiffening and early strength development in order to establish the most effective mix for efficient construction of the frame.

As part of a separate research programme into fire resistance of high strength concrete some columns incorporated polypropylene fibres at triple the normal dosage rate. The effect of the inclusion of this level of fibres on the fresh and hardened properties of the concrete was also investigated.

MATERIALS

The basic premise for the design of the mixes was to use, as far as possible, normally available materials.

One of the primary aims of the mix design process was to produce a mix that would remain workable for approximately 2 hours. This was to allow the superplasticiser to be added at the batching plant and to give time for the concrete to be placed by traditional methods.

Cement

The chemical composition of cement can have an influence on the ultimate strength and setting characteristics of HSC. In particular cement with high C_3A contents can lead to higher strength. However, some researchers [1] have shown that high levels of C_3A can lead to rapid loss of workability. For this reason some countries such as Norway produce specific cement for HSC with C_3A levels below 5.5% [2]. In this project a cement complying with the requirements of BS12 (1996) was used with an average C_3A level of 8%. Cement with a lower C_3A was not readily available and the setting characteristics of the HSC were to be monitored as part of the project.

Pozzolanic Materials

The use of high reactivity pozzolans such as microsilica and metakaolin was necessary to achieve the specified strength without excessive cement contents. Both microsilica and metakaolin were used in separate mixes to provide comparative data on their relative performance.

Microsilica

Microsilica is a by-product of the smelting process used to produce silicon and ferrosilicon. The material is a very fine powder with a mean particle size between 0.1 and 0.2 micron. The microsilica used in this project consisted of 95% Silicon Dioxide (SiO_2) and was pre-mixed with an equivalent weight of water to form a slurry. A slurry is easier to handle than a dry powder and helps the homogenous mixing of the concrete.

Metakaolin

Metakaolin is a highly pozzolanic material formed when china clay, the mineral kaolin, is heated to a temperature between 600 and 800^{0}C. The resulting material is ground to a fine powder with a particle size somewhat finer than pulverised fuel ash. The metakaolin used in this project was added in powder form.

Aggregates

The properties of the aggregates are important in producing high strength concrete and the following properties are particularly significant [3]:

- Mechanical Properties
- Shape
- Particle Size Distribution

For normal "good" quality aggregates the upper limit for concrete strength is around 120-140 N/mm^2. However, these strengths were not possible with the local flint aggregate. Although mechanically strong the smooth surface of the flint leads to bond failure at strengths significantly below 100 N/mm^2. For this reason a crushed carboniferous limestone coarse aggregate was imported into the batching plant for use in HSC. For practical storage reasons a 20-5mm graded limestone was used.

The fine aggregate properties are less significant than the coarse aggregate and a good clean concreting sand is generally considered suitable. The locally available flint sand was used.

Water Reducing Admixtures

High range water reducing admixtures (superplasticisers) were included in the mix to improve the workability of the concrete, disperse the cementitious powder and reduce the overall water content and therefore cement content of the mix. Based on previous experience Grace SP6, a blended polymeric sulphonate, was used as the superplasticiser. This is a powerful water reducing admixture and was used at a dosage rate of 2% by weight of cementitious material equivalent to 8.8 litres/m^3.

MIX DESIGN

In general the potential strength of concrete is governed by the water/binder ratio [4] however, for HSC the influence of aggregate strength and the cement paste/aggregate bond are also important. In order to produce the required strength the mix was designed to minimise the water content without excessive cement contents. Table 1 shows the mix proportions for the preliminary mix designs.

Table 1 Preliminary mix designs

MIX TYPE	MIX PROPORTIONS, kg/m^3							
	PC	Micro-silica	Meta-kaolin	20-5	Sand	Fibres	S/Plast	28 day strength N/mm^2
PC + Micro-silica	400	40	--	1059	754	--	8.8	110.5
PC + Meta-kaolin	400	--	40	1066	737	--	8.8	105.0
PC + Micro-silica + Fibres	400	40	--	1090	710	2.7	8.8	114.0

Although these mixes comfortably exceeded the strength requirement it was evident that the workability loss was rapid with a reduction in slump of 100mm in one hour. In order to counteract the workability loss a standard dose of a plasticiser was added to the mix and the initial workability was increased. Two types of plasticisers were assessed, firstly a standard lignosulphonate type which is still widely used as a water reducing agent. Secondly a polyhydroxy carboxylic acid/lignosulphonate blend which is a more powerful water reducer known to have slight retarding effects. The effects on workability retention of two different types of plasticiser are shown in Figure 1.

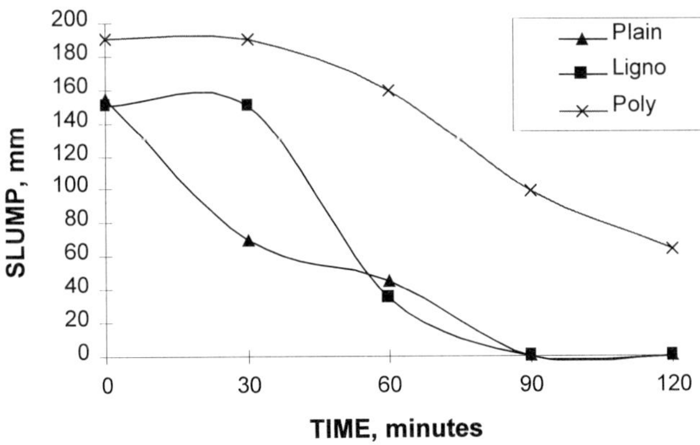

Figure 1 Workability retention

The degree of workability retention was reflected in the early strength development of the concrete. Although the polyhydroxy carboxylic acid blend plasticiser extended the usable life of the concrete, the effect on the 1 day compressive strength was significant as shown in Table 2. However, for practical reasons it was important to maintain the workability at a minimum of 100mm slump for as long as possible to allow all materials to be batched as the batching plant.

Table 2 Early strength development

MIX	MIX TYPE	ADMIXTURES	STRENGTH, N/mm^2			
			1 day	3 days	7 days	28 days
B	PC + Microsilica	SP6	23.2	55.9	74.2	96.6
1	PC + Microsilica	SP6 + Lignosulphonate	11.6	45	54.4	74.2
4	PC + Microsilica	SP6 + Polyhydroxy	5.4	72.7	93.7	108

As a result of these laboratory trials the mixes were modified by increasing workability to a target 550mm flow with a superplasticiser and a polyhydroxy carboxylic acid plasticiser incorporated into the mix. The inclusion of fibres in the mix increased both the water demand and cohesiveness of the mix and as a result the percentage of fine aggregates in the mix was reduced compared to the microsilica mix without fibres. Final mix designs are shown in Table 3.

Prior to final delivery full scale batches were supplied to preliminary works on site and these confirmed that the mixes could be handled and placed by normal construction methods.

Table 3 Final mix designs

MIX	MIX TYPE	FLOW, mm	MIX PROPORTIONS, kg/m^3							FREE W/C
			PC	Meta-kaolin	Micro-silica	20-5	Sand	SP6 l/m^3	P7 l/m^3	
19	PC + Metakaolin	550	400	40	--	1146	717	8.8	1.2	0.32
15	PC + Microsilica	550	400	--	40	1170	732	8.8	1.2	0.25
18	PC + Micro-silica+ Fibres	550	400	--	40	1173	690	8.8	1.2	0.32

RESULTS AND DISCUSSION

Workability Retention

There were no problems with workability retention and all mixes remained workable even when placed by relatively slow methods of placing such as crane and skip. The maximum time from batching to final placement was 1 hour 40 minutes.

The degree of workability loss was much less than that observed in the laboratory. This is attributed to the lower on-site temperatures (concrete was placed during winter) and the agitating effect of the delivery trucks. On the ground to first floor, and first to second floor columns there was no loss in workability during the placing of the concrete. However, the extended workability did not allow the early stripping of columns and it was not possible to strip the formwork until 36 hours after casting.

As a result of the improved workability retention the initial delivered workability was reduced on the second to third floor columns to 400mm flow. This increased the early strength development of the concrete and allowed formwork removal at approximately 24 hours.

Handling Characteristics

Despite the high workability and reduced sand contents of the delivered mixes the concrete still exhibited some thixotropic behaviour. This was particularly evident in the mixes containing polypropylene fibres although once energy was applied to the concrete via internal vibrators the concrete could be placed in the normal way.

The low water content and high level of "fine fines" in the concrete reduced the tendency for bleeding in the mix and there was no evidence of bleed water in any of the HSC supplied.

The lack of bleeding would generally lead to an increased risk of plastic shrinkage cracking. However, there was no evidence of any plastic cracking in any of the members cast. This is attributed to the site conditions – low temperature and an internal location with no wind.

Strength Development

Strength development is a function of the materials, mix design, curing and ambient conditions and Figure 2 shows the strength development of mixes incorporating microsilica, metakaolin and microsilica + fibres up to 28 days for British Standard water cured cubes. From this it can be seen that the mix incorporating microsilica developed strength quicker and reached a higher 28 day strength than the other mixes. This is a reflection of the lower water cement ratio of this mix. The incorporation of fibres into the mix increased the water cement ratio by 0.07 and there was a corresponding reduction of approximately 16 N/mm^2 in the 28 day strength.

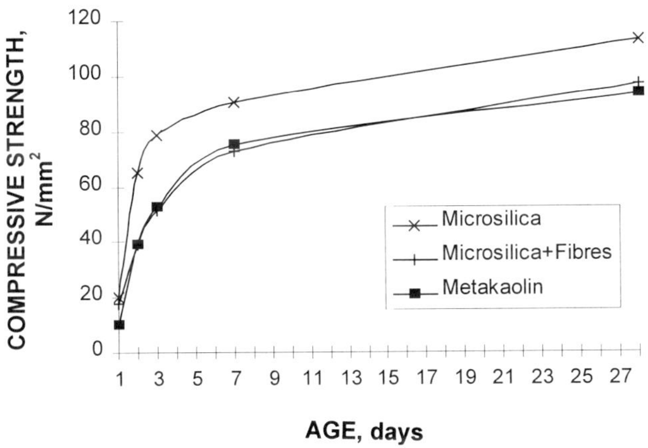

Figure 2 Strength development

All mixes exceeded the specified 28 day strength requirement with an average value of 96.9 N/mm^2. Strength development beyond 28 days was slow with an average gain of 5 N/mm^2 from 28 to 56 days. This is attributed to the lack of free water for further hydration.

Influence of Curing

Very low water cement ratio concrete can be susceptible to self desiccation and some researchers [5] have questioned the validity of water cured cube results in relation to high strength concrete. As part of this study air cured cubes were compared with standard water cured cubes on a certain number of columns and slabs and the results are shown in Table 4.

These initial results confirm that the high strength concrete is more susceptible to lack of curing and significant losses in strength are possible. Further more extensive research is required on the effects of desiccation on high strength concrete which may be particularly significant in relation to longer term strength development of insitu concrete.

Table 4 Comparison of water and air cured cube strengths

ELEMENT	STRENGTH, N/mm^2		STRENGTH REDUCTION, %
	Water Cured	Air Cured	
Column Ga	103.5	89.2	13.8
Column 1c	112.5	100.6	10.6
Column 2a	114.7	102.8	10.4
Slab 1	62.0	60.0	3.2
	63.2	64.0	-1.3
Slab 2	55.6	53.2	4.3
	57.8	54.1	6.4
Slab 3	54.2	49.8	8.1
	56.8	55.0	3.2

Temperature

The ambient temperature within the research facility remained reasonably constant throughout the construction period which was ideal for monitoring the heat development within the concrete.

Thermo couples were attached to a number of columns throughout the structure and Table 5 summarises results for both HSC and normal concrete columns.

With the exceptions of columns B3 and C3 on floors 2-3 there was little difference in the temperature gain between HSC and normal strength concrete. Heat evolution is roughly proportional to the cement content but at a constant cement content less water leads to a reduction in the heat evolved [3]. The results obtained indicated that HSC will not necessarily generate more heat and will not be more sensitive to thermal cracking.

Table 5 Temperature development

FLOOR	COLUMN REF.	MIX	TEMPERATURE, ^{0}C		
			Ambient	Max	Difference
1-2	B3	C85 + Metakaolin	9	24	15
	C3	C85 + Metakaolin	9	24	15
	B2	C85 + Metakaolin	12	30	18
	C2	C85 + Metakaolin	12	29	17
2-3	B3	C85 + Microsilica	14	42	28
	C3	C85 + Microsilica	14	42	28
	B4	C85 + Microsilica + fibres	11	28	17
	C4	C85 + Microsilica + fibres	11	29	18
3-4	C4	C37	10	27	17
	B3	C37	10	28	18
	C3	C37	10	28	18
	B2	C37	10	26	16
4-5	C4	C37	9	26	17
	B3	C37	10	27	17
	C3	C37	10	27	17
	B2	C37	11	26	15

Other Mechanical Properties

Although not specifically required for the design of the structure some other mechanical properties were measured during the course of the research and these are summarised in Table 6.

Table 6 Other mechanical properties

MIX	28 DAY COMPRESSIVE STRENGTH, N/mm^2	TENSILE SPLITTING STRENGTH, N/mm^2	MODULUS OF ELASTICITY, kN/mm^2
C85 + Microsilica + fibres	103.5	5.9	50.0
C85 + Metakaolin	112.5	6.2	54.5
C85 + Microsilica	114.7	7.3	69
C37	50	4.1	37
C37	58	4.0	36
C37	52.5	3.8	29

Tensile strength is an important characteristic in the prediction of cracking and the durability of concrete. These results confirm previous studies [4] which show that tensile strength increases with compressive strength but is not directly proportional.

Some studies [4] indicate that tensile splitting strengths may be as high as 10% of the compressive strengths for normal weight concrete reducing to 5% for HSC. Whilst there is insufficient data in this study to reliably confirm these figures the general trend is confirmed.

The static modulus of elasticity is used to estimate deformation in the structure and BS 8110 gives an equation for estimating the value. This equation under estimated the values obtained in this project by up to 25 kN/mm^2 for HSC. However, in critical structures where deformation is important the standard recommends tests on actual materials to be used and the results of this study reinforce this view.

PRACTICAL APPLICATIONS

This research has shown that high strength concrete can be used effectively in the construction of medium rise concrete frame buildings. HSC allows the frame designer to maintain constant column sizes throughout the full height of the building which has significant benefits in improving construction times, enabling early stripping of the shutters and reducing overall project duration.

Reduced column sizes leads to savings in the cost of material (concrete reinforcement and formwork). The smaller column sizes can result in reduced overall foundation loading and increased lettable floor areas.

When combined with the most efficient forming system and reinforcement the use of HSC can contribute to a significant decrease in construction times. Preliminary estimates from the research indicate that the appropriate combination would reduce frame construction times by up to 50%.

CONCLUSIONS

1. It is possible to produce high strength concrete with normally available materials. However, care is required in the selection of appropriate aggregates and admixtures to ensure optimum performance.

2. The combined use of a superplasticiser and a plasticiser gives workability retention of up to two hours which allows all materials to be added at the batching plant. This removes the need for the addition of superplasticisers on site thereby speeding up construction

3. The use of high dosages of fibres increases the cohesiveness of the concrete and leads to a reduction in strength related to an increase in water demand of the concrete.

4. There was an absence of bleeding in all HSC produced and extreme care would be required to prevent plastic shrinkage cracking in unfavourable conditions.

5. Extending the workability retention of the concrete leads to retardation of the set such that stripping of initial formwork was not possible at less than 24 hours.

6. Early strength development is rapid with typically 80% of the 28 day strength being obtained at 7 days. Strength development from 28 to 56 days was limited with less than 5% increase in water cured specimens in this period.

7. The influence of curing on the compressive strength of HSC is more significant than the effects on normal strength concrete. Reductions of more than 10% in the 28 day strength were observed for air cured specimens.

8. Temperature development of HSC was similar to normal strength concrete.

REFERENCES

1. PERENCHIO, W F. An evaluation of some factors involved in producing very high strength concrete. PCA, 1973.

2. RONNENBURG, S. High strength concrete for North Sea Platforms, Concrete International, 1990, 12, 29-34.

3. SHAH, S P AND AHMAD, S P. High Performance Concretes and Applications. E Arnold, 1994, 391 pp.

4. FIP/CEB, High Strength Concrete : State of the Art Report, FIP, 1990, 61 pp.

5. CORRASQUILLO, P M AND CORRASQUILLO, R L. Evaluation of the use of current concrete practice in the production of high strength concrete. ACI Materials Journal, 1988, 85, 1, 49-54.

PRODUCTION AND MECHANICAL CHARACTERIZATION OF VERY HIGH PERFORMANCE FIBRE-REINFORCED CONCRETE BEAMS

G L Guerrini

L Cassar

Italcementi Group

L Biolzi **G Rosati**

Polytechnic of Milan

Italy

ABSTRACT. In this paper the mix design development of a very high performance fibre-reinforced concrete to be used for the manufacturing of precast beams for a road bridge in Italy is described. Experiments were conducted to characterise the material and to study the flexural behaviour of reinforced beams. In particular, beams with and without a hollow core were considered. The tests were monitored with a moiré grid with 40 lines/mm to evaluate the crack development and the position of neutral axis. This experimental work represents perhaps the first integrated experience in Italy, from the design to the final application of very high performance concrete.

Keywords: Fibre-reinforced concrete, Fracture mechanics, High performance concrete, Mechanical strength, Microfibres, Microsilica, Precast, Quality control, Superplasticizer.

Dr G L Guerrini is a Senior Research Engineer at CTG Italcementi Group Bergamo Laboratories – Italy. He is responsible for the development and application of high performance concretes and other innovative cement-based materials, including fibre-reinforced composites.

Professor L Biolzi is Professor of Structural Engineering Department of Polytechnic of Milan, Italy. His principal fields of interest are fracture mechanics of quasi-brittle materials, size effects, composite materials and constitutive behaviour.

Dr L Cassar is R&D Central Manager of Italcementi Group. He is author of about 60 scientific papers.

Dr G Rosati has been an Assistant Professor since 1992 at Polytechnic of Milan, Italy. His main field of research includes bond, concrete fracture mechanics and very high-strength concretes.

INTRODUCTION

Advanced composites and the fundamental understanding of their behaviour is a rapidly expanding branch within the field of civil engineering materials. In particular, fibre-reinforced cement-based materials have had a great evolution in these years. Large efforts have been made to develop high performance cements and concretes showing further property improvements [1].

Generally speaking, the definition "high performance" is meant to differentiate structural materials from the conventional ones, but also to optimise a combination of properties in terms of final applications related to civil engineering. The most interesting properties are strength, ductility, toughness, durability, stiffness, and thermal resistance. Final cost of the materials must also be considered.

The specific "high performance fibre-reinforced cement-based composites" term refers to high performance cement-based materials, particularly developed for specific applications, for which toughness, ductility and energy absorption are fundamental properties [1, 2]. Furthermore, fibres improve the quality of concrete, in order to obtain an increased homogeneity and to better exploit the high strength matrix.

The experimental work executed for the mix design development, the production and the mechanical testing, is fundamental for the design and production of very high performance concrete precast beams aimed at civil engineering constructions, such as bridges, viaducts and high-rise buildings.

This paper describes the mix design development of a very high performance, fibre-reinforced material to be used for the production of precast, prestressed beams destined to the construction of a small road bridge in Italy.

Industrial manufacturing of high performance concrete possessing compressive strengths more than 100 MPa is now a current practice in many countries, such as U.S.A., Canada, Germany and France [3-5]. On the contrary, in Italy recommendations also concerning the use of high strength concrete (compressive strength up to 110 MPa) have been introduced only recently [6].

The use of fibre-reinforced high performance concrete represents an absolutely innovative experience in Italy, from the industrial point of view: in fact, due to the introduction of fibres in mix design, processing steps (mixing, workability, casting and curing) become more complex and need of high quality production level.

For this special application, fibres were used for the following reasons:

- Increase of tensile strength and ductility, necessary for the pretensioning of concrete beams;
- Better exploitation of high strength matrix as the fibres induce a homogeneous stress distribution in the concrete;
- More homogeneous materials, whose strength standard deviation is reduced with the use of fibres [7];
- Improved fatigue resistance, considering that the final application is a small road bridge heavily stressed by vehicular traffic (particularly trucks).

Other works have demonstrated that fibres do not influence the physical properties (permeability, porosity, freeze-thaw durability), of mixtures similar to those here developed [8].

The setting-up and the optimisation of mix design were performed for a special concrete composed of Portland cement, microsilica, acrylic superplasticizer and steel fibres, having the following basic characteristics: nearly self-compacting concrete (slump more than 20 cm) for almost 1 hour from the batch preparation, a cubic compressive strength of 120-130 MPa and a bending strength of about 18 MPa.

Afterwards, six 3m-long reinforced beams (three of which were lightened with polystyrene cores) were produced in a precasting plant and then, after 28 days from casting, tested in the laboratory.
The two different types of beams were considered to evaluate the contribution of the material in the tensile region. For this reason the tests were monitored with a moiré grid with 40 lines/mm to evaluate the crack development and the position of neutral axis.

For quality control, a complete mechanical characterisation (compressive and bending tests) of the fibre-reinforced concrete was made. The results confirmed the very good reproducibility passing from the laboratory scale to the industrial scale.

Apart from the relatively short length (12 meters) of the beams to be produced for the bridge construction, this study represents the first integrated experience in Italy, from the design to the final application. According to the current Italian design codes and regulations, in order to use these special concretes in structural applications it is necessary to provide an adequate experimental documentation and the adopted design rules, to be examined and approved by technical committee.

On December 1996, a recent document was issued – "Guidelines for structural concretes" by the Italian Consiglio Superiore dei Lavori Pubblici, Ministry of Public Works – containing the recommendations about the possible use of concretes having a characteristic compressive strength from 60/75 MPa up to 100/115 MPa [6].

MIX DESIGN

For this particular application, the selected materials include:

- a Portland cement CEM I 52.5 R (according to ENV 197/1 European Standard), with a Blaine fineness of 4450 cm^2/g;
- a grey microsilica in the form of a dry, undensified powder having a surface area of 20 m^2/g (B.E.T. method), with a SiO_2 content of 95%;
- an acrylic copolymer superplasticizer, 30% solid content;
- a natural sand, of alluvial origin, from Po River (0-3 mm) – fine aggregates;
- crushed Po River stones (3-6 mm) – coarse aggregates;
- carbon steel microfibres, diameter 0.15 mm, length 13 mm.

Both sand and crushed stones have siliceous and calcareous composition.

The mixture components were chosen for workability (time of almost 1 hour for a slump of about 20 cm), and a cubic compressive strength of at least 120 MPa. The optimal microsilica content was also studied [9].

The mixtures were prepared using a 15 litre Hobart planetary mixer. The specimens were prepared in steel moulds through a high vibration frequency table (200 Hz), removed from the moulds after 24 hours and subsequently cured: at least three specimens were tested for the determination of mechanical properties.

After a preliminary particle size distribution study, an evaluation of workability loss and of the optimal curing cycle (see Table 1), the material proportions were defined as follows:

- aggregate/binder ratio of 1.5;
- water/binder ratio of 0.23-0.25;
- superplasticizer (on dry extract basis)/cement ratio of 0.015;
- microsilica/binder ratio of 0.1;
- steel fibres content of 1% by volume
- sand/(total aggregates) ratio of 0.6.

The ASTM C-230 flow table test was performed in order to evaluate workability loss: this test is more sensible than the slump test to evaluate the flow for these special mixtures, which are very thixotropic. Typical values ranges between 160 and 240 mm, as a function of microsilica content.

BEAMS PRODUCTION

The preliminary step before the beam production was the manufacturing of "pilot" beams aimed to ascertain the feasibility of producing the developed mixture in industrial conditions. For this reason, the first test involved the production of six 3 meter-long reinforced beams, three of which here lightened with polystyrene cores, Figure 1.

During the mixture preparation in the precasting plant, particular attention was devoted to the correct sequence of components addition (cement, microsilica, aggregates, liquid components and, above all, steel fibres).

Table 1 Curing conditions and related mechanical properties

	TIME IN MOULD	INITIAL CURING	FURTHER CURING	BENDING STRENGTH MPa	COMPRESSIVE STRENGTH MPa
A	24 hrs	6 days, water 20°C	-	17	116
B	24 hrs	27 days, water 20°C	-	18.6	135
C	24 hrs	6 days, air 20°C (lab)	-	12.5	103
D	24 hrs	24 hrs, water 60°C	-	18	119
E	24 hrs	24 hrs, water 80°C	-	18	137
F	7 hrs	14 hrs, 60°C, 100% RH	-	12	90
G	7 hrs	14 hrs, 60°C, 100% RH	Up to 28 days, water 20°C	14	120

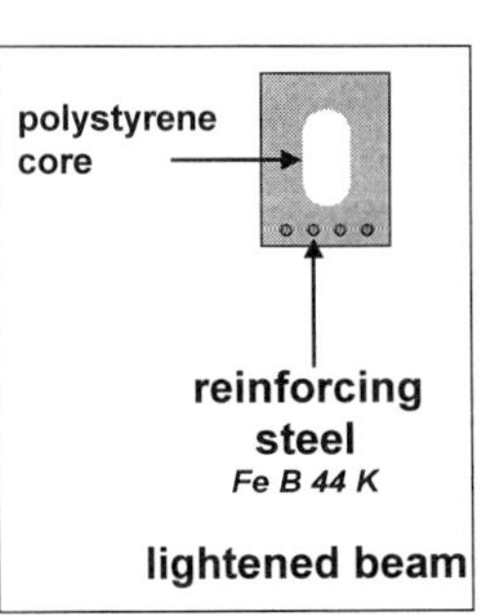

Figure 1 Cross-section of produced beams and view of formworks

Further to this first experience, it was observed that the use of a vibrating fibres feeder is absolutely necessary to avoid fibre "balling" inside the mix.

In addition to the six beams, a number of cubic and cylindrical specimens and small beams were produced for mechanical evaluation.

As a first experiment, it was decided to adopt a natural curing at plant temperature during the first days after the beams demoulding. Steam curing could be the possible and correct alternative, to accelerate the production cycle in a precasting plant.

The beams were protected with wet burlap and plastic heavy clothes to ensure high humidity conditions in the first seven days of curing.

BEAM TESTING

To carry out the bending tests, a suitably-modified 100 kN. Instron testing machine was used. The system is provided with an electromechanical actuator, with a minimum speed of 2 μm per hour. Among the three control channels, one gives the possibility of choosing the feedback signal, in order to allow a stable control of the test; the closed-loop control with integral and derivative gain, makes it possible to remove the effects of the finite axial stiffness of the machine, so as to avoid unstable failures, Figure 2.

In the tests, the feedback signal was the crack opening displacement, which was measured by a clip-gauge placed aside the mid-span section (Figure 3), along the bottom.

To test the reinforced beams representing a portion of the hollow slabs, the load (Figure 3) was applied by means of a hydraulic actuator, and the test was displacement-controlled. A moiré grid with 40 lines/mm was glued to the central part of the specimens, in order to monitor crack formation and propagation (Figures 3 and 4).

The moiré fringe pattern was continuously recorded during the whole test, by means of a video-camera. All the recorded images were analysed by means of a software package, in order to evaluate the displacement field and crack opening. The output of the transducers, as well as that of the load cell were automatically stored in a data logger.

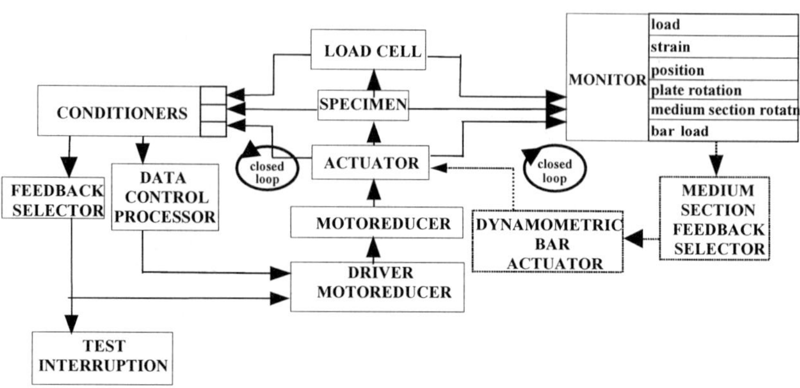

Figure 2 Closed-loop control system

A comparison between two beams with and without the hollow core (lightened) is shown in Figure 5. It may be observed that the qualitative behaviour is similar; the lightened beam has a load carrying capacity smaller by 10%, even though the weight reduction is greater (about 15%).

This difference may be due to the apparent ductility of the fibre-reinforced material that in the cracking zone is able to transmit important stresses. This difference is certainly reduced in materials without fibre reinforcement where the constitutive behaviour tends to be quite brittle.

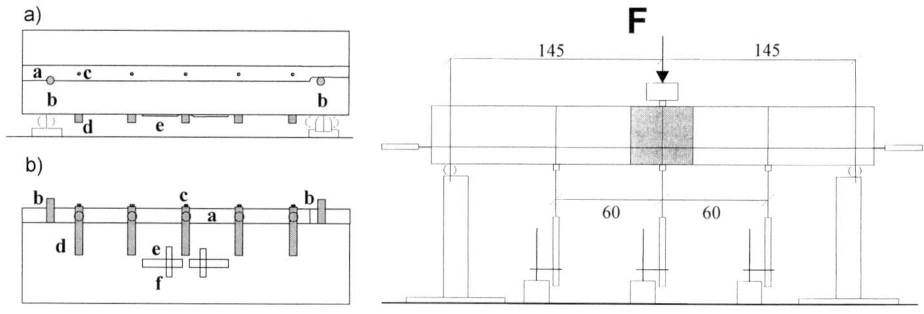

a *support in aluminium for transducers.*
b *pegs in brass fixing the support*
c *screws fixing the transducer*
d *contrast for the transducer*
e *knives supporting the clip-gauge*
f *LVDT ±10 mm*

roller
load device with a spherical hinge
moirè grid
LVDT ±100 mm
magnetic base
LVDT ±10 mm

Figure 3 Instrumentation and loading equipment

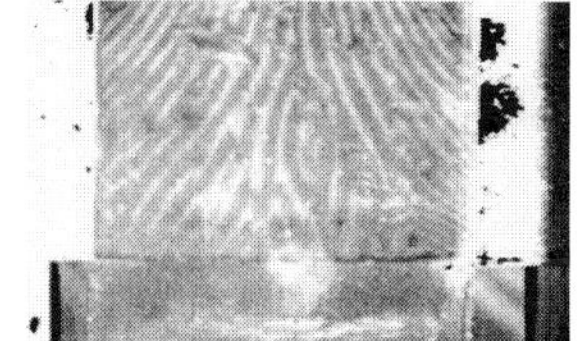

Hollow core beam – about 70% peak load

Full beam – about 70% peak load

Figure 4 Fracture mechanics: beam cracking (left) and moiré fringes (right)

QUALITY CONTROL OF CONCRETE

The trend of both (cubic) compressive and bending strengths for the material are reported in Figures 6 and 7. The direct comparison of these values with the corresponding ones registered during the laboratory mixture development confirms that the production procedures were correctly applied, passing from the laboratory scale to the industrial scale.

Further improvements in strengths could be reached by optimising both mix design and curing conditions, especially with a more accurate humidity control in the first days after production. Table 2 shows the results of quality control for the definitive beams manufactured for the product certification. It is interesting to observe that, in spite of the high cement content, the temperature registered with thermocouples inside the beams was always no more than about 35°C. However, the beam sections were quite limited (17 cm x 20 cm).

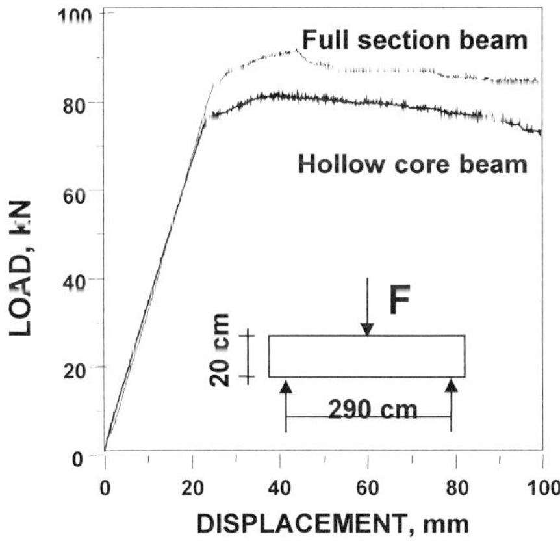

Figure 5 Load-displacement curves for tested beams

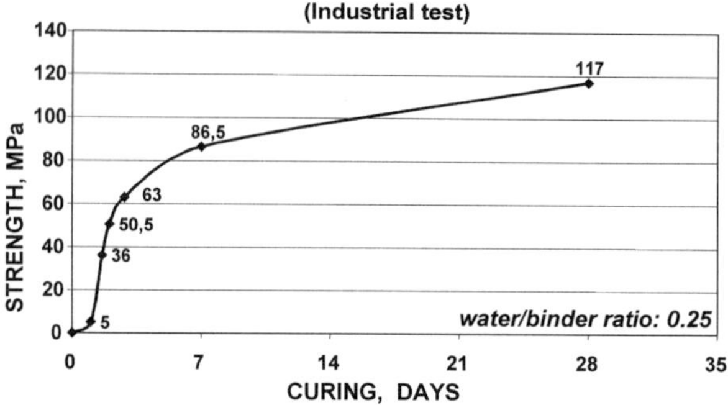

Figure 6 Evolution of cubic compressive strength

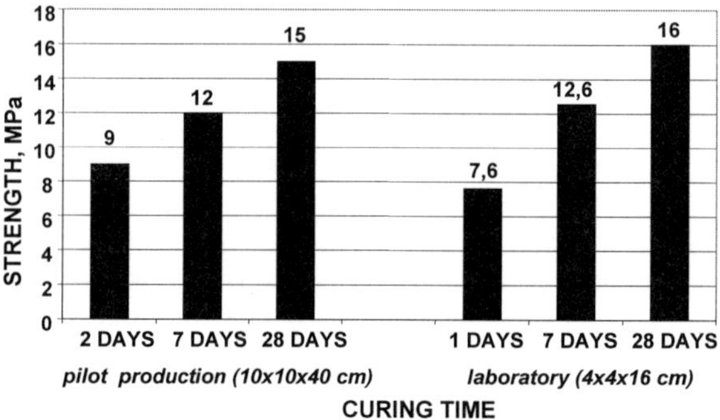

Figure 7 Bending strength: a comparison

CONCLUSIONS

After this pilot trial, according to the recommendations related to the use of high performance/high strength concretes contained in the above-cited "Guidelines for structural concretes", the following program is scheduled:

1. Manufacturing of full-scale beams (12 meters long) for product certification, Figure 8;

2. In-plant characterisation of instrumented beams: static and dynamic loading testing until the breaking point, in presence of cast slab designed for bridge construction;

3. Theoretical elaboration of experimental results and modelling of stress-strain behaviour of the material.

After the submission of complete documentation to the special technical committee designed for the examination, it will be possible to obtain the permission for the manufacturing and the installation of beams for the bridge construction.

Table 2 Main properties of prestressed beams concrete

Specific gravity, kg/m^3	*2500*
Water/binder ratio	*0.23*
Cubic compressive strength (7 days), MPa	103
Cubic compressive strength (28 days), MPa	135
Flexural strength (7 days) MPa,	16.4
Flexural strength (28 days), MPa	19.7
Splitting test (28 days), MPa	8.1
Dynamic modulus, E (7 days), MPa	43300
Dynamic modulus, E (28 days), MPa	45800

The conclusion of this program is forecasted within Summer 1999. This pilot study from the design to the final application would stimulate the use of high performance concrete in structural applications in Italy too, as already has happened in other countries (U.S.A, Canada, France, Japan) [3, 10-12]. Further experimental works are necessary to better understand and model these special materials, so the introduction of very high strength classes as "standard" concrete in the design codes will be possible.

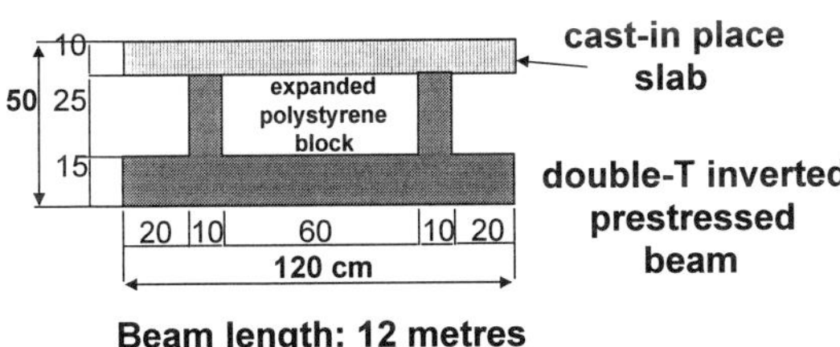

Figure 8 Cross section of beam for bridge construction

The use of fibres in very high strength concrete – more resistant, but also more brittle – will be absolutely necessary so the research would be focused in this direction: the better performance of high performance fibre-reinforced concrete contributes markedly to structural ductility and quality of beams. However, some disadvantages are to be mentioned:

1. Increase of the concrete cost;

2. Increase of the mixing time in the precast plant. for a mix design containing 60-80 % kg/m^3 of steel microfibres, corresponding to 1% about by volume, a duration of 15-20 minutes per batch could be estimated;

3. Need of high quality production level.

Out of these reasons the use of fibres is justified, at this moment, only for special applications.

ACKNOWLEDGEMENTS

This research program about the development and structural applications of high performance concrete is financed by CTG - Italcementi Group. Authors wish to express particular thanks to Precompressi Centro Nord Company (Italy) which is co-operating in the experimental production program.

REFERENCES

1. NAAMAN, A E AND REINHARDT, H W (Eds.). High Performance Fibre Reinforced Cement Composites 2 (HPFRCC 2)", RILEM Proc. No. 31, E & FN Spon, London, 1996

2. BALAGURU, P N AND SHAH, S P. Fibre Reinforced Cement Composites, Mc Graw Hill Inc., New York, 1992

3. AÏTCIN, P C. "High Performance Concrete", 1st edn., E& FN Spon, London, 1998

4. BREITENBÜCHER, R. Developments and Applications of High Performance Concrete, Mat. and Struct., Vol. 31, April 1998, pp. 209-215

5. BONNEAU, O, POULIN, C, DUGAT, C, RICHARD, P AND AÏTCIN, P C. Reactive Powder Concretes: from Theory to Practice. Concrete International, April 1996, pp. 47-49

6. MAURO, M. Guidelines on Structural Concrete of Public Works Ministry, L'Industria Italiana del Cemento, N. 721, May 1997, pp. 398-416

7. ACITO, M, CASSAR, L, GUERRINI, G L AND MIGLIACCI, A. Characteristic Strength of Very High Performance Cement-based Materials (in Italian). Proceedings of 11° Congresso CTE "Nuova Tecnologia Edilizia per l'Europa" - Napoli, 7-9/11/1996, pp. 251-255

8. GUERRINI, G L, BIOLZI, L AND COLOMBET, P. "On the Frost Durability of Very High Performance Cement-based Materials". Proceedings of the Int. Symposium on High-Performance and Reactive Powder Concretes, August 16-20, 1998, Sherbrooke (Quebec) Canada, Vol. 1, pp. 261-276

9. BIOLZI, L, GUERRINI, G L AND ROSATI, G. Influence of Steel Microfibres and Microsilica on Behavior of High Performance Cement-based Composites. Proceedings of ECCM–8 "European Conference on Composite Materials – Science, Technologies and Applications", 3-6 June 1998, Naples, Italy, Ed. I. Crivelli Visconti, Woodhead Publ. Ltd., Proc. Vol. 2, pp. 97-104

10. GOODSPEED, C H, VANIKAR, S AND COOK, R A. High-Performance Concrete Defined for Highway Structures. Concrete International, February 1996, Vol. 18, n.2, pp. 62-67

11. AÏTCIN, P C AND RICHARD P. The pedestrian-bikeway bridge of Sherbrooke. Proceedings of the 4th Int. Symposium on "Utilization of High-strength/High-performance Concrete", F. de Larrard and R. Lacroix (ed.), Presses LCPC, 29-31 May 1996, Paris, France, pp. 1399-1406

12. MITSUI, K, YONEZAWA, T, KOJIMA, M, TEZUKA, M AND KINOSHITA, M. Design and construction of Prestressed Concrete Bridge Using 100 MPa High-Strength Concrete. Proceedings of the 4th Int. Symposium on "Utilization of High-strength/High-performance Concrete", F. de Larrard and R. Lacroix (ed.), Presses LCPC, 29-31 May 1996, Paris, France, pp. 1493-1502

BEHAVIOUR OF COUPLED SHEAR WALLS HAVING REPAIRED COUPLING BEAMS

A Nadjai

A I Abutair

University of Ulster

D Johnson

Nottingham Trent University

United Kingdom

ABSTRACT. The aim of this paper is to investigate the response of planar coupled shear walls before and after the repair of beams to wall at the elasto-plastic joints using the discrete force method. The method is based on the force approach, using as redundants the shear forces not only at the contra-flexural points of the connecting beams but also at the wall junction. The elasto-plastic condition is restricted to the connecting beams using a convenient perfectly elastic/perfectly plastic material model. The work presents the results obtained with the discrete force method and discusses their applicability to some representative problems.

Keywords: Coupled shear walls, Concrete Structures, Structure and Analysis, Repair and Strengthening.

Dr A Nadjai is a Lecturer in Structural and Design Engineering at the University of Ulster, Northern Ireland. His main research interests are, computer modelling and simulation of structural problems using finite elements. He is a joint editor of the Seventh International Conference on Civil and Structural Engineering Computing, Oxford 1999. He is a member of the I Struct E and the MIABSE.

Dr A I Abutair is a Lecturer in Structural Engineering at the University of Ulster. He leads the Sustainable Materials Research Group within the Center for Sustainable Technology at the University. He is a member of the CIB W80 and RILEM 140-TSL committees on Predicting Service Life of Building Materials and Components. His main research interest include the durability, repair and strengthening of concrete structures.

Dr D Johnson is a Principal Lecturer at The Nottingham Trent University. He is the research Director at the department of Civil Engineering. His main interests are finite element modelling in slabs and tall buildings reinforced concrete structures. He is a Fellow of the IStruct E and member of ICE. He has published and travelled widely and serves on many Technical Committees.

INTRODUCTION

In high multistorey, reinforced concrete buildings coupled shear walls (CSWs) can provide an efficient structural system to resist horizontal forces due to wind and seismic effects. CSWs are usually built over the whole height of the building and are laid out either as a series of walls coupled by beams and/or slabs or as a central core structure with openings to accommodate doors, elevator wells, windows and corridors. An idealized coupled shear wall structure and the deformation caused by lateral loading are shown in Figure 1.

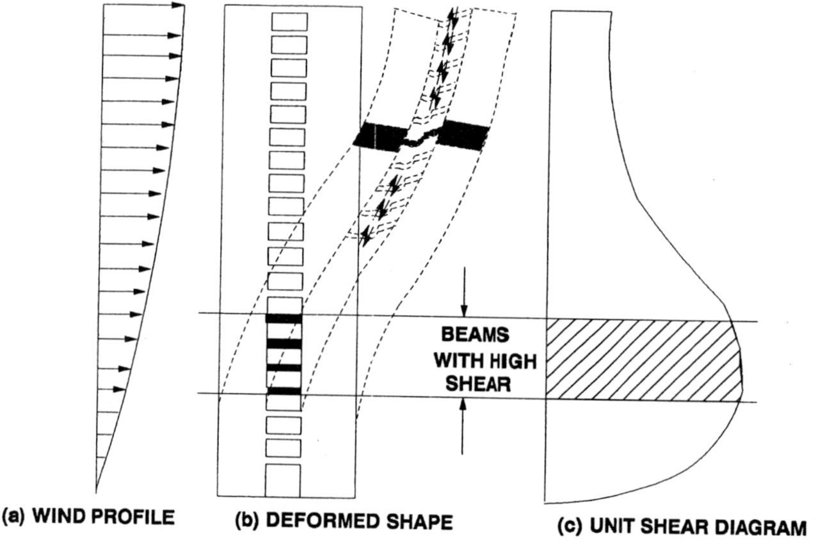

Figure 1 Behaviour of coupled shear walls subjected to lateral load

During major earthquakes, large seismic forces are transferred between individual wall piers through the coupling beams. The coupling beams can suffer stiffness degradation due to cracking and crushing at the joint walls. The mechanical characteristics of the coupling beams significantly affect the behaviour of the system. Damage of coupling beams can provide an increase of the natural period, lateral displacement, and storey drifts. Repair and strengthening of the beams to their original form is essential in multistorey buildings. In recent years, the repair of existing structures has been amongst the most important challenges in civil engineering.

The primary reasons for the strengthening of the structures include: upgrading of resistance to withstand underestimated loads: increasing the loadcarrying capacity; eliminating premature failure due to inadequate detailing. Strengthening of concrete members is usually accomplished by the construction of external reinforced concrete or shotcrete jackets, by epoxy bonding of steel plates to the tension faces of the members, or by external post-tensioning.

DISCRETE FORCE METHOD

The discrete force approach to shear wall analysis has been presented previously for the linear elastic analysis of planar walls by Johnson and Choo [1] and for spatial shear wall systems by Johnson and Nadjai [2]. The method has been extended to the elasto-plastic condition by Johnson and Nadjai [3], assuming that plasticity is restricted to the connecting beams. Further improvement of the method has been developed for the elastic and elastoplastic analysis of planar coupled shear walls with flexible bases by Nadjai and Johnson [46]. In the discrete force formulation, connection points are assumed not only at the contraflexural points of any connecting beams, but also at wall junctions. The beam contraflexural points are assumed to be centrally positioned in the case of coplanar wall connections, but, consistent with an assumption of zero lateral wall bending stiffness, are located at the non-coplanar wall end of the beam in the case of a non-coplanar junction. The wall connections are chosen such that the system is reduced to a set of rectangular wall elements which may have coplanar connecting beam 'stubs' on one, both or neither extremity. The connection points established in the manner described are 'released' and pairs of unknown self-equilibrating shear forces are pressumed to act at each such released point. The stress resultants produced in the released structure due to the external loading and the connection shears are then determined and the effects of these resultants on vertical deformation along the connecting joint positions is established. By enforcing conditions of vertical displacement compatibility between adjacent connection points, the shears at the connections are determined and a full elastic stress resultant and displacement solution may then be accomplished.

ELASTO-PLASTIC ANALYSIS

Connecting Beam Response

The deformation of the coupling beams is a combination of flexural and shear deformation. The flexural deformation, in which the beam bends in double curvature with a point of contraflexure at the centre of the span and the associated forces are shown in Figure 2a. The action of the shear force acting through the point of contraflexure produces maximum bending moments at the wall supports, with the consequent development of flexural cracks. When the applied shear force is increased, the flexural cracking progresses towards the compression corners. Eventually, crushing will take place in the compression corners resulting in the ultimate failure of the beam (Figure 2b).

In extending the discrete force formulation to the elasto-plastic condition, it is assumed that plasticity is restricted to the connecting beams. This assumption is usually warranted since, in practice, flexible connecting beams are desirable for both dynamic response and ductile failure mode considerations. However, several studies [7,8] have shown that plasticity imposes a severe ductility requirement on the connecting beams which is unlikely to be fulfilled in practice. Accordingly, it has been suggested [9,10] that the analysis should be terminated if a specified beam ductility limit is exceeded. Such a procedure has not been adopted here but could be incorporated if so desired. In common with most previous investigations, perfectly elastic/perfectly plastic behaviour is assumed for the connecting beams, since the adoption of more complex models has been shown [7] not to have a major effect on the elasto-plastic response.

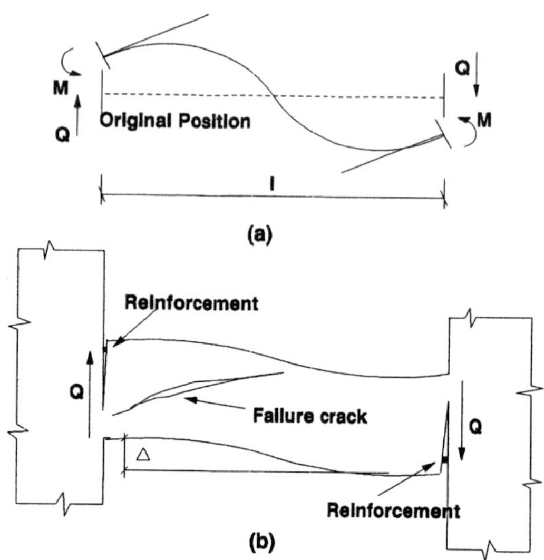

Figure 2 Lintel beam, (a) deformation, (b) cracking pattern

The bi-linear model is particularly convenient for the force formulation, since the perfectly plastic assumption results in the connecting be-am shear forces remaining constant once their elastic limit is exceeded, and hence the number of redundants to be determined is progressively decreased.

Response of Repaired connecting beams

Test specimens were designed as part of a coupled shear wall which was connected to a beam with the latter being cut in the midspan [11]. The samples are part of a shear wall which has dimensions of 400 mm in width, 100 mm in depth and 950 mm in height (Figure 3a,b). After samples have been tested two different types of failure occurred. Firstly shear failure in the beam and secondly tension cracks in the wall section in all samples. To strengthen the beam against shear failure 250xl9x3 mm thick mild steel strips were used (Figure 3c). The strips were placed above the binders and were located using a cover meter, in the same spacing of 120 mm. After repair the initial stiffness was reduced by 40% and the final stiffness was decreased by 15% compared to their original capacities. The mild steel strips provided reduced quite remarkably the crack width and increased the shear capacity of the beam by average of 6 %. Mild steel glued plates were used to strengthen the beam shear capacity. From a structural point of view the stiffening effect had far more influence in reducing the reinforcing bar and steel plate strains than in reducing deflection. Thus the glued plates contributed more to the control of cracking than to control of deflection. The resistance against shear was increased, as none of the repaired samples failed due to shear loads applied.

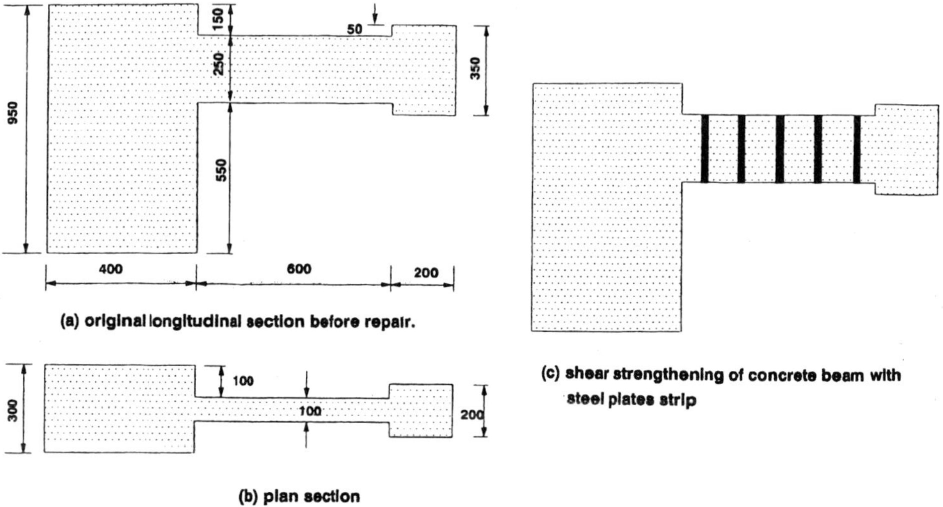

Figure 3 Structural layout and details of testing specimens

Analysis

Following an initial elastic analysis, the most severely loaded connecting beam, i, is identified and the load factor, λ_1, at which this beam reaches its plastic shear value, Q_{ui} ~ is obtained by linear extrapolation. Since the shear force in beam i subsequently remains constant at Q_{ui}, the unknown Q_j may be deleted from {q} and the elastic compatibility equation relating to storey i becomes both superfluous and inappropriate in view of the subsequent plastic deformation of beam i. Also, the compatibility equation for storey i+1 will now refer to the compatibility of deformations between storeys i+l and i-1 (Figure 4a) rather than to storey i+1 alone. This modification may be effected if the flexibilities previously associated with storey i are accrued to those of storey i+l.

The analysis is now repeated. Since the process is piecewise linear, at any stage, s, the load factor λ_s, may be obtained by accumulation of the previous factor, λ_{s-1}, with an addition, obtained by extrapolation, required to increase the most severely loaded beam, i, to its ultimate shear capacity. It is necessary to retain a note of the beams which have achieved their plastic capacity, since, at a general stage (Figure 4b) several beams at their ultimate load may intervene between beam i and the nearest superior, j, and lower k, beams which are still elastic. The principle remains the same however, in that the addition to storey j of the flexibilities presently associated with i will ensure that the compatibility condition for storey j subsequently refers to deformations between storey levels j and k. The complete procedure is repeated until all the connecting beams have achieved their ultimate capacity, at which stage the system is statically determinate. thus, this force approach becomes simpler as plasticity spreads rather than requiring additional plastic rotation variables as is the case with the stiffness method.

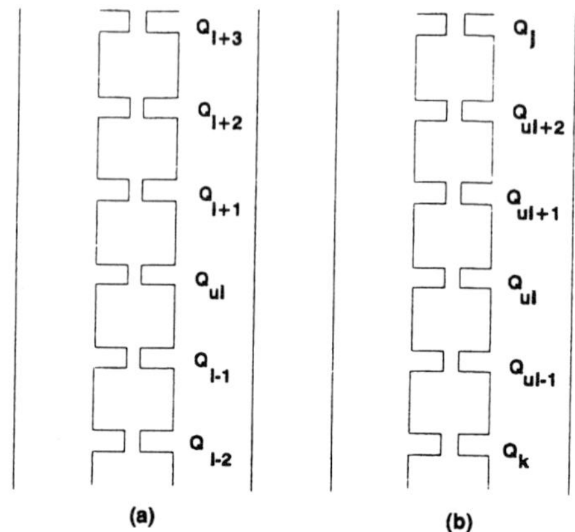

Figure 4 Spread of plasticity, (a) isolated beam, (b) within a plastic zone

NUMERICAL EXAMPLE

To demonstrate the application of the above procedures, an example previously considered by Pekau and Cistera [12] is analysed by the discrete force method. Figure 5 shows a 26storey coupled shear wall subjected to triangular loading, in which W is the total load resultant.

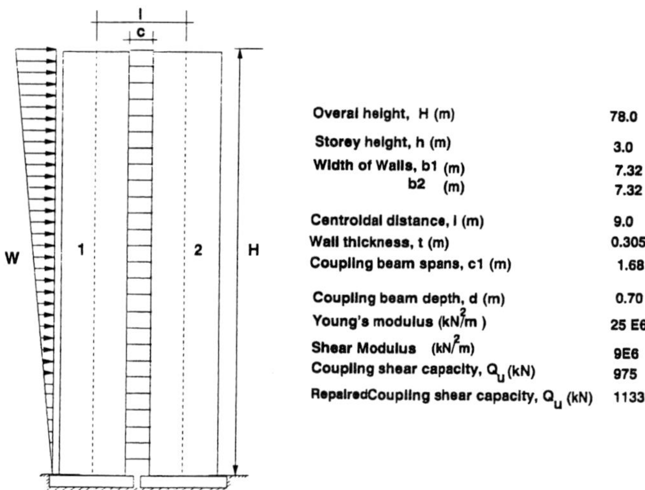

Figure 5 Coupled shear walls

The structure is studied under monotonically increasing lateral loading up to the ultimate load. The stiffness degradation of the repaired coupled beams obtained by tests [11] was implemented and the structure was reanalysed by the discrete force method. A comparison between the original and strengthened coupled shear walls was made.

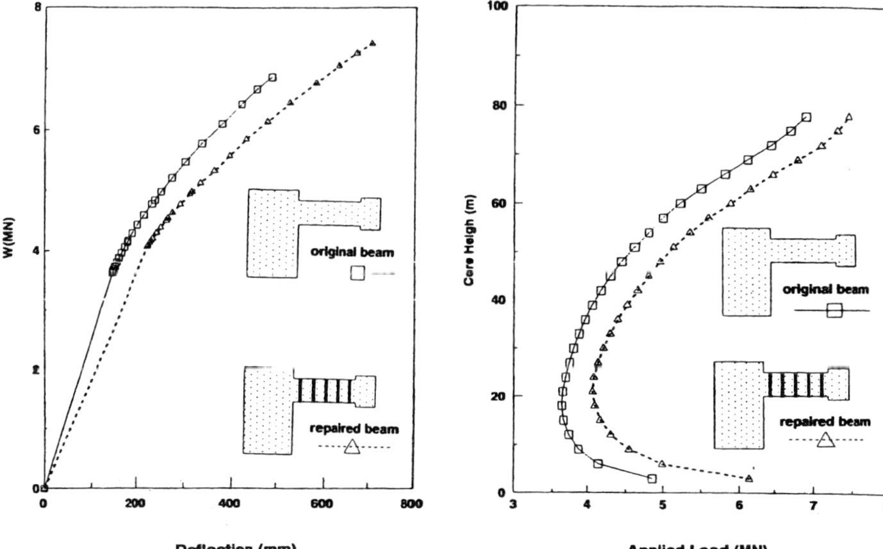

Figure 6 Load-displacement curve

Figure 7 Variation of limit of plasticity with the core height

The load-displacement curves for the original and repaired coupling beams are shown in Figure 6 plotted up to the ultimate state where plastification of all coupling beams has occurred. It can be seen that strengthening the beams results in a 10% increase in the ultimate load. However, the top deflection is more sensitive to the strengthening, showing a corresponding increase of 40%. Also the first yield of the strengthened coupling beams occurs at only a slightly increased load. It is thus clear that the strengthening affects primarily the lateral stiffness of the coupled walls, with relatively small influence on lateral strength.

When the structure is loaded beyond the elastic range, plastic behaviour develops and plastic regions are formed over the height of the stnucture. Figure 7 shows the variation of the limit of plasticity with the core height for different beam stiffnesses. The coupling walls before and after repair result in the formation of a middle plastic zone which rapidly extends into the lower part of the wall while the upper part remain an elastic zone until later in the loading history.

Figure 7 shows that the strengthened coupling beams can withstand more loading before reaching its ultimate load capacity compared to its original form.

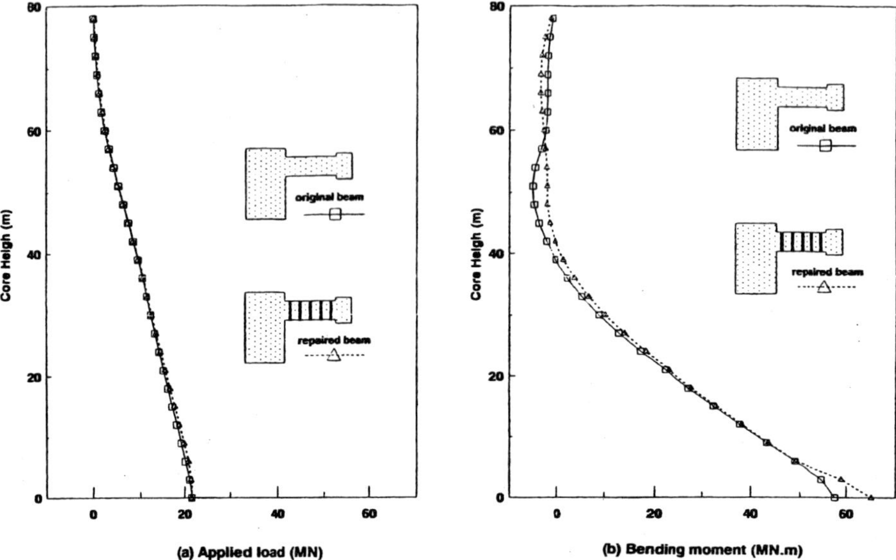

Figure 8 Distribution of axial force and bending moment in walls

Internal stresses within the structure for the two different shear capacities of the coupling beams at a load level equal to 4985 kN are examined to show the distribution of axial forces and moments in the walls. Figure 8 shows the distribution of the axial forces and bending moments over the height of the structure with different strength and stiffness of beams. It is interesting to note that the axial forces and the bending moments in the walls behaves almost the same way compared to its original form. It is clear that the stresses at the base walls will not be increased when beam strengthening with bonded steel plates strip technique is used.

CONCLUSIONS

Based on the discrete force method, an analysis has been presented for planar shear walls before and after repairing the coupling beams and subjected to any type of practical lateral loading. The discrete force method requires the use of customised software, rather than standard plane frame software. The method essentially embodies its own pre-processor, so that input data requirements are kept to a minimum and are related and directly to the wall layout. A typical example showed strengthening of beams with plate bonding plates can:

1. increase the ultimate load
2. extend the life of the beam with an increase of shear capacity
3. reduce the lateral stiffness of the total building
4. keep constant stresses at the base compared with its original form

The enhanced efficiency of the discrete force is expected to be of greater significance in the implementation of other repair techniques to increase the lateral stiffness of the building, which is the subject of the current research effort.

REFERENCES

1. JOHNSON, D and CHOO, B S. The static and dynamic analysis of coupled shear walls by a discrete force method, The Structural Engineer, 71, 1993, pp.10-14.

2. JOHNSON, D, and NADJAI, A. Static analysis of spatial shear wall systems by a discrete force method, Int. J. of Structural Engineering Review, Vol. 8, No.2/3, 1996, pp.133-144.

3. JOHNSON, D, and NADJAI, A. The Elasto-plastic analysis of spatial shear wall systems by a discrete force method, Int. J. of Structural Engineering Review, Vol. 8, No.4, 1996, pp.345-356.

4. NADJAI, A, and JOHNSON, D. Elastic and elasto-plastic analysis of planar coupled shear walls with flexible bases, Int. J. of Computers & Structures, 68, 1998, pp. 213229.

5. NADJAI, A, and JOHNSON, D. Elasto-plastic analysis of spatial shear wall systems with flexible bases, Third Canadian Conf. On Computing in Civil and Building Enaineerina, Montreal, Canada, August 1996, pp. 422-431.

6. NADJAI, A, and JOHNSON, D. Elastic analysis of spatial shear walls with flexible bases. Int. J. of The Structural Design of Tall Buildings, Vol. 5, 1996, pp. 55-72.

7. NAYAR, K K, and COULL, A. Elasto-plastic analysis of coupled shear walls, Proc. ASCE, J. Struct. Div., 102,1976, pp. 1845-1869.

8. PAULAY, T. Elasto-plastic analysis of coupled shear walls, ACI Journal, 67, 1970, pp.9 1 5-922.

9. GLUCK, J. Elasto-plastic analysis of coupled shear walls, Proc. ASCE, J. Struct. DIV., 99, 1973, PP.1743 1760.

10. COULL, A, and CHOO, B S. Simplified elasto-plastic analysis of coupled shear walls, Proc. ICE, Part2, 73, 1982, pp.365-381.

11. ABUTAIR, A, NADJAI, A, COUSIN, W, and MANFRED, G. Behaviour of epoxy-repaired beams concrete walls, Congress Proceeding, Creating with Concrete, Dundee, 1999.

12. PEKAU, O A, and CISTERA, V. Behaviour of nonlinear coupled shear walls with flexible bases, Can. J. Civ. Eng., 16, 1989, pp. 45-54.

FIREPROOF CONCRETE FOR LINING OF SOAKING PITS

S I Pavlenko

A A Permjakov

Siberian State University of Industry Novokuznetsk

A Yu Pronjakin

West-Siberian Steel Works

Russia

ABSTRACT. The Siberian State University of Industry and the West-Siberian Steel Works have developed compositions and technology for fireproof concrete made from local materials and industrial by-products. The concrete is designated for the production of unburnt blocks and bricks to be used for lining heating facilities of metallurgical enterprises. Concrete for producing blocks for soaking pits consists of the following components: quartzite's of three grading fractions and a binder (technical grade liquid glass) with a density of 1.42 to 1.44 g/cm^2. The technical grade liquid glass developed increased the strength of concrete and its heat resistance by 35 to 45 % and 25 to 30 %, respectively, as compared with the commonly used liquid glass. The service life of the blocks manufactured at the pilot plant of the West-Siberian Steel Works in walls of a soaking pit was 18 months while that of the blocks produced by the Pervouralsky plant ranges from 6 to 12 months. The construction of a department for producing blocks and bricks from the above concrete is in the stage of completion at West-Siberian Steel Works.

Keywords: Quartzite, Liquid glass, Fireproof concrete blocks, Strength and heat resistance of concrete.

Professor S I Pavlenko is Head of the Department of Civil Engineering, Siberian State University of Industry, Corresponding Member of the Academy of Engineering Sciences RF, Member of ACI and CIA, active Member of NYAS, the author of more then 200 of scientific works, Novokuznetsk, Russia.

Professor A A Permjakov is Head of Ecology and Mineral Resources Laboratory, Geology Department, Siberian State University of Industry, the author of many research works on mineralogy of ores, sinters, firebricks, etc. His interests also include the complex use of industrial mineral waste Novokuznetsk, Russia.

Mr A Yu Pronjakin is Head of Shop of Quartzite Fireproof Concrete of the West- Siberian Steel Works, Novokuznetsk, Russia.

INTRODUCTION

The paper describes the development and the organization of the production of fireproof materials for lining heating facilities of four large metallurgical enterprises located in Novokuznetsk. To reduce expenditures, locally procurable materials and waste products from local industries had to be used. The problem was solved by developing the composition and technology for fireproof concrete made of quartzite from Antonovka deposit and silica fume from Kuznetsk Ferroalloy plant.

MATERIALS

Quartzite

Quartzite from the Antonovka deposit consists of 97.5 to 98.2 % quartz, 1.4 to 1.6 % goethite and hydrogoethite, 0.5 to 0.7 % illit, smektit and very small amounts of magnetite, fluorite, apatite, zircon, kaolinite, feldspar, calcite and dolomite. The chemical composition of quartzite is given in Table 1. The DTA showed the endothermic effect of reversible transformation of low-temperature quartz into high-temperature quartz at 575 °C. The reformation of the crystalline structure began at 530 °C and ended at 590°C. Quartzite were ground at the crushing plant to three grading fractions: 3 to 8 mm, 0.09 to 3 mm and powder fraction < 0.09 mm.

Table 1 Chemical composition of quartzite, % by mass

SiO_2	Fe_2O_3	Al_2O_3	CaO	$K_2O + Na_2O$	MgO
97.64 - 98.74	0.97 - 1.24	0.23 - 0.46	0.07 - 0.20	0.33	0.20

Technical Grade Liquid Glass

Since conventional liquid glass produced in Russia [1] is expensive and far away from the user, the technology for producing technical grade liquid glass from silica fume of Kuznetsk Ferroalloy plant has been developed and tested at a pilot plant. The technology was based on direct solution of silica fume in sodium hydroxide. Silica fume collected by filters is a finely dispersed material with a fineness of 60 to 90 m^2/g determined according to the method of low-temperature sorption of nitrogen. About 75 % of the particles of silica fume are less than 5 μm in size (Table 2).

Table 2 Granulometric composition of silica fume

SIZE, mm	> 10	5-10	0.5-5.0	0.2-0.5	0.1-0.2	< 0.1
GRADING FRACTION, %	5-10	10-15	20-28	32-50	10-15	10-15

Bulk density and density in a caked state are 180 to 220 and 440 to 530 kg/m^3, respectively. Silica fume tends to becoming caked and to be suspended in hoppers, while "fresh" silica fume can be sufficiently aerated. Taking this into consideration, the West-Siberian Steelworks and Kuznetsk Ferroalloy plant supply silica fume in a "fresh state" in hermetic containers equipped with aerators. The XRD analysis of the silica fume showed that silica was mainly in the amorphous state (77 to 90 % by weight of silica fume). The chemical analysis of the silica fume is given in Table 3. Desolving active silica in a concentrated solution of caustic soda took place in a vertical steel apparatus with a cover. The apparatus was provided with a mixing device. The reaction lasted 1 hour at 90 to 95 °C.

Table 3 Chemical composition of silica fume, % by mass

SiO_2	Al_2O_3	MgO	FeO	CaO	K_2O+Na_2O	SO_3	LOI
85 - 94	1.0 -1.5	0.8 - 1.2	1.5 - 2.0	0.7 - 2.5	0.6 - 1.5	0.3 - 0.5	2.3 - 2.5

The technical grade liquid glass produced meets the requirements of State Standard 13078-81 for liquid soda glass. The density, SiO_2 content, silicate modulus and Na_2O content were 1.36 to 1.45, 24 to 31, 2.6 to 3.0 and 8.7 to 12.2, respectively. The content of solid insoluble sediment (up to 5%) was higher than the standard (0.2 %). As shown by the investigation, this solid sediment increased the strength and heat resistance of fireproof concrete by 35 to 45 and 25 to 30 %, respectively, as compared with concretes containing conventional liquid glass. The technical grade liquid glass exhibited the same binding properties as the conventional liquid glass.

Ferrochrome Slag

Ferrochrome self-decomposing slag from Chelyabinsk Electrometallurgical Works was used as a hardener for the technical grade liquid glass (3 to 6 % by weight). This material was taken because of a high content of olivine in an active form. It fell into the program of developing a fireproof concrete from locally procurable quartzite and industrial by-products (technical grade liquid glass and ferrochrome slag). Unlike conventional metallurgical slag, ferrochrome slag had high contents of magnesium oxide and very low contents of calcium oxide (Table 4). The slag contained small amounts of metal. The MgO-to-SiO_2 ratio was 1.7 which provided the formation of Mg_2SiO_4. Testing of the slag showed the necessity of its grinding, classification and separating of metallic inclusions.

Table 4 Chemical composition of ferrochrome slag, % by mass

SiO_2	Al_2O_3	Cr_2O_3	FeO	MgO	CaO
27 - 36	16 - 22	3 - 8	1 - 2	38 - 45	1 - 3

MIXTURE PROPORTIONS AND PROPERTIES OF FIREPROOF CONCRETE

Proportioning of Concrete Mixtures for Blocks of Soaking Pits

To determine the optimum mixture proportions, 36 x 50 mm cylinders were cast by pressing in accordance with the recommendations [2]. The unit pressure was 30 Mpa. Mixture proportions determined by the method of experimentation [3] are given in Table 5.

Table 5 Mixture proportions

MATERIALS	AMOUNT, %
1. Quartzite (3 - 8 mm)	30 - 50
2. Quartzite (0.09 - 3 mm)	40 .- 60
3. Quartzite (< 0.09 mm)	10 - 20
4. Technical grade liquid glass (density (1.40-1.44) x 10^3 kg/m^3)	10 -. 14
5. Ferrochrome slag	3 - 6

The specimens were first cured at room temperature for 24 h and then in a drying chamber at 105 ± 5 °C for another 24 h. After drying, they were burnt in a laboratory furnace using the cycle (temperature rise period, holding and temperature rise rate) corresponding to that of soaking pits of the West-Siberian Steel Works. The optimum mixture proportions for the fireproof concrete determined by computer technique are given in Table 6.

Table 6 Mixture proportions (quartzite from Antonovsky deposit)

MATERIALS	AMOUNT, %
1. Quartzite (3 -. 8 mm)	30
2. Quartzite (0.09 -. 3 mm)	45
3. Quartzite (< 0.09 mm)	12
4. Technical grade liquid glass (density (1.40-1.44) x 10^3 kg/m^3)	9
5. Ferrochrome slag	4

Properties

Physic-chemical properties of the concrete developed were compared to the requirements of specification 14-8-129-74 [4] for silica concrete (Table 7).

Table 7 Properties of concrete as compared with specification 14-8-129-74

CHARACTERISTICS	SPECIFICATION	CONCRETE DEVELOPED
Chemical analysis, %		
SiO_2	≥ 92	≥ 94
$Na_2O + K_2O$	≤ 2.5	≤ 2.9
Fire resistance, °C	≥ 1610	≥ 1680
Porosity, %	≤ 25	≤ 21
Ultimate Compressive Strength, N/mm²	≥ 12.5	≥ 28.3
Apparent density, 10^3 kg/m³	≥ 1.80	≥ 1.51

As can be seen from Table 7, the values for the concrete developed are higher than the requirements of the specification 14-8-129-74 which indicates that the concrete can be used for lining of soaking pits.

Concrete for Use in Products for Iron and Steel Industries

Physic-chemical requirements (Specification 14-8-184-75) [5] for unburnt quartzite products are higher than for concretes used for soaking pits (Table 8). The concrete developed had insignificant deviations from the values required: SiO_2 content must not be lower than 95 % (92 % for soaking pits). It was 94 % for the concrete developed. The content of $Na_2O + K_2O$ must not exceed 1.2 % (2.5 % for soaking pits). It was 1.9 % for the concrete developed, fire resistance must not be lower than 1690°C (1610 °C for soaking pits), it was 1780 °C for the concrete investigated; the open porosity must not exceed 19 % (25 % for soaking pits), it was 21 % for the concrete investigated; the ultimate compressive strength must not be lower than 25.0 N/mm² (12.5 for soaking pits), it was 28.3 N/mm² for the concrete investigated; the apparent density must not be lower than 2.14 g/cm³ (1.80 for soaking pits), it was 2.51 g/cm³ for the concrete investigated. It is evident that the SiO_2, $Na_2O + K_2O$ contents, porosity did not meet the requirements of the specification. Therefore, new concrete mixtures were made having the following mixture proportions:

1. Aggregate (quartzite with a particle size of 0 to 3 mm) 92.0 %
2. Technical grade liquid glass 4.3 %
3. Ferrochrome slag (hardener) 3.7 %

The results of testing of concrete specimens cast from these mixtures as compared with the requirements of specification 14-8-184-75 are given in Table 8. As can be seen from Table 8, the concrete developed meets the requirements of the specification for use in products for iron and steel industries.

Table 8 Properties of concrete as compared with specification 14-8-184-75

CHARACTERISTICS	SPECIFICATION	CONCRETE DEVELOPED
Chemical analysis, %		
SiO_2	≥ 95	≥ 96
$Na_2O + K_2O$	≤ 1.2	≤ 1.1
Fire resistance, °C	≥ 1690	≥ 1705
Porosity, %	≤ 19	≤ 17
Ultimate Compressive Strength, N/mm^2	≥ 15	≥ 27.6
Apparent density, g/cm^3	≥ 2.14	≥ 2.40

TESTS

Blocks for Soaking Pits

In June - July 1992, 30 blocks from the concrete developed were produced at the pilot plant of the West-Siberian Steel Works. The blocks were installed in the lining of a soaking pit No 9-4 along with dinas-quartzite blocks brought from Pervouralsk plant (the aggregate of the dinas-quartzite blocks consists of 80 % broken dinas and 20 % quartzite). Testing commenced in August 1992. The blocks from the concrete produced stood 18 months versus 12 months of those from Pervouralsk plant [6].

Fireproof Products

To mould quartzite fireproof concrete blocks in traditional equipment by vibration or vibroshock is difficult because of a poor workability of fireproof mixtures attributed to physic-mechanical properties of quartzite and the rheology of the alkaline binder having a density of 1.4 to 1.44 g/cm^3. The technology of preparation of an alkaline-silicate binder entailed employing a method of vibrowave impulse for mixing followed by applying a small compressing effort. The fireproof mixture was prepared in a rotary-drum concrete mixer. The components (quartzite, ferrochrome and technical grade liquid glass) were mixed for 5 minutes until uniform dry mass was obtained.

During the vibrowave impulse pressing, the working-face of a die does not compress the entire surface of a product, while the impulse continuously travels along the surface of the moulded product creating a high unit pressure at a relatively small compressive effort. After the continuous treatment of local zones at a frequency of impulses f = 10 Hz, compressive and shear deformations occur in the material in a press-form which accelerates the movement of the mixture and intensifies packing. The materials are characterized by higher homogeneity and density. The blocks were moulded by a single-spindle vibrowave press. A series of 500 blocks, 260 x 130 x 90 mm, were made at a pilot plant of the West-Siberian Steel Works. On March 31, 1995, a nozzle of an iron chute of a blast-furnace was lined with those blocks. Over a period of April 4-18, 1995, 27582 tons of iron (85 taps) passed through the nozzle of the chute. After the contact of 100 h of the lining and iron, no injury to the lining was observed. By using the concrete developed the service life of the lining was increased by half. A department for the production of fireproof materials using this new method is in the stage of completion at the West-Siberian Steel Works. Four-spindle vibrowave presses have been installed

CONCLUSIONS

1. Compositions and technology for fireproof unburnt units made from locally procurable materials (quartzite, silica fume and ferrochrom slag) have been developed.

2. Concrete developed meets the requirements of specifications and technique guides (202 - 10 - 86 and 202 - 11 - 86) for use in the production of silica blocks for lining of heating facilities of iron and steel industries. It produced better performance and required less expenditure than dinas - quartzite concrete.

3. Technology for producing technical grade liquid glass from silica fume by its direct solution in sodium hydroxide at 90 to 95^0 C has been developed.

4. The use of technical grade liquid glass as a binder increased the strength and resistance of concrete by 35 to 45 and 25 to 30%, respectively.

REFERENCES

1. STATE STANDARD 13078 - 81, Moscow, Gosstandart, 1981.

2. GUIDES FOR THE PREPARATION OF A FIREPROOF CONCRETE WITH QUARTZITE AGGREGATE, UralNII -Stromproyekt, Chelyabinsk, 1986.

3. VOZNESENSKY, V A. Statistical Methods of Planning Experiment in Technicomechanical Investigations, Moscow, Izdatelstvo Statistic, 1974, 192 pp.

4. PRODUCTION OF UNBURNT HEAT INSULATING ARTICLES AND CRUSHED QUARTZITE MATERIALS, Technique Guides 202-10-81, Minchermet USSR, 'Soyuzogneupor', Pervouralsk, 1987, 9 pp.

5. PRODUCTION OF UNBURNT FIREPROOF QUARTZITE MATERIALS FOR LINING TEEMING LADLES, Technique Guides 202-11-86, Minchermet USSR, Soyuzogneupor', Pervouralsk, 1987, 19 pp.

6. ПРОНЯКИН, Ю Н, КУСТОВ, Б А, ПЕРМЯКОВ, А А, ПАВЛЕНКО, С И, ПРОНЯКИН, А Ю, Перспективы производства огнеупорных бетонов на минеральном сырье Кузбасса, в “Черная металлургия Кузбасса: пути преодоления кризиса”. Издательство СибГИУ, Новокузнецк, 1998. С. 43-46.

BEHAVIOUR OF MASONRY WALLS SUBJECT TO ELEVATED TEMPERATURES

A Nadjai

M E O'Gara

D J O'Connor

University of Ulster

United Kingdom

ABSTRACT. A finite element model called MasSET has been developed which is capable of predicting the structural behaviour of single leaf masonry walls subject to elevated temperatures. The analysis models a slice through the wall as a column strip in plane stress, and also includes material with geometric non-linearity. The model has been previously validated by comparison with experimental results [1] and is used in this paper to conduct a parametric study investigating the effects of slenderness ratio, load eccentricity and boundary conditions. The results of the investigation are presented by way of failure temperatures for each condition, and show conclusive findings to the effects of each parameter investigated.

Keywords: Masonry, Fire conditions, Finite element model, MasSET, Parametric study, Slenderness ratio, Load eccentricity, Boundary conditions.

Dr A Nadjai is a lecturer with the University of Ulster in the School of the Built Environment. His research interests included fire engineering, shear wall analysis, tall building analysis, and soil structure interaction problems.

Mr M E O'Gara is a research student at the University of Ulster in the School of the Built Environment, and is due to complete his DPhil research on the subject of 'the finite element modelling of masonry subject to elevated temperatures', in October 1999.

Dr D J O'Connor is a senior lecturer at the University of Ulster in the School of the Built Environment. He has numerous publications concerning the thermo-structural response of building elements, the thermal dynamics of a model test methodology, and heat flux measurement techniques.

INTRODUCTION

The role of a single leaf masonry wall in a fire situation is generally seen to be three fold:

1. Firstly a load bearing wall must provide structural stability to prevent collapse. This provision is termed structural adequacy.

2. The wall is required to maintain integrity, i.e. preventing the passage of flame from one side of the wall to the other, in an attempt to compartmentise a fire.

3. The wall must provide adequate insulation properties to prevent excessive temperature on the unexposed face creatine further spread of flame.

The increasing emphasis on the use of masonry walls as a load bearing element has lead the primary interest of this study to the first criterion of structural adequacy. The maintenance of structural stability may also in part fulfil the requirements of the second criterion, while also providing safe egress for any building occupants.

In fire separating elements, such as masonry walls, heat is usually exposed to one side of the element only. This is particularly important in the case of brick masonry due to its low thermal conductivity, producing high thermal gradients over the cross section [2]. Thermal bowing is therefore produced due to differential thermal expansion. With the hot face of the wall expanding more rapidly than the cool one, the wall will tend to bow towards the fire. The fire exposed face of the wall will also experience a considerable reduction in mechanical material properties which effectively can be represented as a reduction in thickness of the hot face [3]. As a result of the change in thickness any applied load will have moved towards the fire exposed face and must be recognised as having the advantageous effect of maintaining structural stability by counteracting thermal bow

Current design code does not provide concise calculation models for the determination of fire resistance times. Some design guidance is available based entirely on isolated standard fire tests, which related minimum periods of fire resistance to required wall thickness' [4]. Important parameters such as load, load eccentricity, material type and slenderness ratio are not accounted for.

The finite element model called MasSET (Masonry Subject to Elevated Temperatures) presented in this paper has been employed to conduct a parametric study investigating several influential factors on the behaviour of sinale leaf compartment walls in fire.

DESCRIPTION OF MasSET

MasSET is a two dimensional finite element model written in displacement formulation. It models a slice through a masonry wall as a column strip using eight noded isoparametric elements, and is capable of predicting thermal response under various restraint and loading conditions. The effects of both material and geometric non-linearity have been accounted for and a smeared cracking model was incorporated to simulate the brittle behaviour of a concrete type material.

The model assumes the structure to firstly receive an applied load, followed by successive temperature increments. At present MasSET is not coupled with a thermal analysis program and explicitly defined temperature distributions are directly assigned to the nodal points by the user as input data. The standard or modified non-linear Newton Raphson iteration procedure is adopted to reach a converged solution, whereby convergence is based on the norm ratio of the external and internal nodal force vectors.

Material Modelling

The two phase masonry material of brick unit and mortar joint exhibits distinct non-linear stress - strain behaviour. Each material component may be modelled separately providing a more accurate description of the macro level behaviour between the two constituents. Alternatively an equivalent homogenised material may be employed, giving more flexibility in the finite element mesh descritization.

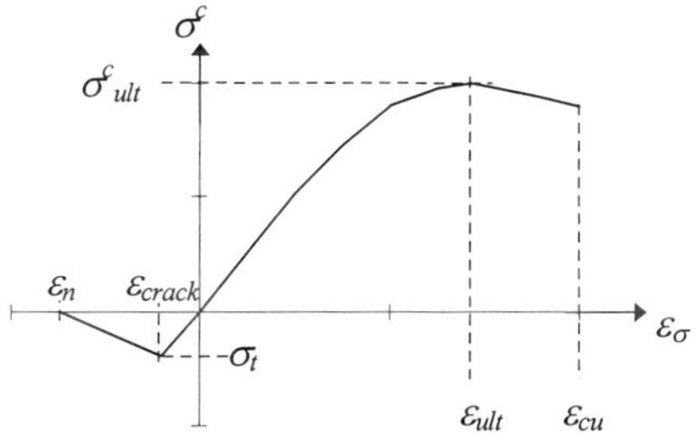

Figure 1 Stress-strain relationship of masonry materials in compression and tension

$$\frac{\sigma^c}{\sigma^c_{ult}} = \frac{\varepsilon_\sigma}{\varepsilon_{ult}} \cdot \frac{3}{2 + \left[\dfrac{\varepsilon_\sigma}{\varepsilon_{ult}}\right]^3} \quad (1)$$

Where; σ^c_{ult} = Compressive strength
σ_t = Tensile strength
ε_{ult} = Ultimate strain
ε_{cu} = Crushing strain
ε_{crack} = Cracking strain
ε_n = Crack softening parameter

Material non-linearity was approached using the tangent modulus method. The temperature dependent parabolic stress-strain relationship for a concrete material proposed in Eurocode 4 [5] was adopted with a linear strain-softening branch, as illustrated in Equation (1), Figure 1. In order to relate the uniaxial stress-strain relationship to a two dimensional continuum material, the well established biaxial failure envelope for concrete [6] was employed.

The brittle tensile nature of the masonry materials was accounted for by the inclusion of a smeared cracking model. If the maximum principle stress at any gauss point exceeded the user specified tensile failure stress then a crack was assumed to develop in a plane normal to the direction of the maximum principle stress over the associated gaussian integration area. The crack was simulated by setting the Young's modulus in the direction normal to the crack as zero. The stress in the direction normal to the crack was not set to zero immediately but reduced according to a linear crack softening relationship.

The effects of crack closure were also incorporated by restoring the compressive strength of the material in the relevant direction. A certain amount of shear capacity was retained, but reduced as the crack width increases. The σ-ε relationship of the cracking model employed is shown in Figure 1. Crushing of a material was modelled by reducing all strength properties and stresses in an integration area to zero when the effective biaxial strain at that particular gauss point exceeded the user specified crushing strain.

P-Δ Effects

It was recognised that the problem at hand contained certain aspects of geometric non-linearity. Changes in geometry due to deformation significantly influence the predicted response of a masonry wall [8]. The total langrangian descriptor was used to redefine the structural geometry, and account for the additional stresses experienced in a laterally deforming axially loaded masonry wall. Green's strain tensor was used to include higher order strain terms. The membrane stresses, which exist in an axially loaded slender structure, may actually have a stiffening effect and have been accounted for by the inclusion of the geometric stiffness matrix.

Elevated Temperature Effects

The effects of thermal expansion were included by the application of thermal strains £,h, which were defined by the ε_{th}-temperature relationship adopted by Eurocode 4 [5]. Thermal strains are indeed an initial strain component, and are therefore independent of stress. The nature of the finite element method written in displacement formulation prevents a direct explicit displacement-strain relationship. Therefore any deflection sought after (in this case thermal bowing) must be achieved from an equation of the form;

$$P^j = K^j a^j + f^j \tag{2}$$

where P is the vector of externally applied nodal loads, K is the stiffness matrix, a is the vector of nodal displacements end f is the vector of loads due to element body forces. Superscripts i and j denote iteration and increment respectively.
Temperature changes result in an initial strain vector of the from;

$$\varepsilon^j{}_{th} = \begin{Bmatrix} \varepsilon_{xth} \\ \varepsilon_{yth} \\ \gamma_{xyth} \end{Bmatrix} \tag{3}$$

where in an isotopic material $\gamma_{xyth} = 0$, as no shear strains are caused by thermal dilation. The thermal strains must be converted to equivalent nodal loads to act as body forces, f_{th}, which are capable of creating the state of thermal strains in the element, when applied to that element.

$$f^{j}{}_{th} = -\int_{Ve} B^{i^{T}} D^{i}{}_{T} \varepsilon^{j}{}_{th} dV \tag{4}$$

The thermal stains are converted to imaginary thermal stresses, σ_{th}, using the current tangential elasticity matrix, DT The thermal stresses are then integrated over the element volume, therefore the thermal strains must be evaluated at the gauss points to facilitate the gauss quadrature integration. Temperature, T_i, is assigned to each nodal point, and with use of the element shape functions N_i; the temperature at the gauss points T_{gp}, is determined,

$$T_{gp} = [T_i]\{N_i\} = [T_1, T_2, \ldots\ldots]\begin{Bmatrix} N_1 \\ N_2 \\ : \end{Bmatrix} \tag{5}$$

then using the explicit ε_{th} - temperature relationship employed, the thermal strains are evaluated. Rearranging Equation (2) gives;

$$Pj - fj = Kj\,aj$$

and by substituting Equation (4);

$$P^{j} + \int_{Ve} B^{i^{T}} D^{i}{}_{T} \varepsilon^{j}{}_{th} dV = K^{i} a^{i} \tag{6}$$

The equilibrium equations are solved for nodal displacements, a, and it is apparent that when the external applied nodal loads are zero, displacements are entirely due to thermal effects. Upon determining the nodal displacements, the total strains £`o~, are calculated;

$$\varepsilon^{i}{}_{tot} = B^{i} a^{i} \tag{7}$$

The individual components of the total strain were recognised in order to determine the stress related strain component for the calculation of internal forces. Equation (8) describes each strain

$$\varepsilon^{i}{}_{tot} = \varepsilon^{i}{}_{\sigma} + \varepsilon^{j}{}_{th} + \varepsilon^{j}{}_{tr} \tag{8}$$

Where ε_{tot} = total strain
ε_{σ} = stress related strain
ε_{tr} = transient strain
ε_{th} = thermal strain

The transient strain component has been reported to be significant in concrete materials that were loaded and experiencing heating for the first time. This strain component has the effect of reducing and redistributing thermal strains in loaded elements, preventing excessive damage by thermally induced stresses [8].

In the present analysis the transient strain model proposed by Anderberg [9], Equation (9), has been employed, and is dependent on both thermal strain and stress level.

$$\varepsilon^{j}_{tr} = -2.35\frac{\sigma^{j-1}}{\sigma^{c}_{ult}}.\varepsilon^{j}_{th} \quad [T < 500°C]$$

$$\Delta\varepsilon^{j}_{tr} = -0.0001.T.\frac{\sigma^{j-1}}{\sigma^{c}_{ult}} \quad [500°C < T < 800°C] \qquad (9)$$

Where; σ^{j-1} = stress in the element from the previous increment.
σ^{c}_{ult} = Compressive strength at ambient temperature.
T = element temperature.
ε_{th} = thermal strain.

From Equation (9) as the stress level increases, so does the transient strain component, thus reducing the effective thermal strains.

Modification of material strength and stiffness with increase in temperature was incorporated which is particularly influential in the case of loaded masonry walls. Upon the application of temperature the material properties of compressive strength σ^{c}_{ult}, ultimate strain ε_{ult}, crushing strain ε_{cu}, and tensile strength σ_{t} were modified for each integration area according to the temperature at the gauss point. Both σ^{c}_{ult}. and ε_{ult} were modified according to the Eurocode 4 recommendations. A tri-linear decay model for σ_{t}, based on test results by Thelanderson [10] was employed, while the linear crushing strain model used by Anderberg [9] was implemented.

PARAMETRIC STUDY

MasSET was used to conduct a parametric study, investigating the effect of slenderness ratio X, load eccentricity *e*, and boundary conditions, on the critical temperature of fire walls. The range of values considered are illustrated in Table 1.

The thickness of the wall was taken as 50 mm due to the availability of temperature distribution information from associated experiments on 1/2 scale masonry wall panels subject to a model BS476 fire curve on one side [11]. This model fire curve was specifically developed [12] to produce similar temperature profiles in a reduced scale wall as a standard BS476 fire curve would create in a full scale wall.

Table 1 Range of parameters investigated

PARAMETERS	P=150.0N
λ = h/t	7.5, 10.5, 12.5, 17.5, 20.0, 25.0, 30.0, 40.0, 50.0, 60.0
e/t	0.0, 0.25 (See Figure 3)
Boundary Conditions	Wall Type A Wall Type B Wall Type C (See Figure 2)

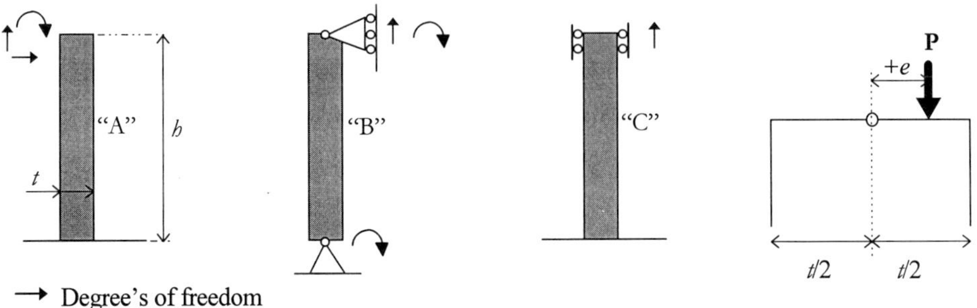

Figure 2 Boundary conditions of type A, B and C walls

Figure 3 Load eccentricity

An applied load for all scenarios of 150.0 N was chosen and reflected approximately an average of the BS5628 design resistance load.

A homogenised material was assumed and the specific values of properties employed where;

Compressive strength σ^{c}_{ult}, N/mm²	18.0
Tensile strength σ_t, N/mm²	1.0
Ultimate strain ε_{ult}	0.0021
Crushing strain ε_{cu}	0 0035

The finite element mesh density was 3 elements wide yielding seven nodal points over the wall thickness enabling the curvi-linear temperature distribution to be simulated. Steel plates were introduced at the loading and concentrated reaction points to prevent early failure by numerical instability caused by local crushing.

RESULTS AND DISCUSSION

Effect of Slenderness Ratio

In all cases of boundary conditions and load eccentricity the failure temperatures decreased as the slenderness ratio increased. However, wall type C did not follow this trend and within the range of practical λ no failure was experienced until temperatures in excess of 1000 °C (see Figure 4). It was also observed that in C, no significant thermal bowing occurred during the course of heating due to rotational restraint at both boundaries. The eventual failure mode of all type C walls was due to material property failure.

The predominant failure mode of the type A and B walls was due to lateral instability (buckling) and therefore the failure temperatures where largely affected by λ, however in the lower values of (7.5, 10.0, 12.5) of wall type B with e/t= 0.0, reverse bowing was observed and failure occurred at high temperatures due to reduction of material properties (see Figure 5).

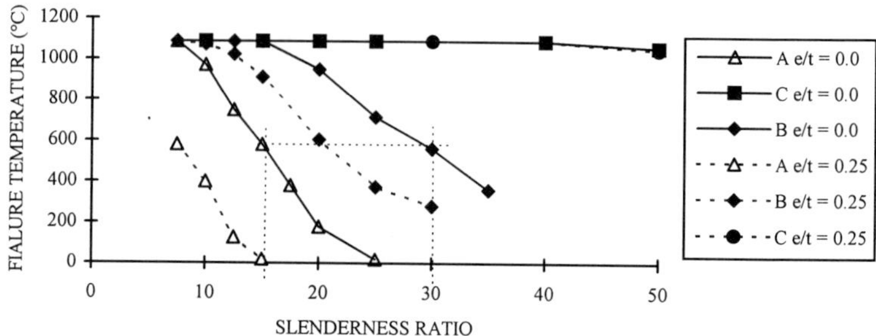

Figure 4 Slenderness ratio V's Exposed face temperature for all wall types and load positions

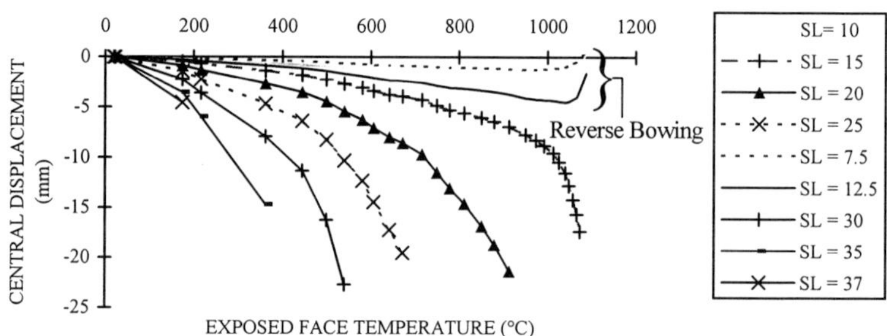

Figure 5 Exposed face temperature V's Central displacement for wall type B, e/t = 0.0

Effect of Load Eccentricity

Two situations of load position were considered, that of uniaxial loading *(e/t* = 0.0) and eccentric loading away from the fire occurring exposed face *(e/t* = 0.25). In an internal single leaf compartment wall the possibility of fire generally exists on either side. Since the effect of load eccentricity towards the fire would be expected to increase failure temperatures the decision was made to investigate eccentric loading away from the fire as it was considered to be the more critical condition.

Upon examination of Figure 4 it was apparent that the eccentric load *(e/t* = 0.25) had the effect of reducing failure temperatures in wall types A and B, while no significant difference was observed in wall type C.

The uniaxial load produced uniform compression through the thickness of the wall and therefore upon the application of temperature the transient strain component was significant and resulted in reduced thermal strains.

The application of an eccentric load in wall types A and B created an additional moment ($M_{ec} = P\ e$), producing an initial bending synonymous with the direction of thermal bowing. The initial bending resulted in reduced compressive stresses on the fire exposed face, leading to a reduced transient strain component on the exposed face and hence increased thermal strains inducing additional thermal bowing than in the uniaxially loaded case. This is supported by Figure 6 which compares the central displacement of a sample wall type B, and Figure 7 reveals the various strain components of the two walls at the arbitrary point indicated. Upon examination of Figure 7 the lesser transient strain component in the $e/t = 0.25$ loaded wall was apparent.

No initial bending due to an eccentric load was observed in wall type C due to complete rotational fixity, thus restraining the moment MeC, ensuring uniform displacement in the Y - direction and therefore uniform initial compressive stresses irrespective of load eccentricity.

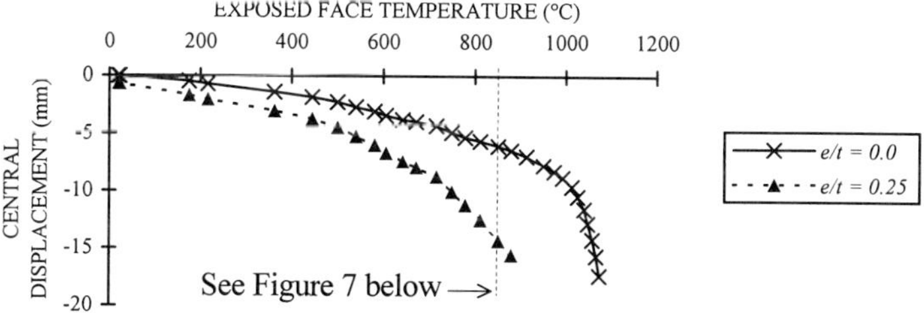

Figure 6 Exposed face temperature V's Central displacement for wall type B, $\lambda = 15$

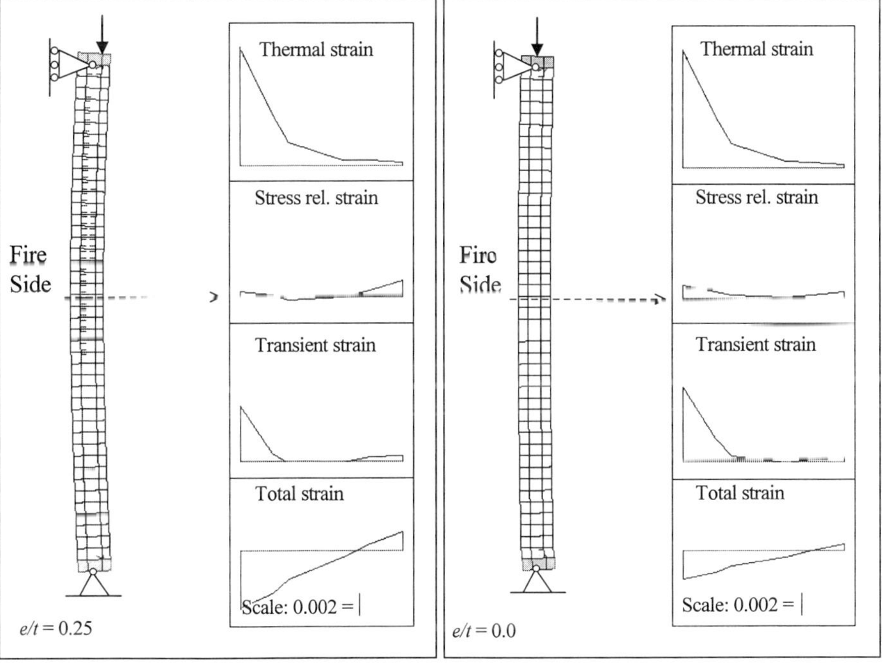

Figure 7 Comparison of strain components of wall B, $\lambda = 15$, between e/t = 0.0 and e/t = 0.25

Effect of Boundary Conditions

Wall type A is seen to be the most critical, followed by type B, and finally (within practical limits) type C maintains structural stability until temperatures in excess of 1000 °C. Figure 8 reveals the various strain components of a sample wall type C nearing failure. Since the thermal deflections were restrained the thermal strains were converted to stress related strains, resulting in considerably high compressive stresses on the fire exposed face. The high compressive stresses hence produced significant transient strains serving to reduce the thermal strains.It was also noted that in the centrally loaded walls the failure temperatures of type A corresponded with similar failure temperatures of type B with twice the λ value, thus upholding the fact that the effective height of a cantilever wall (type A) is 2.0h, while that of a pinned top and bottom wall (type B) is 1.0h. Figure 4 illustrates this point.

Wall type A is assumed to be representative of actual fire separating structural cantilever walls, however, it must be recognised that both wall types B and C are unlikely to represent practical structural situations, and indeed complete rotational restraint or freedom would be difficult to achieve. In real structural circumstances partial rotational restraint is more likely to be the case, and therefore the failure domain would exist intermediate of the B and C failure envelopes. In order to simulate a representative structural situation the immediate structure surrounding the boundary conditions may be included in the numerical modelling at the cost of significantly increasing computational time. Alternatively, and perhaps more favourably, the moment-rotation relationship of actual wall boundary conditions may be numerically applied in analysis.

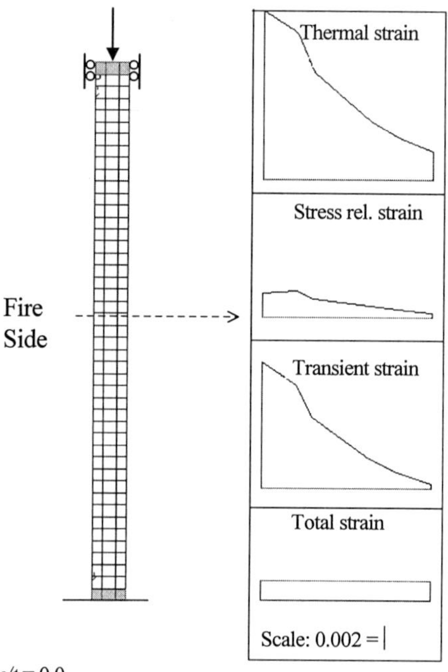

Figure 8 The individual strain components of a type C wall nearing failure

CONCLUSIONS

MasSET has been used to conduct a parametric study investigating the effects of slenderness ratio, load eccentricity and boundary conditions.

For wall types A and B as λ increases the failure temperature decreases, and failure temperature is further reduced as the applied load moves away from the exposed face. Both λ and load eccentricities have an insignificant effect on the failure temperatures of wall type C, and failure is not experienced until material properties have sufficiently degraded.

The information obtained may assist in the compilation of design guides. However, in order for real structural situations to be investigated the model must be developed to account for partial rotational restraint at the boundaries.

REFERENCES

1. O' CONNOR, DJ, NADJAI, A, AND O'GARA, ME. A Numerical Model for the Behaviour of Masonry Under Elevated Temperatures. Proc. Fourth Int. Conf. Comp. Struct, Edinburgh, 1998, Cp7.1, pp 229-237.

2. COOKE, GME. Thermal bowing and how it affects the design of fire separating constructions. Interfam '88 - Fourth Int. Fire Conf. 22-24, March, Church College Cambridge, 1988, pp230236.

3. LAWERENCE, SJ AND GNANKRISHNAN, N. The fire resistance of masonry walls. National Building Technology Centre, Technical Record No. 531.

4. CP 121. Code of Practice for Walling, Pt 1: Brick and Block Masonry. British Standards Institution, London, 1973.

5. EUROCODE NO.4. Design of Composite Steel and Concrete Structures. Pt 1.2: Structural fire design (second draft). Commission of the European Communities, 1992.

6. KUPFER, H, HILSDORF, KH AND RUSH, H. Behaviour of concrete under biaxial stresses. Proc. ACI, Vol 66, No 8, 1969, pp 656-666.

7. O'MEAGHER, AJ AND BENNETTS, ID. Modelling of Concrete Walls in Fire. Fire Safety Journal, Vol 15, No 4, 1991, pp 315-335

8. KHOURY, GA, GRAINGER, BN AND SULLIVAN, PJE. Transient Thermal Strain of Concrete: Literature Review, Conditions Within Specimen and Behaviour of Individual Constituents. Magazine of Concrete Research, Vol 37, No 132,1985, ppl3 1-144.

9. ANDERBERG, Y AND THELANDERSSON, S. Stress Deformation Characteristics of Concrete at High Temperatures. Division of Structural Mechanics and Concrete Construction, Lund Institute of Technology, Bulletin 54, 1976.

10. THELANDERSSON, S. Effect of High Temperature on Tensile Strength of Concrete. Lund Institute of Technology, Division of Structural Mechanics and Concrete Construction. Lund. NeostYled pp27~ 0ct. 1971.

11. LAVERTY, D. Private Communication. University of Ulster at Jordanstown, 1998.

12. O'CONNOR, DJ, SILCOCK, GWH AND MORRIS, B. Furnace Heat Transfer Processes Applied to a Strategy for the Fire Testing of Reduced Scale Structural Models. Fire Safety Journal, Vol 26, No 1, 1996, pp l -22.

STEEL/CONCRETE COMPOSITES - AN UPDATE ON THE EXTERNALLY REINFORCED OPTION

P G Lowe

University of Auckland

New Zealand

ABSTRACT. This paper follows on from a paper of somewhat similar title presented at the first conference in this series in 1993. The theme in both papers is a description of a steel/concrete composite in which all the reinforcement for the concrete is provided as an external steel case rather than as traditional bar reinforcement. There are many new aspects and features which flow from such a change. A primary feature of all the proposals made in this and related papers is that the steel is provided as a thin walled, closed, hollow, convex box, or outer shell, inside which the concrete is cast. Partial cases such as open topped, three sided rectangular shapes are not included. The steel case is a permanent form for the concrete. Because the steel is thin, a few millimetres only for even quite large components, fabrication is essentially a sheet metal operation. This means that the shaping of the sheets can be undertaken by numerically controlled equipment and other simplifications follow. The tubular nature of the steel boxes gives great freedom to devise simple and effective jointing systems. The torsional and shear properties are very favourable. Failure is by flexure. This is an ideal outcome.

Several extensions and new aspects have been experimented with since 1993 including damage control, fire properties and protection of such a material and the application of prestress. Selected outcomes from these studies, and some proposed future studies, are described.

Keywords: Externally reinforced concrete (ERC), Steel/concrete composites, Composite action, Simplification of construction.

Peter G Lowe is Professor of Civil Engineering in the University of Auckland, New Zealand. His other interests include structural mechanics, structural optimisation, engineering heritage and the use of waste materials in the building industry.

INTRODUCTION

Our traditional material, reinforced concrete, has remained relatively unchanged for a long period. Advances have been made in performance, and much progress has been achieved to render the material less brittle and hence more suitable for use in situations where ductility is important. But there are respects in which much less progress has been made, and there seems little prospect of this situation changing in the foreseeable future if the traditional material is retained. For example, automation of the whole reinforcement operation is essentially not possible. There are also respects in which achieving certain desirable features in the product are too costly or complicated for practical use. An example is to reinforce a beam-column intersection so that in the ultimate state the plastic hinge forms in the beam and well away from the face of the column. The current congestion of reinforcement in a heavily reinforced junction between a column and a beam is already such that increasing the steel content still further to move the hinge into the beam is, generally, not a practical choice.

One of the immediate benefits from considering the external reinforcement option is that virtually all the present barriers which prevent further progress with traditional reinforced concrete are removed, and new opportunities present themselves which allow further innovative uses for the composite.

Much of the motivation for wishing to consider other options for steel/concrete composites comes from the desire to reduce the cost of the product. Important cost savings can be achieved by adopting fabrication and construction procedures capable of being automated and of simplifying site operations. Traditionally design has taken precedence over construction of the product. Here the other alternative is followed and the primary decisions have been made with ease of construction taking precedence over the design considerations. Such a difference of approach could extend as far back as the teaching of the young engineer.

In the 1993 paper [1], the emphasis was on discussing the basics of the externally reinforced option. Here we shall give a survey of the externally reinforced programme content in the Auckland laboratory since that time.

FLEXURAL BEHAVIOUR

The emphasis in the programmes continues to be on flexural behaviour: strength and ductility. The observation is that full composite action between the steel case and concrete contents is achieved without the need to provide bonding of one material to the other at the interface. The composite action arises from the steel case and the concrete contents being bent as a single unit with a common curvature. An essential feature is that the case must be complete: three sided or partial enveloping of the concrete will not provide the actions sought. Versions of the commonly used models and calculation methods for section properties and ultimate moment capacities are sufficient to explain the experimental observations.

Maximum ease of reinforcement fabrication, with scope to automate most or all of this set of operations, has been given a high priority in the programmes. This feature has been demonstrated in the many experimental beams of various sizes and capacities [2]. Progress made in this direction can be measured by the relative ease and economy with which the experimental work in the Laboratory can be funded and worked on, compared with what has been possible in the past when working with the traditional material. Broadly speaking, the costs and time requirements are substantially less for the ERC scenario than for conventional reinforced concrete. A newer part of the programme has been the incorporation of pre-stress along with the external reinforcement. Some of these data are presented below.

One of the features of ERC members is that the effects of shear and other secondary features are much less likely to cause a premature failure than is the case with traditional reinforced concrete. For example the following data relate to an ERC beam cut from a frame and tested as a simple beam. The section was B=300 mm, D=350 mm, T=2mm, there was no inner duct, b=d=0, see Figure 1. When cut from the frame of which it was part, pieces of column were left attached to the beam. These pieces were of similar section to the beam, and were connected to it by simple flange plates through bolted connections. The beam was 'stocky', with a length of less than 2m. With a depth of 350mm the span/depth was less than six. Tested as beam in simple bending over a 2 m span, under a central concentrated load, the limit was reached still in flexure, with no signs of shear weakness near the supports or at the beam/ column inter-face. Recently much larger components have been made.

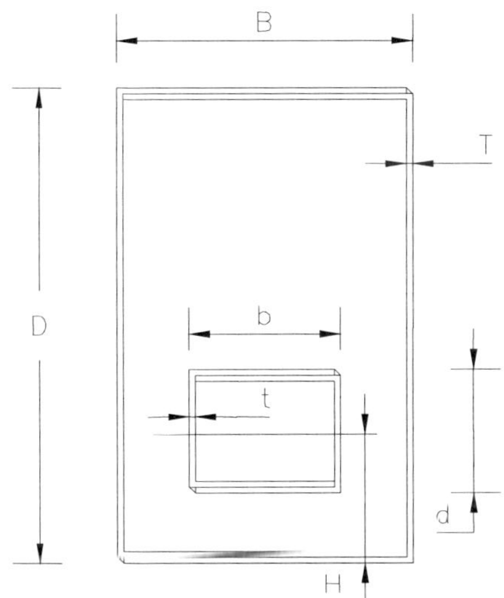

Outer edges of flanges are welded

Figure 1 Typical beam section

JOINTING OF BEAMS AND COLUMNS

Because the reinforcement used in the externally reinforced option is usually tubular, there is always the *inside* to use for connection purposes, as well as the *outside*. This simple observation is the key to greatly simplifying the jointing of beams to columns or other elements. Typical steelwork connections require endplates on one member which are bolted or welded to prepared surfaces or plates attached to the second. The result is that quite frequently all the load carrying actions must pass through the connecting medium, the bolts or welds, and this is usually all arranged to be in a single plane. It is often the situation that the total strength of the connecting medium is less, even substantially less, than the strength of the members being connected.

In the typical reinforced concrete application the joining of one member to another is usually achieved by main bars passing from one member into the other. The typical member, the beam say, is of uniform strength in the part of the member adjacent to the member it is being joined to. Under extreme loadings, such as in earthquakes, the moments are likely to be increasing monotonically as the joint is approached. The usual result is that beam hinges form very close to the column face, or may even migrate some distance into the column. This is usually very undesirable, and may be difficult to avoid.

The externally reinforced option can offer several alternatives. Some at least of these alternatives can be made to be very simple as well as very strong. The arrangement favoured in the Auckland programme has been to consider the joint proper to consist of a joint *module*, a local component housed within the cases of the members framing into the joint. An important feature of this module is that each arm of it is of greater strength than the member casing which contains it. Then what can be achieved in a routine manner is a hinge, for example, which is designed to occur in the beam at a prescribed distance from the column face. An example is shown in the photograph, Figure 2. Typically, this distance is between

Figure 2 An early example of a beam hinge, which had formed at the left side where the flame cut has been made: the hinge being well clear of the column face. This is a portion of the frame shown in Figure 3 [1]

one and three beam depths along the beam from the column centre-line. A mechanism of collapse with hinges in such positions leads to a significant increase in collapse load, compared with the more usual situation of hinges at the ends of the beams and very close to the column faces. A related feature of such jointing systems is that the transfer of moment from member to member does not occur at *a section* but along *a length* of the component. A consequence of this is that many fewer connectors, be they bolts or runs of weld, are needed to achieve the necessary strengths. Also, all circumferential welds can be avoided which means that the remaining welds are essentially non-structural. Some examples of these features are described elsewhere [3].

With the connection component housed inside the case simple connection details can be evolved. In addition there is scope to achieve clean lines and aesthetic qualities which alternatives do not usually provide. Several frame structures with these features have been built and tested [4]. The best of the jointing systems studied have been those which employ an internal connection, termed a module. Such modules can be made both simple and strong. The incorporation of other materials such as glue laminated timber for beams, for example, is also simple to achieve and may provide a useful alternative in particular applications [5].

PROTECTION FOR COVER CONCRETE AND DAMAGE CONTROL IN STRUCTURES.

The *cover concrete* in most reinforced concrete structures is the protection for the bar reinforcement, but is not itself protected. In seismically active parts of the world the first requirement of any structure should be that it survive the earthquake. A further requirement, which is not at present a high priority, is that the damage to the structure should be minimized. In damage terms the cover concrete is very vulnerable. The ideal outcome would be if both collapse resistance and damage limitation can be provided simultaneously in the same construction. The external reinforcement approach achieves some measure of this combination of qualities.

For damage control some protection will be needed for the cover concrete, if present. The ERC approach achieves protection by eliminating the cover concrete. But more than that, the external case ensures that any and all the concrete (filling) is protected in some measure by retaining the concrete in place, even when broken. Broken concrete, if prevented from spalling or dislodging, is quite capable of continuing to bear compressive stress. This is one of the key features of the external reinforcement concept.

Observation has been that when an ERC member is eventually cut open for inspection after test, it is not immediately evident that the concrete contents have been severely damaged. Often there are few signs. Only after a time, when retained water has evaporated and may be the contents have been disturbed in some way, is it evident that there has been extensive breakage. Major cracks may develop, but the more common observation is multiple smaller cracks. Loss of stiffness is observed as must occur after the steel yields in tension. Compression buckling of the case is often quite evident but rarely is strength lost prior to eventual tensile failure of the casing. All of these features point to an efficient retaining of the concrete by the very thin casing.

So far as repair of damaged components or structures is concerned, the concrete contents may be severely broken but repair or replacement of broken concrete may not be necessary. The steel case on the other hand can both stretch and buckle and may require repair if full initial stiffness is to be regained, and for visual reasons. Experience has been that repair of the case is best made by straightening as necessary and perhaps adding some minor amounts of steel. What should be avoided is cutting the case open and doing 'open-heart surgery' on the contents. Repairs have been made as described and the results have met requirements [6].

Repair of damaged conventional reinforced concrete components or structures can also be undertaken by applying external reinforcement principles [6]. In such situations there is often loss of concrete from the body of the member, following after loss of cover concrete. Adding a thin case of the type advocated here can be shown to be an effective repair option [7]. The case is a permanent form for the added concrete, which makes good the earlier loss. It is often quite straightforward to regain the lost strength and even add to the strength of the member, in addition to regaining the lost stiffness.

In the future, damage limitation in structures experiencing extreme events will probably become a much more discussed topic. There are no doubt many ways of proceeding, the external reinforcement option being just one. Another, less radical, use of external reinforcement in new construction is to place a short section of case in the formwork of conventional reinforced concrete members at the sites of potential hinges. Since most damage in such structures is at or close to hinges as they develop in the extreme event, the 'cased' hinge could be expected to show much less damage and will require much less repair.

FIRE TESTING OF ERC COMPONENTS

Some data have been accumulated on the effects of fire on ERC components [8]. The test method has been unconventional but the data have usefulness. Conventional fire testing is expensive and there is a need to explore the effects of fire-type loads on ERC structures. Rather than the usual gas-fired furnace arrangement, in this programme known amounts of electrically generated heat were supplied to parts of the structure and the temperatures and softening characteristics studied.

Most of this experimentation has been with relatively small section beams. The beams were loaded by mechanical means during the heating cycle. The typical section was D=200mm, B=150mm, see Figure 1. Several were constructed using an inner as well as the outer steel casing. The inner tubular casing, termed a duct, was empty of concrete and was instrumented with thermocouples. In practical use such ducts might house services as well as providing the tension capacity of the section which is well protected from the heat of the fire. Some much larger section beams are at present being prepared for testing in a similar fashion.

With the data already accumulated it is possible to analyse and design ERC sections in the fire engineering sense. The thinking is that passive fire resistance can be built into sections through the use of one or more inner steel ducts which are well protected by virtue of the thickness of concrete surrounding these ducts., and which act with the cool concrete as a member in bending during the fire. The outer case, if not protected, will lose strength, as will some 20 - 40 mm of outer concrete, and not contribute any resistance. There is also always the option to protect the whole structure by active means, such as sprinklers, or by passive means such as intumescent paints. A selection of fire test data is given elsewhere [8]. There is a fuller account of the fire-related experimentation in an unpublished internal report [9].

Figure 3 One of several ERC frames [4] and a typical ERC beam [2]
The unpainted segments on the frame were the portions heated to study fire properties [9]

APPLICATION OF PRESTRESS

One of the more exciting prospects currently being investigated is the application of pre-stressing to externally reinforced members. There is scope to pre- or post- tension the ERC member, either before or after the component has left the factory environment. This is in contrast to conventional reinforced concrete for which pre-tensioning is a factory-only operation. The pre-tension might, typically, be used to provide pre-camber in the members as well as adding to the flexural strength.

If pretension is to be applied, then the empty steel case, with suitable end plates, can be fitted with the tendon or bar which can then be stressed by reacting against the case itself. When the concrete filling is added this can either grout the tendon or not as desired. For post tensioning there is less need for an endplate, and separate grouting of the bar or wire tendon is required.

Experience has been that the pre-stress is straightforward to install and can greatly add to the strength of the member. Pre-tensioning will generally not permit as much initial or primary stressing force to be applied as compared with post-tensioning, since there is only the empty case available to resist the pre-stress. But there may be scope to develop greater ultimate capacity with pre-tensioning than with post tensioning. This can result when the pre-tensioned strand or bar is grouted but the post-tensioned strand or bar is not.

An Experimental Beam

A typical moment of resistance strength increase due to prestress is 50%, which roughly equates with an initial pre-compression in the tension flange of around half the yield stress. A typical laboratory specimen was D=200mm, B=200mm, T=3mm, see Figure 1. Typically, by test, the yield stress was 430MPa for the steel and concrete cylinder strength of 65MPa. Such beams have been observed to develop a moment of resistance of about 250kNm when pre-stressed with a 32mm bar, placed 45 mm up from the lower flange and stressed initially to 280kN. At maximum moment this force was estimated to have increased to about 750kN. The moment strength increase due to the pre-stress in this example was more than 50%.

A Proposed Experiment

A much larger beam, with D=600mm, B=320mm, T=t=3mm, b=d=200mm, H=160mm, see Figure 1, has maintained a moment of 600kNm for some months and has been used in various Class related exercises. Later it is intended to post-tensioned this member, with the prestressing cable passing through the hollow inner duct, when it is expected the beam will develop an ultimate moment capacity of about 1500kNm, up from an estimated 1000kNm without the post-tension. In due course this 6.4m long beam will be studied under a 'fire' load regime. The middle third of the span will be heated over a 90 minute period while the beam has a working load applied.

FABRICATION OF COMPONENTS

Because the steel sheet being employed is very thin, there is ample scope to produce misshapen members through folding, welding and assembly, if care is not taken. But as experience is accumulated and adequate control is achieved, accurate and clean-lined components can be achieved at very favourable cost. Because all the welds can be MIG, with much less heat employed than with all-steel construction of similar capacity, distortion from this source is controllable and probably results in much lower levels of residual stress than in thicker plated structures. Reduced residual stresses probably also follow from the use of internal modules to connect components, as described earlier.

The bulk of the welds are longitudinal rather than circumferential: such welds are essentially non-structural. There is considerable scope to automate fabrication, since design data can be supplied to the fabricator on floppy-disc or direct on-line. If the quantities are sufficiently large, roll forming of the shape is an option. An important consideration in some applications is possible distortion of the shape of the casings as they are filled with wet concrete. The normal filling operation would be with a concrete pump, and the filling rate may need to be carefully specified.

A FINISHED STRUCTURE

Recently a permanent structure constructed in ERC has been built. The ERC components are twin 23m high columns, spaced 5m apart. These support a 5m wide by 10m high, double sided, advertising sign. The columns present the appearance of a simple rectangular shape, 380mm x 760mm, but the section of each column actually consists of three cells. Referring to Figure 1, each of the columns consists of an outer and an inner casing, but here the inner case is turned through a right angle. The dimensions are B=380mm, D=760mm, T=t=3mm, b=170mm, d=374mm, and the section has two axes of symmetry. The 23m length was split into a 13m and a 10m length, and these lengths were in turn built up from 3.6m long sheets, butt welded together, these welds being staggered.

Once connection was made through the web of the larger, outer case to the flanges of the inner case, the maximum unsupported plate length was reduced from 760mm to about 300mm. This was beneficial, both to stiffen the assembly for handling and transporting to site, and to ensure that the pressure of the fresh concrete during the filling process could be accommodated. On site the 13m lengths were stepped into a 950mm steel caisson foundations and carefully positioned before being concreted both outside and in. Each splice joint between the 13m and 10m lengths consists of a pair of 2m long cases, similar in all respects to the main cases, and sliding fits into the main cases. The column filling was limited to 3m high lifts, at four hourly intervals, to ensure no distortion of the section.

Finally the sections of the sign proper were positioned between, and bolted to, the cantilever columns. The close tolerances achieved, with some of the fixings made in the shop and the column positioning on site, meant that the final assembly of the sign was straightforward and met all the requirements. One of several benefits that the ERC solution provided was that there was no need to clad the columns. The steel was zinc sprayed and painted in the shop before transport to site. There was an overall cost saving compared with a more conventional, all steel alternative. The typical ultimate major axis moment strength of each column is about 1400kNm: the minor axis strength is about half this value. The major axis stiffness is around 250,000 kNm^2.

OTHER FEATURES

In the future, supplies of the best quality aggregates are likely to become scarcer and more costly. Use of re-cycled or waste materials as the aggregate for the concrete may be forced upon the Industry. Such lower grade fillers can be easily accommodated in the ERC scenario, provided that these materials are not reactive, and do not degrade with time. Another option available with ERC is to use un-bonded fillers. Aggregate, compacted in the case without the cement, is one option which has been experimented with, and might be a choice if there was no available cement, or in an emergency situation. The strength attainable will be lower than if bonded, but the general features are still retained.

Lightweight aggregates could be used. At the other extreme, high strength or ultra high strength concretes would provide additional compression capacity, when for example pre- or post- stressing was being incorporated and additional compression capacity was needed without increasing the member size.

Alkalinity levels for concrete used in traditional reinforced concrete must be maintained to protect the embedded steel against lifetime changes due to carbonation. In the ERC scenario, with the exclusion of contact with atmospheric CO_2, there may be scope to reduce the alkalinity levels without prejudicing the corrosion resistance. This in turn could save on CO_2 emissions in the manufacture of the cement.

CONCLUSIONS

Conferences such as the present one bring together a wide range of interested parties whose particular interests and expertises may be very specialized. With regard to the subject of the present paper, there is an interest group not represented, and this is the other half, the Steel-related Industry and Technology. This is understandable, given the title of the Conference. But there is, in our view, a pressing need to bring these two sectors, the Concrete and the Steel related, together in a suitable discussion mode, in order to achieve a proper dialogue, and furnish the environment where cross-benefits for both materials can be more easily achieved.

Externally reinforced concrete could equally well be described with a name which featured steel, but it owes more to concrete technology than it does to steel, and hence the present name. Our experience has been that neither established Industry grouping is willing to concede anything to the other and this makes progress, when using the two together, more difficult than it need be.

Even so progress can be made investigating new materials combinations on small budgets and in the current environment. The physical appearance of externally reinforced components is of a steel component, and the first two concerns may be corrosion and fire performance. These are two legitimate concerns and must be resolved to the satisfaction of the user of any such technology. There is confidence, however, that neither corrosion nor fire present major technical difficulties which cannot be resolved satisfactorily.

Alternatives to present established technology should always be under investigation, especially in the teaching and research environment. Many other aspects of the whole picture of construction and the impacts on society come into focus when alternatives are examined. The use of construction waste in a value-added manner and where appearance is not an issue, is another option which opens for study when the tubular nature of external reinforcement is available.

By analogy with the motor industry it could be argued that present structural concrete technology is analogous to the motor industry practices before the introduction of stressed skin concepts into body design. A change to a stressed skin approach will not be wholly advantageous. The possible advantages are however very substantial.

ACKNOWLEDGEMENTS

It is a pleasure to acknowledge assistance, useful discussion and aid in various forms, which have all contributed to the research programme relating to external reinforcement. N. A

Charman, A. J. Van Erp and S. G. Kanji have each written Master of Engineering theses on the topic. It was a pleasure to supervise and work with them. All three when graduate students were recipients of the BHP/New Zealand Steel Scholarship for Innovative Use of Steel. This scholarship assistance was particularly valuable. Several friends have been generous with their time and have discussed aspects of the programme: R.M.Thompson and Dr. J.W. Butterworth particularly. This is much appreciated. The technical staff in the departmental laboratories, especially H. Mooy, P. Thomas, K. Gotts and B. Mehaffy, have made important contributions to the programme, and are thanked for their assistance.

The financial support for this programme has been provided by Industry. The main contributors have been: Milburn (NZ) Ltd., McConnell-Dowell Corp (NZ) Ltd., Mainzeal Group Ltd., and BHP/NZ Steel Ltd. Had this financial support not been available the programmes described here would not have proceeded.

REFERENCES

1. LOWE, P G. Externally reinforced concrete: a rethink of steel/concrete composites. Concrete 2000, Eds. R K Dhir and M R Jones, E & F. N. Spon, 1993, pp.1717 - 1726.

2. LOWE, P G. Externally reinforced concrete composites. Advances in Concrete Technology, Proceedings of Third CANMET/ACI International Conference, Auckland, 1997, Ed. V M Malhotra, ACI Special Publication SP - 171, pp. 551 - 568.

3. VAN ERP, A J. Externally reinforced concrete : further jointing systems, Thesis for Master of Engineering, University of Auckland, 1993, pp. 164.

4. CHARMAN, N A, LOWE, P G AND VAN ERP, A J. Externally reinforced concrete framed structures. Australasian Structural Engineering Conference, Auckland, Ed. J W Butterworth, 1998, pp. 151 - 158.

5. LOWE, P G. A role for innovation in structural engineering programmes? Structural Stability and Design. Eds. S Kitipornchai , G J Hancock & M A Bradford , Balkema, Rotterdam, 1995, pp.455 - 460.

6. LOWE, P G AND KANJI, S G. Externally reinforced concrete a report on some recent studies. N.Z. Concrete Society Conference '96 , Wairakei, 1996, pp. 115 - 119.

7. KANJI, S G. Seismic strengthening and repair of existing reinforced concrete members using external reinforcement principles. Thesis for Master of Engineering, University of Auckland, 1996, pp. 235.

8. CHARMAN, N A AND LOWE, P G. Experimental performance of loaded beams in a simulated fire. N.Z. National Society for Earthquake Engineering, Annual Conference, New Plymouth, 1996, pp. 185 - 190.

9. CHARMAN, N A AND LOWE, P G. Thermal loading of externally reinforced concrete beams, Auckland Uniservices Ltd. Unpublished Report, 1993, pp. 57

PERFORMANCE OF VOLCANIC PUMICE CONCRETE WITH SPECIAL REFERENCE TO HIGH RISE COMPOSITE CONSTRUCTION

K M A Hossain

Papua New Guinea University of Technology

Papua New Guinea

ABSTRACT. This paper will highlight the experimental investigation on volcanic pumice concrete (VPC) and its performance in thin walled composite (TWC) beams and columns. Preliminary investigations confirmed that it is possible to produce a VPC of strength 30 MPa and 30% lighter than normal concrete. Comparative performance study of VPC in normal and confined environment will be described based on tests on TWC sections. This paper will also highlight the feed backs from manufacturing and construction industry regarding the practicality, viability and economy of this novel form of construction in the context of Papua New Guinea. The successful completion of this research will result in the formulation of design guidelines for this type of structural elements and their application in the construction of high rise structures.

Keywords: Volcanic pumice, Concrete, Beams, Columns, Buckling, Strain, Composite, Strength, Chemical, Mix design.

Dr Khandaker M Anwar Hossain is a Lecturer in the Department of Civil Engineering, The Papua New Guinea University of Technology. His research interests include the properties of concrete, small-scale modelling of structures, thin walled composite construction and non-linear finite element analysis of reinforced and composite structures.

INTRODUCTION

Pumice is a natural material of volcanic origin produced by the release of gases during the solidification of lava. Volcanic pumice has been used as aggregate in the production of light weight concrete in many countries of the world. It is found that satisfactory concrete [1,2] 30% to 40% lighter than normal concrete having good insulating characteristics with high absorption and shrinkage can be manufactured using volcanic pumice. This paper will present the properties of volcanic pumice concrete (VPC) and its performance in thin walled composite (TWC) structural elements. TWC sections is a new idea for beams and columns [3,4] comprising cold formed steel elements with an infill of concrete that are suitable as replacement for hot-rolled steel or reinforced concrete in small to medium sized buildings. Typical TWC beams and columns are shown in Figure 1. The inherent advantages of this system are derived from its structural configurations. Open box sections for beams will allow easy casting of in-fill concrete. TWC sections do not require temporary form work for infill concrete as the steel acts as form work in the construction stage and as reinforcement in the service stage. The infill concrete in TWC sections is less likely to be affected by adverse temperature and winds as experienced in the case of reinforced concrete. The in-fill concrete is generally cured quickly and in any case, the load capacity of the steel alone may be relied upon for most construction loads.

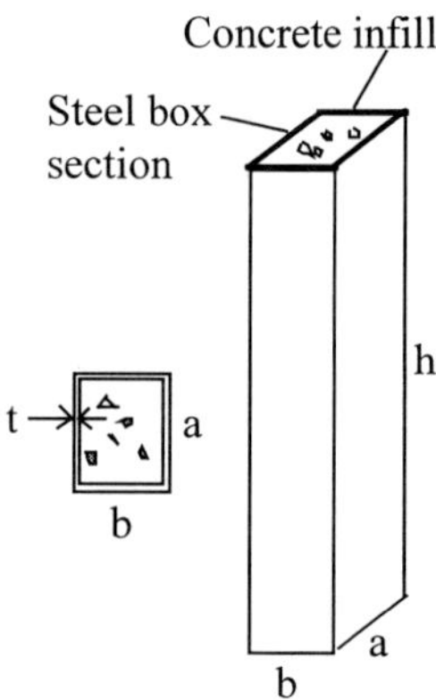

a,b: cross-sectional dimensions of column
h: height of column; t: thickness of sheeting
b,d: Crossectional dimension of beam
s: width of opening

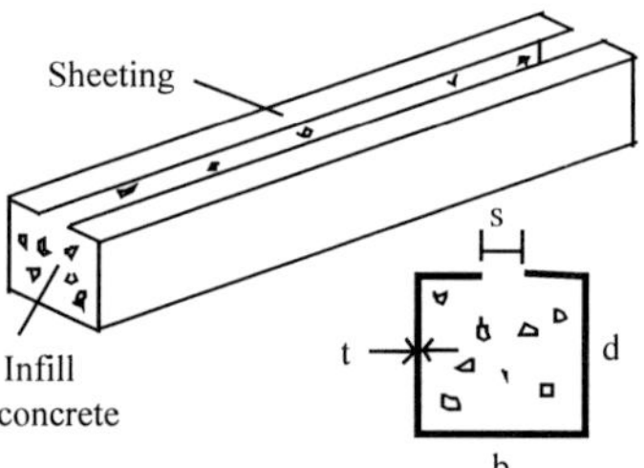

Figure 1(a) TWC columns

Figure 1(b) Typical custom folded Beam section

RESEARCH SIGNIFICANCE

Research is now ongoing, to investigate the behaviour and potential use of TWC sections as structural elements with particular emphasis on the use of locally available material, economy and local structural needs. As a consequence, volcanic pumice which is a natural waste from volcanic activities in Papua New Guinea is used to manufacture infill light weight concrete for TWC sections. Advantage of this system includes a very light construction weight, excellent surface finish, relatively slender dimensions, enhanced ductility and the potential for semi-rigid connections. The improved performance of the VPC in a confined environment will allow the use of comparatively low strength VPC in this form of construction as the main function of the concrete infill is to prevent local buckling of the thin sheeting. The successful completion of this project will explore the possible utilisation of VP in low cost concrete production and potential construction of medium to high-rise buildings in PNG using TWC elements. The use of VP in construction will also decrease the environmental hazard in the volcanic areas of Papua New Guinea.

EXPERIMENTAL INVESTIGATION

Chemical And Physical Properties of VP And VPA

Volcanic pumice can be found on various coastal islands of Papua New Guinea. The investigation was based on VP obtained from Tavurvur and Vulcan craters located in the Rabaul area of the East New Britain province. Chemical analysis [5] as shown in Table 1 indicated that the VP is principally composed of silica (about 61%).

Table 1 Chemical and physical properties

(a) CHEMICAL PROPERTIES OF VP

Calcium oxide (CaO): 4.44% Silica (SiO_2): 60.82% Alumina (Al_2O_3): 16.71%
Iron oxide (Fe_2O_3): 7.04% Magnesia (MgO): 1.94% Sodium oxide(Na_2O): 5.42%
Potassium oxide (K_2O): 2.25% Sulphur trioxide (SO_3): 0.14%
Loss on ignition: 1.52% Chloride : <0.012% Sulphate: 0.07- 0.78% pH: 5.3

(b) PHYSICAL PROPERTIES OF VP AND AGGREGATES

Item	Stone aggregate	Sand	VPA
Bulk density, kg/m^3	2470	2610	763
Absorption %	2.86	2.04	37.00

The properties of normal aggregate are compared with those of VP aggregate (VPA) in Table 1. The bulk density results suggest that the VPA is much lighter than normal aggregate and also has high water absorption which indicates high degree of porosity. The high abrasion value [2] of VPA compared to normal aggregate also indicates its low strength.

Investigation on Volcanic Pumice Concrete (VPC)

The VPC was manufactured from locally manufactured Portland cement called 'Paradise'. The aggregates used were 10 mm maximum size crushed gravel, 26 mm maximum size volcanic pumice and local river sand. A series of tests were performed to investigate the effect of different percentages of volcanic pumice as aggregate on the strength of VPC. The concrete mixes had the overall ratio of 1:2:3 (mix 1) and 1.2:4 (mix 2) on volume basis. Each of the two mixes are classified into five sub-mixes according to the % of VPA as a replacement of normal coarse aggregate (by volume). The coarse aggregate consisted of 10 mm maximum crushed gravel and VPA with river sand as fine aggregate. The mix parameters and some characteristics of the fresh concrete for mix 1 are presented in Table 2. The first numeric in the mix designations represents % of VPA of total coarse aggregate by volume, second numeric represents % of VPA of total aggregate by weight and the third numeric represents total aggregate cement ratio by weight. The specimens were 100 x100x100 mm cubes and 100 x 200 mm cylinders for compressive strength. Four cubes and four cylinders were cast from each sub-mix. The samples were compacted by using a table vibrator. The specimens were demoulded after 24 hours and cured under water at a temperature of 23 ±2°C until testing at 28 day.

Table 2 VPC mix details

MIX DESIGNATION	VPA kg/m^3	10mm AGG. kg/m^3	NW/C (BY WEIGHT) *	CUBE STRENGTH MPa	CYLINDER STRENGTH MPa
VPC Mix-1: 1:2:3 , Cement =814 kg/m^3					
100-36.9-2.38	358	0	0.465	27.1	21.6
90-25.6-2.55	322	119	0.44	28.8	24.1
75-19.4-2.80	268	297	0.39	31.34	26.43
50-11.3-3.22	179	594	0.376	35.3	27.53
0-0-4.07	0	1188	0.36	40.21	35.42

* Net water excluding water absorbed by VPA

A study on fresh VPC mixes shows that more water is needed to get a workable mix as the % of VPA is increased from 0 to 100%. The total water requirement is much higher due to high water absorption capacity (about 37%) of VPA as well as the presence of higher quantity of fines in the VPA compared to replaced 10 mm aggregate. For 100% VPC, a slump of 50 to 60 mm represents satisfactory workability compared to 80-82 mm slump of 0% VPC. The variation in the 28-day compressive strength of VPC with different percentages of VPA is shown in Table 2. As expected, strength decreased with the increase of VPA due to the replacement of normal stone aggregate by relatively weak pumice aggregate. Results show that by using 100% VPA, it is possible to obtain a VPC of 27 MPa (1:2:3) and 22 MPa (1:2:4). The use of 100% replacement of coarse aggregate by VPA (designated as 100% VPC) can produce a VPC of 25% lighter than the normal concrete. A study is now under progress to study the other properties including permeability, shrinkage, corrosion resistance and fire resistance of VPC. Research is also under progress towards the development of design charts for mix designs of VPC.

Investigation on TWC Beams And Columns With VPC In-fill

A comprehensive series of experimental investigation is now under progress to study the performance of TWC beams and columns with VPC infill. The experiments are designed to provide information on the load-deformation response, stress-strain characteristics and failure modes.

Detail Description of Tests

The TWC beams and columns were manufactured using hollow steel sections available in the market and also using flat sheet folded to hollow sections as shown in Figure 1 with thickness of the steel varies from 0.45 mm to 3 mm. The infill concrete for TWC beam-columns either VPC or normal concrete (NC) was made from 10 mm maximum size aggregates. The specimens were cast and then air cured at room temperature until tested. The behaviour of thin steel sheeting under wet concrete was observed during the casting operation. The concrete casting was much simpler than reinforced concrete. The strain gauges were installed at key locations of the specimens as shown in Figure 2. Axial deformation for column and central deflection for beam were measured by dial gauges. The load was applied in increments and at each load increment strains and axial and central deformations were measured to get complete load-deformation response of both columns and beams.

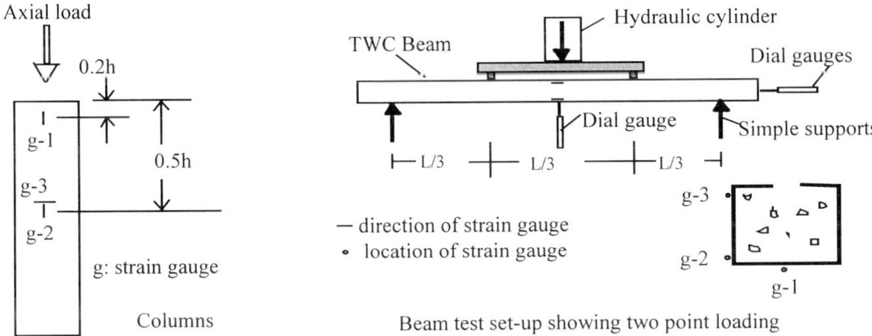

Figure 2 Instrumentation and testing of TWC columns and beams

Test Results and Discussion

The typical load-deformation responses from NC and VPC infill TWC columns are compared to each other in Figure 3. The VPC columns shows higher deformation than NC columns. This is reasonable as the modulus of elasticity of VPC is found to be lower than NC as reported by Hossain [2]. Taking into consideration the effect of different cube strength of concrete, the VPC columns are found to have similar strength when compared with NC columns. It seems to be that the confinement of concrete in TWC sections improve the performance of VPC and the failure load is dictated by the performance of steel section. It is interesting to note that the VPC columns are more ductile than NC columns of the same strength. The strength is found to be affected by the slenderness of the columns. For comparatively slender columns with h/b ratio of 24, the strength is reduced by about 15% compared to columns with h/b ratio of 6 as the failure is dictated by global buckling.

The load-deflection response for NC and VPC beams are compared in Figure 3. Both beams show similar behaviour with strength seems to be not affected by the type of concrete. After initial stages of debonding, the lateral separation between steel and concrete starts due to compression in the open top flange where the steel plate is free to deform laterally. As load increases, the increased compression force causes the steel plate at the top open flange, to buckle at about 90% of the ultimate load. The transition between start of buckling and failure is quick and finally the beam fails due to excessive outward buckling and cracking of infill concrete.

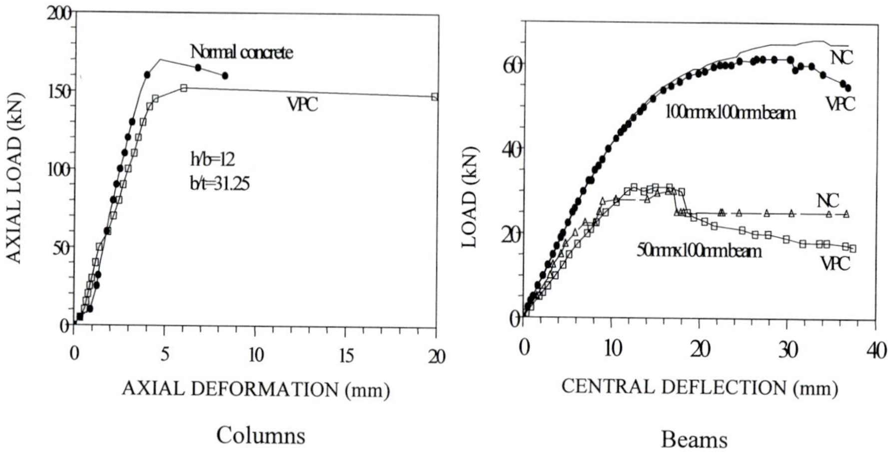

Figure 3 Comparative study of load-deformation response for beams and columns

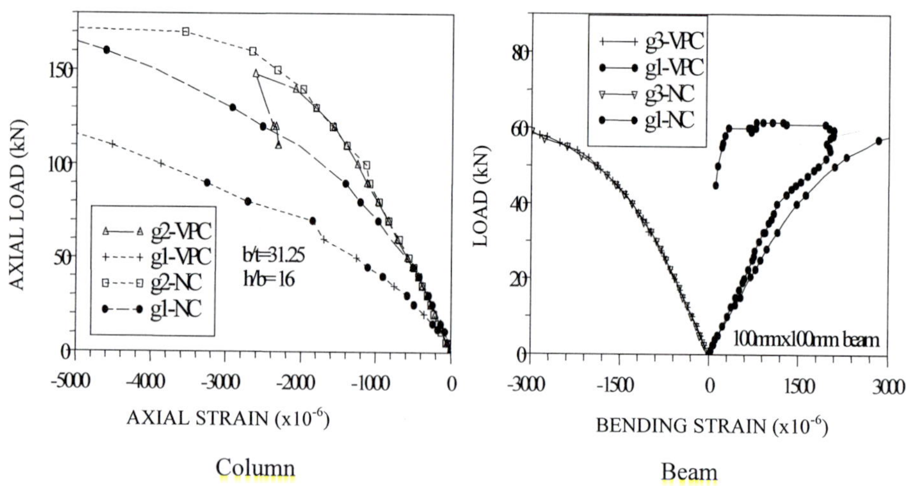

Figure 4 Effect of VPC infill on strain

The typical variation of axial strains for both NC and VPC infill columns is shown in Figure 4. The strains are found to increase steadily with the increase of load. The higher strain at the top end of the column (gauge-1) compared to the centre (gauge 2) was observed for comparatively short column with slenderness, h/b=6 while for comparatively slender column with slenderness, h/b=24 the strain in the middle of the column is found to be higher than that at the top for both NC and VPC columns due to the global buckling mode of failure. The strains in VPC columns are found to be higher than those in NC columns. The bending strains in VPC and NC beams are also compared to each other in Figure 4. The compressive strain in gauge 3 seems to be similar in both type of beams although the tensile strain in gauge 1 shows higher values for VPC beams. But it seems to be that the type of concrete has less influence on the strain than that observed in columns.

STUDY OF VPC IN NORMAL AND CONFINED ENVIRONMENT

The axial capacity of TWC columns is found to be based upon the increased strength of the concrete due to confinement and the increased plate buckling capacity due to composite action. It is found that [4] the axial capacity of VPC column is 30% higher than the sum of individual capacity of concrete core and sheeting. Tests were also performed varying the strength of VPC infill in TWC columns having the same geometric dimensions and steel properties. The strength of VPC varied from 10 MPa to 35 MPA. It is found that only 7.5% (compared to 24%, based on BS 5400 [6] using actual concrete capacity) increase in axial capacity is possible by increasing the strength of VPC by almost 200%.

POTENTIAL APPLICATION OF VPC IN THE CONTEXT OF PNG

VPC can be used in the construction of traditional building with TWC beams, columns and walls as shown in Figure 5. The TWC walls are supposed to consist of double skins of steel sheeting with an infill of VPC. The following structural advantages can be gained from this form of construction:

- Very light construction weight generally and can be reduced further by 25-35% using VPC ; ductile beam-column frame with possible semi-rigid connections; infill panel of TWC wall with frame can resist the earthquake forces in building; speedy construction due to simple fabrication and erection; excellent potential for pre-cast construction.

In the context of PNG, the research should be intended to answer the viability of these type of construction where the country is susceptible to earthquake. The viability of the structures in the light of structural performance, feasibility in fabrication and economy should be assessed. In this regard, the TWC sections are suitable as structural elements as confirmed from their advantages. However, structural performance should also be related to feasibility and economy. For feasibility assessment, the co-ordination between manufacturing and construction industry is needed. With this system the steel strip is reasonably cheap to import and the cost of setting up a rolling mill is acceptable in quite poor countries. The manufacturing of sheeting locally will provide economy. In this systems, infill concrete does not need high strength as its main job is to prevent local buckling of the sheeting in addition to ensure mechanical interlock between steel and concrete. It is possible to use low strength, low cost and light weight concrete manufactured from locally available materials. This will reduce the cost and weight of the TWC sections.

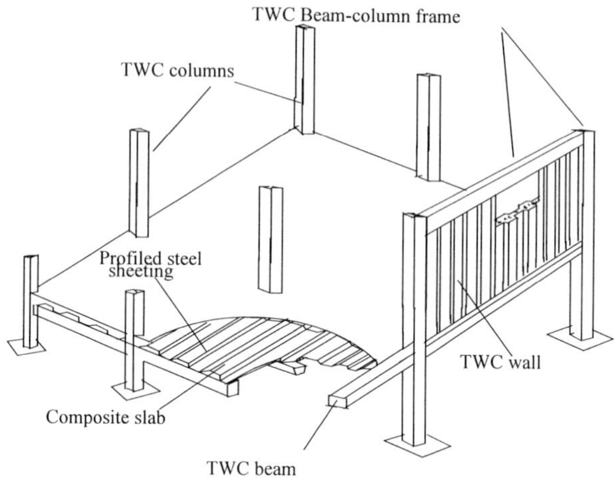

Figure 5 Schematic of a typical future TWC building

FEED BACK FROM MANUFACTURING AND CONSTRUCTION INDUSTRY

The building of an elegant structure satisfying structural (strength associated with ductility and lightness), construction (feasible and simple) and economic (low cost) requirements is vital for PNG and in the future, it is possible that TWC sections will meet those requirements and can lead to the construction of high rise buildings in Papua New Guinea. At the moment, desk studies are ongoing where a set of questionnaires are sent to various manufacturing and construction industries in PNG. The various industries welcome and appreciate the type of construction as confirmed from the preliminary feedback. The completion of this project will provide a foundation for the use of VPC and will introduce TWC construction in PNG.

CONCLUSIONS

This paper presents the results of investigations on the suitability of using Volcanic Pumice as a coarse aggregate in light weight concrete production. It is found that a VPC of strength 27 MPa can be obtained by replacing 100% (volume basis) of coarse stone aggregate by VPA. The performance of VPC in TWC sections is found satisfactory and in future, TWC elements can be an alternative economical solutions for housing construction in PNG. This paper also demonstrates how the use of technology can transform the cause of social and environmental disasters into a natural resource and hence can be used in post-disaster rehabilitation projects.

ACKNOWLEDGEMENTS

The author is grateful to the Papua New Guinea University of Technology for providing financial assistance in this project. The author is also grateful to the Technical staffs of the materials laboratory of the Department of Civil Engineering.

REFERENCES

1. NEVILLE, A M. Properties of Concrete. 3rd Edition, Pitman Publishing Limited, London, 1981, pp. 779.

2. HOSSAIN, K M A. Properties of volcanic pumice based cement and light weight concrete. (to be published)

3. WRIGHT, H D. The behaviour of Composite Rectangular Tubes under Bending and Axial Loading. Proce. Nordic Steel Construction Conference, Malmo, Sweden, June 19-21, 1995, pp. 523-530.

4. HOSSAIN, K M A. Behaviour of thin walled composite sections as structural elements. Proce. ASEC Conference, Auckland, 31-2 October, Vol. 1, 1998, pp. 175-180.

5. HOSSAIN, K M A. Volcanic ash and pumice based blended cement. Proce. of 23rd OWICS Conference, CI-Premier Conference Organisation, Singapore, 24-26 August, 1998, pp. 297-302.

6. BRITISH STANDARDS INSTITUTION. BS 5400. Steel, Concrete and Composite Bridges: Part 5: Code of Practice for design of composite bridges., 1979.

ELASTOPLASTIC BEHAVIOUR OF HYBRID RC COLUMNS LATERALLY CONFINED BY PRESTRESSING AND SQUARE TUBE

T Yamakawa

University of Ryukyus

M Kurashige

Neturen Co Ltd

Japan

ABSTRACT. Lateral confinement for RC columns has become of great interest to the structural design, because of preventing the brittle shear failure and improving load-carrying capacity, ductility and energy absorption capacity. The confinement by PC bar prestressing and square steel tube is newly proposed as both passive and active confinement methods in this paper. The paper summarises the test results and discusses the elastoplastic behaviour of high confined RC columns by PC bar prestressing and square steel tube in comparison with ordinary RC column and non-hooped or hooped RC columns laterally confined in square steel tube.

Keywords: Confinement, Ductility, Elastoplastic behaviour, Hoop, PC bar prestressing, RC column, Seismic design, Square steel tube.

Professor Tetsuo Yamakawa is a Professor in the Department of Civil Engineering and Architecture, University of the Ryukyus, Okinawa, Japan. His main research interests include structural concrete, seismic repair and retrofitting, hybrid structures and structural mechanics. He is a member of AIJ, JCI and NZSEE.

Mr Masayoshi Kurashige is a Chief Engineer and Manager in Neturen Co., Ltd., Tokyo, Japan. He specializes in the prestressing design and development of the bridges and buildings. He is a member of AIJ.

INTRODUCTION

It has been widely accepted that the strength and the ductility of concrete are improved by transverse reinforcement. Recently hybrid or composite structures which consist of concrete and steel or new materials have been developed and applied to buildings and bridges. The author proposed the hybrid RC columns doubly transversally confined by square steel tube and hoops in 1991[1]. The axial compression tests and cyclic lateral loading tests under constant vertical load were carried out according to the research programme [1-2]. As a result, it was concluded that the combination of high strength concrete (35MPa-50MPa) and double confinement by square steel tube and hoops remarkably improved the seismic performance of RC columns [3].

On the basis of the research, the high confined method for hybrid RC columns was proposed in 1997[4]. This method consisted of both passive and active confinement systems. In addition to square steel tube, transverse PC bar prestressing was applied into the column on both directions. The transverse PC bar prestressing prevents the wall of square steel tube from bending deflection and also gives the active lateral pressure through the wall of steel tube. As a result, superior seismic performance for columns may be expected because of the high confined effect even if RC columns are likely to yield to shear failure. The purpose of this paper is to discuss the elastoplastic behaviour of hybrid RC columns highly confined by square steel tube and PC bar prestressing.

COLUMN SPECIMENS

Total number of specimens is four. Among them three specimens are jacketed with square steel tube except for the reference RC column specimen. Wall thickness of the steel tube is 6mm and small gaps (about 10mm) between the steel tube and stubs are provided so that axial compressive force may not be transmitted through the steel tube. The details of these column specimens are listed in Table 1.

The shear span to depth ratio M/(VD) is 1.5. The amount and arrangement of deformed rebars are common in all specimens. The mechanical properties of square steel tube and reinforcement bars are shown in Table 1. The ratio of total area of 12 longitudinal deformed bars (ϕ19mm) to gross area of the section is 5.51% in all specimens. This value is close to the highest limit. The yield strength of rebar is 490MPa and its value is higher than those of the other reinforcements except for PC bars. The concrete cylinder strength is 37.4MPa and 50.2MPa. Therefore, the reference RC column specimen is shear failure type due to bond splitting failure.

PC bar is a manufactured pre-tensioned pipe whose outer diameter is 23mm and inner diameter is 15.8mm. The reaction steel round bar for prestressing is inserted in the pipe previously. The thickness of the pipe is 3.6mm and its section area is 219.4mm^2. The yield strength is 931MPa and its strain is 0.475%. The value of the prestress is 147.1kN per one PC bar. This value is 72% of the yield strength. The prestressing arrangement is referred to Figure 1.

Table 1 Column specimens

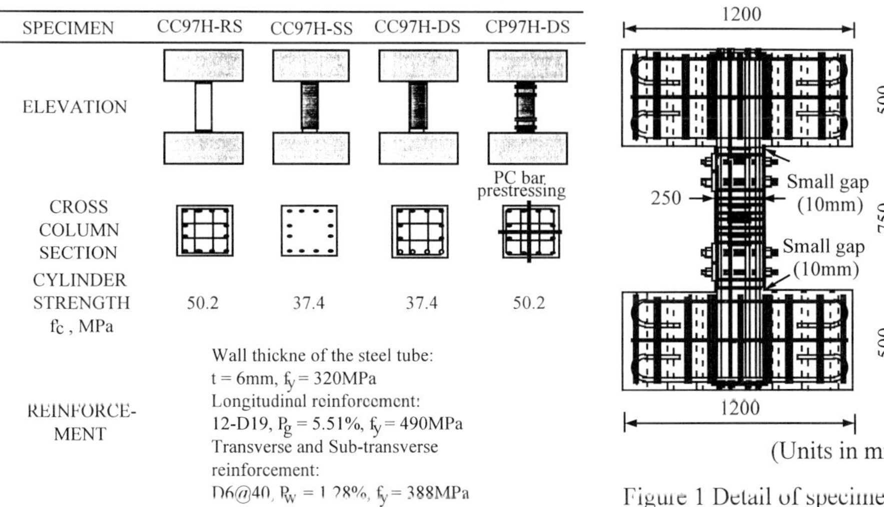

SPECIMEN	CC97H-RS	CC97H-SS	CC97H-DS	CP97H-DS
ELEVATION				PC bar prestressing
CROSS COLUMN SECTION				
CYLINDER STRENGTH f_c, MPa	50.2	37.4	37.4	50.2
REINFORCEMENT	Wall thickne of the steel tube: t = 6mm, f_y = 320MPa Longitudinal reinforcement: 12-D19, P_g = 5.51%, f_y = 490MPa Transverse and Sub-transverse reinforcement: D6@40, P_w = 1.28%, f_y = 388MPa			

Figure 1 Detail of specimen

EXPERIMENTAL RESULTS

The test setup of Ken-ken type as shown in Figure 2 was used in this experiment. A couple of servohydraulic actuators figured in Figure 2 always maintained the horizontal loading beam in parallel to the strong floor and sustained a constant vertical total load during cyclic lateral loading test. The instrumentation methods were designed to obtain the following data: (1) applied forces (2) horizontal and vertical displacements (3) strains of the reinforcement bars and the steel tube.

The lateral loading programme is shown in Figure 2. At first, several cyclic small lateral forces were applied to all specimens in order to obtain the initial lateral stiffness. Then the loading pattern was a cyclic type with alternating displacement reversals as shown in Figure 2. The peak drift angles were increased stepwise until R=3% with an incremental drift angle of ΔR=0.5% after three successive cycles at each displacement level. On the other hand, the axial load which was kept constant during the test was applied to the column by means of a couple of servohydraulic actuator. The magnitude of the constant axial load was $A_g f_c'/2$, where f_c' is the compressive strength of the concrete cylinder and A_g is the cross sectional area of RC column.

Crack patterns for the reference RC column specimen CC97H-RS were ones due to bond splitting failure. Figure 3 shows measured V-R hysteretic loops, where V is lateral force, and Figure 4 shows measured strains of the rebars, namely, the longitudinal reinforcement at the top of each column against the drift angle R. The bottom stub was fixed to the strong floor and the top stub was restrained by a couple of servohydraulic actuators so that its rotation might not occur.

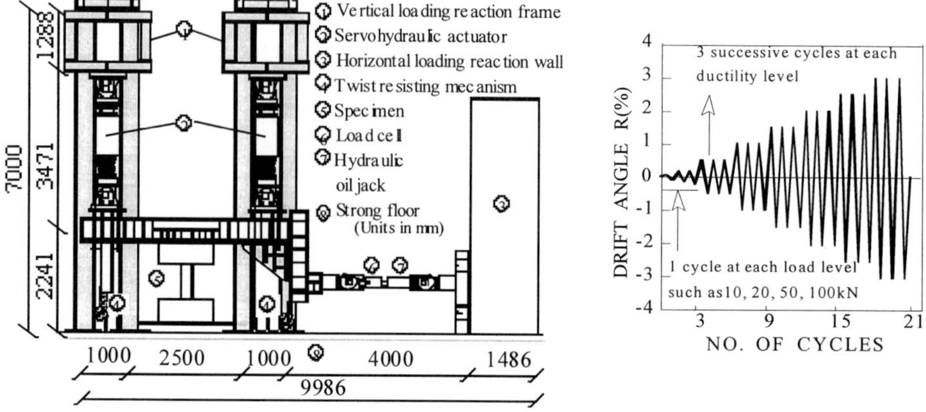

Figure 2 Detail of test setup and loading programme

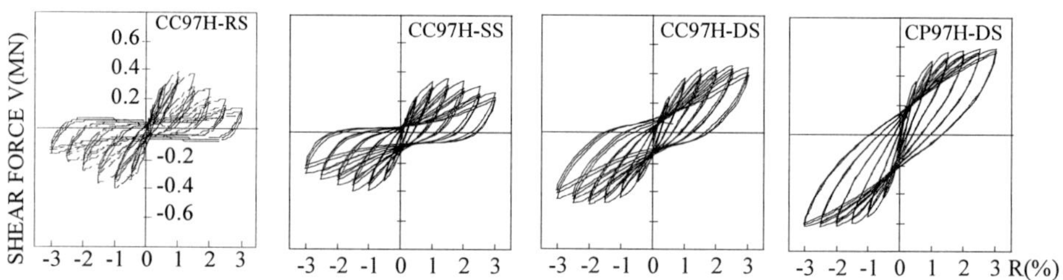

Figure 3 Measured shear force versus story drift angle relationships for columns

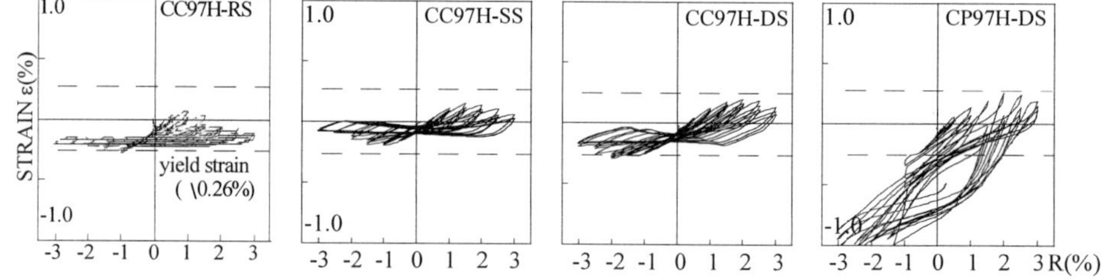

Figure 4 Measured strains versus story drift angle relationships at the top of rebars in columns

In the reference RC column specimen CC97H-RS, a pinching phenomenon, that is to say, a bond slip was observed in its hysteretic loops as shown in Figure 3. As a result, the lateral capacity was degraded gradually as the story drift angle increased and its ductility couldn't be expected. Although the jacket by square steel tube prevented collapse of RC column from shear failure, the bond slip was observed through measured hysteresis loops of the specimen CC97H-SS (see Figure 3). This fact is also recognised from the measured strain of the rebar as illustrated in Figure 4. The strain doesn't reach yield strain. The specimen CC97H-DS doubly confined with square steel tube and hoops improved the hysteresis loops. However, a little bond slip phenomenon was observed in hysteresis loops and the rebars didn't yield (see Figures 3 and 4). On the other hand, the hooped column specimen CP97H-DS doubly confined with square steel tube and PC bar prestressing demonstrated on excellent seismic performance and bond slip didn’t yield at all. The rebar of the specimen CP97H-DS only reaches yield strain as shown in Figure 4. Its mean shearing stress in the column section is about 10MPa at a peak shear capacity.

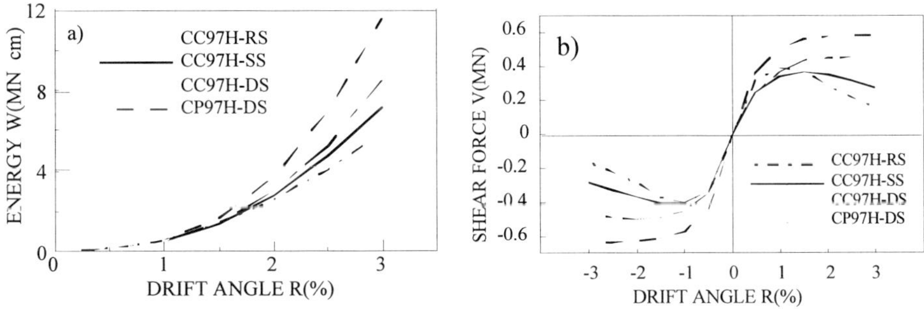

Figure 5 a) Accumulated absorbed energy and b) measured skeleton curves for columns

Accumulated absorbed energy and measured skeleton curves based on experimental results are illustrated in Figure 5. Through Figure 5 the specimen CP97H-DS is far superior to the other specimens. The higher the transverse confinement become, the better the seismic performance is improved. The experimental results under high axial load ratio $N/(A_g f_c')=0.5$, where N is a constant axial load, present that it is possible to produce an excellent seismic performance by adding high confined effect even if the shear span to depth ratio $M/(VD)$ is 1.5, the longitudinal reinforcement ratio p_g is 5.51% and its yield strength is high, namely, 490MPa.

THEORETICAL INVESTIGATION

In this theoretical investigation, a stress-strain curve model proposed by Sakino and Sun [5] is adopted for the confined or unconfined concrete. The compression strength of singly confined concrete in square steel tube was proposed by Matsumura and Itoh [6] based on some experimental results. The compressive strength of doubly confined concrete in square steel tube and hoops is obtained by adding the strength, namely, by using the simple superposed strength method. PC bar strengthens the passive confinement by square steel tube. In addition to improvement of the passive confinement, the lateral pressure due to PC bar prestressing may be expected as an active confinement through the square steel tube. The proposed equation by Richart [7] is applied as a fluid pressure. Therefore the compressive strength of confined concrete is given by Equation (1).

In Equation (1) the first term is the cylinder concrete strength, the second term is an incremental strength due to the passive confinement by square steel tube and PC bar, and the third term is one due to the active confinement by PC bar prestressing.

$$\frac{f'_{cc}}{f'_c} = 1 + \frac{46\,(n+1)\,{}_s\sigma_y\,(t/B)^2}{f'_c} + 4.1\frac{\sigma_r}{f'_c} + \text{an incremental strength by hoops} \qquad (1)$$

where,
f_{cc}' = compressive strength of confined concrete (MPa)
f_c' = cylinder concrete strength (100 ϕ x 200mm) (MPa)
n = number of PC bars inserted in each direction across the column section
${}_s\sigma_y$ = yeild strength of wall of steel tube (MPa)
t = thickness of wall of steel tube (mm)
B = width of square steel tube (mm)
σ_r = average lateral pressure due to PC bar prestressing through square steel tube (MPa)

Numerical examples with regard to the constitutive law for the high-strength concrete are illustrated in Figure 6. As the confined effect increases, the stress-strain curves are improved in strength and ductility. Flexural analysis was carried out by using the fiber model.

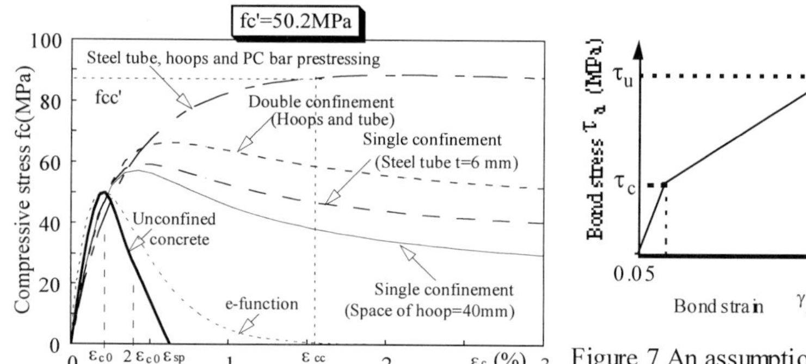

Figure 6 Calculated stress-strain curves for concrete

Figure 7 An assumption of mean bond stress τ_a versus mean bond strain γ_a relationships

Numerically calculated skeleton curves based upon the flexural analysis were obtained by assuming curvature distribution along the member axis together with the constitutive law for a confined concrete. On the other hand, the proposed equation for shear strength by AIJ (Architectural Institute of Japan) consists of both truss and arch mechanisms [8]. Steel tube was converted into hoops as the truss mechanism. The compressive strength of the confined concrete was adopted in arch mechanism. Also a modified Arakawa's Equation [8] was applied in order to calculate the shear strength as well as the AIJ Equation [8].

Since occurrence of bond slip was observed from the experimental results, the flexural analysis was also carried out for reference by assuming the constitutive law with regard to the bond stress and bond strain which is defined by the amount of bond slip divided by the column height h. The bond stress-strain relationship is assumed as illustrated in Figure 7 with reference to the experimental study on bond behaviour between deformed reinforcement bars and confined concrete in square steel tube [9].

In Figure 8 measured *V-R* skeleton curves and calculated ones with perfect bond or bond slip by flexural analysis are illustrated together with calculated shear strength due to shear failure based on two proposed equations. Through Figure 8 both flexural and shear strengths are compared. As the transverse confinement is getting higher, the shear strength is improved significantly. Then shear strength is superior to flexural strength. As a result, the flexural failure goes ahead of the shear failure and the ductility can be maintained in case of high transverse confined column specimen. This fact is supported by both experimental and calculated results as shown in Figure 8.

Ultimate strength by experimental results and calculated *N-M* interaction diagrams with perfect bond or bond slip are illustrated in Figure 9. As the transverse confinement becomes higher, both experimental and calculated results on flexural strength are getting larger, because the bond slip may be ignored. However, the experimental results show a lower level than the calculated *N-M* interaction diagrams for column specimens CC97H-SS, CC97H-DS and CP97H-DS. Especially in the column specimen CP97H-DS, the opening holes in the square steel tube for PC bar prestressing are too large. Since the influence of the opening holes is not taken account of, the calculated results are likely to be overestimated.

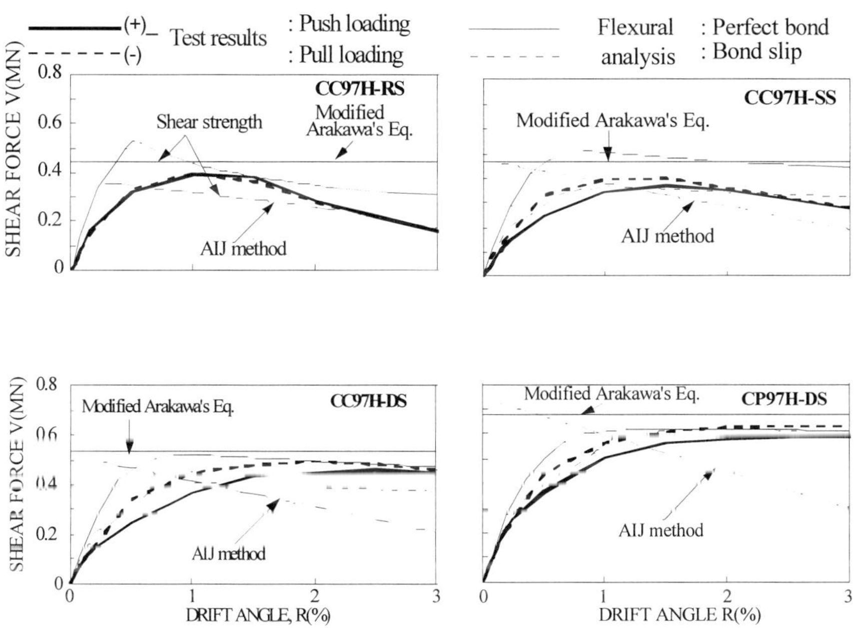

Figure 8 Comparison on shear or flexural strength and measured skeleton curves

The relationships between the increasing ratio of flexural strength due to transverse confinement and axial load ratio are shown in Figure 10 through calculation. The increasing ratio of flexural strength is defined by dividing the increment of the flexural strength due to

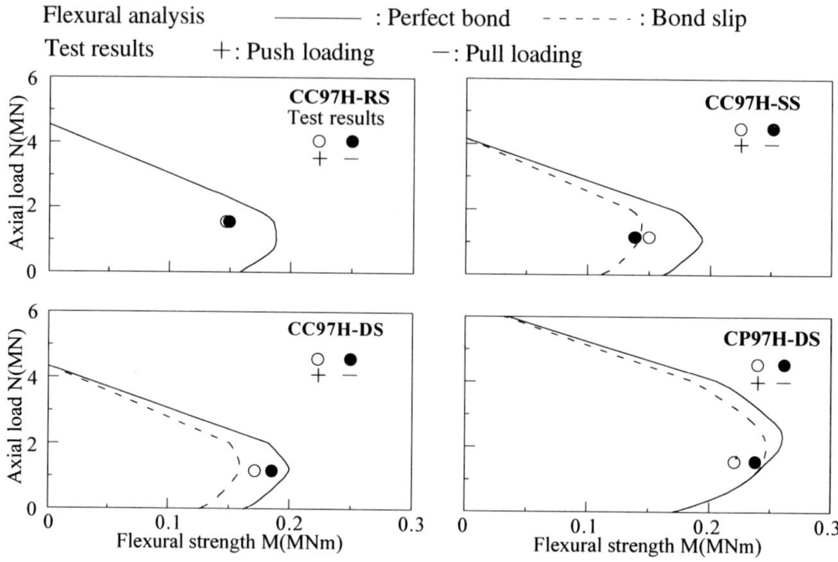

Figure 9 Calculated *N-M* interaction diagrams and experimental results

transverse confinement by the flexural strength for unconfined concrete whose stress-strain relationship is presented by exponential function (see Figure 6). In this calculation the bond slip is ignored. As the axial load ratio increases, the flexural strength also increases. The higher the transverse confinement and the axial load ratio become, the more the calculated flexural strength increases.

In order to discuss the ductility of the specimens, both experimental and calculated results are compared with respect to the lateral capacity deterioration. In this calculation, the bond slip is

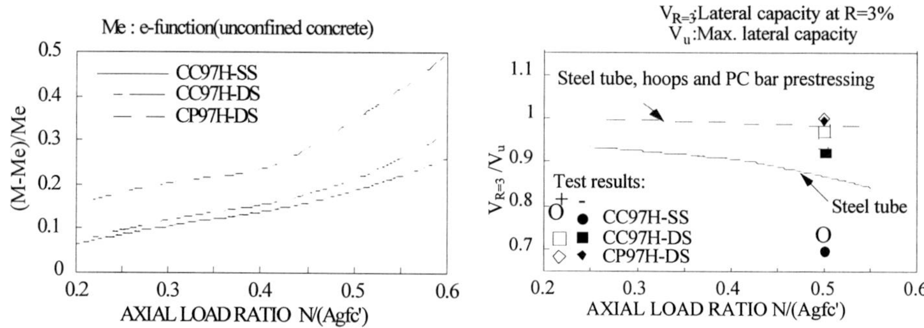

Figure 10 Calculated flexural strength

Figure 11 Lateral capacity deterioration

ignored and the lateral capacity is equal to the flexural strength. The lateral capacity deterioration $V_{R=3}/V_u$ is defined by dividing the lateral capacity $V_{R=3}$ at a drift angle R=3% by the maximum lateral capacity V_u. Figure 11 represents the lateral capacity deterioration $V_{R=3}/V_u$ versus the axial load ratio $N/(A_g f_c')$. The solid and broken curves in Figure 11 show calculated results and black or white marks show the experimental results respectively. From Figure 11, the calculation results almost agree with the experimental results except for the specimen CC97H-SS. This figure suggests that high transverse confinement by both the square steel tube and prestressing or hoops produces superior ductility even if large amount of high strength longitudinal reinforcement is arranged in a RC column in which the bond splitting failure is likely to happen.

CONCLUSIONS

As the transverse confinement increases, the seismic performance of a hybrid RC column is improved. Transverse prestressing through the square steel tube is very effective for enhancement of strength and ductility.

ACKNOWLEDGEMENTS

The authors wish to thank Mr. Muranaka, K, from Shimizu Corporation, and Messrs Ogawa, K, and Kamogawa, S, graduate students at the Univ. of the Ryukyus, for their assistance in the experiment and preparation of this paper.

REFERENCES

1. YAMAKAWA, T, SAKINO, K, AND YAMADA, Y. A study on elastoplastic of R/C short columns doubly confined in steel tube and hoops, Proc. of the 3rd International Conference on Steel-Concrete Composite Structures, Fukuoka, Japan, 1991, pp 665-670.

2. YAMAKAWA, T. An experimental study on axial compression behaviour of concrete doubly confined in steel tube and hoops, Proc. of 4-ICCS International Conference on Steel-Concrete Composite Structures, Kosice, Slovakia, 1994, pp 131- 134.

3. YAMAKAWA, T, HAO, H T, AND MURANAKA, K. Elastoplastic behaviour of doubly confined R/C columns in steel tube and hoops, Journal of Structural and Construction Engineering, Architectural Institute of Japan, AIJ, No. 500, 1997, pp 83- 89.

4. YAMAKAWA, T, MURANAKA, K, AND KURASHIGE, M. An Experimental study on seismic behaviour of high confined R/C columns by steel tubes and prestressing, Proc. of the Japan Concrete Institute, JCI, Vol.19, No.2, 1997, pp 1437-1442 (in Japanese).

5. SAKINO, K, AND SUN, Y. Stress-strain curve of concrete confined by rectilinear hoop, Journal of Structural and Construction Engineering, AIJ, 1994, pp 95-104 (in Japanese).

6. MATSUMURA, H, AND ITOH, S. Compressive strength of filled concrete in square steel pipe, Summaries of Technical Paper of Annual Meeting of Architectural Institute of Japan, AIJ, 1989, pp 1627-1628 (in Japanese).

7. RICHART F E, ET AL. A study of the failure of concrete under confined compressive stresses, Univ. of Illinois Eng. Experimental Station, Bulletin No.185, 1928.

8. ARCHITECTURAL INSTITUTE OF JAPAN. AIJ Design Guidelines for Earthquake Resistant Reinforced Concrete Buildings Based on Inelastic Displacement Concept (Draft), AIJ, 1997 (in Japanese).

9. MORISHITA, Y, TOMII, M, AND SAKINO, K. Experimental studies on bond behaviour between deformed reinforcement bars and concrete confined in square steel tube, Transactions of Japan Concrete Institute, JCI, Vol.9, 1987, pp 335-342.

STRENGTH AND STRAIN OF CONCRETE UNDER REPEATED STATIC LOADS

A Y Barashikov

A D Zhuravsky

Kiev State Technical University

Ukraine

ABSTRACT. As a result of generalizing of numerous experimental researches it has been established that at not repeated static loading of concrete by compressing load there occur changes of its strength and strain characteristics. Data are given on influence of repeated static loads on concrete prism strength, concrete deformation module and ultimate compressibility of concrete.

Keywords: Repeated static loads, Prism strength, Concrete deformation module, Levels of cyclic loads.

Professor Arnold Y Barashikov is a Head of the Department of Reinforced Concrete Structures, Kiev State Technical University of Construction and Architecture, Ukraine. He specialises in the repeated static loads on concrete, shrinkage and creep of concrete. Professor Barashikov has published widely and serves on many Technical Committees in Ukraine.

Dr Alexander D Zhuravsky is a Lecturer at the Department of Reinforced Concrete Structures, Kiev State Technical University of Construction and Architecture, Ukraine. His main research interests include the concrete creep under two-axial compression, the investigation of two-axially prestressed slabs.

INTRODUCTION

Repeated static compressing stresses while in service test many concrete and reinforced concrete elements of buildings and structures. Practically all loads besides the own weight, pressure of a ground and constantly installed equipment are variable and varied in a time. Investigation of buildings shows large changes of stresses in structures in a consequence of a adverse combination of variable force and temperature effects.

In view of modern lines to reduction of sections of bearing structures danger of destruction of elements owing to low-cyclic fatigue, increase of share of plastic deformations, influence initial imperfections and etc. is increased. According to existing methods of calculation the slow changes of effects are not taken into account, and the structure is designed on a to the utmost possible effort, arising during exploitation.

INFLUENCE OF REPEATED STATIC LOADS ON THE STRENGTH AND STRAIN OF CONCRETE

Experimental and theoretical researches, conducted for the last years [1-9], have shown essential influence of repeated static loads on the physical and mechanical characteristics of concrete. During the effect of repeated compressing loads, strength and strain of concrete can be increased or decreased. Such changes depend on a number of factors: structure of concrete, concrete inundation and the character of the load.

There were many proposals [1-9] under the description of influence low-cyclic static loading on strength and strain of various concretes. The common fault of offered ratios should be thought necessity of the introduction in the settlement formula of factors, received as a to statistical way for each kind of a concrete and character changes of loads.

Dependences on determination of influence of repeated compressing loads on strength and strain of a concrete are below offered. These dependences in certain degree are universal and capable of approximately to take into account repeated effects practically for frequently used of kinds of concrete.

Naturally, accuracy of results then a little below, than at approximation of properties by formulas for each kind of a concrete and load. However, taking into account, that characteristics of materials and loads, and, hence, the initial data for designing of buildings have large changeability, the offered approach can be thought justified.

Is now already proven, that at low levels of cyclic loads ($\sigma_{b,max}$=(0.2-0.5)R_b) of a concrete prism strength R_b can be increased up to 10-12 %, at average levels ($\sigma_{b,max}$=(0.6-0.8)R_b) - does not practically change, at high levels ($\sigma_{b,max}$=(0.8-1.0)R_b) decrease of strength occurs, which causes to low-cyclic fatigue. Account of change of a concrete strength during apply not frequent of repeated compressing loads is offered to make on formula:

$$\widetilde{R}_{bi} = R_b(1 + \varphi_i), \tag{1}$$

Where:

$$\varphi_i = \begin{cases} K_i^+ \sqrt{1-\vec{\eta}_i} - \eta_{b1} \ldots\ldots\ldots\ldots at \ldots \vec{\eta}_i \le \eta_{b1}; \\ 0 \ldots\ldots\ldots\ldots at \ldots \eta_{b1} < \vec{\eta}_i \le \eta_{b2}; \\ K_i^- \sqrt{1-\vec{\eta}_i} - \eta_{b2} \ldots\ldots\ldots\ldots at \ldots \vec{\eta}_i > \eta_{b2}. \end{cases} \quad (2)$$

Here $\vec{\eta}_i = \sigma_{bi} / R_b$ - maximum level of loading in a cycle relatively of prism strength.

In ratio (2) two levels of stress in a concrete - concerning bottom $\eta_{b1} = R_{b1} / R_b$ and top $\eta_{b2} = R_{b2} / R_b$ parametrical points are used [5]. The introduction in the settlement formulas of parametrical points permits indirectly taking into account a structure of a concrete, its strength and other characteristics of a concrete mix and hardened concrete. At loading of a sample below a level η_{b1}, as a rule, the prism strength is increased, and the excess of a top level of microcracks η_{b2} causes to low-cyclic fatigue of a concrete.

Coefficients K_i^+ and K_i^- reflect the influence of quantity of cycles loading for a constant amplitude of cyclic tests can be calculated on formulas:

$$K_i^+ = \frac{2(i+3)}{3i}; K_i^- = \frac{3(i+4)}{2i}; i = 1,2,\ldots,n. \quad (3)$$

For determination of parametrical levels for various kinds of concrete dependences, received in [4, 5, 7] can be used.

In Figure 1 schedules are constructed at η_{b1}=0.509 and η_{b2}=0.834, reflecting change increase of a concrete prism strength at repeated static loading. On schedules points by results of experimental researches of various concrete, including on wastage of enrichment of iron ores are put. As it is visible from Figure 1 experimental points are enough well stacked in the field of change of prism strength.

In difference from of prism strength utmost strain of a concrete $\tilde{\varepsilon}_R$ at high, average and low levels of a repeated compressing load is always increased. It is thus possible to find out rather sharp dependence between value of deformations of compression and maximum level of a load in a cycle $\vec{\eta}_i$. This dependence is good approximation square parabola:

$$\eta_\varepsilon = \frac{\tilde{\varepsilon}_R}{\varepsilon_R} = 1.25 - c(\vec{\eta}_i - \eta_{b2})^2 \quad (4)$$

where ε_R - utmost deformations of a concrete at standard tests of monotonous loading; c - experimental parameter of cyclic strain, located for various concrete within the limits of $1.3 < c < 1.8$ [3, 4].

In Figure 2 relative change of utmost deformations of a concrete for value c=1.583; η_{b2}=0.6 is shown [3].

Simultaneously with an increase of utmost deformations of a concrete at compression decrease of a initial module of elasticity E_b is established.

Relative reduction of initial module of elasticity of a concrete with an increase of a level of a load and number of cycles is well described by family exponential functions:

$$(\widetilde{E}_b) = \widetilde{E}_b / E_b = \alpha_{lim}(1 - me^{-\gamma}), \qquad (5)$$

where $\widetilde{E}_b$ - initial module of elasticity on a cycle of loading; $\alpha_{\lim}$ - relative utmost significance of a initial module of elasticity; m, γ - experimental constants, dependent from a level cyclic of loading and kind of concrete.

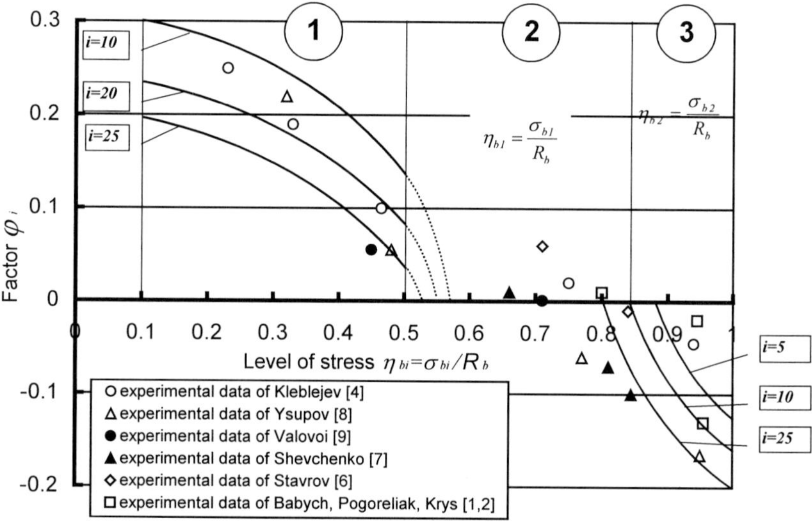

Figure 1 Change of factor φ_i for determination of prism strength at low-cyclic loading: 1-areas of hardening; 2-areas of stable deformation; 3-areas of loss strength

As have shown researches [3, 6, 8] factors γ and $_{lim}$ for operational levels of a load can be accepted by sizes constant γ=0.05, $_{lim}$=0.55, and m - linearly depends from:

$$m = a - b\alpha_{lim.} \qquad (6)$$

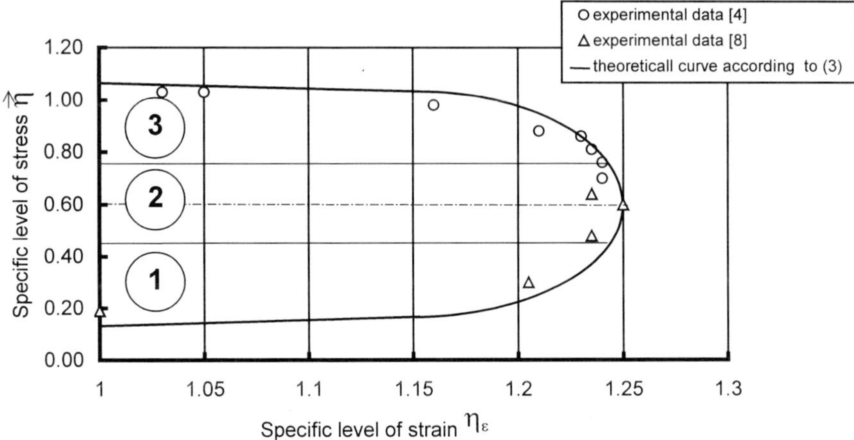

Figurc 2 Chagc of utmost deformation of a concrete at cyclic loads (the borders of the areas are established approximately): 1, 2, 3 - see Figure 1

CONCLUSIONS

Such as above ratios characterize change stress and strain of properties of a concrete at repeated static loads. They allow more exactly to calculate reinforced concrete constructions attached to action of variable loadings.

REFERENCES

1. BABYCH, Y AND KRUS, Y. Peculiarities of concrete behavior under the low cyclical static loading. Proceedings of the XLI International Conference "Krynica 95". Vol.5, Krakow-Krynica, 1995, pp 13 20.

2. BABYCH, Y M, POGORELIAK, A P. Investigation of influence of not repeated compression for deformation and cracking of heavy-weight concrete // Hydromelioration and hydrotechnical construction. Vol.5, Lviv, 1977, pp 120-124, (in Russian).

3. BARASHIKOV, A Y. Calculation of designs of buildings and structures with consideration of real conditions of construction and operation : Reliability and durability of machines and structures. 1988, Vol.14. pp 23-32, (in Russian).

4. BARASHIKOV, A Y, KLEBLEJEV E K. Physical and mechanical properties of concrete under repeated loading. Scientific, theoretical and practical investigation at the sphere of construction and architecture. Samarkand, 1991, pp 70-74, (in Russian).

5. BERG, O Y, SCHERBAKOV, E N AND PISANKO, G N. High-strength concrete. Moskow: Stroyizdat, 1971, 208 pp, (in Russian).

6. STAVROV, G N, RUDENKO, V V AND FEDOSEEV, A A. Stress and strain of concrete under repeated static loading. Mag. Concrete and reinforced concrete. 1985, Vol.1, pp 33-34, (in Russian).

7. SHEVCHENKO, B N. Structures from concrete on wastage of enrichment iron ore. Kiev: Higher school, 1989, 192 pp, (in Russian).

8. YSUPOV, Z. Reinforced concrete behaviour under not repeated loading at the condition of Central Asiatic climate. Tashkent, Fan, 1988, 130 pp, (in Russian).

9. BARASHIKOV, A Y, SHEVCHENKO, B N, VALOVOI, A I. Low-cyclic fatigue of concrete under compression. Mag. Concrete and reinforced concrete. 1985, Vol.4, pp 27-28, (in Russian).

MEMBRANE FORCES IN SLABS AND PUNCHING SHEAR

A W Beeby

University of Leeds

United Kingdom

T Zaib

CHASNUPP PAEC

Pakistan

ABSTRACT. A study is described into the effects of compressive membrane forces on the punching resistance of flat slab – column connections. Previous experimental work is reviewed which shows that membrane forces may increase the punching resistance by around 30%. Experimental work is presented which shows that the relationships between axial force and shear strength given in BS8110 is excellent and that in Eurocode 2 is reasonable. Methods of calculating the expansion of members under loads below that required to yield the reinforcement are considered and compared with experimental data. It is shown that a method based on Eurocode 2 provides a safe lower bound estimate. The stiffness of the concrete slab surrounding a slab – column connection is calculated using thick cylinder theory. The above elements are combined to calculate the punching resistance of slabs published in the literature. The method is shown to give a safe estimate but up to 25% higher that methods ignoring membrane effects.

Keywords: Punching shear, Membrane forces, Flat slab design

Professor Andrew W Beeby is Professor of Structural Design at the University of Leeds, UK. He has specialised in the development of design methods for reinforced concrete and is a member of both UK and European code drafting committees. Over the past 35 years he has carried out research into many aspects of both the ultimate and serviceability behaviour of reinforced concrete elements. Since 1991 he has chaired Permanent Commission 2 'Member and Material Modelling' of the Comite Euro-International du Beton. He is currently a member of the council of the Institution of Structural Engineers.

Dr Tariq M Zaib is currently employed by CHASNUPP PAEC in Pakistan. He studied for his PhD in the Department of Civil Engineering at the University of Leeds where he investigated the influence of membrane forces on shear strength.

INTRODUCTION

Compressive membrane forces develop in restrained slabs as soon as flexural cracking commences but develop much more quickly when yield of the tension steel initiates. The forces arise because both cracking and yield lead to a movement of the neutral axis towards the compressive face and a consequent tendency to expand at the level of the uncracked section centroid. If this expansion is restrained, compressive forces in the plane of the slab are induced. These forces can lead to a very substantial increase in the flexural capacity of lightly reinforced restrained slabs. This issue has been the subject of extensive research over the years and is well understood (see, for example, reference 1). Except in some very limited areas, design methods have yet to take advantage of this potential increase in flexural capacity. Another area where the presence of membrane forces could have a beneficial effect on behaviour is in the field of punching shear. High bending moments develop in a limited area around concentrated loads or slab-column connections. The cracking in this limited region and also possibly the initiation of yield locally will result in a tendency to expand. This expansion will be restrained by the concrete outside this local zone and could be expected to result in the development of in-plane compressive forces. Punching shear strength is known to be enhanced by the presence of in-plane compressive forces and design formulations for shear, such as those in Eurocode 2 (2) make allowance for this.

The possibility that punching shear strength could be enhanced by compressive membrane forces has been investigated to a limited degree by a number of workers over the years. Taylor and Hayes (3) published the results of punching tests on restrained slabs as long ago as 1965 and Aoki and Seki (4) carried out test in 1971. Both these investigations showed significant increases in punching capacity as a function of the degree of restraint provided to the slab. A theoretical study by Hewett and Batchelor (5) showed that Kinnunen and Nylander's theory of punching shear could be modified to take account of membrane effects if the restraint could be quantified. More recently, Rankine and Long (6) showed that punching strength was increased by 30 – 50% if tests were carried out on full panel specimens rather than the type of specimen normally used for punching shear tests where the edge of the specimen roughly corresponds to the ring of contraflexure around the column. This enhancement was ascribed to membrane forces developed due to restraint of the local area by the concrete outside the ring of contraflexure. This was demonstrated by measurements of the strains in the slabs and finite element analyses. Chana and Desai (7) have published results from even larger specimens (9m square) which resulted in similar enhancements. Kuang and Morley (8) carried out 12 tests on square slabs restrained by edge beams of varying stiffness and with different percentages of reinforcement which resulted in enhancements of punching strength of up to about 30%. These various experimental programmes were not able to measure the membrane forces that developed so no clear relationship has been proved between restraint and the in-plane force developed at failure and hence the enhancement in shear capacity as a function of the in-plane force.

If a clear experimental and theoretical basis can be developed for the prediction of compressive membrane forces on punching shear strength then it appears that it would be much easier to introduce this into design practice than is the case for the effects of membrane forces on flexural strength. The reason for this is that it is difficult to characterise the restraint present over the whole of the slab, which affects the overall flexural strength but it may be much easier to establish the presence of adequate restraint locally to a potential punching shear failure.

The project described in this paper is an attempt to establish the critical steps necessary for the development of a design method. The stages in the project are as follows:

1. Establish, for slab elements, the relationship between shear strength, in-plane expansion and axial force.

2. Develop a means of calculating the load – axial force – axial expansion relationship for slab elements and validate this.

3. Develop a means of calculating the stiffness of the slab surrounding a failure zone under the action of in–plane forces.

4. Combine 2 and 3 above to enable the axial force and hence the enhancement in punching shear strength to be predicted and test this against the available experimental data.

The paper will consider each of these stages. Much of the work is based on research carried out by T. Zaib for his PhD studies (9).

SHEAR STRENGTH AND AXIAL COMPRESSION

Various codes of practice include equations to predict the shear strength of members subjected to bending and axial load. Two such are Eurocode 2 (2) and BS8110 (10). The formulae from these codes are given below.

Eurocode 2:

$$V_c / b_v d = 0.0452 f_{cu}^{2/3} [k(1.2 + 40 A_s/bd) + 0.15 N/A_c] / \gamma_m$$

BS8110:

$$V_c / b_v d = 1.25 k (f_{cu} A_s / b_v d) 1/3 / \gamma_m + 0.6 N V h / A_c M$$

Where:

V_c	=	the shear capacity of a section without shear reinforcement
f_{cu}	=	the cube strength of the concrete
A_s	=	the area of tension reinforcement at the section considered
b_v	=	the breadth of the section at the level of the tension steel
d	=	the effective depth of the section
A_c	=	the area of the concrete section
k	=	a coefficient allowing for the effect of section depth
N	=	the axial load
M	=	the moment at the section considered
γ_m	=	partial safety factor

For ease of comparison of the equations, the notation in BS8110 has been adopted, at least two equations in the actual documents have been combined and the Eurocode formula has been expressed in terms of cube strength rather than cylinder strength. It will be seen that the equations give a very different account of the significant variables.

Recently, 42 tests on axially loaded slab elements were carried out by Zaib (9). The variables considered were concrete strength, reinforcement ratio and axial load. All specimens were 200 mm wide, 155 mm overall depth and with a shear span of 425 mm giving a ratio of shear span to effective depth of 3. Measurements were recorded of longitudinal expansion and of the shear force at failure. All but four of the beams are reported to have failed in shear. The ultimate shear strengths from these tests are compared with the two code formulae in Figures 1 and 2. In making the comparisons, partial safety factors have been removed and the maximum cube strength of 40 N/mm^2 implied in BS8110 has been ignored. It will be seen that, though both formulae give reasonable predictions, the BS8110 formula is somewhat better; indeed, the coefficient of variation of the errors for the BS8110 formula is only 8.7% while that for Eurocode 2 is 17.8%.

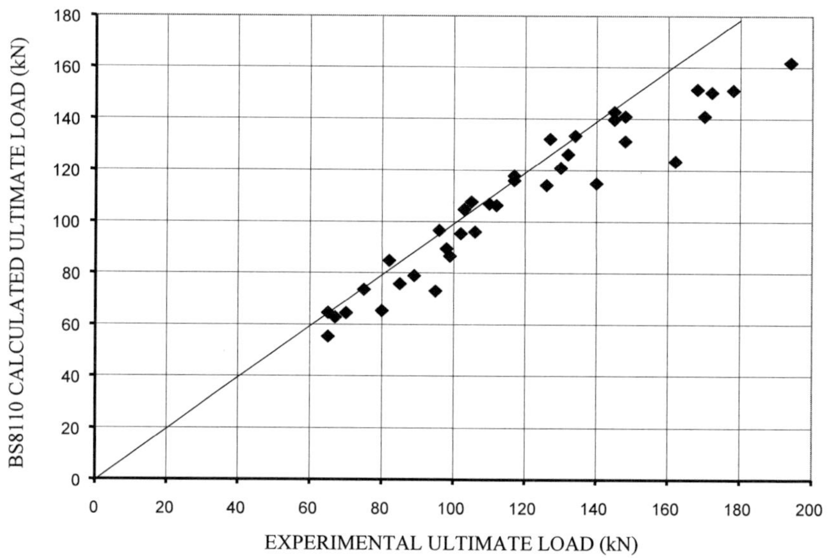

Figure 1 BS8110 prediction of the influence of axial load

On the basis of these comparisons, it appears that the BS8110 equation provides an excellent prediction of the influence of axial force on shear strength. In passing, it is interesting to note that neither BS8110 nor Eurocode 2 include any safety factor in the estimation of the additional shear capacity resulting from the axial load.

PREDICTION OF AXIAL EXPANSION

Equations exist which relate flexural capacity to axial force and overall expansion (eg. see (1)). Unfortunately, such relationships are not helpful for considering shear since the main factor leading to expansion close to flexural failure is yield of the reinforcement and, when considering shear failure, by definition the reinforcement will not all have yielded.

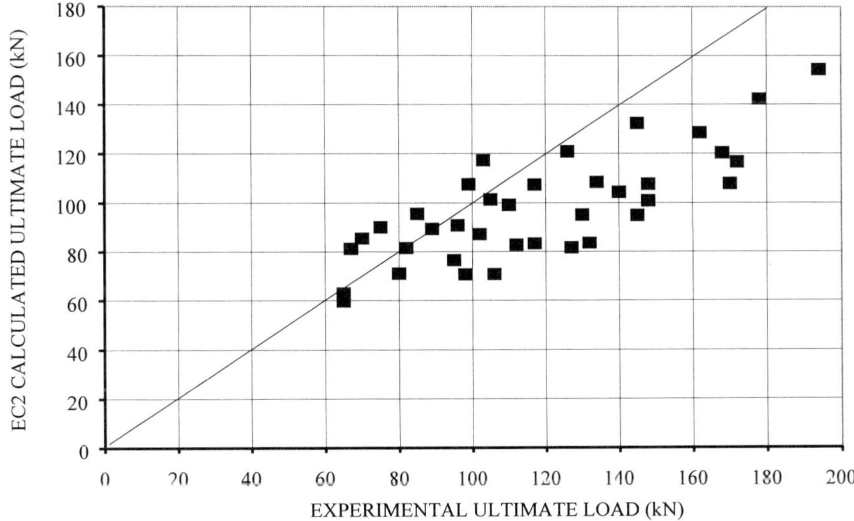

Figure 2 Eurocode 2 prediction of the influence of axial load

In punching situations it may well be that some yield develops prior to failure but the full yield mechanism will not have developed. The approach adopted in this study is therefore to attempt to predict the expansion due only to the development of flexural cracking assuming no yield. This approach will either be correct or conservative.

Both BS8110 and Eurocode 2 provide methods of calculating the location of the neutral axis and curvature of a cracked section while making allowance for tension stiffening. Both methods have weaknesses from the point of view of the calculation of member expansions but that in Eurocode 2 is considered both the more convenient to use and technically preferable. The approach is as follows.

Elements of structure loaded above the cracking load behave in a manner intermediate between the behaviour of an uncracked section and a fully cracked section where the concrete in tension is assumed to carry zero stress. This may conveniently be idealised by considering a proportion of the member under the particular loading considered to be uncracked and a proportion to be fully cracked. If the fraction of a short length δL of the member considered to be fully cracked is ζ and hence the fraction which is fully cracked is $(1 - \zeta)$, then, if the deformation parameter we wish to calculate is χ_e (which may be an effective curvature, strain or deflection) then χ_e will be given by the relationship:

$$\chi_e = \zeta\chi_2 + (1 - \zeta)\chi_1$$

where:

χ_e = effective value of deformation parameter, allowing for tension stiffening.
χ_1 = value of deformation parameter calculated for an uncracked section
χ_2 = value of deformation parameter calculated assuming a fully cracked section.
ζ = distribution coefficient

Eurocode 2 defines the distribution coefficient, ζ, as:

$$\zeta = (1 - \beta_1\beta_2(\sigma_{sr}/\sigma_s)^2)$$

where:

β_1	=	coefficient taking account of effect of bond
β_2	=	coefficient taking account of duration of loading
σ_s	=	stress in reinforcement calculated on the basis of a cracked section
σ_{sr}	=	stress in reinforcement calculated on the basis of a cracked section under the loading that just causes cracking.

If the deformation parameter that is required is the strain at the level of the centroid of the uncracked section, then, by definition, χ_1 is zero and hence:

$$\varepsilon_{cen} = (1 - \beta_1\beta_2(\sigma_{sr}/\sigma_s)^2)\varepsilon_{2cen}$$

where :

ε_{cen}	=	the effective strain at the level of the centroid of the uncracked section
ε_{2cen}	=	the strain at the centroid of the uncracked section calculated on the basis of a cracked section.

The overall expansion of a member may now be found by calculating the effective centroidal strain at sufficient sections along the member and then using numerical integration. Figure 3 shows the calculated and experimental expansions as a function of applied load for a typical specimen. It will be seen that agreement is reasonable up to close to failure where the expansion is underestimated. This underestimate is clearly due to some degree of non-linearity developing at loads close to failure and it occurs in almost all the tests. The proposed approach thus generally underestimates the expansion and will hence lead to an underestimate of the membrane force. This is clearly a safe state of affairs. Figure 4 shows the expansion at a load stage close to failure plotted against the calculated expansion. During the tests, the axial load was applied first followed by the transverse loading in stages. It is clear from the results that some bedding in had occurred of the bearing plates to which the axial load was applied and between which the overall expansion was measured. The axial deformation under zero transverse load has therefore been corrected to the calculated axial shortening under the axial load. It will be seen that the calculated expansions provide a good lower bound to the experimental expansions. The experimental expansions greater than 1 mm are almost all cases where significant non-linearity has clearly occurred. Many of the cases are for specimens with zero axial load where the non-linearity is due to the reinforcement exceeding the elastic limit. The reinforcement, being of small diameter bars, had no definite yield.

PREDICTION OF THE STIFFNESS OF THE SURROUNDING CONCRETE

If it is assumed that the restraint to local failure around an internal column is provided by the concrete outside the ring of contraflexure, what is required is a relationship between pressure and radial expansion at the inner surface of the ring. If it is further assumed that the stiffness of the concrete outside the ring of contraflexure against in-plane radial force is unaffected by flexural cracking in the slab then the stiffness's may be calculated from the classical elastic formulae for thick cylinders as:

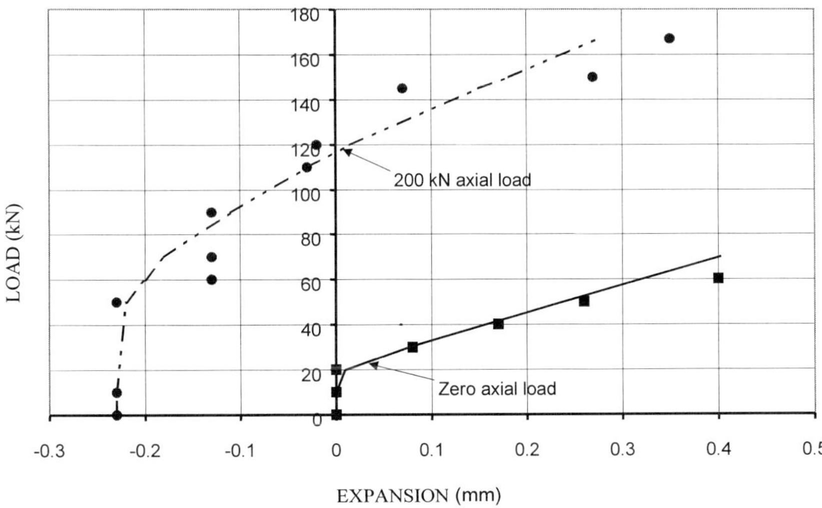

Figure 3 Typical calculated and experimental load-expansion curves

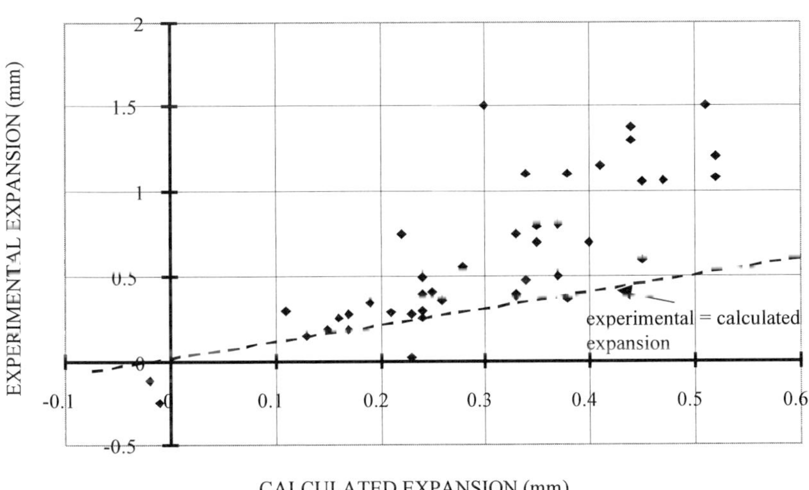

Figure 4 Comparison of calculated and experimental expansions

$$S = \frac{E}{2a\left[\upsilon + \frac{b^2 + a^2}{b^2 - a^2}\right]}$$

where:

a = the distance from the centre of the column to the ring of contraflexure
b = the distance from the centre of the column to the edge of the slab
υ = Poissons ratio.

In most practical situations, b may be assumed to be infinite and the expression reduces to:

$$S = \frac{E}{2a(1 + \upsilon)}$$

PREDICTION OF PUNCHING STRENGTH

The process adopted may conveniently be illustrated graphically. Chana and Desai's Specimen FPS1 (7) has been used to illustrate the procedure. This is the only specimen that they tested which did not include shear reinforcement and it actually failed at a load of 1225 kN. Figure 5 shows axial load plotted against axial expansion for various levels of column loading. The expansion has been calculated on the basis of a distribution of moments in the region of the column obtained from a finite element analysis. Also plotted is the axial load – expansion relationship for the surrounding concrete. The points where the axial load-expansion relationship for the surround intersect the relationship for a particular punching load gives the actual membrane stress generated under that load. Taking the intersection points permits a graph to be drawn of membrane stress against punching load (Figure 6). It is now necessary to plot on Figure 6 a relationship between axial stress and punching strength. The point where these two lines intersect will then give the failure load.

Application of the BS8110 formula is not straightforward as it is far from clear how the term 0.6Vh/M should be assessed in a punching shear situation. For a beam with a point load a distance a_v from the face of the support, this can be written as $0.6h/a_v$. If it is assumed that approach may be used for punching then a_v could be taken as the distance from the face of the column to the ring of contraflexure, giving a value for the specimen considered of 0.15, corresponding to an increase in resistance of 130 kN/(N/mm^2). The strength ignoring membrane effects according to BS8110 for this specimen is 773 kN. This relationship is shown as a broken line on Figure 6 and will be seen to lead to a predicted strength of 925 kN under a membrane stress of 1.42 N/mm^2; an increase over the value ignoring membrane stresses of 20%. There are other possible interpretations of the term 0.6Vh/M. For example, it could be argued that it should be calculated at the critical perimeter where M is the moment per unit width at the perimeter and V is the shear per unit length of the perimeter. This results in a very much higher contribution from the axial stress of 467 kN/N/mm^2 which leads to a totally unrealistic punching strength. It therefore seems reasonable to use the first interpretation.

The approach may be used to predict the strength of the slabs tested by Rankine and Long (6). The comparison of calculated and experimental results is shown in Figure 7. It will be seen that the proposed method provides a very reasonable prediction of the strength.

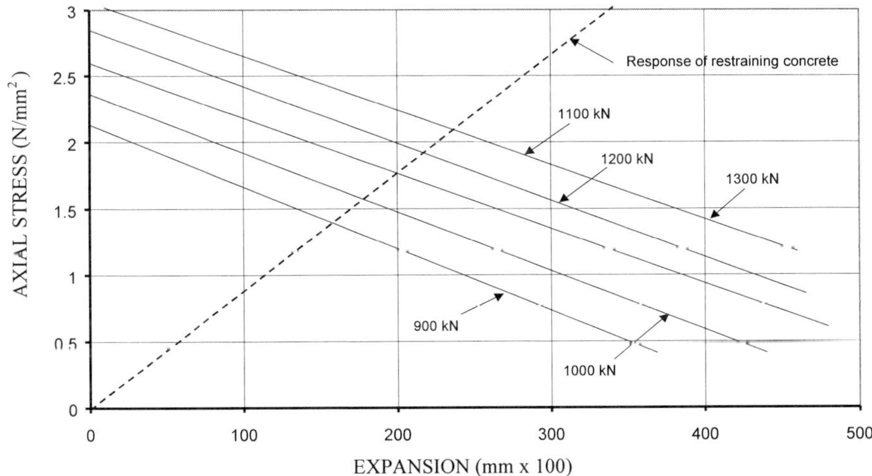

Figure 5 Expansion plotted against axial stress

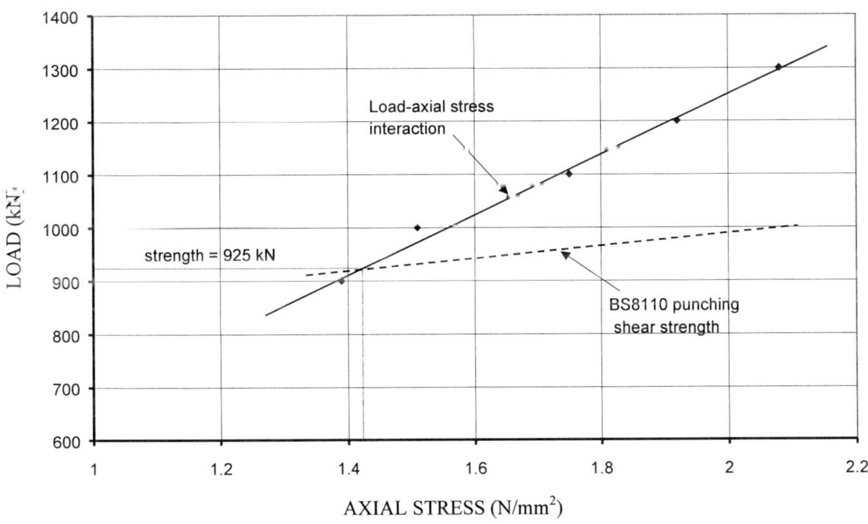

Figure 6 Axial stress plotted against punching load

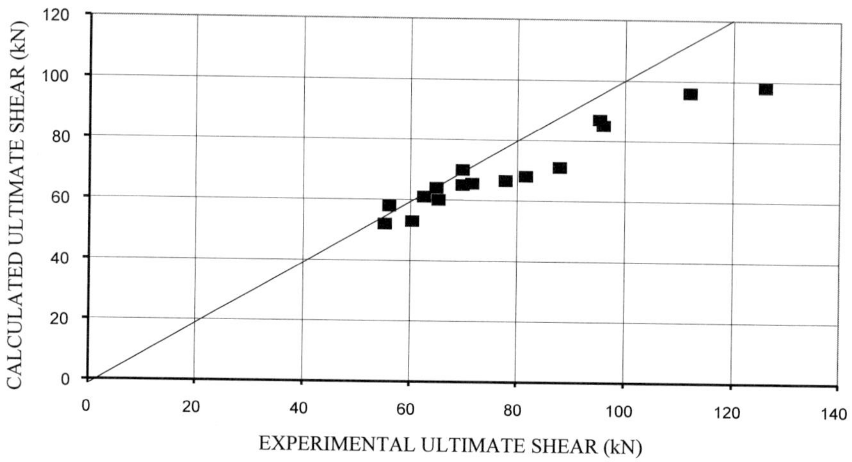

Figure 7 Comparison of calculated and experimental punching strengths for the tests by Rankine and Long

DISCUSSION

A method has been developed and validated which will permit the punching strength of internal flat slab – column connections to be calculated making allowance for the beneficial effects of compressive membrane forces. The method is based on provisions already written into current codes of practice, which will be an advantage in trying to justify its practical application. As it stands, the method is clearly not a practical hand design tool, though it can quite easily be programmed. It could, however, be used to run a parameter study covering the expected range of values of span/depth ratio, reinforcement ratio around the column, ratio of column size to slab thickness and concrete strength. The results from this study could be used to establish punching strength enhancement factors which could be used in practice.

Such an approach should be able to be used safely in practice provided there was a reasonable area of slab surrounding the column without holes of such a size as to inhibit the development of the hoop tension in the concrete surrounding the ring of contraflexure. Simple rules could be formulated to guard against this.

CONCLUSIONS

1. Compressive membrane effects can lead to a significant increase in the punching shear strength of internal flat slab – column connections.

2. Equation 6 in BS8110, which predicts the influence of axial load on shear strength is shown to fit the data from tests on axially loaded slab elements excellently. The Eurocode 2 formula is also reasonable.

3. Use of the assumptions given in Eurocode 2 for the prediction of the expansion of members loaded beyond their cracking load is shown to give a reasonable but conservative estimate of the expected expansion.

4. A method has been demonstrated whereby the effect of axial load on shear and knowledge of the expansion of the area of slab around the slab-column connection may be used to predict the compressive membrane force and the punching shear strength. The method gives reasonable estimates of the punching.

5. This method could be used to develop a design approach which could lead to considerable economies in the design of internal slab – column connections.

REFERENCES

1. BRAESTRUP, M. Dome effect in RC slabs: rigid-plastic analysis. Journal of the Structural Division, Proc. ASCE Vol. 106 No. ST6, June 1980.

2. BRITISH STANDARDS INSTITUTION. DD ENV 1992-1-1:1992 Eurocode 2: Design of concrete structures. Part 1. General rules and rules for buildings. BSI 1992.

3. TAYLOR, R and HAYES, B. Some tests on the effect of edge restraint on punching shear in reinforced concrete slabs. Magazine of Concrete Research, Vol. 17, No. 50, March 1965.

4. AOKI, Y, and SEKI, H. Shearing strength and flexural cracking capacity of two-way slabs subjected to concentrated load. Cracking, deflection and ultimate load of concrete slab systems, American Concrete Institute Publication SP30, 1974.

5. HEWITT, B E and BATCHELOR, B, deV. Punching shear strength of restrained slabs. Journal of the Structural Division of ASCE, Vol. 101 No. ST9, Sept 1975.

6. RANKINE, G I B and LONG, A E. predicting the enhanced punching strength of interior slab – column connections. Proceedings of the Institution of Civil Engineers, Part 1, Vol.82 December 1987.

7. CHANA, P S and DESAI, S B. membrane action and design against punching shear. The Structural Engineer, Vol. 70 No. 19, 6th October 1992.

8. KUANG, J S, and MORLEY, C T. Punching shear behaviour of restrained reinforced concrete slabs. ACI Structural Journal, Vol. 89 No. 1, Jan/Feb 1992.

9. ZAIB, T M. Effect of compressive membrane action on shear strength. PhD thesis, University of Leeds, October 1996.

10. BRITISH STANDARDS INSTITUTION. BS8110: part 1: 1997: Structural use of concrete. Part 1. Code of practice for design and construction. BSI 1997

TOWARDS THE TOTAL DESIGN OF REINFORCED CONCRETE

S B Desai

Department of the Environment Transport and the Regions

United Kingdom

ABSTRACT. Recent research in the production of concrete has often concerned its expected performance, strength and durability, and the best usage of natural minerals and resources for achieving economical and sustainable construction. However, developments in structural design do not appear to have accounted for the potential influence of various concrete constituents on its physical properties. Similarly, compressive strength has continued to be used as the only measured parameter in general methods and in rigorous analyses capable of using nonlinear properties of materials. This paper seeks to investigate realistic properties and attributes of various concretes, in the interest of optimising levels of safety and competitiveness of concrete construction as the first practical step towards the total design of reinforced concrete. The proposals presented in this paper could lead finally to the whole life design of concrete structures, incorporating integrated measures to deal with long term actions of the environment and normal and accidental loading.

Keywords: Concrete, Structural design, General design, Rigorous analyses, Tensile strength, Nonlinear properties.

Dr Satish Desai is a Principal Civil Engineer with the Building Regulations Division of the Construction Directorate, Department of the Environment, Transport and the Regions. His main research interest is in the fields of reinforced concrete design and collaboration between research in concrete technology, structural design and construction process for enhancing competitiveness and sustainability of concrete construction.

INTRODUCTION

The concept of simultaneous consideration of the different factors contributing to the stability and durability of structures is not new. It has been reported[1] that *Vastushatra*, the ancient and prehistoric Indian science of dwellings, had recognised the need for an overall approach and some of its principles are relevant even today:

- Use of conical shapes of buildings for providing stability and protection against rain.
- Vulnerability of joints and binding materials in construction. (The ultimate solution was to use stones cut to perfect shapes and assembled to form a building without mortar.)
- Building towns on the convex side of the river, so that a deep pool of water is available during most of the year and the foundations of buildings on the embankment sloping up from the river bank could avoid problems associated with water.

It is conceivable that the master builders of olden times could have adopted some similar strategies in other parts of the world. The success of their work is visible in the form of many monuments, built with a combination of their knowledge of various materials, including the earlier versions of concrete, and their craftsmanship and skill in building. At some stage in the history of concrete construction, research in the role of chemical reaction in hardening of concrete might have parted company with the development in structural analysis and design based on approximations of physics and mathematics. The difference has become more apparent during the past hundred years of research in reinforced concrete. As a result, concrete technology and structural design have become separate fields receiving separate sponsorship from various public or private sectors. The sponsors seem to believe that research in the individual fields could be initiated, planned and funded separately and that the progress can be achieved by simply combining the outputs sequentially. This approach would appear to ignore the advantages of integration and collective consideration of research in different fields, which is arguably the most effective way to serve the concrete industry as a whole.

Prediction of the service life of a building would depend on clear definitions of many factors, e.g. intended use, service conditions, performance criteria, effectiveness of maintenance, etc. It is also recognised that the present experience and statistical data are insufficient for providing general guidance on quantitative prediction of service life. However, in the interest of progressing towards service life prediction and whole life design of concrete structures by intent, quality of the overall design process has to improve sufficiently. The design procedures should reflect the real properties of modern materials and the existing design rules, which have been based on tests carried out some years ago, should be reviewed. Such improvements should aim subsequently at designed provision of mutually compatible and complementing measures against all foreseeable actions, natural and manmade, e.g. carbonation, heat, wind, chemicals, fire, service loads, accidents through impact and explosions, etc. For design of large and specialist projects, an integrated defence should be sought against the long term environmental deterioration and short term action of fire and accidents. The design solution should be based on understanding of the response to these actions provided by concrete as a material and that afforded by the structural frame members as a unit. For use in the general design, simplified rules should be developed with clear definition of their limitations.

Such an approach could avoid incorrect applications of prescriptive methods, which often mask other design and construction deficiencies and the true safety margin, through uneconomical and extra provisions, and they may even lead to unsafe solutions in some projects.

As the first step in the process of improvement in design, a current research project on production of concrete is adapted to serve as a part of a "Sustainability Portfolio". This project concerns optimised binders made with waste materials, e.g. pulverised fuel ash (PFA) and granulated blast-furnace slag (GGBS), in conjunction with Portland Cement (PC). Other projects are similarly brought under the portfolio for carrying out coherent tasks, e.g. investigation of properties of different concretes (additional to the compressive strength) and their use in rigorous design (based on the study of various issues as described in this paper), beam tests and their analyses for assisting a review of the existing general design rules, etc.

MERITS OF PRODUCING CONCRETE WITH OPTIMISED BINDERS

Production of concrete using Portland cement (PC) on its own could be regarded as excessively expensive and energy intensive, unless it is really essential, e.g. in components requiring an early gain in strength. Cements containing PFA, GGBS and such other products are allowed in the concrete mixes conforming to British and European standards and, as far as possible, they should be considered for common applications. This objective could be served more effectively through research on the performance of a variety of optimised binders, appropriate for making concrete suitable for its intended use in a structure during its service life.

Current concrete standards and design codes specify measures for durability and guidance on the mix design and materials, which are judged to provide adequate performance. The chosen options depend on local conditions of cost, availability, specifier preference, etc. However, a specifier should avoid any unreasonable restriction on the concrete producer's choice of cement type or such measures as the minimum cement content and the grade of concrete. In certain cases with modest demand on durability, use of pure PC as a binder may be wasteful and, also, concrete could be of a grade lower than that recommended generally. On the other hand, durability is very important in the construction of bridges and other civil engineering structures and alternatives to concrete made with PC alone may have to be considered for providing adequate defence against aggressive conditions of exposure to chemicals (e.g. sulfates or chlorides) to suit local conditions of cost, availability, specifier preference, etc. In such situations, optimised binder combinations could be considered, e.g. binders containing PFA and GGBS, where advantage can be taken of their physical and chemical influences in providing durable concrete. Additionally, the use of such binders can help in reducing an early age temperature rise and cracking. Further options are provided by using water-reducing admixtures and fine fillers, which help in developing a denser microstructure of concrete.

MAIN ISSUES CONCERNING THE STRUCTURAL DESIGN

The Role of Concrete as a Structural Material

The work described above is meant to show the cost benefits and enhancement in durability afforded by the blended binders and their role in sustainable concrete construction.

However, it is also important to show that these binders can be effectively used in structural concrete with reliable properties for effective and efficient design of structures against all in-service actions.

Simultaneously, design of concrete structures, both general and rigorous, should account for the fact that concrete is a multi-phase material[2] comprising a multi-phase matrix, weak in tension, strong in compression and brittle. Its low tensile strength is mainly due to the presence of voids (about 50 to 100 microns) and it depends on its micro-structure, which differs with different constituents. The concept of reinforced concrete is based on the addition of steel bars as another phase to overcome the weakness in tension and brittleness of concrete. For achieving composite action between steel and concrete, however, concrete must have bond strength, which is essentially related to the post-elastic behaviour of concrete and its tensile strength. Besides its role in the normal design, bond strength is vital in providing ductility, which is an important factor in the survivability of framed structures subjected to accidental actions. The following sections include an examination of various issues linked with general and rigorous design, arising from the influence of constituents of concrete on its role as a structural material.

Approximations in the General Design of Reinforced Concrete

The current codes of practice give general and semi-empirical rules for flexural design of reinforced concrete elements. These rules assume that an elastic analysis of a frame can be carried out first and the post-elastic behaviour (leading to the ultimate limit state) can provide some redistribution between mid-span and support bending moments. This is essentially a compromise assuming a direct link between the two different phases of response of concrete to the increasing external actions, elastic and non-elastic, which depend on linear and nonlinear properties of concrete respectively.

Moreover, design rules employ only the compressive strength of concrete and they imply that the nonlinear properties have the same direct and unique relationship with the compressive strength for all types of concrete. This may not be valid for concretes made with different constituents and binders, which could have different microstructures, sand-binder matrixes, percentages of voids etc.

Shear failure is attributable to excessive principal tensile stresses in the neutral axis region and, therefore, the tensile strength of concrete is vital to the shear resistance of a reinforced concrete member. Slabs can resist punching shear without any provision of links, only because of the tensile strength of concrete. However, tensile strength of concrete does not feature explicitly in any design rules in current codes of practice.

The general design methods assume a simple relationship between tensile strength and compressive strength, which may not be valid for all types of concrete. It is essential, therefore, to examine these rules to verify the influence of constituents of concrete on these parameters. This study will be significantly important for examining the general validity of prescriptive rules for shear design in case of members made with different types of concrete, particularly the Eurocode EC2 "Variable Strut Inclination Method" where the contribution of concrete to the shear resistance is apparently ignored[3].

Emergence of Interest in Serviceability Design

For some considerable time, development in design methods has mainly involved the strength or load-carrying capacity of reinforced concrete elements as the primary criterion, complemented by serviceability checks in certain cases. However, commercial pressures on economy of construction at present seem to lead to reduced member sizes and towards reduction in partial factors for material strength. Consequently, the global factor of safety may reduce significantly and serviceability design criteria could become critical, e.g. limits on cracking, deflection, vibration, etc. This potential development is attracting the interest of researchers**[4]**. Serviceability design concerns perceptible signs of distress to the structure (deformations and cracking), which enable decision-making, whether to continue the use of the building or to repair or demolish it. This is an important consideration in the management of buildings and in the process of whole life design of concrete structures.

For serviceability design, precise information of the various parameters is essential, including the modulus of elasticity of concrete, and an estimate of only the compressive strength of concrete could not suffice. Such data are also important for the integrated design of deformation-sensitive structures, combining superstructure design with foundation design and employing principles of soil-structure interaction.

Data Required for Provisions against Post-construction Actions

The properties of concretes made with different constituents to suit the durability criteria should also feature in an integrated design for provisions in the framed structures against fire and accidental actions. In such cases, difference in types of concrete could affect the post-construction response of structures to these actions, especially due to any potential difference in ageing of concretes made with different constituents. A realistic assessment of nonlinear analysis parameters, based on accelerated ageing and exposure tests, could provide more reliable design estimates of residual strengths of structures and measures for avoiding disproportionate collapse, based on assessment of the ductility of concrete members.

TEST METHODS FOR TENSILE STRENGTH

Identification of Test Methods

The cylinder-splitting test and the flexural test are identified as suitable for this exercise. Previous research on these tests is reviewed in this section.

Cylinder-splitting tests

Concrete specimens were tested by Chapman**[5]**, made with minimum cement content of 154 kg/m^3 and with W/C ratio 0.68 (28 days cube strength between 20 and 26 N/mm^2) and 0.57 (28 days cube strength between 28 and 34 N/mm^2). The following rule was used for estimating the cylinder-splitting strength.

$$\text{Cylinder-splitting strength} = \frac{2\ (\text{machine load at failure})}{(\text{perimeter x length of the cylinder})}$$

Chapman has noted influence of the following factors on results of cylinder-splitting tests.

i) Correct alignment of packing strips is essential for reducing stress-concentration under the applied load. Cylinders provided with plywood packing strips may give 7% higher test results than those provided with hardboard strips. For 100 mm dia. cylinders, the influence of size is negligible for strips larger than 12 mm wide and 3 mm thick

ii) 100 mm dia. cylinders give higher splitting strength compared with that given by 150 mm dia. cylinders and, on average, the difference is 13% for flint concrete and 5% for limestone concrete.

iii) Steel-moulded cylinders may give marginally higher splitting strength compared with that given by cardboard tube moulded cylinders, but the difference is not significant.

iv) W/C ratios do not influence "splitting/compressive" strength ratios.

v) An increase in the "aggregate/cement" ratio gives greater splitting strength for the same compressive strength.

vi) Limestone concrete gives comparatively higher cylinder-splitting strength (about 8% higher than flint concrete) and higher compressive strength for equivalent concrete mixes. Ratios of "Splitting/compressive" strength are also higher than the corresponding granite or flint aggregate concrete for given compressive strengths.

vii) "Splitting/compressive" strength ratio could be higher for a specimen which fails in compression as a result of failure of the aggregate, compared with a specimen made with an equivalent grade of concrete and which fails as a result of matrix failure.

viii) Size of aggregate has no influence on "splitting/compressive" strength ratios, as shown by tests using a range of aggregate from 10 mm to 38 mm in size.

Comparison Between Cylinder-splitting Strength and Flexural Strength

Malhotra et al**[6]** have estimated cylinder-splitting strength (150 dia. and 400 mm long) and flexural tensile strength (also known as modulus of rupture) using tests on beams (89 x 100 x 400 mm long) loaded at the centre or at third points.

The specimens represented a ranged of W/C ratios (by weight) from 0.31 to 1.00, "aggregate/cement" ratio ranged from 2.69 to 9.24 and 28 days cylinder crushing strength (150x300 cylinders) range from 11 to 53 N/mm^2. All specimens were covered with glass plates after casting. After 24 hours, they were placed in a moist-curing room for 27 days with 75° F temperature and 100% RH.

Malhotra has concluded that the values of flexural strength are higher than the corresponding cylinder-splitting strength (Figure 1) since the formulae for flexural strength assume linear distribution of stress, which is theoretically incorrect. He has also noted that the results for flexural tensile tests depend on the number of loading points (one at the centre or two at third points), span-depth ratio and moisture condition of test beams.

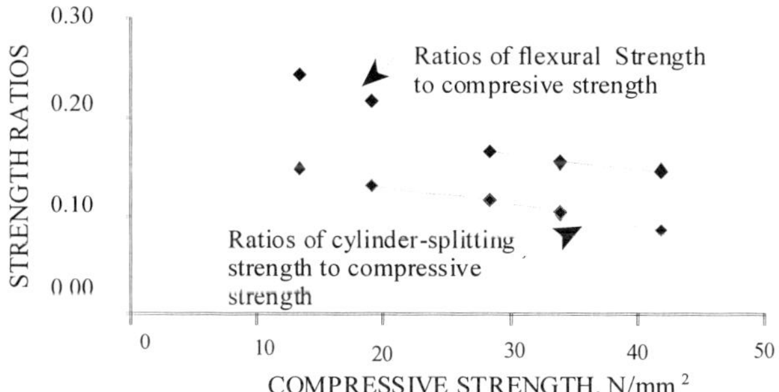

Figure 1 Comparison between flexural and tensile strength (Malhotra et al)

REALISTIC INPUT FOR RIGOROUS STRUCTURAL ANALYSIS

The Role of Rigorous Analysis

It is proposed to review the design rules given in the current codes of practice with tests on beams made with different types of concrete and loaded to failure. The failure load of a beam will be validated against its load-carrying capacity given by rigorous analyses, using realistic linear and nonlinear properties of the concrete derived from tests, as far as possible. ABAQUS, a nonlinear finite element analysis computer program, has been identified as suitable for this exercise and for analysis of complex structures. Other programs are also available for such analyses but the need for realistic input of properties is expected to be important for their use as well. A case for improving the ABAQUS input is discussed here, as an example, in relation with the author's past experience[7].

Input for ABAQUS Program[8]

ABAQUS analysis depends on the ability of individual elements of a member to reach a condition of equilibrium between the actions and reactions, without exceeding the assumed strength and properties of materials. Options for input information have a wide range, to suit the nature of the problem, e.g. types of elements, properties of materials, support conditions, load increments and number of iterations for convergence of a solution. While the input can be chosen to suit the complexity of the problem, the properties of concrete are derived within ABAQUS in relation with the compressive strength of concrete, as described below.

Stress-strain relationship

The stress-strain curve given in BS8110: Part 1**[9]** is a parabolic curve corresponding to the modulus of elasticity as 5.5 $\sqrt{f_{cu}}$ k/N/mm2 (f_{cu} = characteristic compressive strength). This rule does not agree with Table 7.2 of BS8110: Part 2**[9]**, which gives values of "E" as (20000 + $200f_{cu}$) N/mm^2, for $60>f_{cu} > 20$ N/mm^2 (f_{cu} = the actual 28 days cube strength, which is used in test data in place of the characteristic strength). A new stress-strain curve for concrete is proposed, based on the following conditions:

- The curve starts at the origin, with value for the stress and the strain being zero.
- At the origin, the slope of the curve is "E" (modulus of elasticity of concrete)
- Limiting longitudinal stress is taken as $0.67f_{cu}$ corresponding to the strain at peak stress

  of $2.4(\sqrt{f_{cu}}) \times 10^{-4}$, where the slope of the curve is zero.

(The author has proposed that tests should be used to provide better information on "E" and nonlinear properties of different concretes, as explained later in this paper.)

ABAQUS requires the part of stress-strain input in the elastic range, which is assumed to correspond to the range of stresses from zero up to 5 N/mm^2. This could be chosen as a convenient point on the stress-strain curve plotted up to the maximum stress $0.67f_{cu}$. For stresses higher than 5 N/mm^2, ABAQUS requires only the difference between the total strain and the elastic strain. ABAQUS adds these input values of strains to the limiting elastic strain to obtain the total strains. ABAQUS requires that the input should include the horizontal part of the curve at the peak stress. The curve could, therefore, be extended up to the strain value of 0.0035, the same as the BS8110: Part 1 curve as shown in Figure 2.

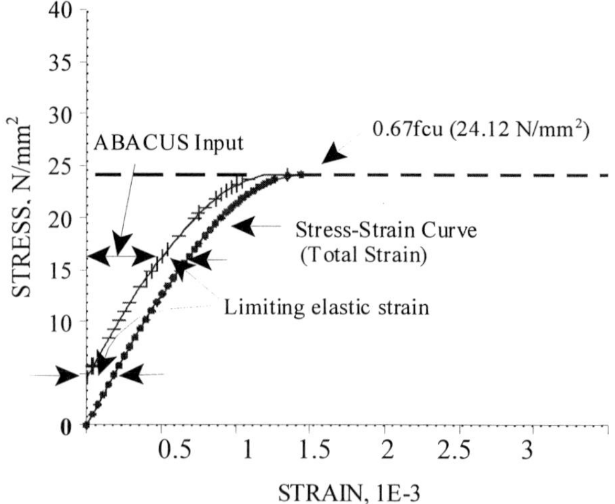

Figure 2 Stress-strain relationship for concrete (f_{cu} = 36 N/mm^2)

Failure ratios

Failure ratios is an option used for defining the shape of the failure surface, as a function of the predefined field variables. The sketch in Figure 3 (not-to-scale) shows the yield and failure surfaces in plane stress, as described overleaf:

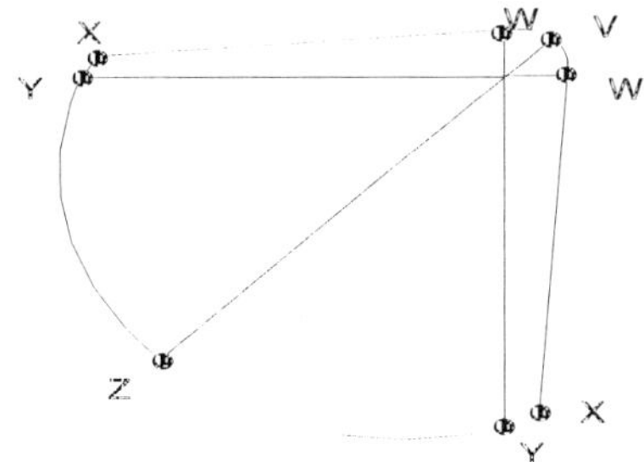

Figure 3 Yield and failure surfaces in plane stress

i) Lines "wx" and line "yz" represent the crack detection surface and the compression surface respectively.

ii) Points "w" and "y" show the uniaxial tension and compression respectively.

iii) Points "v" and "z" show the biaxial tension and compression respectively.

As defined within ABAQUS, the following four values are entered in the input file:

i) The ratio of the ultimate biaxial compressive stress to the ultimate uniaxial compressive stress = 1.16

ii) The absolute value of the ratio of uniaxial tensile stress at failure to the ultimate uniaxial compressive stress = 0.12

iii) The ratio of the magnitude of a principal component of plastic strain at ultimate stress in biaxial compression to the plastic strain at ultimate stress in uniaxial compression = 1.28

iv) The ratio of the principal tensile stress at cracking (in-plane stress, when the other non-zero principal stress component is at the ultimate compressive stress value), to the tensile cracking stress under uniaxial tension = 0.33

Of these ratios, item (ii) concerning the relationship of uniaxial tensile stress with the compressive stress would be different for different concretes and should be investigated. The other properties should then be reviewed accordingly, as described later in this paper.

Tension stiffening effect

When a section cracks, it undergoes a discontinuity in the distribution of stresses and the cracks affect the material stiffness associated with the integration point. In a zone where cracks are present, the uncracked concrete between the cracks can carry a certain proportion of the tensile force. This proportion depends on the properties of constituents of the section and, also, on the density and distribution of reinforcement in the section. This phenomenon is called the tension stiffening effect, which enables the program to simulate an interaction between the reinforcement crossing the crack and the concrete, to allow "smearing" of cracking over the finite volume associated with an integration point.

Among the options provided by ABAQUS, "TYPE=STRAIN" is normally chosen as a suitable option for reinforced concrete members. This input requires two pairs of parameters. The first pair, for example (1.0, 0.0), relates to the number of field-variable dependencies included in the definition of compressive yield stress besides the temperature and these default values should be chosen. Otherwise the program will assume that the post-cracking behaviour depends only on temperature.

The second pair of parameters, e.g. (0.0, 3.1×10^{-3}), defines the total strain at which the tensile stress normal to a crack will be zero. This strain is a multiple of the strain at which concrete cracks, called "the strain at failure". Its "typical" value within ABAQUS is 10^{-4}. However, this should be taken as only an indication since the actual value of this strain for normal grade concrete is nearer to 3×10^{-4} or 0.0003. ABAQUS recommends that the total strain (for zero tensile stress across a crack) should be 10 times this strain as a starting point. It is also suggested that a calibration should be done to arrive at an optimum value for analysing a particular type of beam. Such calibrations should consider the density and distribution of steel, strength of concrete, etc. and the values chosen after some initial trial runs may fall within a narrow range, say between 0.003 and 0.0035. This is considered as acceptable, although the aim should be to derive a methodology to evaluate the tension stiffening for various grades of concrete made with different constituents and the reinforcement distribution. This could be followed up in future phases of the research.

Shear retention

The shear retention input is used to describe the reduction of shear modulus associated with the crack surfaces. This is taken as a function of the tensile strain across the crack. The shear stiffness of open cracks is assumed to reduce linearly to zero as the crack widens and the tensile strain is, say, 0.0075 or 0.009. The shear retention input values are typically chosen to be within a narrow range, between (1.0, 0.0075, 1.0, 0.0075) and (1.0, 0.009, 1.0, 0.009), after some initial trials. However, future research could narrow down a preselection to suit the type of concrete and the corresponding tensile strength.

The shear retention factor accounts for deterioration effect of cracking on the shear modulus of the initial uncracked concrete (G), which depends on its modulus of elasticity (E) and Poisson's Ratio (ν).

$$G = \frac{E}{2(1+\nu)}$$

This function applies strictly to concrete within its elastic range. In the test programme under this project, it is intended to measure "E", "v" and the full stress-strain curve. It is expected that these values can give the shear modulus but the problem could lie in measuring the lateral strain during the post-elastic stage. This measurement could become impractical for the portion of stress-strain curve beyond the "peak stress", owing to the speedy progress towards failure of the specimen. A solution may be sought through studying the trend of variation in "G" with "E", treating "v" as a constant, and extrapolating the relationship to reflect a calibration against data provided by beam tests of this exercise.

Values of Poisson's Ratio depend on the compressive strength and the density of concrete as shown by Klink**[10]**. Tests included in this exercise would reveal whether the ratio depends on the constituents of concrete. Klink has shown that "v" is higher in the inner part of the specimen, 55% higher at the centre than at the surface. He has recommended that this be the "actual" Poisson's ratio. For the purpose of this exercise, "v" is proposed to be measured on the surface of specimens, for the reasons of simplicity and ease of comparison with other available data.

PLAN FOR RESEARCH AND EXPECTED OUTPUT

Research in the production of concrete with optimised binders is proposed to be combined with research in structural design, to form a "Sustainability Portfolio" of projects of three year duration starting in June 1998. The portfolio includes the following main tasks:

- Development of suitable test methods for measuring all properties (besides the compressive strength) of various grades of concretes made with different binders, and
- Investigation into a practical assessment of realistic properties, based on the review of relevant research and experience as described above

The output is intended to serve rigorous analyses capable of using nonlinear properties of concrete, in the form of full stress-strain curves and the relationship of compressive strength with tensile strength, modulus of elasticity and other relevant parameters. It is proposed to use the data in nonlinear finite element analyses for obtaining estimates of load carrying capacity of beams and compare them with test results on beam specimens with provisions for strain measurements.

The general design rules may have to be modified depending on these beam test results and the results of rigorous analyses. The modified rules may include additional parameters to represent the constituents and binders in typical categories of concretes suitable for general use.

The outcome of the portfolio exercise, therefore, would maintain ease of using concrete specifications and normal structural design rules, without introducing unnecessary complications, and the use of general specifications and design rules could continue where it is appropriate.

It is hoped that these findings could serve as an input for reviewing the recommendations for strength design and serviceability design given in BS8110 and Eurocode EC2. Finally, a guidance document is planned to cover all common issues concerning concrete technology, structural design and construction process.

Objectives of the portfolio are limited to the initial steps in developing the concept of total design of competitive and sustainable concrete structures, based on using realistic properties of concrete for general design or rigorous analyses as appropriate. Future research could make the whole life design of concrete structures a practical proposition and extend the procedures to include design of mutually compatible and common provisions against environmental actions and in-service loading, including accidental actions and fire.

ACKNOWLEDGEMENTS

The author would like to thank the Department of the Environment, Transport and the Regions for sponsoring the research work referred to in this paper. The author would also thank Professor George Somerville and the staff of the British Cement Association and Professor Ravindra Dhir and Dr Nutan Subedi of the University of Dundee for their views expressed during various discussions on the subjects covered by this paper.

REFERENCES

1. KAND, C V. Structural Concepts in Vastushastra. Proc. Seminar on Innovative Applications and Trends in Structural Concrete, FIP Day, 1998, Mumbai, pp 197-204.

2. HEWLETT, P C. Concrete: The Material of Opportunity. Proc.Int. Conf. Concrete 2000 Eds. R K Dhir and M R Jones, E & FN Spon, 1993, Vol.1, pp 1-22.

3. DD ENV 1992-1-1: Eurocode 2: Design of concrete structures, Part 1: General rules and rules for buildings. 1992. British Standards Institution, London.

4. BEEBY A W AND FATHILBITARAF F. The design of framed structures - A proposal for change, Proc. Int. Conf. Concrete in the service of mankind, Eds. R K Dhir and M J McCarthy, E & FN Spon 1996, pp 495 - 506.

5. CHAPMAN, G P. Cylinder splitting tests on concretes made with different natural aggregate, Research Note, Sand and Gravel Association of Great Britain, SfB Dpl: Df2:Db, SR 6701, August 1967, London

6. MALHOTRA, V M, AND ZOLDNERS, N G. Comparison of Ring-Tensile Strength of Concrete with Compressive, Flexural and Splitting Tensile Strengths, American Concrete Society, Journal of Materials, Vol. 2, No 1, March 1967, New Jersey, USA.

7. DESAI, S B. Shear Resistance at normal and high temperatures of reinforced concrete members with links and central bars, Thesis for the degree of Doctor of Philosophy, City University, London 1995.

8. ABAQUS Manual Version 5.2, Hibbitt, Karlsson and Sorensen (UK) Ltd, Warrington, Cheshire, UK.

9. BRITISH STANDARDS INSTITUTION, BS8110: Structural Use of Concrete, Part 1: 1985: Code of practice for design and construction. Part 2: 1985: Code of practice for special circumstances.

10. KLINK, S A. Actual Poisson's Ratio of Concrete, Title no. 82-74, ACI Journal, American Concrete Society, November-December 1985, New Jersey, USA.

STRUCTURAL BEHAVIOUR OF MID-RISE SHEAR WALL UNDER LATERAL LOAD

A A Tasnimi

Building and Housing Research Centre (BHRC)

Iran

ABSTRACT. This paper reports the results obtained from tests conducted on four reinforced concrete structural shear wall specimens SHW1, SHW2, SHW3 and SHW4 which were subjected to seismic-type lateral forces. Scale ratio of 1 to 8 was adopted for all specimens, which were identical in size and reinforcement ratios. These walls were cast with constant thickness (t) and a height-to-length ratio (h/l) equal to 3 and were considered to represent the critical structural elements with a rectangular cross-section. All specimens were experimentally subjected to different slow cyclic horizontal loading, and instrumented with electrical strain gauges on the steel surface so that the strain distributions were used for prediction of forces within the section. The type of failure and behaviour of such structural elements were investigated. The results obtained experimentally are discussed and compared with the recommendations provided by the current code of practice. The results also indicate fairly good flexural strength but not sufficient ductile capacity for all specimens.

Keywords: Ductility, Cracking, Lateral cyclic load, Earthquake, Seismic.

Dr Abbas Ali Tasnimi is the President of Building and Housing Research Centre (BHRC), Tehran, I.R. Iran. He is an associate professor in structural engineering of Tarbiat Modarres University (TMU), Tehran, I.R. Iran. He received his Ph.D degree in structural engineering from Bradford University,U.K. His main research interests includes, seismic behaviour of R/C structural components, non-linear behaviour of R/C structures, strengthening and retrofiting of existing R/C structures. He was formerly the President of TMU (Tarbiat Modarres University) Tehran, I.R. Iran.

INTRODUCTION

The design of shear walls in high seismicity regions is governed by the ultimate limit state (ULS) behaviour recommended by codes of practice [1]. In general shear wall is expected to exhibit stable non-linear hysteretic load-versus-deformation behaviour under cyclic seismic loads. In order to assure this behaviour, the design philosophy is, to induce flexural yielding by providing a flexural strength less than the shear strength of the shear wall's critical section. A comprehensive review of experimental research on shear walls indicates that stable hysteretic wall behaviour can not be fully guaranteed using the current design codes of practice. These studies suggest changes in current design procedures to provide ductile behaviour modes under seismic action [2-4]. The main objective of this paper is to investigate the behaviour of structural walls used in mid-rise buildings comprising frames and shear walls, and designed according to the Iranian code for seismic design [5]. According to this code if the building's height is limited to 8 storeys or 30 meters, shear walls can resist 100% of the seismic lateral forces. In this case comparison of stiffness of resisting elements may be neglected provided that frames in addition of resisting 100% of gravity load are capable of resisting at least 30% of seismic lateral forces.

EXPERIMENTAL PROGRAMME

Wall Details

The walls selected for testing purposes had geometry of 500mm length, 1500mm height and 50mm thichness. In all cases they were monolithically connected to a foundation, which was utilized, to fix down the wall to the laboratory's strong floor, simulating a fully fixed footing. The nominal dimensions of the specimens, together with the arrangement of flexural vertical reinforcement and both horizontals as well as vertical shear reinforcement are shown in Figure 1. The vertical and horizontal shear reinforcement comprised deformed steel bars of 3-mm diameter. Four vertical reinforcements comprised deformed steel bars of 6mm diameter were used as flexural steel at each corner of each wall. Table 1 summarizes the yield and ultimate strength characteristics of all steel bars. All reinforcement was designed in compliance with the recommendations given by the ACI [1]. The concrete for all specimens were made from Type (I) Portland cement, river sand, and 10mm maximum size crushed gravel. Measured slumps ranged from 60mm to 180mm for all specimens. From the six 150x300mm concrete cylinders, three were tested in compression at 28 days and the remaining three used for the tensile splitting test. Also compression tests were carried out on 150mm cubes. The modulus of elasticity of concrete was calculated on the basis of data obtained from cylinder compression tests. The concrete strain corresponding to its strength was measured for each specimen. Full details of the concrete mix used and the concrete properties are summarized in Tables 2 and 3 respectively.

Loading History and Testing Procedures

To simulate loading sequence that might be expected to occur during an earthquake, four simplified types of horizontal cyclic loading history were adopted. Since no standard cyclic test procedures has been introduced for testing reinforced concrete structures, the horizontal load was applied at a quasistatic rate in displacement controlled cycles with different patterns which corresponded to three major states namely cracking state, yielding state and ultimate state. In all cases this horizontal load was applied at the centroid of the distributed lateral forces which were evaluated analytically at each storey level. Figure 2 provides the loading programmes used for the tests in terms of horizontal load. All specimens were incrementally loaded with a hydraulic actuator having 50kN capacity through a load cell. At each incremental loading state the load was maintained constant for a few seconds in order to measure and record the load, displacement response of the walls and the steel strain via an electronic data logger. Two types of electrical strain gauges (FLA-5-11) and (FLA-10-11) were used for measuring the strain of vertical wall reinforcement. The position of these strain gauges is shown in Figure.1. In addition, electrical strain gauges of type KYOWA:KC-20-A1-II were used to measure the surface concrete strains for specimen SHW1 only. Linear transducers of type LVDT were used in monitoring the in-plane horizontal displacements at 1100mm height and 50mm above the foundation of the wall. In order to insure that the foundation is fixed to the laboratory's strong floor vertical displacement of the top of the foundation was measured. Providing special arrangements using ball bearings as shown in Figure 3 prevented the out-of-plane displacement of all specimens. The measured values of load, displacement and strains were recorded by a computer data logger capable of measuring to sensitivity ranges of ±0.1N, ±0.001 mm, and ±12 microstrain, respectively, with speed of about 0.08 seconds per channel

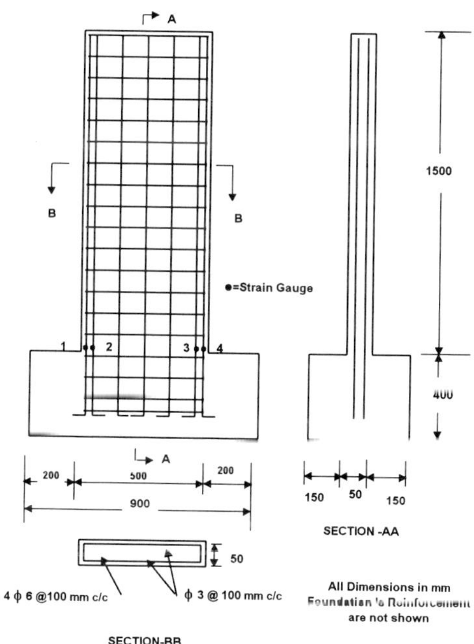

Figure 1 Geometry and reinforcement details of wall specimens

Table 1 Properties of reinforcing bars

TYPE	YIELD STRENGTH f_{sy}(Mpa)	ULTIMATE STRENGTH F_{su}(Mpa)
3mm dia. Bar	216	317
6mm dia. bar	276	475

Table 2 Concrete mix proportion by weight

PROPERTIES BY WEIGHT	SPECIMENS SHW1 & SHW2 (kg)	SPECIMENS SHW3 & SHW4 (kg)
10mm Aggr.	825	827
River sand	950	954
Cement	320	325
Free water	250	254

Table 3 28 days characteristics strength of concrete

SPECIMENS	COMPRESSION STRENGTH		SPLITTING TENSILE STRENGTH f_{ct}(Mpa)	MODULUS OF ELASTICITY E_c (Mpa)	CONCRETE STRAIN RELATED TO $f'_c \varepsilon_0$ * E-3
	Cube f_{cu}(Mpa)	Cylinder f'_c(Mpa)			
SHW1	25.85	21.60	2.19	20.98	1.75
SHW2	23.95	21.60	2.22	20.98	1.75
SHW3	26.10	22.45	2.26	22.46	1.96
SHW4	26.97	23.45	2.24	22.46	1.96

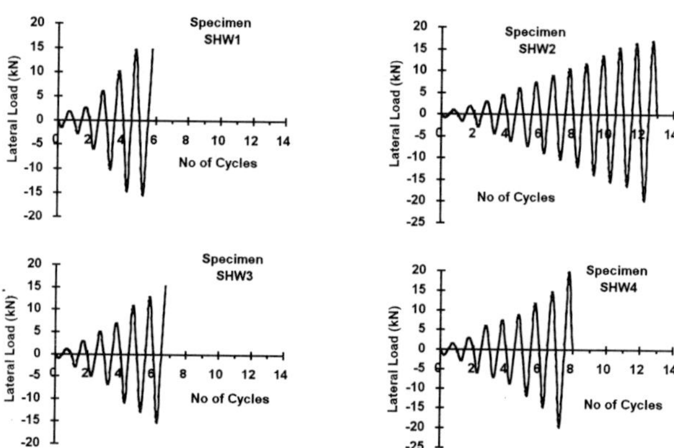

Figure 2 Loading history applied cyclically

Table 4 Tested and calculated results in successive loading cycle, Specimen SHW1

CYCLE NO	HORIZONTAL LOAD (kN) AND DIRECTION ⇒	⇐	AVERAGE TOP DEFLECTION (mm)	AVERAGE MOMENT OF RESISTANCE (kN-m) (1)	AVERAGE APPLIED MOMENT (kN-m) (2)	MOMENT RATIO (kN-m) (1)/(2)
1	1.57	1.77	0.520	1.69	1.72	0.97
2	2.95	2.55	0.870	2.67	2.79	0.96
3	5.99	6.00	2.390	5.92	6.18	0.96
4	10.11	10.11	4.480	9.79	10.41	0.95
5	14.34	14.63	13.46	14.03	14.92	0.94
6	15.42	14.53	16.45	14.87	15.88	0.95

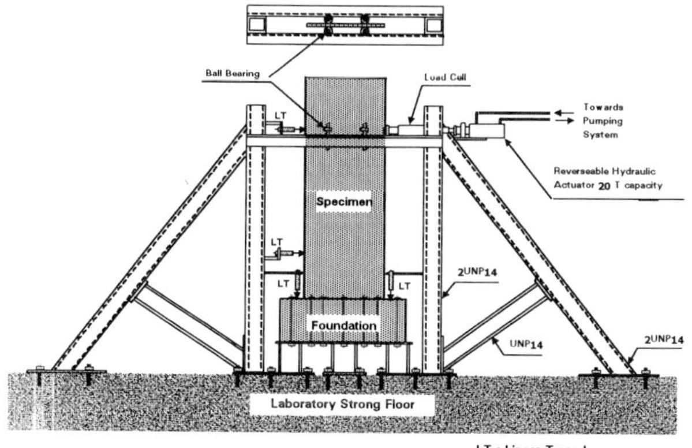

Figure 3 Schematic view of test setup with the positions of strain gauges and transducers

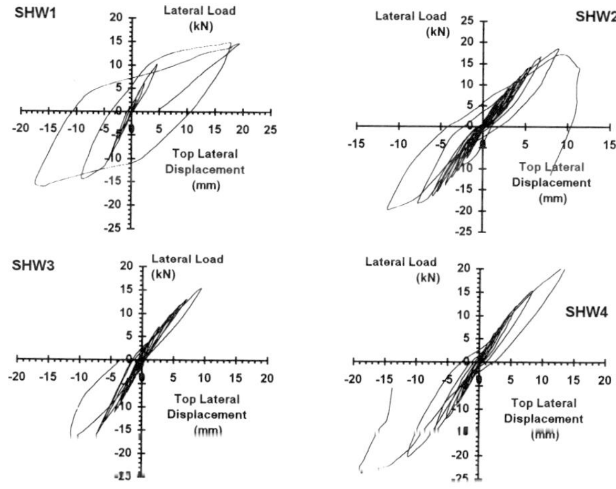

Figure 4 Lateral load versus top horizontal displacement loops for the tested specimens.

Testing Results

The main results obtained from the tests, together with information necessary for their interpretation are given in Tables 4 to 7 and Figure 4. These tables provide the information of the experimental data and the principal results of the specimens tested in the programme. All forces within critical section of the walls at any state of loading were evaluated on the basis of the method given in refrences 6 and 7. The applied and resisting moments at the base and their ratio indicating good agreement are given in the same tables.

Figure 4 illustrates the lateral load versus the top displacement curve estblished from the tests for all specimens.

Table 5 Tested and calculated results in successive loading cycle, Specimen SHW2

CYCLE NO	HORIZONTAL LOAD (kN) AND DIRECTION ⇒	HORIZONTAL LOAD (kN) AND DIRECTION ⇐	AVERAGE TOP DEFLECTION (mm)	AVERAGE MOMENT OF RESISTANCE (kN-m) (1)	AVERAGE APPLIED MOMENT (kN-m)(2)	MOMENT RATIO (kN-m) (1)/(2)
1	0.90	0.90	0.20	1.04	0.95	1.12
2	1.57	1.86	0.41	1.85	1.91	0.93
3	2.95	2.95	0.72	3.03	3.11	0.99
4	4.42	4.42	4.48	5.98	4.66	0.92
5	6.10	6.00	1.62	5.98	6.38	0.94
6	7.37	7.37	2.13	7.28	7.78	0.94
7	8.85	8.95	2.74	8.68	9.39	0.93
8	10.32	10.62	3.39	11.04	11.05	1.00
9	11.89	11.80	4.08	12.97	12.49	1.04
10	13.66	13.66	4.81	13.65	14.36	0.95
11	16.02	16.02	6.40	15.53	17.00	0.91
12	17.79	18.19	8.25	18.17	18.98	0.95
13	17.79	19.56	11.25	19.85	20.87	0.94

Table 6 Tested and calculated results in successive loading cycle, Specimen SHW3

CYCLE NO	HORIZONTAL LOAD (kN) AND DIRECTION ⇒	HORIZONTAL LOAD (kN) AND DIRECTION ⇐	AVERAGE TOP DEFLECTION (mm)	AVERAGE MOMENT OF RESISTANCE (kN-m) (1)	AVERAGE APPLIED MOMENT (kN-m) (2)	MOMENT RATIO (kN-m) (1)/(2)
1	1.09	1.09	0.26	1.05	1.15	0.88
2	3.05	2.95	0.75	2.30	3.11	0.95
3	5.01	5.01	1.55	4.89	5.26	0.93
4	6.90	7.08	2.58	6.87	7.34	0.94
5	10.91	11.30	5.01	10.75	11.67	0.92
6	13.56	12.78	6.60	12.97	13.83	0.95
7	15.53	15.14	8.30	15.11	16.11	0.94
8	16.42	17.50	15.50	17.44	17.81	0.97

Table 7 Tested and calculated results in successive loading cycle, Specimen SHW4

CYCLE NO	HORIZONTAL LOAD (kN) AND DIRECTION ⇒	HORIZONTAL LOAD (kN) AND DIRECTION ⇐	AVERAGE TOP DEFLECTION (mm)	AVERAGE MOMENT OF RESISTANCE (kN-m) (1)	AVERAGE APPLIED MOMENT (kN-m) (2)	MOMENT RATIO (kN-m) (1)/(2)
1	1.47	1.47	0.41	1.41	1.54	0.91
2	2.95	2.95	0.89	2.92	3.01	0.93
3	5.90	6.00	2.13	6.00	6.25	0.95
4	7.27	7.37	2.63	7.15	7.69	0.93
5	8.85	8.95	3.28	9.03	9.34	0.96
6	11.80	11.80	5.89	11.75	13.32	0.94
7	14.84	14.94	7.88	15.15	15.64	0.97
8	19.76	19.76	13.62	19.63	20.75	0.94

DISCUSSION

Cracking

For all specimens, flexural cracks initially appeared near the bottom part of the tensile zone of the wall, when only 10% of the wall capacity were reached. As the cyclic lateral load was appraoched nearly 62% of its maximum value, significant inclined cracks initiated at the tension zone of the wall during consecutive load reversals. These cracks continued to penetrat deeply into the centre of the wall towards the compressive zone. Some of these cracks formed a diagonally crisscrossing crack pattern for all specimens. Beyond a load level of about 85% of the failure load, the crack patterns underwent insignificant changes. It should be noted that two load cycles prior to failure, narrow flexural cracks were formed on previous cracks and hence the width of the major flexural cracks became considerable. This may be considered to indicate that the shear resistance of the wall is inadequate. Figure5 shows that, as in all cases, horizontal crack were developed at the vicinity of the foundation which eventually caused the sudden drop of the lateral load-carrying capacity of the specimen.

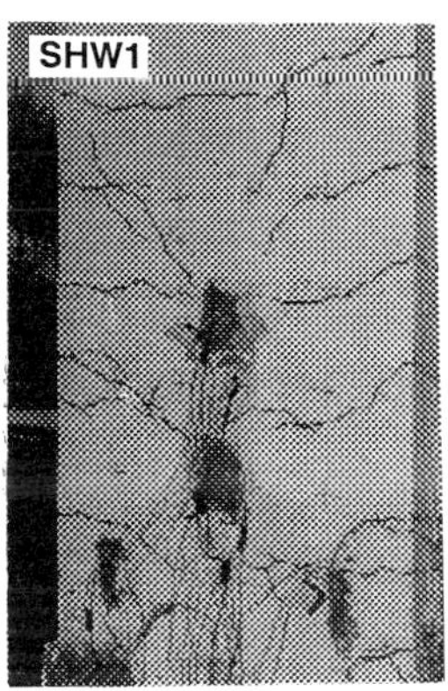

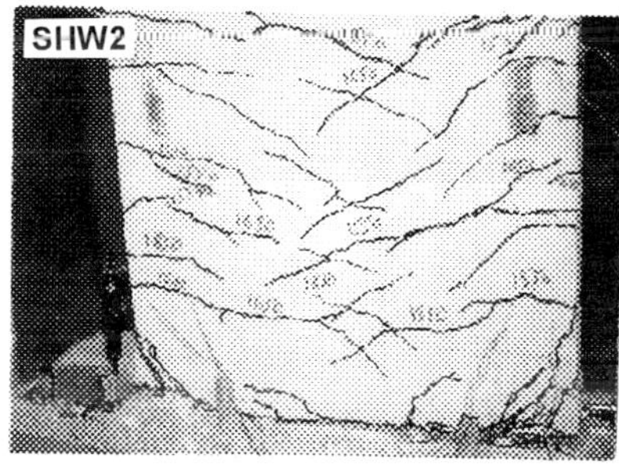

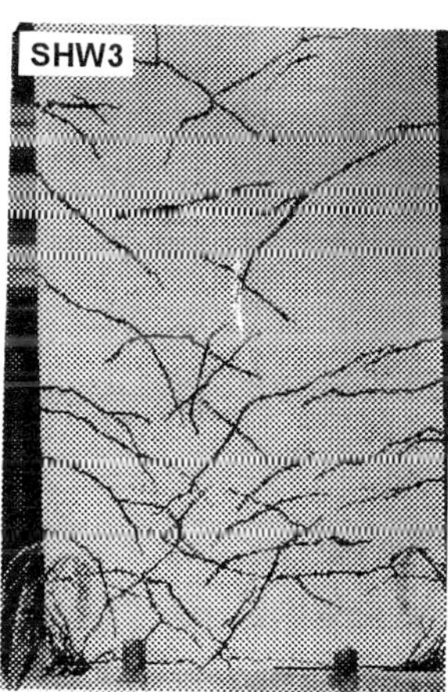

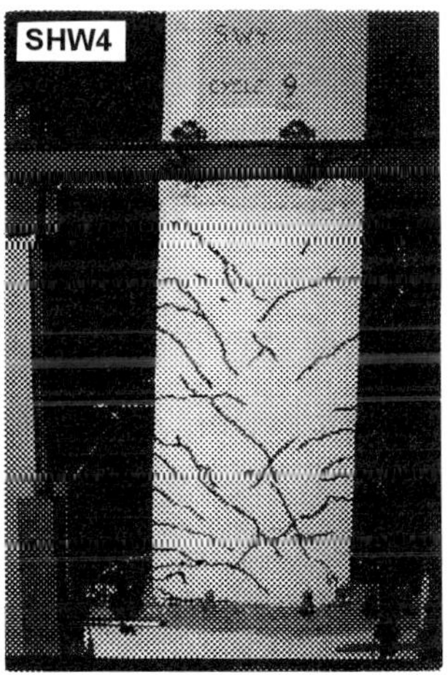

Figure 5 Behaviour and cracking process, for all specimens

Strength Degradation

Tables 4 to 7 indicate that the horizontal load-carrying capacity of specimens SHW2 and SHW4 were almost identical, where specimens SHW1 and SHW3 exhibited lower capacities of about 22 and 12 percent respectively. Since the strength of the concrete used for all specimens differ only slightly, the variation in wall strength is considered to predominantly reflect the effect of loading history on the strength of the vertical reinforcement. The cyclic loading applied to the specimen SHW1 had a relatively higher intensity in early stages which caused more decrease in stiffness which may be due to rapid strain-softening response exhibited by the reinforcement resulting in a reduction in wall strength. However, it is seen from these tables that the cyclic loading regime employed to all specimens appeared to have had an insignificant effect on the ultimate strength of the walls.

Deformation and Stiffness

Variations of lateral deflection in successive cycles are presented in Tables 4 to 7. The values of horizontal top displacement were evaluated theoretically on the basis of the measured displacement at the level 110mm above the wall base. The displacement relating to the first cracking and the first yielding states corresponded to 3 and 29 percent of that at the ultimate state. This shows the sharp reduction in the stiffness of the walls after yielding. The horizontal load versus horizontal top displacement of all specimens, shown in Figure.4, indicates a distinctly non-linear deformational response. Another interesting feature of the curves is the pinching effect which is mainly due to the low shear strength of the specimens. Also, as the load-displacement curves indicate that for all speciments the stiffness in elastic range is approximately constant, while at elasto-plastic and plastic a considerable reduction was noticed. This may indicate that the increase in the width and depth of cracks will cause a reduction in the stiffness of the section. It is also remarkable that increasing the maximum displacement of the loading cycle increases the stiffness deterioration.

Ductility

The capacity of the structural elements to deform beyond yield or elastic limit with minimum loss of strength and stiffness depends upon their ductility. Table 8 gives the ductility value for all specimens based on Δ_u/Δ_y and ϕ_u/ϕ_y. ratios. The measured displacement at yield state was controlled through electrical strain gauges, ie; when the strain of vertical reinforcement reached the value of 0.2%, the load and displacement reading were recorded. Also curvature at ultimate state was calculated based on the ACI method described earlier in this paper. The value of ductility in Table 8 indicate that ductility based on defromation was considerably higher than that of curvature. As it was stated previously, this is mainly due to the reduction of stifness and low shear capacity of the specimens.

CONCLUSIONS

The results of the tests conducted on shear wall specimens loaded laterally were used in this investigation. Based on the experimental study in this paper the following conclusions may be drawn:

1. The crack patterns and failure mode of all specimens are not fully in compliance with the concept of the compressive force path reported in reference [3]. Since they indicate that the wall capacity is affected by the flexural as well as shear strength of the specimen.

Table 8 Curvature and displacement ductility of the specimens

SPECIMEN	DUCTILITY BASED ON ACI μ_ϕ (1)	CALCULATED DUCTILITY BASED ON TEST RESULTS						RATIO OF CALCULATED DUCTILITY TO ACI (4)/(1)
		Based on wall curvature			Based on wall displacement			
		ϕ_u*10^{-6} (2)	ϕ_u*10^{-6} (3)	ϕ_u*10^{-6} (4)	Δ_y (5)	Δ_u/Δ_y (6)	Δ_u (6)	
SHW1	3.98	4.49	8.980	2.00	4.48	16.45	3.67	0.51
SHW2	3.45	3.60	8.140	2.26	3.39	11.26	3.32	0.66
SHW3	4.36	4.29	8.980	2.10	5.05	15.50	3.10	0.46
SHW4	4.69	4.89	15.15	2.10	4.76	18.20	3.82	0.45
AVG	4.12	4.32	9.060	2.12	4.42	15.35	3.48	0.52

2. For all specimens the plastic hinge was formed at the extreme fiber of the wall section and at the vicinity of the base.

3. The strength and deformational responses of the specimens were found to be independent of the cyclic loading sequence.

4. The observed horizontal crack at the base (just above the foundation) prior to ultimate state may be due to sliding of the vertical reinforcement. This caused considerable reduction in the strength, stiffness and energy dissipation of the specimens.

ACKNOWLEDGEMENT

The work described in this paper was financially supported by "Building and Housing Research Center" (BHRC), and was carried out in the structural laboratory of BHRC. The author gratefully acknowledges the assistance of Mr. Shokrzade, laboratory staff and financial support of BHRC.

REFERENCES

1. Building Code Requirements for Reinforced Concrete, ACI-318M-89, 1991

2. CARDENAS A.E AND RUSSELL H.G AND CORLEY W.G., Strength of Low-Rise Structural Walls, ACI Committee 442, Publication (SP-63), 1980, pp.221-242.

3. LEFAS I.D., KOTSOVOS MICHAEL D., AND AMBRASEYS, NICHOLAS N., Behaviour of Reinforced Concrete Structural Walls Strength, Deformation, Characteristics, and Failure Mechanism, ACI Structural Journal, 1990, Vol.87, January, No.1, pp. 23-31.

4. IOANNIS D.LEFAS AND MICHAEL D.KOTSOVOs, Strength and Deformation Characteristices of Reinforced Concrete Walls Under Load Reversals, ACI Structural Journal, 1990, Vol. 87, No.6, pp. 716-726.

5. Iranian Code for Seismic Resistant Design of Buildings, IBCS-2800, 1988

6. OESTERLE, R.G, FIORATO, A.E ARISTIZABAL-OCHOA J.D.AND CORLEY W.G., Hysteretic Response of R.C Structural wall, ACI Committee 442, Publication SP-63, 1980, pp. 243-274.

7. TASNIMI, A.A, Prediction of forces within prestressed sections, Ph.D Thesis, Dept. of Civil and Sturctural Engineering, University of Bradford, UK. 1988.

8. TASNIMI A.A, AND MOSSAYEBI A., Cyclic Behaviour of Exterior Beam-Column joints in Precast Concrete Frames, 3rd International Kerensky Confence on Structural Engineering, 20-22 July, Singapore, 1994.

9. SALE, E.A.B. AND FINTELA M., Strength, Stiffness and Ductility Properties of Slender Shear Walls , ACI Committee 442, Publication Sp -63,1980.

10. JUN-ICHI T., AKENORI S. AND TOSHIO S. Crack Indices of Reinforced Concrete Shear Walls for Seismic Damage Evaluation, Proceedings of Ninth World Conference on Earthquake Engineering, August 2-9, Vol.4, 1988, Japan.

11. PARK R., Ductility Evaluation from Laboratory and Analytical Testing, Proceeding of Ninth World Confrence on Earthquake Engineering. Augost 2-9, Vol. 8,1988, Japan.

12. PAULAY T., Design of Ductile Reinforced Concrete Structural Walls for Earthquake Resistance, Earthquake Spectra, Vol.2, No.4, 1986, pp. 783- 82.

ANALYSIS AND DESIGN OF CONCRETE COUPLED CORE SUBSTRUCTURES SUBJECT TO GRAVITY AND LATERAL LOADS

C K Manatakos

M S Mirza

McGill University

Canada

ABSTRACT. The results of a research program involving linear elastic and nonlinear inelastic two- and three-dimensional finite element analyses, and design of a 21 storey reinforced concrete core-slab-frame structure subjected to gravity loads and lateral forces until failure are reported. Stage 1 studied the elastic response of the structure, while Stage 2 focussed on the design of the core substructure following the existing building code provisions and taking into consideration the three-dimensional building response. Design procedures and appropriate steel reinforcement details are proposed for the cores, coupling and lintel beams, slabs enclosed within and surrounding the cores, and the core wall-slab-beam connections. Stage 3 examined the nonlinear response of the core-slab substructure as designed, considering the effects of cracking and crushing of the concrete, strain-hardening of the steel reinforcement, and tension-stiffening. These results were incorporated in evaluating and modifying the design.

Keywords: Core-slab-frame structure, Cracking and crushing of concrete, Ductility of core wall-slab-beam connections, Elastic behaviour, Load-deflection response, Nonlinear behaviour, Three-dimensional finite element modeling, Yielding of steel reinforcement.

Dr Charles K Manatakos is a Post Doctoral Fellow n the Department of Civil Engineering and Applied Mechanics at McGill University, Montreal, Canada. He specializes in analysis, design and behaviour of reinforced concrete buildings, bridges and other structures.

Professor M Saeed Mirza, teaches in the Department of Civil Engineering and Applied Mechanics at McGill University, Montreal, Canada. His research focusses on the behaviour and design of concrete structures, and durability of concrete infrastructure. Professor Mirza has been an author and coauthor of several publications, has been a member of many related committees, and has been frequently invited to be an expert witness. He is a past president of the Canadian Society of Civil Engineering (1985-86) and is presently Chair of the CSCE Technical Committee on Rehabilitation of Infrastructure.

INTRODUCTION

Present practice of analysis and design of reinforced concrete core-slab-frame structures is based on a simplified structural system derived by decoupling the three-dimensional building into the different core and frame substructures,which are further subdivided into individual planar walls, frames and slabs. Detailed linear elastic and nonlinear inelastic finite element analyses~[1,2] show beneficial three-dimensional interaction between the various structural components when subjected to gravity loads and lateral forces, resulting in an overall increased lateral and torsional stiffnesses of the building, by a factor of as much as four times in this case, compared to the aggregation of the responses of the individual planar components. Current building codes provide some guidance for design of planar ductile frames and structural walls, along with empirical tools to ensure serviceability. However,the codes are deficient and do not take into consideration the three-dimensional response of core-slab-frame structures at the ultimate limit state. An equal emphasis on energy dissipation and ductility at the higher load levels is lacking. Design procedures and detailing of the steel reinforcement need to be developed for open-, closed- or partially closed-section cores, for slabs enclosed within and surrounding cores, for coupling and lintel beams, and to ensure properly detailed core wall/column-weak beam/slab connections.

OBJECTIVES

The objectives of this study were to provide an improved understanding of the complex behaviour of reinforced concrete core-slab-frame structures subjected to gravity loads and lateral forces until failure, investigating the influence of the various structural components on the overall building response [3,4], and to develop practical design procedures for the different elements and their connections[5].

DESCRIPTION OF THE STRUCTURE

The core-slab-frame structure is 21 storeys in height including a basement and has a non-rectangular plan (Figure 1) of 4 by 6 bays in the X- and Y-directions, respectively. The core substructure consists of an infilled-slab core with an enclosed slab (E-Slabs), an elevator core with one-third storey height deep lintel beams, and a stairwell core with partially enclosed slabs. In the building X-direction, the frame substructure is composed of columns and slab-band girders (SB1, SB2, SB3) forming two exterior frames of two-bays each, two interior frames of four-bays each and three core frames of one-bay each on either side of the cores. In the building Y-direction, the cores are connected by one-third storey height deep coupling beams (labelled as C-E and E-S) spanning between core sections flange walls. A flat slab, termed the surrounding slabs (S-Slabs), joins the core and frame substructures.

ELASTIC AND NONLINEAR ANALYSES

A detailed three-dimensional finite element model of the building was developed for the various analyses. Figure 2 illustrates the mesh details for a typical floor slab and the cores.

Elastic analyses of the core-slab-frame structure [3] were undertaken using the SUPERSAP (enlarged version of the SAP IV) computer program [6].

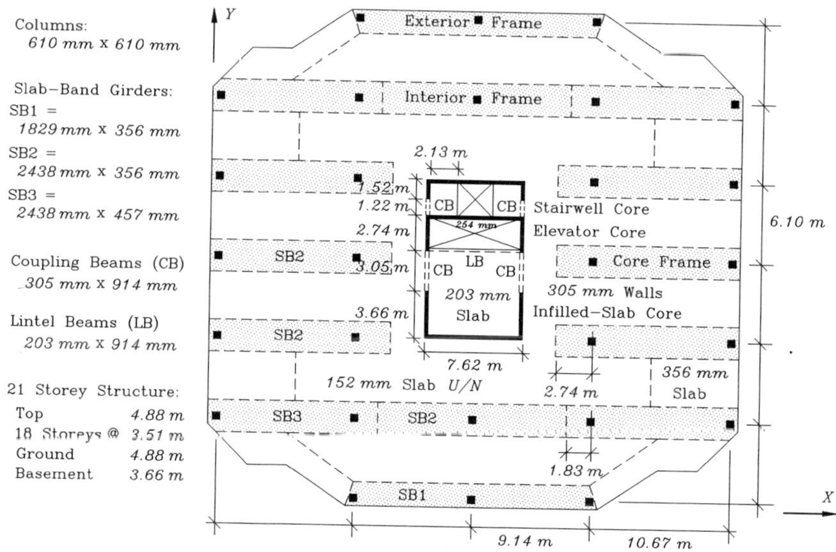

Figure 1 Floor plan of core-slab-frame structure

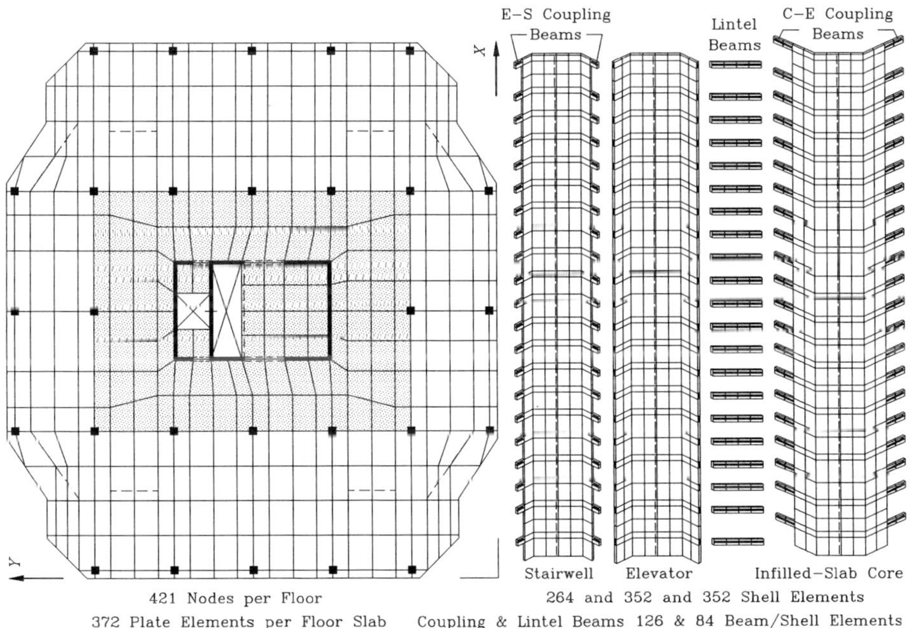

Figure 2 Finite element mesh for slab and cores

The model consisted of 9525 nodes, 8780 quadrilateral shell/plate elements modeling the cores, slabs and slab-band girders, and 3310 beam elements modeling the columns, coupling and lintel beams and auxiliary elements. Boundary conditions consisted of 6 degrees of freedom permitted at all nodes except at the ground level perimeter nodes where the horizontal translations were restrained, and at the base level where all translations and rotations were restrained.

Nonlinear analyses of the core-slab substructure [4] (Figure 2), were performed using the NONLACS {NON Linear Analysis of Concrete and Steel structures) computer program [7]. A portion of the surrounding slabs extending to one-half of the panels around the core sections perimeter, is incorporated in the model due to the limitation of the number of elements imposed by the NONLACS program. Constitutive material properties include strain-softening, crushing, cracking, tension-stiffening and biaxial stress responses of the concrete, and yielding and strain-hardening of the steel reinforcement. Quadrilateral facet shell plate elements are used, each divided into 8 layers across the concrete thickness. Uniformly distibuted vertical and horizontal reinforcement in the cores, top and bottom reinforcement in the slabs, basket and shear-torsion transverse hoops and longitudinal bars in the coupling and lintel beams are modeled as smeared steel layers. Concentrated reinforcement at the core wall end and corner regions, diagonal reinforcement in the coupling beams, and top and bottom flexural reinforcement in the lintel beams are modeled as bundled bars. The model is characterized by 2238 nodes, 2186 shell elements with 6 concrete layer systems, 20 smeared reinforcing steel layer systems and 18 steel reinforcement bar element systems. Appropriate boundary conditions are imposed at the cut slab edges by the addition of boundary elements as springs incorporating the vertical deflections Δ_z and rotations θ_x and θ_y to anticipated values at the ultimate load, taken as 3.5 times the values obtained from the uncracked elastic analysis.

Several loading combinations for serviceability and ultimate limit states were considered according to the CSA Standard A23.3-M94 [8]. For the elastic analyses, the equivalent static earthquake forces Q_X and Q_Y, in the principal X- and Y-directions of the structure, respectively, are determined from the base shear based on a force modification factor of $R = 3.5$ for reinforced concrete ductile coupled flexural core wall and frame lateral load resisting systems. For the nonlinear analyses, the earthquake forces on the core-slab substructure were determined from the elastic interactive shear force distributions in the core substructure. The gravity loading was applied first, then the earthquake forces were applied as monotonically increasing loads in 30 load steps by incrementing the value of α_Q from 0.1 to values of 3.0 and 5.0 for the Q_X and Q_Y earthquakes, respectively, to obtain an accurate determination of the ultimate load and a well defined load-deflection response.

DESIGN PROCESS

Current concrete codes lack appropriate procedures for design of cores of open-, closed- and partially closed-sections, slabs enclosed within and surrounding cores, coupling and lintel beams, and core wall-beam-slab connections. For the core-slab-frame structure, design of the various components [5], takes into consideration the three-dimensional building response and is implemented following Clause 21, Special Provisions for Seismic Design of the CSA Standard A23.3-M94 [8]. Assumptions and limitations underlying the conventional design procedures and code clauses are reviewed.

In areas where the CSA Standard is not applicable, the ACI Standard 318M-95 [9] Building Code Requirements for Reinforced Concrete, the New Zealand Standard NZS3101:1982 [10] Code of Practice for the Design of Concrete Structures and the available research data are incorporated in the design and detailing of the steel reinforcement.

For the design of cores, the presence of coupling and lintel beams combined with enclosed and surrounding slabs, alter the core response to that of a partially closed- or closed-section. Concentrated reinforcement regions in cores at wall ends, corners and junctions, confined compression wall regions, and regions adjacent to beam connections, must be designed as equivalent columns using hoops and cross-ties stabilizing the longitudinal reinforcement and confining the concrete to preserve structural integrity at higher load levels. Design recommendations for cores are developed relating to the dimensional requirements, plastic hinging regions, effective core wall and slab widths for lateral load resistance, confined core wall-beam-slab regions, and concentrated and uniformly distributed steel reinforcement.

For the design of two-way slabs enclosed within and surrounding cores, joining to frame substructures, these slabs do participate in the lateral load resistance along with the core flange and web walls. Design recommendations for two way slabs are provided relating to the flexural steel reinforcement for spacings within and outside the plastic hinging regions, ductility requirements for the maximum tension reinforcement ratio and limits of the average negative to positive ultimate moment ratios throughout the slab, and the ultimate compressive strain in the concrete. Design and detailing of the slab-core wall connections, and corner and end regions, is examined for the maximum slab reinforcement ratio and spacings, serviceability criteria and confinement of the concrete.

The coupling beams are designed using diagonal reinforcement to maintain their strength and develop plastic hinges while being subjected to lateral load reversals. Analysis results demonstrate a coupled core substructure response, showing that portions of the slabs participate in the lateral load resistance resulting in a T-shaped coupling beam behaviour. Design recommendations for coupling beams are provided relating to the development length and hoop stabilization of the diagonal reinforcement, and basket reinforcement.

The lintel beams do not satisfy the CSA Standard dimensional requirements to qualify as ductile lateral load resisting elements. However, the elevator core response shows that the lintel beams participate in the lateral load resistance, developing significant flexural and torsional actions, and must be designed as ductile members. Portions of the enclosed and the surrounding slabs participate in the lateral load resistance, resulting in an L-shaped lintel beam response. Design recommendations for lintel beams are presented relating to the flexural, shear, torsional and skin reinforcement, and stability and confinement requirements.

Two types of deep joints are encountered in the core substructure, the core wall-coupling beam joints and the elevator flange wall end-lintel beam joints. Detailing of these joints is presented ensuring strong wall-weak beam ductility requirements.

CORE-SLAB-FRAME STRUCTURE ELASTIC RESPONSE

Elastic analysis results of the three-dimensional core-slab-frame structure [3], demonstrate that for the core sections, the bending moment distributions due to earthquake (Figure 3) are largest in value in the lower 20% to 25% of the core height, with moment reversals occurring in the bottom two storeys. The largest moments are in the ground storey, showing very irregular force distributions with local concentrations of about 2 to 4 times larger than the average values, basically due to shear lag effects at the slab-core wall corners and ends. The earthquake effects are as significant as the gravity load effects, and the shear and torsional effects are comparable with the axial and flexural actions. Hence, the critical region for seismic design of cores is in the lower 10\% to 25\% of the core height. For the enclosed and surrounding slabs (E-S-Slabs), the bending and twisting moment distributions due to the earthquake and dead loads (Figure 4 for level 1) demonstrate that the slabs experience load reversals and are subjected to considerable twisting moments. Values of the moments due to dead loads vary between 0.5 to 5 times the correponding values due to earthquake. The resulting slab forces due to the earthquakes are largest in the lower 10\% of the structure height with local increases of 8 to 10 times the average values and significant reversals occurring in the bottom two storeys. Bending and twisting moment distributions display concentrations of 3 times the average values at the stiffer slab-core wall corner and end regions. These irregular slab forces are caused by the complex slab support conditions and are located within a slab band width of approximately 3 to 5 times the core wall thickness adjacent to the slab-core wall junctions around the core perimeter. More importantly, the results show clearly that slabs enclosed within and surrounding the cores do participate in the lateral load resistance, transferring out-of-plane forces between the core and frame substructures.

For the coupling beams, maximum shear forces and bending moments at the beam ends occur between the one-third and the mid-height of the structure. The earthquake effects are greater than those due to the gravity loads. Bending moments due to Q_Y earthquake in the C-E coupling beams (Figure 5), are up to 8 times larger than those due to dead loads, which are as much as 20 times the forces due to Q_X earthquake. For the lintel beams, the bending and twisting moment distributions along the structure height show that the forces due to Q_Y earthquake are 3 to 6 times larger than those due to Q_X earthquake. The resulting forces due to dead loads are 2 to 4 times larger than those due to earthquake. Analysis results demonstrate that as a result of the three-dimensional structure response, the governing lateral forces for design are not necessarily in the direction of the beam span.

CORE SUBSTRUCTURE NONLINEAR RESPONSE

The lateral load-top deflection response of the core-slab substucture is presented in Figure 6, for the DQX and DQY models idealizing the three-dimensional coupled core substructure considering the coupling and lintel beams, the membrane, flexural and torsional actions of the enclosed and surrounding slabs for loading conditions of [1.0D + 0.5L + α_Q\,Q_X] and [1.0D + 0.5L + α_Q\,Q_Y]. The curves are divided into four regions. Region 1 is the linear elastic portion of the response up to initial cracking of the concrete. Region 2 represents an almost linear response where the steel reinforcement is in the elastic range and cracking of concrete has occurred. Region 3 is the nonlinear portion of the response where yielding of the steel reinforcement and further cracking of concrete occurs with tension-stiffening. Region 4 is the final nonlinear portion of the response to failure, with yielding and strain-hardening of the reinforcement, and strain-softening and subsequent crushing of concrete. For the DQX model response, Region 2 extends to load level 1.55Q_X with a 88mm top drift, Region 3 extends to 2.50Q_X with a 494mm top drift and Region 4 predicts a failure load of 2.65Q_X with a top drift of 871mm, providing a deflection ductility ratio of 9.90.

For the DQY model response, Region 2 extends to load level $1.85Q_Y$ with a 56mm top drift, Region 3 extends to $4.45Q_Y$ with a 852mm top drift and Region 4 predicts failure at $4.60Q_Y$ with a 1190mm top drift, providing a deflection ductility ratio of 21.25.

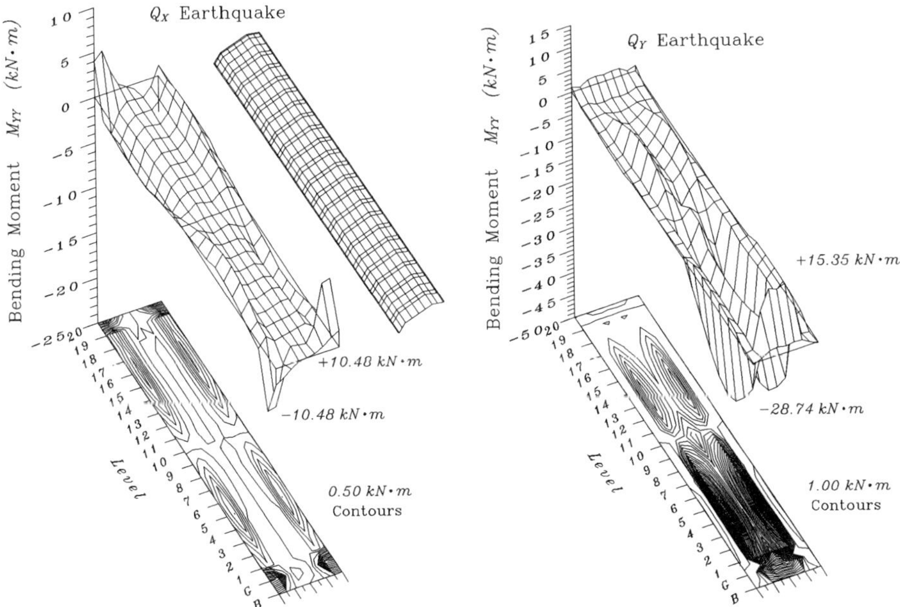

Figure 3 Distribution of elastic moments in the elevator core

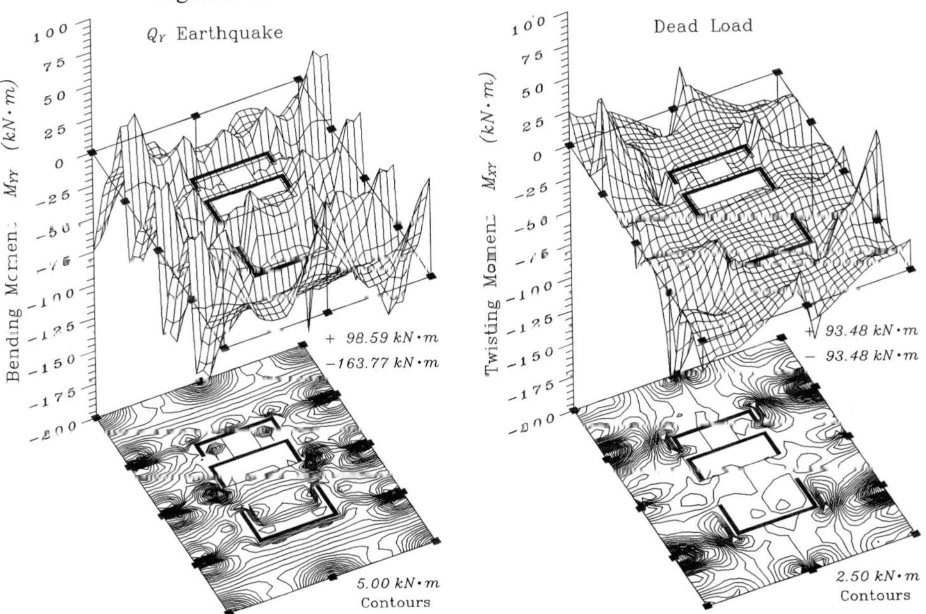

Figure 4 Distribution of elastic moments in the E-S-Slabs

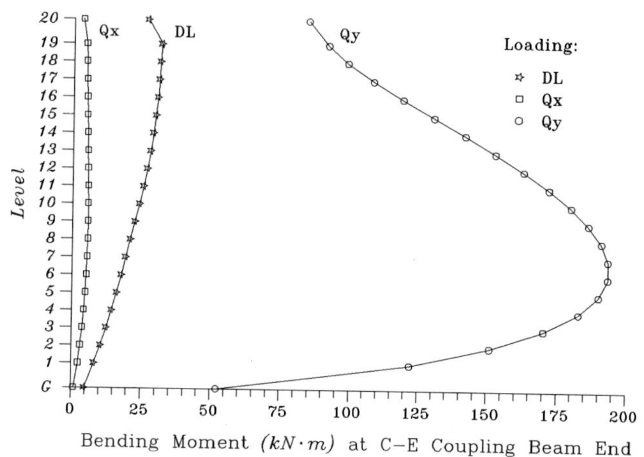

Figure 5 Bending moment distribution along the structure height

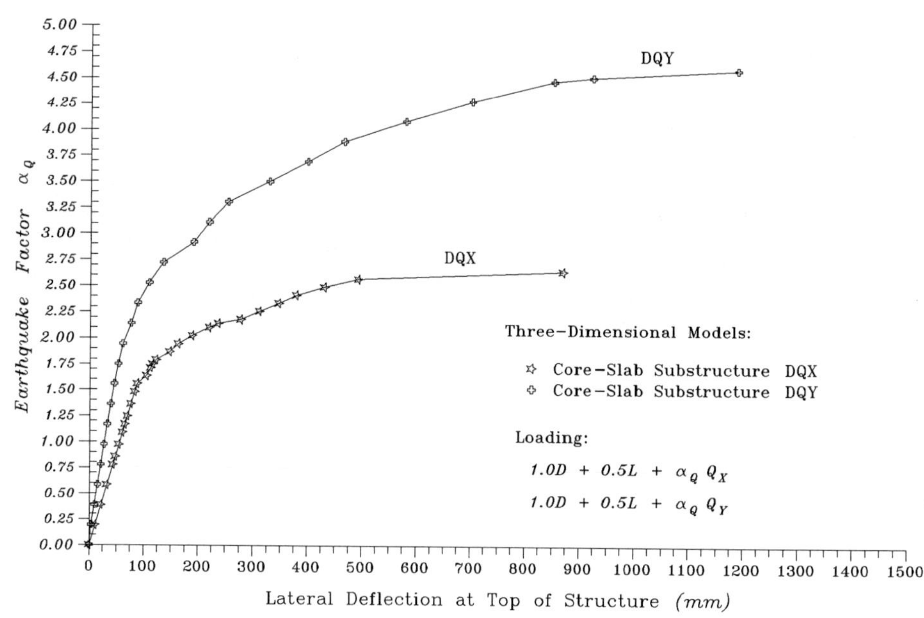

Figure 6 Earthquake vs lateral deflection response to failure

Failure mode of the core-slab substructure, as predicted by the response of the DQX and DQY models, includes concrete crushing in the core section corner wall regions and flange walls due to the high stresses, together with rupture of the concentrated tension steel reinforcement in the cores, attaining steel strains of 12.5% at failure. This causes tearing of the concrete and separation at the slab-core junctions, up to a distance of one-half of the flange wall and one-third of the web wall lengths at the ground level, with an upward deflection of the slab of 50mm to 100mm at the core-slab corners. This leads to the deterioration of the core stiffness and the associated instability problems, resulting in local buckling of the core sections wall corner regions in the lower two storeys. Structural integrity of the lower 10\% of the structure height. must be preserved for inelastic response at the higher load levels.

Contributions of the slab and beam elements to the lateral load response of the substructure, were also examined by eliminating their effects in the DQX and DQY models. The results showed that accounting for the flexural and torsional actions of the surrounding slabs with the cores and at the slab-core wall connections, predicted the same ultimate load but with a 33% decrease in the top drift, the energy absorption capacity and the ductility, as a result of the stiffening of the core substructure. Consideration of the flexural actions of the enclosed and the surrounding slabs, the slab-core wall connections and the lintel beams showed a 20% increase in the lateral stiffness of the substructure in the post-cracking and yielding regions, with a 25% to 65% reduction in the top drift, and a 15% to 20% increase in the ultimate load. As a result of the overall stiffer structural response, due to the consideration of these elements, the ductility and energy absorption capacity of the ubstructure decrease by about 20% to 45% in the X- and Y-directions, respectively.

SUMMARY AND CONCLUSIONS

Research findings [1,2]demonstrate that consideration of the three-dimensional response of core-slab-frame structures subjected to gravity loads and lateral forces, results in a considerably improved lateral and torsional stiffness characteristics with an increased energy absorption capability and ductility until failure. In this case, by factors of about two to four times the stiffness of the aggregation of the planar components. Coupled core substructures demonstrate complex interaction between the various structural elements. Core sections combined with enclosed slabs, coupling and lintel beams, and surrounding slabs, respond as closed- and partially closed-sections. Core flange and web walls and slabs participate in the lateral load resistance. Combined with the fixed conditions at the base region, a 40% increase in the overall stiffness of the structure occurs in the lower 10% to 15% of the height, compared with the planar response of the structure.

Elastic analyses of core-slab-frame structures subjected to gravity load and lateral forces are not representative of the actual behaviour of the structure, which becomes inelastic in the higher load range. Nonlinear response of the structure due to lateral forces shows that concrete cracking, yielding of steel reinforcement and the plastic hinging regions can extend over the lower 20% to 25% height of the structure. The majority of the inelastic action and damage due to lateral forces is concentrated in the lower 10% to 15% of the structure height. Near the ultimate load a considerable reduction of the stiffness of the core sections and slabs occurs, especially at the slab-core wall-beam connections, with the stiffness of the core walls and slabs decreasing to values of 15% to 25% and 40%, respectively, of the uncracked stiffness values at the base region where plastic hinging occurs [2, 11, 12, 13].

Analysis and design methods must take into consideration the three-dimensional behaviour of the structure, at the higher load levels to ensure ductility and structural integrity for appropriate energy dissipation characteristics, with careful detailing for these elements and regions in the bottom storeys.

ACKNOWLEDGEMENTS

The authors would like to acknowledge the support of the Natural Sciences and Engineering Research Council of Canada to the continuing research program at McGill University, the staff members at the Computing Centre at McGill University for their help and very generous grant of computer time and storage facilities to conduct all of the analyses reported in this paper, and Professor A.G. Razaqpurof the Department of Civil Engineering, Carleton University, Ottawa, Ontario, for providing the NONLACS computer program used in the nonlinear analyses. Special thanks are extended to Miss Palmina D'Alelio for her encouragement.

REFERENCES

1. MANATAKOS, K. Analysis of low- and medium-rise buildings. M.Eng. Thesis, 1989, McGill University, Montreal, Quebec, Canada. 1050pp.

2. MANATAKOS, K. Behaviour and design of core-slab-frame structures. Ph.\,D. Thesis, 1997, McGill University, Montreal, Quebec, Canada. 355pp.

3. MANATAKOS, K and MIRZA, M S. Elastic behaviour of reinforced concrete core-slab-frame structures. Departmental Report No.95-4, 1995,a, McGill University, Montreal, Quebec, Canada. 260~pp.

4. MANATAKOS, K and MIRZA, M S. Nonlinear behaviour of reinforced concrete core-slab-frame structures. Departmental Report No.~95-5, 1995\,b, McGill University, Montreal, Quebec, Canada. 460~pp.

5. MANATAKOS, K and MIRZA, M S. Design of reinforced concrete core-slab-frame structures. Departmental Report No.~95-6, 1995\,c, McGill University, Montreal, Quebec, Canada. 240~pp.

6. BATHE, K J, WILSON, E L and PETERSON, F E. A structural analysis program for static and dynamic response of linear systems. National Science Foundation EERC Report No. 73-11, 1973, University of California, Berkeley, California, USA. 122~pp.

7. RAZAQPUR, A G and NOFAL, M E. Transverse load distribution at ultimate limit states in single span slab-on-girder bridges with compact steel sections. Research and Development Branch, Report No. MISC-88-01, 1988, Ontario Ministry of Transportation, Toronto, Ontario, Canada. 281pp.

8. CANADIAN STANDARDS ASSOCIATION. Design of concrete structures with explanatory notes CAN3-A23.3-M94, 1995, Canadian Standards Association, Rexdale, Ontario, Canada. 220~pp.

9. AMERICAN CONCRETE INSTITUTE COMMITTEE~318. Building code requirements for structural concrete ACI~318-95 and commentary ACI~318R-95, 1995, American Concrete Institute, Farmington Hills, Michigan, USA.369~pp.

10. NEW ZEALAND STANDARD NZS~3101. Code of practice and commentary for the design of concrete structures NZS~3101:82, Parts 1 and 2, 1982, Standards Association of New Zealand, Wellington, New Zealand. 283~pp.

11. PARK, R and PAULAY, T. Reinforced concrete structures. John Wiley and Sons, Inc., 1975, New York, New York, USA. 769~pp.

12. PARK, R and GAMBLE, W L. Reinforced concrete slabs, John Wiley and Sons, Inc., 1980, New York, New York, USA. 618~pp.

13. PAULAY, T and PRIESTLEY, M J N. Seismic design of reinforced concrete and masonry buildings. John Wiley and Sons, Inc., 1992, New York, New York, USA. 744~pp.

CREEP EFFECTS IN SLENDER, BIAXIALLY LOADED STRUCTURAL CONCRETE COLUMNS

N C Mickleborough

K W Lee

Hong Kong University of Science and Technology

Hong Kong

ABSTRACT. The principal objective of this work is to present a study into the time dependent behavior of slender structural concrete columns subjected to biaxial loading. A non-linear procedure is developed and described for the analysis of these columns where both the material and geometric non-linearities are considered. The solution is based on an iterative procedure to predict this behaviour. Throughout the analysis, bending of the column is assumed to occur about the principal axis of the cross-section and these sections are analysed using the age-adjusted effective modulus method to include the effects of creep and shrinkage. The time scale is divided into increments to permit a study of the structural behaviour, including the on-set of cracking, as lateral deflections increase with time. To consider the validity of this analytical behaviour, an experimental investigation has been conducted and reference made to previously published data. The experimental investigation described involves the testing of columns subjected to varying degrees of biaxial loading. The axial force and initial eccentricities about the two axes are the major variables considered. The analytical model and analysis show good correlation with the experimental results obtained in the study.

Keywords: Biaxial, Slender column, Cracking, Creep properties, Serviceability, Shrinkage, Slenderness ratio, Structural concrete, Buckling.

Dr Neil C Mickleborough is an Associate Professor in Civil Engineering at the Hong Kong University of Science and Technology and has an active research interest in structural dynamics and serviceability behaviour of tall buildings.

Mr Ka Wai Lee is a post graduate student in Civil Engineering at the Hong Kong University of Science and Technology and is currently studying the effects of creep, shrinkage and temperature effects in the behaviour of slender structural concrete columns.

INTRODUCTION

Slender concrete columns are becoming increasingly popular in structural design because of the advanced development of high strength concrete materials and the economical advantages of the smaller cross sectional dimensions of these column. The time dependent behaviour of slender concrete members under sustained loads is often not well understood in structural design and oversimplifying assumptions are often made.

In most structural members, creep and shrinkage of concrete cause increases of deformation and redistribution of stresses, but do not adversely affect strength. In some situations, however, the deformations caused by creep and shrinkage can lead to an increase in the internal actions in the structure and therefore, a reduction in strength carrying capacity. A slender column carrying a sustained, eccentric, compressive load is such an example.

To predict these deformations due to creep and shrinkage accurately, models for the creep and shrinkage of concrete are required. Details of creep models have been developed for many years [1-3].

Normally, when a concrete column is subjected to load, its response is both immediate and time-dependent. Under sustained load this deformation of the member gradually increases with time, and may eventually be many times greater than its instantaneous value, because of the P-Δ effect. Consider a slender, pin-ended column subjected to a compressive force P applied at a constant initial eccentricity about the principle axis, e_{ox}, e_{oy}, the instantaneous deflections of the column under this load P are δ_{ix} and δ_{iy}. For long and slender columns, the secondary moments ($P\delta_{ix}$ and $P\delta_{iy}$) may be many times greater than the initial primary moments Pe_{ox} and Pe_{oy}, and these moments in turn course further deformations. The load carrying capacity may be considerably less than that of a short-column with the same cross-section. For very long columns, an instability failure may occur under relatively small compressive loads. Buckling may initiate a failure prior to the strength of the cross-section being reached.

To more accurately predict these deformation models, the creep and shrinkage characteristics of concrete are required and analytical procedures using these models must be developed. In the analysis presented, an iterative computer based procedure for the analysis of slender concrete columns subjected to uniaxial and biaxial bending is described. Analytical predictions are compared with measurements from test data for pin-ended columns. The study of uniaxially loaded slender columns has been previously conducted and reported [4].

METHOD OF SLENDER COLUMN ANALYSIS

Overview

The method of analysis of slender columns presented uses an iterative technique to account for the secondary effects of bending due to the non-linear behaviour of load and lateral deflection. To determine the effects of loading, both cross-section analysis and column behaviour need to be considered. With time dependent effects both short-term and long-term analyses are required.

Short Term Analysis

Considering a pin-ended column in Figure 1, a concentrated load P is applied with an eccentricity of e_{ox} and e_{oy} to the x-axis and y-axis respectively. The initial primary moments acting on the column at the time of first loading τ_o are Pe_{ox} and Pe_{oy}.

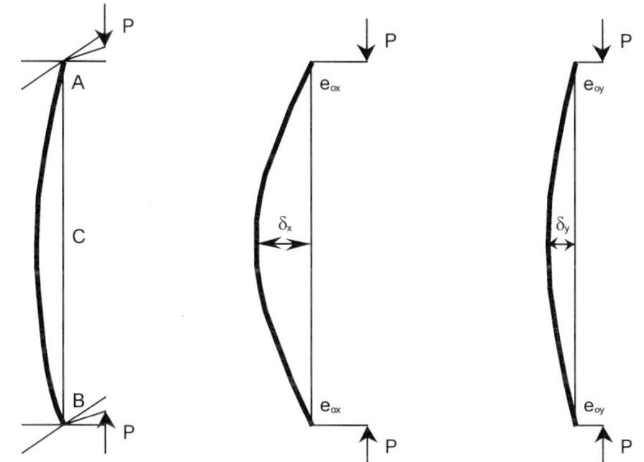

(a) Biaxially loaded column (b) x-direction configurations (c) y-direction configuration

Figure 1 Configurations of column under biaxial bending

The strain distribution and curvature at sections along the length of the column caused by the external load P can be determined from basic mechanics. The secondary effects of the axial force P acting on the column require consideration in the iteration procedure to determine the total deflection, which can be achieved by assuming a deflected shape function. A parabolic variation has shown to give a good prediction with only small variations in the prediction for different shape functions [4]. The central deflection in x and y direction can be expressed as a function of curvature at the ends and centre of the column by:

$$\delta_x = \frac{L^2}{96}(\kappa_{Ay} + 10\kappa_{Cy} + \kappa_{By}) \quad (1a)$$

$$\delta_y = \frac{L^2}{96}(\kappa_{Ax} + 10\kappa_{Cx} + \kappa_{Bx}) \quad (1b)$$

Through iteration, if the column does not reach its buckling instability load under initial loading, the total moments are:

$$M_{total,x} = Pe_{oy} + P\delta_{oy} \quad (2a)$$

$$M_{total,y} = Pe_{ox} + P\delta_{ox} \quad (2b)$$

In this analysis the curvature at central section C now needs to be re-analyzed, at the same time, the magnitude of curvature at C will increase, which implies the increment of δ_x and δ_y. This procedure will continue until the lateral displacement at the mid-height converges to δ_{ix}, δ_{iy} after the ith iteration.

To determine a more accurate estimate of behaviour, further sections can be considered along the length of the column and used in the iteration process to determine a deformed shape function.

Time Analysis

Time dependent deformation occurs and changes in cross-section affect the behaviour of the column. The age-adjusted effective modulus method is used to determine the effects of creep and shrinkage at any time on each cross-section.

Linear creep, being load dependent, gradually increases the curvature of section with time. The magnitude of the secondary moment increases with increased lateral deflection, hence causing additional increments of secondary moments.

The analysis presented considers response at successive time intervals with each interval considering an iterative solution procedure to determine the lateral deflection [4]. Iteration continues until, at each time instant, convergence is reached on a lateral deflection. Instability failure is deemed to have occurred when convergence on a lateral deflection cannot be reached.

Cracking of the cross section may occur during the iteration process, at a time instant, and the depth to the neutral axis is adjusted at the end of each iteration when the secondary moment has been determined.

CROSS-SECTIONAL ANALYSIS

Short-term Analysis – Uncracked Sections

The strain distribution at any section can be represented as:

$$\varepsilon=\varepsilon_o+\kappa_x y+\kappa_y x \tag{3}$$

Where κ_x and κ_y are curvatures in the yz and xz plane respectively. The stress distribution of the cross section can be represented as:

$$\sigma=E_c\varepsilon \tag{4}$$

The internal force and moments can be found from:

$$\begin{aligned} N &= \int \sigma dA & \qquad & N = E_c(A\varepsilon_o + B_x\kappa_x + B_y\kappa_y) \\ M_x &= \int \sigma y dA & \text{or} \qquad & M_x = E_c(B_x\varepsilon_o + I_x\kappa_x + I_{xy}\kappa_y) \\ M_y &= \int \sigma x dA & \qquad & M_y = E_c(B_y\varepsilon_o + I_{xy}\kappa_x + I_y\kappa_y) \end{aligned} \tag{5a}$$

Where A, B, I, I_{xy} are area, first moment of area, second moment of area and product of moment inertia respectively. These section properties are determined from the transformed section.

Equation (5) can be expressed in matrix form:

$$\begin{bmatrix} \varepsilon_o \\ \kappa_x \\ \kappa_y \end{bmatrix} = \frac{1}{E_c} \begin{bmatrix} A & B_x & B_y \\ B_x & I_x & I_{xy} \\ B_y & I_{xy} & I_y \end{bmatrix}^{-1} \begin{bmatrix} N \\ M_x \\ M_y \end{bmatrix} \tag{5b}$$

For any applied moment and axial force, the short-term deformation can be determined from the relationships expressed in Equation (5a) or (5b). N, M_x and M_y are the axial force and applied moment at the section.

Short-Term Analysis – Cracked Sections

For a combined bending and axial force (M_x, M_y and N) sufficient to cause cracking, the initial strain can be expressed as given in Equation (3).

The uncracked concrete stress is acting on the area above the line shown in Figure 2,

$$\sigma = E_c(\varepsilon_o + \kappa_x y + \kappa_y x) \tag{6a}$$

For the section area subjected to cracking which occurs below the line shown in this figure,

$$\sigma = 0 \tag{6b}$$

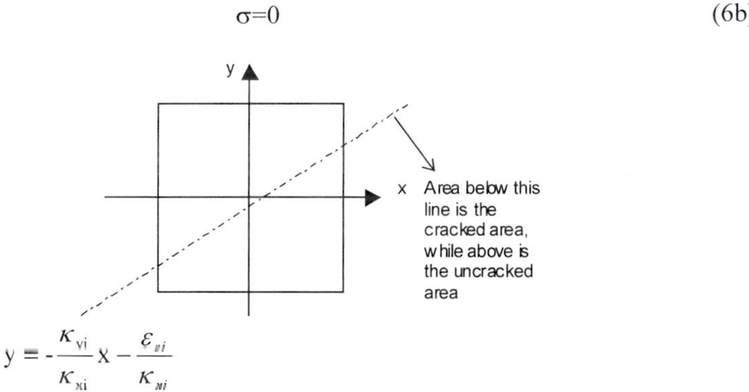

Figure 2 The corresponding cracked and uncracked area of cross-section

The position of the neutral axis in axial force and biaxial bending is given by putting ε in Equation (3) equal to zero,

$$0 = \varepsilon_o + \kappa_x y + \kappa_y x \tag{7}$$

The steel stress in the kth level of reinforcement can be expressed as:

$$\sigma_{sk} = E_s(\varepsilon_o + \kappa_x d_{sky} + \kappa_y d_{skx}) \tag{8}$$

From basic equilibrium, the axial force N is equal to the integral of the stress distribution over the uncracked area of section.

$$N= \iint \sigma dA + \sum_{k=1}^{m} \sigma_{sk} A_{sk} + \sum_{k=1}^{m} \sigma_{ck} A_{sk} u_k(t) \quad (9a)$$

$u_k(t)=-1$ when $t=1$ if the steel bar is located in the compression zone and $u_k(t)=0$ when $t=0$ if the steel bar is located in the cracked zone.

The resultant moment is:

$$M_x= \iint \sigma y dA + \sum_{k=1}^{m} \sigma_{sk} A_{sk} d_{sky} + \sum_{k=1}^{m} \sigma_{ck} A_{sk} d_{sky} u_k(t) \quad (9b)$$

$$M_y= \iint \sigma x dA + \sum_{k=1}^{m} \sigma_{sk} A_{sk} d_{skx} + \sum_{k=1}^{m} \sigma_{ck} A_{sk} d_{skx} u_k(t) \quad (9c)$$

Using Equations (9a), (9b) and (9c) the neutral axis can be determined numerically. Hence, the short term values of ε_o, κ_x and κ_y under initial loading can be calculated.

TIME ANALYSES

As described for uniaxial loading of columns and for flexural members [5], the time dependent analysis for a cross section to determine the change of strain at any particular time period can successfully be determined by relaxation solution procedure proposed by Bresler and Selna [6].

At any time interval that strain can be expressed as:

$$\varepsilon_{total} = \varepsilon_{cr} + \varepsilon_{sh} + \varepsilon_{el} \quad (10)$$

During the time interval, the strain condition is assumed to be unchanged. With ε_{total} constant, and $\varepsilon_{cr} + \varepsilon_{sh}$ changing, then ε_{el} must change by an equal and opposite amount. As the instantaneous strain changes, so too does the concrete stress. The stress on the cross section is therefore allowed to vary freely due to relaxation. The procedure effectively allows the internal actions to change, however the equilibrium is therefore not maintained. To restore equilibrium, an axial force ΔN and bending moments ΔM_x and ΔM_y must be applied to the section. When ΔN, ΔM_x and ΔM_y are applied to the section, the restraining actions are removed and equilibrium is restored. Creep and shrinkage may then be considered to be artificially introduced by external loads ΔN, ΔM_x and ΔM_y.

The restraining forces ΔN, ΔM_x and ΔM_y can be calculated. If creep were not restrained in anyway, the strain and curvature would increase to $\phi(t,\tau_o)\varepsilon_{oi}$ and $\phi(t,\tau_o)\kappa_i$, respectively, during the time interval $(t-\tau_o)$. The restraining forces required to prevent this deformation are obtained from Equation (5a).

$$\begin{aligned} -\Delta N_{cr} &= -\bar{E}_e\phi(A_c\varepsilon_o + B_{c,x}\kappa_x + B_{c,y}\kappa_y) \\ -\Delta M_{cr,x} &= -\bar{E}_e\phi(B_{c,x}\varepsilon_o + I_{c,x}\kappa_x + I_{c,xy}\kappa_y) \\ -\Delta M_{cr,y} &= -\bar{E}_e\phi(B_{c,y}\varepsilon_o + I_{c,xy}\kappa_x + I_{c,y}\kappa_y) \end{aligned} \quad (11a)$$

Since only the concrete is subjected to creep, the properties of A_c, B_c, I_c, $I_{c,xy}$ refer the concrete component of the section (ignoring the steel).

Similarly, if shrinkage is uniform over the depth of the section and completely unrestrained, the shrinkage induced strain which develops during the time interval $(t-\tau_o)$ is $\varepsilon_{sh}(t-\tau_o)$. The restraining forces required to prevent this uniform deformation are:

$$\begin{aligned} -\Delta N_{sh} &= -\bar{E}_e\,\varepsilon_{sh}A_c \\ -\Delta M_{sh,x} &= -\bar{E}_e\,\varepsilon_{sh}B_{c,x} \\ -\Delta M_{sh,y} &= -\bar{E}_e\,\varepsilon_{sh}B_{c,y} \end{aligned} \quad (11b)$$

The total restraining forces are the sum of the creep and shrinkage components. By substituting the total restraining forces to the equations shown below, the change of the strain distribution with time is established.

$$\begin{bmatrix} \Delta\varepsilon_o \\ \Delta\kappa_x \\ \Delta\kappa_y \end{bmatrix} = \frac{1}{\bar{E}_e} \begin{bmatrix} \bar{A}_e & \bar{B}_{e,x} & \bar{B}_{e,y} \\ \bar{B}_{e,x} & \bar{I}_{e,x} & \bar{I}_{e,xy} \\ \bar{B}_{e,y} & \bar{I}_{e,xy} & \bar{I}_{e,y} \end{bmatrix}^{-1} \begin{bmatrix} \Delta N_{cr+sh} \\ \Delta M_{cr+sh,\,x} \\ \Delta M_{cr+sh,\,y} \end{bmatrix} \quad (12)$$

Where $\bar{A}_e$ is the area of age adjusted transformed section, $\bar{B}_e$ $\bar{I}_e$ and $\bar{I}_{e,xy}$ are the first, second and product moment of area of the age-adjusted transformed section. The properties of the age adjusted transformed section are calculated using the age adjusted modular ratios $\bar{n}_s = E_s/\bar{E}_e$. $\bar{E}_e$ is the age-adjusted effective modulus [7] defined as:

$$\bar{E}_e = \frac{E_c}{1 + \chi\Delta\phi} \quad (13)$$

The term $\Delta\phi$ refers to the increment of the time-dependent concrete creep coefficient associated with the time interval under consideration and χ is the aging coefficient for concrete during the same interval.

ANALYTICAL INVESTIGATION

Previously reported pin ended reinforced concrete tests [8] have been analysed by the biaxial model and is reported.

Consider the column is 6000 mm long, with 150mm square cross section containing four 12mm diameter reinforcing bars in two layers.

A_{s1}=220mm^2 at d_{s1} = 26mm; A_{s2}=220mm^2 at d_{s2} = 124mm
E_s=200000MPa, E_c=24000MPa, f_c'=27MPa
$\phi(40)=1.45$, $\varepsilon_{sh}(40)=0.00018$, $\chi(40)=0.86$

Values of e_{ox}=30mm and e_{oy}=15mm with P=58kN are considered

First loading, the calculated lateral deflection of the column at central section is δ_x=7.9mm and δ_y=6.6mm. This is compatible for values determined for the uniaxial case. Finally, the total deflection including the time dependent deformation is determined as $\Delta_{total}=\delta_x+\Delta_x$=19.5mm and $\Delta_{total}=\delta_y+\Delta_y$=6.5mm respectively. Figure 3 represents the stress and strain resultants of the loaded column at mid-height.

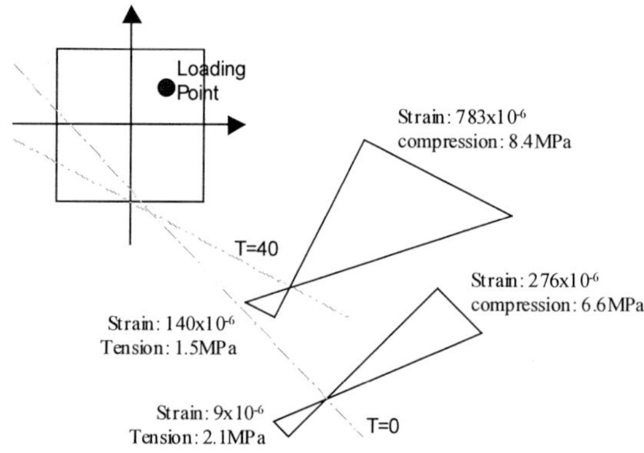

Figure 3 Stress Strain Relationship at the middle section at time T=0 and T=40

EXPERIMENTAL PROGRAMME

Four pin-ended short reinforced concrete columns have been subjected to combined compression and biaxial bending. Each column was 1000mm long, pin-ended, with 120mm square cross section containing four 8mm diameter reinforcing bars in two layers, details can refer to Figure 4.

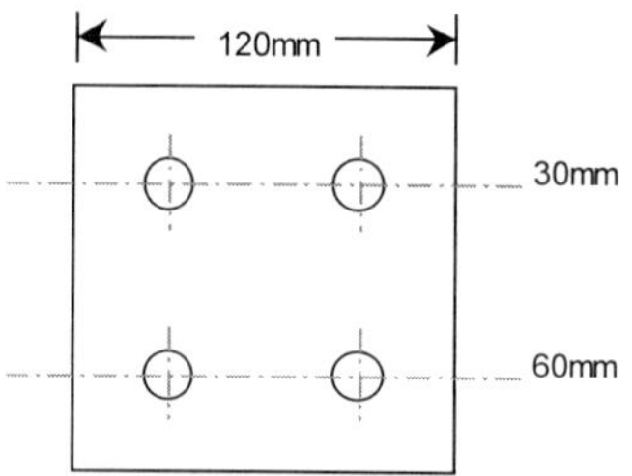

Figure 4 Column cross section used in experiment

The columns were all manufactured from the same concrete mix, with measured material properties as follows. Figures 5, 6 and 7 show photographic details of the column tests.

E_s=227×10^3 MPa; Ec=14×10^{3}MPa
ε_{sh}(30)=0.00015; ϕ(30)=1.5

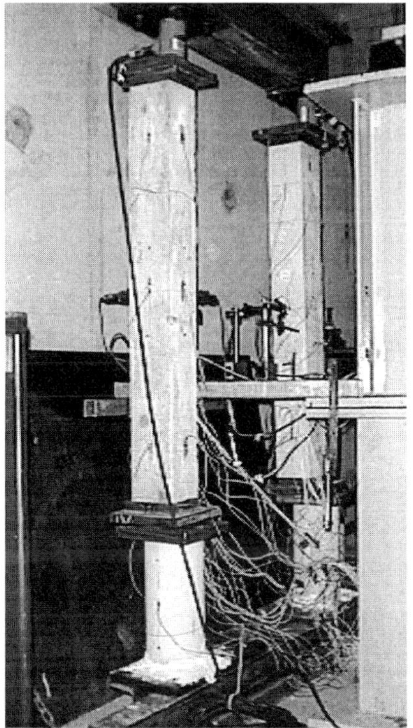

Figure 5 Test columns

Figure 6 Strain gauge at the concrete surface

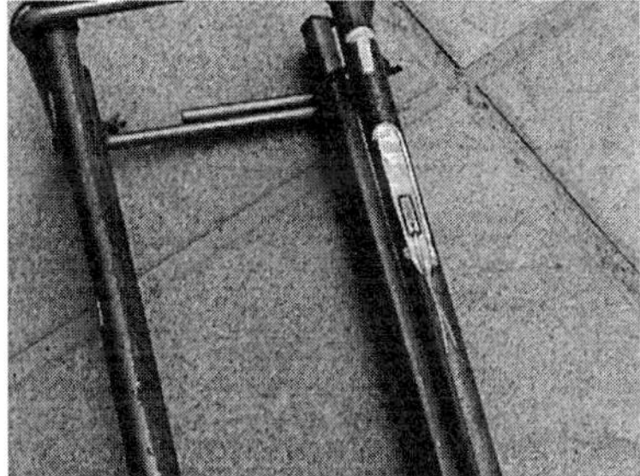

Figure 7 Strain gauge at the reinforcement

The initial eccentricities e_{ox}, e_{oy} and the sustained compressive load P are different for each column and are outlined below, together with the measured and predicted midlength deflections at first loading and at 30 days for the comparison.

CONCLUSIONS

The procedure for the prediction of time dependent behaviour of slender structural concrete columns under biaxial loading has been developed. The method uses the age adjusted effective modulus to determine sectional properties, which in turn are used in an iterative procedure to calculate the structural deformations. The method has been shown to be accurate in the limiting cases of uniaxial loading for previous published slender column deflections. Experimental work on time dependent behaviour of short concrete columns has been conducted and these results indicate that the model also gives good predictions on the time dependent effects under biaxial loading.

Table 1 Summary of test result compare with the prediction from the model

COLUMN NO	TEST 1a		TEST 1b		TEST 2a		TEST 2b	
e_{ox} (mm)	30		10		15		30	
e_{oy} (mm)	40		15		40		30	
P (kN)	60		75		65		75	
δ_i (mm)	δ_x	δ_y	δ_x	δ_y	δ_x	δ_y	δ_x	δ_y
1. Measured	1.03	1.31	0.42	0.51	0.48	1.34	1.05	1.01
2. Predicted	0.95	1.26	0.38	0.57	0.48	1.28	1.16	1.16
3. Meas/Pred	1.08	1.04	1.10	0.89	1.00	1.05	0.91	0.87
$\delta_i + \Delta\delta(30)$ (mm)	x	Y	x	y	X	y	x	y
1. Measured	1.94	2.74	0.9	1.32	1.55	4.82	3.59	3.44
2. Predicted	1.95	2.59	0.79	1.18	1.65	4.42	3.39	3.39
3. Meas/Pred	0.99	1.06	1.13	1.12	0.94	1.09	1.06	1.02

REFERENCES

1. GILBERT, R I., Time effects in Concrete Structures, Elsevier Science Publishers, Amsterdam, 1988, 321 pp.

2. NEVILLE, A M., DILGER, W H, and BROOKS, J J. Creep of plain and structural concrete, Construction Press (Longman Group Ltd.), 1983, 361 pp.

3. GHALI, A, and FAVRE, R. Concrete Structures: Stress and Deformations, Chapman and Hall, London, 1994, 444 pp.

4. GILBERT, R I, and MICKLEBOROUGH, N C. Creep effects in slender reinforced and prestressed concrete columns, ACI Special Publication, SP-129, No. 5., 1991, pp. 77-100

5. GILBERT, R I, and MICKLEBOROUGH, N C. Design of prestressed concrete, Unwin Hyman, London, 1990, 504 pp.

6. BRESLER, B, and SELNA, L. Analysis of time dependent behaviour of reinforced concrete structures, symposium on creep of concrete, ACI special publication SP-9, No.5, Mar 1964, pp.115-128.

7. BAZANT, Z P. Prediction of concrete creep effects using age-adjusted effective modulus method, ACI Journal, 69, April 1972, pp. 212-217

8. MICKLEBOROUGH, N C, and GILBERT, R I. Creep buckling of uniaxially loaded reinforced concrete columns, ACI Special Publication, SP-129, No.3, 1991, pp39-53

PROMOTED INVESTIGATION ON REVISION OF SEISMIC DIVISION INTO DISTRICT TERRAIN ON THE BASE OF MODELLING IN THE SLIDING MODE OF RESULTS SENSITIVE SINGLE-LINE CHECKPOINT MEASUREMENT

V Mkrttchian **N Krmoyan**

State Engineering University of Armenia

R Sinanyan

Yerevan Institute of Architecture and Construction

Armenia

ABSTRACT. The State Engineering University of Armenia (SEUA) and the Yerevan Institute of Architecture and Construction (YIAC) have done research on anti-seismic construction. The measurements done upon the Spitak fault generated novel solutions for estimating the magnitudinal limits of the forthcoming, which is directly associated with planning the anti-seismic construction and regional development. Terrain deformation is presumed to start prior to the actual earthquake, therefore, the sensors established along the fault extension will help determine the modified pattern of seismic activity in anticipation of the tectonic action. The network of sensors will provide integrated data on the origin length and magnitude of the future earthquake. The database and the electronic map of seismic hazard for Yerevan City will be used in combination with the sliding-mode techniques to make determinations of the limits for structure reinforcements and to reduce the expenditure of materials with a relevant improvement of anti-seismic features.

Keywords: Sensitive single-line checkpoint measurements, Terrain deformation, Sliding mode technique, Discontinuous control and setting adjustment (DC&SA), Expenditure of cement.

Professor Vardan Mkrttchian is the President of Armenian Automatic Control Council and Head Professor of Electrical Apparatus Division State Engineering University of Armenia. He specialises in application of Sliding Mode Technique at modelling in anti-seismic areas and in Discontinuous Control and Setting Adjustment (DC&SA).

Mr Norayr Krmoyan is a Ph. D Student in State Engineering University of Armenia. He specialises in databases and the electronic mapping of Yerevan City, including information of the seismic hazards

Dr Rafael Sinanyan works at the Scientific Research Centre of Yerevan Institute of Architecture and Construction (YIAC). He specialises in development and research of high precision means of linear measurements for earthquake prognosis.

INTRODUCTION

With regard to the ongoing urban development, concentration of population in seismically active regions, the problems of developing the methods of reliable earthquake predictions receive the highest priority.

The initial and most important link in estimating the seismic hazard is seismic zonation (SZ). The principle of seismic zonation is based upon two interconnected predictive models - the Model of Zones of Origin (MZO) and the Model of Seismic Effect (MSE). However, the structuring and the physical substance of both models may be non-coincidental with the earlier approaches.

A NEW METHOD OF SEISMIC ZONATION

The method [1] of earthquake prediction and magnitude assessment is based upon 3 premises:

1. The hypothesis of precursors, emerging at the stages of earthquake preparation and creeping. Locations of earthquake preparation and creeping are determined along the fault.

2. G. F. Reid's theory of kickback and its resultant predictive characteristics formulated by A. K. Pevnev. According to this, the locations of earthquake preparation is displayed by exponential curving of the elastic rock formations perpendicular to the fault.

3. The principle of creating the local networks designed for determining the characteristics of the deformation curve. Each cross line will have its local system of orthogonal co-ordinates, to avoid incremental measurement errors and to secure measuring accuracy.

The suggested method will enable a determination to be made of the deformation curve characteristics, yielding the potential original length and the magnitude of the forthcoming earthquake. To be marked upon the adjacent wings of each fault are the geodesic networks in the form of quadrangles, repeated each 10 km along the fault. The marked vertexes of the quadrangles should be fixed at a depth of 1.5 - 2.0 m in the earth of upon rocks using displacement sensors.

Periodic measurements within 1-2 mm accuracy of the quadrangle sides and their diagonals will help determine locations of creeping relative displacements of lithospheric units, if any.

SLIDING MODES IN MODELS OF CHECKPOINT MEASUREMENTS

The sliding mode [2] is a technique used for discontinuous control and setting adjustment (DC&SA) [3] for optimisation model of checkpoint measurements. Non-linear systems of equations in mathematical solutions generate models intended to obtain the object-transitive functions of models using the operator-type forms yielded through consecutive analysis of spatial variables. The models of optimisation on measurements will include transport delays within the system.

The delays prevent HF switching, necessitating improvements to be made to the sliding-mode techniques. Control within sliding mode enables us to decrease the sensitivity to variations of measurements characteristics making them independent upon the environment.

The mentioned problems can be overcome by using asymptotic observers of state and anticipatory devices eliminating delays.

The physical properties of controlled flow does not allow to use the object parameter identification. Developed in this paper is a statistical method based upon the optimal rating[3].

Our purpose is to generate an optimisation model of a model of checkpoint measurements wherein the discontinuity of control in control components would be formulated preceding the stages of selecting the specifications.

Discontinuous-control systems is an efficient tool for solving an entire family of control problems and optimisation problems for dynamic groups of checkpoint measurements. Discontinuity of control results in a discontinuity of the right-hand parts of the differential equations describing the system dynamic properties. Deliberate introduction of measurement into the sliding-mode operation will necessitate a continuous monitoring of the sliding-mode occurrence and stability.

A MEASUREMENT MODEL

It is our purpose to achieve a model suitable for use with all types of circuit simulations. Therefore it should be presented as a simple equivalent circuit. The circuit elements are derived from physical concepts that more often than not get lost whenever a complex model for a packaged device is extracted from an optimisation model.

For the Finite Difference Time Domain computation of the package ports are defined on the coaxial lines as it is used in most simulators by observing wave ratios. The ports for measurement connection are concentrated or internal ports defined by voltage and current The resulting S-parameters of the ports are used to extract equivalent circuit elements leading to well defined values with physical sense.

The essential components of the packaging model are the sliding-mode indicator, observer and the anticipatory device.

As was already noted, the approach used is oriented toward a deliberate introduction of sliding modes over the intersection of surfaces on which the control vector components undergo discontinuity. Realisation of such approach implies the knowledge of the conditions of the occurrence of sliding mode. Designed for this purpose was the indicator of sliding modes. From the point of view of mathematics, the problem may be reduced to that of finding the area of attraction to the manifold of the discontinuity surfaces intersection. On the indicator input two signals x and g from package-mode chip are supplied,

$$g = Clx + C2\ dx/dt \qquad (1)$$

where: C1 ,C2 = constant of checkpoint measurements
x = input parameter of checkpoint measurements
t = time

The indicator compares of those two signals recording the moment of changing the signs of function x and g. Similar signs of functions x and g at the output of the indicator show zero potential (sliding mode does not occur in the package), differing signs on the output shows negative potential (sliding mode does occur).

The state vector of the package is not given. The given data include only certain dimensions of the vector. The problem of identifying the state vector is solved based on data provided by the observer (Equation 2).

$$dy/dt = Ay + Bx - L (z - Ky) \qquad (2)$$

where A, B, L, K are the known matrixes

y = is output parameter of package
z = is the vector composed of linear combinations of the intended and actual parameters of the package

In the Equation (2) the state vector of the system can be directly measured, and the last member of the Equation (2) characterises the gap between the intended and actual parameters of the state vector of the package. The value of the gap will be the solution of the homogeneous differential equation with any desired distribution of roots of the characteristic equation. We can receive components of matrix L for each case by solving the Equation (3) and transforming the differential integrating links. The link of delay in the package is disregarded in the observer model (Equation 2).

The technique of anticipatory compensation is based upon the preliminary supply of input signal with regard to the duration of delay. Considering the delay, the problem of controlling a multi-level package can be presented as follows:

$$dy/dt = Ay + Bx (t - T) \qquad (3)$$

where T is delay.

Integrating the Equation (3) and replacing the integral values by the sum yielded the model of anticipatory device making part of the general model of the checkpoint measurements.

PRACTICAL APPLICATION OF RESULTS

Being currently established in Armenia is a Regional Geo-informational System. The initial stage will include a generation of the electronic map of Yerevan City to the scale of 1: 2000, with a relevant database. With regard to this, operations are in progress on enhancing the accuracy of the map of seismic hazard for Yerevan City. Calculations are being based upon the physical and mathematical models[3].

CONCLUSIONS

The experimental results have shown that using the sliding mode simulation techniques will enhance the accuracy of the existing models. Modelling of measurement having identical sizes and topology will generate a range of device models, such as a sliding-mode indicator, sliding-mode observer, and an anticipating device.

On the basis of the resulting seismic map calculations were made of the limits for reinforcing the structures having differing hazard characteristics.

The calculated results have also been used to design the new buildings and urban structures with 15-20 percent reduction in the expenditure of cement and with an exclusion of pre-fabricated panels.

REFERENCES

1. MOVSESYAN, R A Patent # 470 Republic of Armenia. 1998

2. UTKIN, V I. Sliding Modes in Control and Optimisation. Springer Verlag. Berlin. 1992

3. MKRTTCHIAN, V S. Discontinuous Control and Setting Adjustment (DC&SA) in Non-linear Systems. Hayastan. Yerevan. 1995

RESPONSE OF MICROCONCRETE PANELS TO SIMULATED BLAST LOADING

H Achyutha

K Radhakrishna

S Santhakumar

Indian Institute of Technology Madras

India

ABSTRACT. Microconcrete panels have been subjected to near-field simulated blast pressure loads to study their response. The panels are mounted on a target support facility. A blast pressure pulse is applied at the centre of the panels using a compressed air gun. Three parameters , the thickness of the panels, the percentage reinforcement and the microconcrete mix (strength) are varied. Design of experiment technique is used to determine the parametric values of each of the panels. Thirty one panels with thickness varying from 20 to 40mm, reinforcement varying from 0.16 to 0.64 % and microconcrete strength varying from 11.7 to 47.75N/mm^2 are tested to failure. Response is recorded in the form of displacement-, acceleration- and strain-time histories at the centre of the panels. From the response, load-displacement curves are generated and the strength of the panels at various stages of behaviour is determined.

Keywords: Microconcrete, Panel, Blast pulse, Response, Near-field blast pressure load, Design of Experiment, Dynamic loading.

Professor Dr H Achyutha is Professor of Structural Engineering in the Department of Civil Engineering, Indian Institute of Technology Madras, India. His interests are in the areas of safety of structures against blast and impact loads, experimental techniques and behaviour of masonry infilled frames.

K Radhakrishna is a Research Scholar in the Department of Civil Engineering, Indian Institute of Technology Madras, India, working in the area of blast resistance of structures. He also specialises in ferrocement and concrete structures.

Professor Dr S Santhakumar is a Professor in the Department of Aerospace Engineering, Indian Institute of Technology Madras, India. He specialises in flight dynamics, aerodynamics and instrumentation. He has developed a low cost compressed air missile launcher. He is associated with research institutions of the aviation industry.

INTRODUCTION

Blast resistance of structures is an important criterion for the design of military structures, structures of strategic importance and structures which are potential targets for terrorist attack like nuclear reactor structures [1]. Codes of practice [2] recommend analysis and design procedures which consider the far-field effect of an explosive blast where a whole structure or structural panel is subjected to blast loading. Research is now being carried out to study the near-field effect of blast loads [3]. This paper describes investigations on the response of microconcrete panels to simulated near-field blast pressure loads using a simple experimental facility.

EXPERIMENTAL FACILITY

The experimental facility consists of a compressed air gun [4] and a target support facility shown in Figure 1. Panels of size 1000 x 1000mm can be mounted on the target support facility. The gun is supported on a low level gun carriage and moved into position so that its muzzle is close to the centre of the target panel. The gun is used to simulate near field blast pressure loads at the centre of the panel. The load is varied by varying the gun chamber pressure.

Figure 1 Compressed air gun and target support facility

Tests are carried out with mild steel panels to determine the pressure-time history and the pressure distribution on the panel surface using piezoelectric pressure transducers inserted in the panels at various points [5] . A typical pressure-time history at the centre of the panel with gun chamber pressure at 150psi (1.035 N/mm^2) is shown in Figure 2. From the pressure-time histories at various locations on the panel surface, the pressure distribution on the panel surface is determined. Tyical peak pressure distribution on the panel surface for the blast pulse generated with the gun chamber pressure at 150psi (1.035 N/mm^2) is shown in Figure 3. There is a significant difference in the magnitude of the peak pressure but no significant change in the blast pulse duration when the gun chamber pressure is varied.

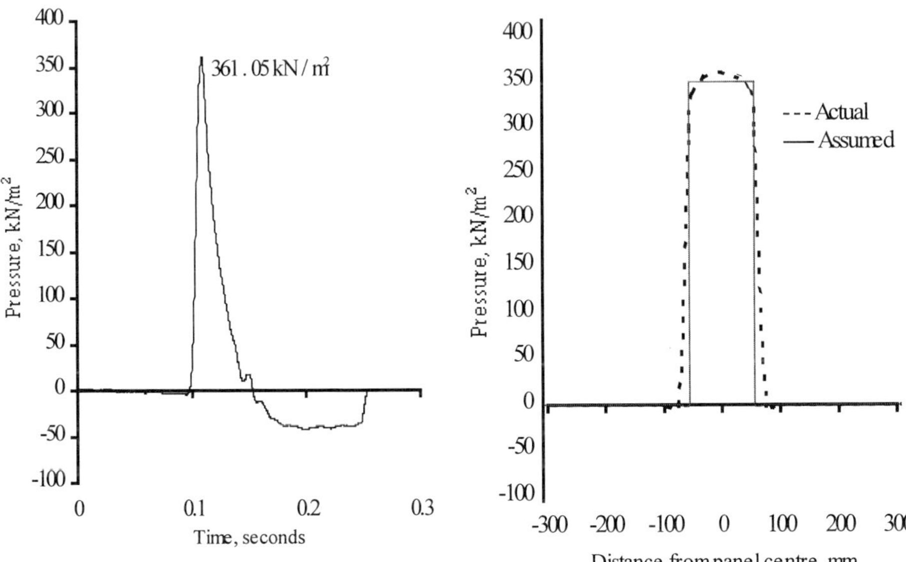

Figure 2 Typical pressure-time history at the centre panelof the panel

Figure 3 Pressure distribution on the surface

INSTRUMENTATION

Instrumentation for studying the reponse of the panels includes inductive displacement transducers, accelerometers and electrical resistance strain gauges. Data is collected by the voltage mode at the rate of 3925 points per second per channel. Data from the various instruments is stored on a computer after passing the signals through an A/D converter. Data analysis is carried out using the DaDISP (Data Analysis and Digital Signal Processing) program.

DESIGN OF EXPERIMENT

Feasibility tests have shown that microconcrete panels of thickness 40mm or less only can be tested to failure using the present experimental facility. Investigations are carried out to study the effect of three parameters, the thickness of the panel (T), percentage reinforcement (S) and microconcrete mix / strength (M) on the response of the panels. The number of panels to be tested and their parametric values are determined in accordance with the design of experiment [6].

A full 3-parameter study is carried out with the magnitude of the parameters at 2 extreme values (levels). From preliminary tests [7], the two extreme values of the parameters are determined as : thickness of 20 and 40mm; reinforcement of 0.16 and 0.64 % and microconcrete mix 1:5 (strength 11.70N/mm^2) and 1:1 (strength 47.75 N/mm^2).

A total of thirty one panels are tested (Table 1).

Table 1 Nomenclature of microconcrete panels

SUCCESSIVE INCREMENTAL BLAST PRESSURE LOADING				
		Reference panels		
$T_{20}S_{0.16}M_{1:5}$	$T_{20}S_{0.64}M_{1:5}$		$T_{40}S_{0.16}M_{1:5}$	$T_{40}S_{0.64}M_{1:5}$
$T_{20}S_{0.16}M_{1:1}$	$T_{20}S_{0.64}M_{1:1}$		$T_{40}S_{0.16}M_{1:1}$	$T_{40}S_{0.64}M_{1:1}$
		Central value panels		
	$T_{30}S_{0.40}M_{1:3}$		$T_{30}S_{0.40}M_{1:3}$	
		Other panels		
$T_{30}S_{0.40}M_{1:5}$	$T_{30}S_{0.16}M_{1:3}$	$T_{20}S_{0.40}M_{1:3}$	$T_{20}S_{0.16}M_{1:3}$	$T_{40}S_{0.16}M_{1:5}$
$T_{30}S_{0.40}M_{1:1}$	$T_{30}S_{0.64}M_{1:3}$	$T_{40}S_{0.40}M_{1:3}$	$T_{20}S_{0.64}M_{1:3}$	$T_{40}S_{0.64}M_{1:5}$
		Panels with chicken mesh reinforcement		
	$T_{20}S_{0.24}M_{1:5}$	$T_{20}S_{0.48}M_{1:5}$	$T_{20}S_{0.72}M_{1:5}$	
		Single shot blast loading		
		Reference panels		
$T_{20}S_{0.16}M_{1:5}$	$T_{20}S_{0.64}M_{1:5}$		$T_{40}S_{0.16}M_{1:5}$	$T_{40}S_{0.64}M_{1:5}$
$T_{20}S_{0.16}M_{1:1}$	$T_{20}S_{0.64}M_{1:1}$		$T_{40}S_{0.16}M_{1:1}$	$T_{40}S_{0.64}M_{1:1}$

TESTING OF PANELS

The microconcrete panels are mounted on the target support facility. The natural frequency of each panel is determined by a hammer test.

In the successive incremental blast pressure loading tests (Table 1), each panel is first subjected to a peak blast pressure load of 432N corresponding to a gun chamber pressure of 20psi (0.138N/mm^2). Response of the panel is recorded in terms of the displacement- and acceleration-time histories at the centre of the panel. Strain-time histories are recorded from strain gauges fixed to the steel reinforcement near the centre of the panel during casting. The damaged area and cracks if any, are marked and the peak pressure load is noted. The gun chamber pressure is then increased and the panel is subjected to the corresponding peak blast pressure load. The incremental loading is continued until failure of the panel.

Some of the panels do not fail even at the peak blast pressure load of 4062N corresponding to the maximum imposed limit of 170psi (1.173N/mm^2) of gun chamber pressure. In such cases , the panels are repeatedly subjected to the peak blast pressure load of 4062N until failure. In a few cases, the peak blast pressure load is increased up to 4494N corresponding to a gun chamber pressure of 190psi (1.311N/mm^2).

In the single shot blast pressure loading tests, the panels are directly subjected to a single blast pressure load corresponding to the load at failure of identical panels subjected to successive incremental blast pressure loading.

TEST RESULTS

A comparison of natural frequencies of the panels with theoretical calculations [8] shows that the experimentally determined natural frequencies range from 0.74 to 0.88 times the theoretically calculated values, with a mean of 0.80 and a standard deviation of 0.05. Thus the simply supported boundary conditions are adequately simulated.

The displacement-time histories are combined with load (pressure)-time histories to generate load-deflection curves (Figure 4) for each panel at different stages of the successive incremental blast pressure loading tests. The peak displacements (Figure 5) at the different stages of loading are joined together to obtain the load-deflection envelope (Figure 6) for each panel.

The strength of the panels at significant stages of loading like formation of first crack, elastic limit and at failure is determined by actual observations and from the load deflection envelope of each panel.

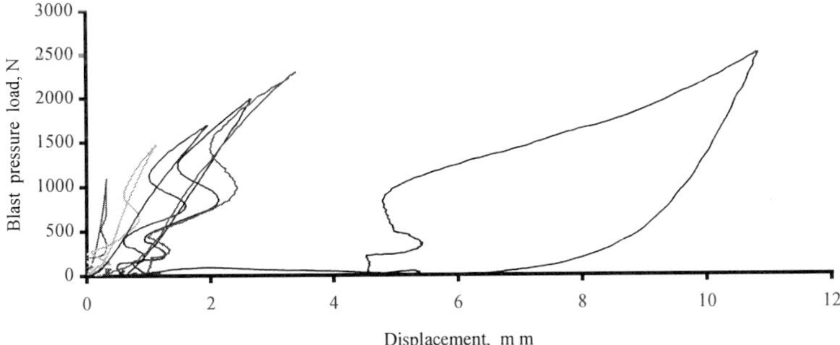

Figure 4 Load-deflection curves for microconcrete panel T_{20} $S_{0.40}$ $M_{1:3}$

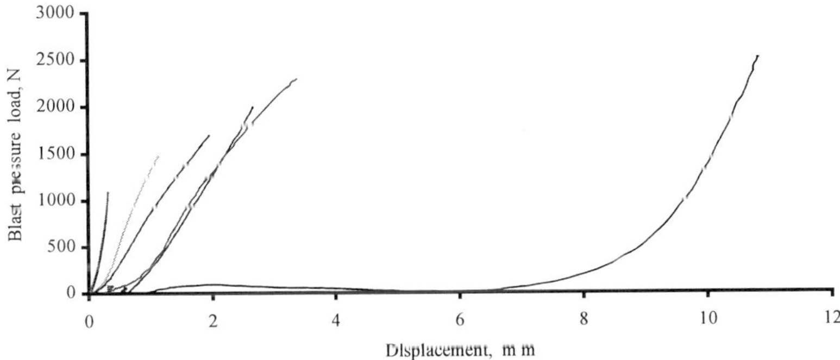

Figure 5 Peak displacements at the centre of the microconcrete panel T_{20} $S_{0.40}$ $M_{1:3}$

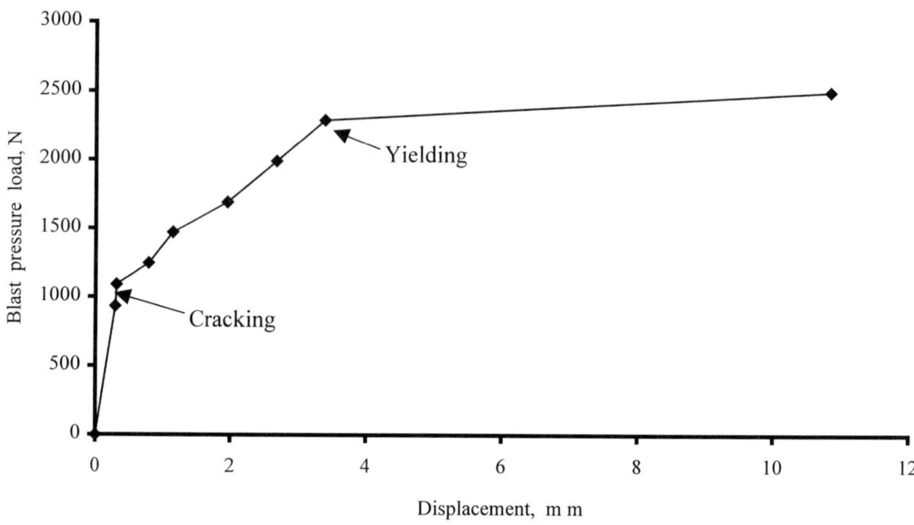

Figure 6 Load-deflection envelope of the microconcrete panel T_{20} $S_{0.40}$ $M_{1:3}$

DISCUSSION OF TEST RESULTS

Significant stages in the behavior of the microconcrete panels like formation of the first cracks, elastic limit and failure of the panels are clearly identified and the corresponding loads are determined from the load-deflection envelopes of the panels (Figure 6).

The microconcrete panels fail by a complex phenomenon of cracking and punching, by yielding of reinforcement, by crushing of microconcrete or by a combination of the basic failure modes. Typical failures are shown in Figures 7, 8 and 9.

Panels with low thickness, percentage reinforcement and microconcrete strength exhibit a unique mode of local failure. On loading upto failure, the reinforcement at the centre of the panel ruptures and the microconcrete breaks off along the reinforcement, leaving an opening of a size corresponding to the reinforcement spacing, around the centre (Figure 7). This local failure of the panels must be taken into consideration in the design of the panels, especially in limiting the spacing of reinforcement.

CONCLUDING REMARKS

The ratio of the failure load sustained by the microconcrete panels subjected to successive incremental blast pressure loading and of the failure load sustained by identical panels subjected to single shot blast pressure loading is in the range 0.87 to 1.48, with a mean of 1.09 and standard deviation of 0.23 [8]. Apparently, the damage sustained by the panels does not influence the strength of the panels at failure. Results from the experimental investigations have been compared with theoretical calculations and a regression analysis has been carried out to generate regression equations and parametric curves. The comparison of the results and the regression analysis are presented in reference [8].

Figure 7 Yielding and rupture of reinforcement

Figure 8 Crushing of microconcrete

Figure 9 Punching failure

REFERENCES

1. MUSACCHIO, J M AND ROZEN, A. Architectural and structural engineering aspects of protective design for nuclear power plants against terrorist attack. Transactions of the 9th International Conference on Structural Mechanics in Reactor Technology, Lausanne, Vol J, August 17-21, pp 153-160.

2. BUREAU OF INDIAN STANDARDS, IS 4991-1968 : Indian Standard Criteria for Blast Resistant Design of Structures for Explosions Above Ground, New Delhi, 1969, pp 38.

3. EIBL, J AND OCKERT, J. Shock wave propagation in concrete, Proceedings of the Symposium on Advances in Structural Dynamics, Department of Civil Engineering, Indian Institute of Technology Madras, Chennai, India,1993.

4. SANTHAKUMAR, S. Design and Development of a Low Cost Missile Launcher, Closing Report, Department of Aerospace Engineering, Indian Institute of Technology Madras, Chennai, India, 1990, pp 88.

5. ACHYUTHA, H, RADHAKRISHNA, K AND SANTHAKUMAR, S. Response of mild steel plate elements to simulated blast loading, Proceedings of the International Conference on Experimental Model Research and Testing of Thin-Walled Structures, Academy of Sciences of the Czech Republic, Institute of Theoretical and Applied Mechanics, Ed. Milos Drdácký and Teoman Peköz, Prague, Czech Republic, September 1997, pp109-115.

6. KAFAROV, V. Cybernetic Methods in Chemistry and Chemical Engineering, Translated from Russian to English by Kuznetsov, Mir Publishers, Moscow, 1976, pp 484.

7. ACHYUTHA, H, RADHAKRISHNA, K AND SANTHAKUMAR, S. Some preliminary studies on microconcrete slab models subjected to simulated blast loading. Trends in Structural Engineering - Towards the 21st Century, Ed. A. Rajaraman, Proceedings of the Structural Engineering Convention 1997, Indian Institute of Technology Madras, Chennai, India, February 12-14, 1997, Tata McGraw-Hill Company, New Delhi, pp 555-563.

8. RADHAKRISHNA, K. Experimental Investigations and Analysis of Steel and Micrconcrete Panels to Simulated Near-field Blast Pressure Loading, Doctoral Thesis (to be submitted), Indian Instiitute of Technology Madras, Chennai, India.

9. YOUNG, W C. Roark's Formulas for Stress and Strain, McGraw-Hill Book Company, Singapore, 1989, pp 763.

THEME FOUR:

DEEP BASEMENTS

CRACK PREVENTION IN WALLS AND SLABS – THE INFLUENCE OF RESTRAINT

M Nilsson

J-E Jonasson

K Wallin M Emborg

S Bernander L Elfgren

Luleå University of Technology

Sweden

ABSTRACT. The avoidance of through cracking in young concrete structures is of utmost importance with regard to waterproofing and durability. Restraint and restraint-transferring casting joints are very important factors influencing the risk of cracking. One method to minimise the restraint is presented below. It is applicable to concrete structures founded on resilient material. The influence of restraint-transferring casting joints between young and older concrete structures is also discussed. Finally, some test results are presented from a medium scale study of casting joints loaded by restraint stresses - a wall 6x2x0.15 m cast on a slab 8x1.4x0.12 m.

Keywords: Concrete, Structures, Restraint, Cracking, Casting joints, Young concrete, Resilient foundation

M Eng M Nilsson is a PhD-student in Structural Engineering at Luleå University of Technology, Sweden. His field of research is casting joints, loaded by restraint stresses, and their influence on the risk of cracking of young concrete structures.

Dr J-E Jonasson is an Associate Professor in Structural Engineering at Luleå University of Technology, Sweden. His research speciality is modelling of thermal and moisture conditions and associated structure behaviour in concrete structures.

Mr K Wallin is a Research Engineer in Structural Engineering at Luleå University of Technology in Sweden.

Dr M Emborg is an Associate Professor in Structural Engineering at Luleå University of Technology, Sweden. His research specialities include modelling of young concrete, fibre reinforcement and fatigue of concrete structures.

Professor S Bernander is an Adjunct Professor Emeritus in Structural Engineering at Luleå University of Technology, Sweden.

Professor Dr L Elfgren is Head of the Division of Structural Engineering, Luleå University of Technology, Sweden. His research specialities include fracture mechanics, fatigue, fasteners, and combined torsion, bending and shear

INTRODUCTION

During the hydration phase of a concrete structure, the temperature varies within the cross section. Normally, surface layers are generally cooler then centre parts. This temperature difference leads to larger motions in the centre parts then on the surfaces. The larger expansion in centre parts during the heating phase of the hydration induces tensile stresses in the surface layers. If these stresses are larger then the actual tensile strength, surface cracks will occur. The surface layers in turn counteract the larger expansion in the centre parts, and therefore cause compressive stresses. When the temperature passes its maximum, naturally the expansion turns into contraction. By the same reason as before, temperature-induced movements in the centre parts are larger than in the surface layers. Thereby, the surface cracks tend to close. The surface layers also counteract these motions, but this time they cause tensile stresses in the centre parts. Although the concrete is now stronger, but also stiffer and more brittle, cracks will occur if the stresses are larger than the actual tensile strength. The large difference compared with cracks occurring during the expansion phase is that these cracks are both through and lasting cracks, see Figure 1. In [3] - [7] and [10] the problem is more carefully described and discussed.

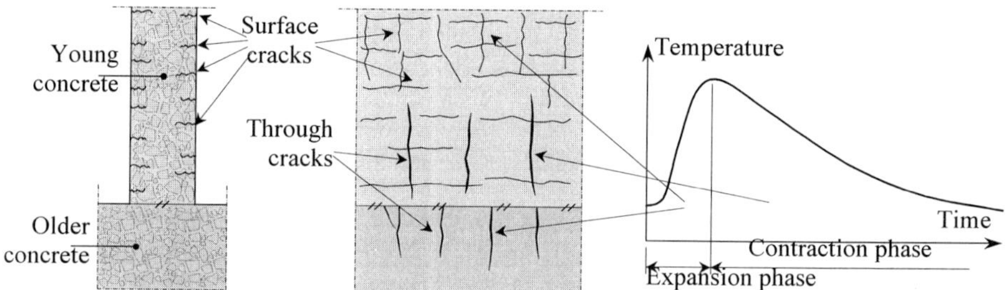

Figure 1 Examples of early age expansion cracks and contraction through cracks in a symmetrical wall cast on older concrete, Bernander [3]

All adjacent structures and materials counteract/restrain temperature-induced free movements of young concrete structures, especially when the temperature of the young concrete decreases and approaches the temperature of the surrounding. This is of major concern. At this state when the concrete is stronger, but also more brittle, it is much more sensitive to restraint-induced stresses and strains. If these stresses exceed the tensile strength of the young concrete, through cracking is possible. Such cracking is mostly common in mass-concrete structures such as deep basements and tunnel walls. Several parameters are influencing the counteraction/restraint, e.g. the amount of through reinforcement, the strength and surface conditions of casting joints as well as the stiffness of adjacent structures and materials. In addition, the temperature difference between young concrete and adjacent structure, and the length-height ratio of the structure are other parameters of great importance.

One example of lasting damages of through cracks is leakage through tunnel walls under one-sided water pressure. In such cases, it is of utmost interest to study the risk of cracking in newly cast concrete structures.

The effects mentioned in the last passage caused by different temperature distributions and counteractions are sometimes designated as *internal* and *outer* restraint, respectively. One must however bear in mind what is in the model or not, otherwise the terms *internal* and *outer* restraint may lead to misunderstandings and uncertainties. Therefore, it is instead preferable to classify cracks caused by restraint as, see [1]:

1. Early cracking during the expansion phase. These cracks arise during the hydratation phase shortly after casting (one or a few days) and tend towards closing by time. Their influence on the static capacity, the function and strength of a structure must be judged from case to case.

2. Cracking during the contraction phase. Cracks arising during the cooling phase are usually through cracks. Depending on dimensions and other factors, they do not arise until weeks, months and in extreme cases even years after casting. Cracks formed during the cooling phase are as a rule permanent.

STRUCTURES ON RESILIENT FOUNDATION

In the following, a young concrete wall is studied. The wall is cast on an older slab, which in turn is founded on a resilient material of some kind. During the cooling phase when the contraction takes place, an internal tensile force at total restrained conditions, N_{RI}, can be estimated. Normally, an internal eccentricity will be present between the force and the centroid of the combined (slab + wall) cross-section. This causes a bending moment endeavouring to rotate the structure. This bending moment is denoted bending moment of internal loading, M_{RI}. See also Figure 2. The actual rotation corresponds to an outer bending moment, M_{RO}. This moment is caused by the counteraction/restraint from the adjacent slab and the resilient material on the free movements of the young concrete.

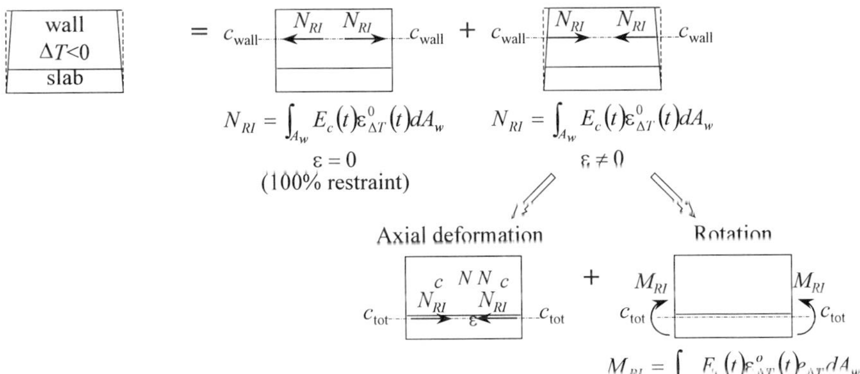

Figure 2 Description of internal forces and bending moments loading a young wall cast on an older slab, Nilsson [8]

How much the construction is permitted to bend depends partly on its length, partly on the stiffness of the foundation. The length influences in that way that the weight of the construction in combination with its length result in a bending moment counteracting the temperature induced bending. The stiffness of the foundation material determines how much the construction may bend down into the ground. A soft material offers almost no resistance to the free deformation of the structure. Thereby the concrete is not subjected to any, or very little, restraint. On the other hand, if the construction is founded on a stiff ground, for example rock or dense gravel, it can hardly bend at all if full cohesion exists to the ground. This type of restraint is also a factor that can cause cracking.

Determination of Restraint

It is possible to estimate the amount of restraint acting on a concrete structure that is founded on a resilient foundation during for example the contraction phase. First, the contraction from the temperature decrease in the young structure must be known, here expressed as free strain, $\varepsilon_{\Delta T}^{0}$. This strain can then be used to describe the bending moment of internal loading, M_{RI}. Furthermore, the stiffness of the foundation material must be known to calculate the outer bending moment corresponding to the rotation, M_{RO}.

The free strain during the contraction phase can be expressed as:

$$\varepsilon_{\Delta T}^{0}(t) = \Delta T(t - t_1) \cdot \alpha_c \tag{1}$$

where

$\Delta T(t\text{-}t_1)$ is change of temperature from the state of zero stresses, [°C]

α_c is thermal contraction coefficient, [1/°C]

The bending moment of internal loading is expressed as:

$$M_{RI} = \int_{A_w} E_c(t) \cdot e_{\Delta T} \cdot \varepsilon_{\Delta T}^{0}(t) \cdot dA_w \tag{2}$$

where

$E_c(t)$ is the actual Young's modulus, [MPa]

$e_{\Delta T}$ is the eccentricity of the internal tensile forces, N_{RI}, to the centroid of the construction, [m]

A_w is the area of young concrete, [m^2], under temperature decrease, ΔT

The stiffness of the foundation material, k [kN/m^2], is calculated as:

$$k = \frac{K_j}{f} \tag{3}$$

where

K_j is the modulus of compression of the ground, [kN/m^2], depending on the type of soil and its grading,

F is a factor describing the relation between the length and width of the structure resting on the ground.

The outer bending moment corresponding to the rotation in the mid span of the structure can be calculated according to, see [2] and [8]:

$$M_{RO} = \frac{2M_{RI}}{\sin\frac{L}{L_e} + \sinh\frac{L}{L_e}} \left(\cos\frac{L}{2L_e}\sinh\frac{L}{2L_e} + \sin\frac{L}{2L_e}\cosh\frac{L}{2L_e} \right) \tag{4}$$

where

L is the length of the structure, [m]

$L_e = \sqrt[4]{\frac{4E_c I_y}{k}}$ is the elastic length, [m] (5)

I_y moment of inertia, [m^4]

The derivation of Equation (4) is cited from Timoshenko [9] for a beam on a resilient foundation bent by two equals and opposite couples at its ends. Here the couples are equal to the bending moment of internal loading, M_{RI}.

The estimation of the restraint can be done in the following way. A structure founded on a material not giving any noticeable resistance against bending due to a bending moment of internal loading, M_{RI}, is not subjected to any restraint, $R_O = 0$. The outer bending moment corresponding to the rotation is then equal to M_{RI}, ($M_{RO} = M_{RI}$). If, on the contrary, the structure is not permitted to bend at all, then the restraint is defined as 100 %, $R_O = 1$. The curvature of the construction is then equal to zero, implying that the outer bending moment corresponding to rotation is also equal to zero.

The restraint can generally be defined as:

$$R_O = \frac{M_{RI} - M_{RO}}{M_{RI}} = 1 - \frac{M_{RO}}{M_{RI}} \tag{6}$$

When using Equation (4) in Equation (6) the restraint in the mid span of the structure, R_O, can then be calculated as:

$$R_O = 1 - \frac{2}{\sin\frac{L}{L_e} + \sinh\frac{L}{L_e}} \left(\cos\frac{L}{2L_e}\sinh\frac{L}{2L_e} + \sin\frac{L}{2L_e}\cosh\frac{L}{2L_e} \right) \tag{7}$$

By use of this equation, it is for example, possible to determine the length of a section being cast for which the restraint can be neglected in practice, (e.g. $R_O < 0.05$).

CASTING JOINTS

Cracking during the contraction phase of a young concrete member on an older one very much depends on the connecting casting joint between the members. If the casting joint is very strong and contains a lot of through reinforcement, it is then capable to transfer counteractions/restraint on the free movements from older members to younger ones. On the other hand, if there is a slip failure in casting joints, the restraint stresses are reduced as well as the cracking risk, see Figure 3. A considerable amount of money can thereby be saved during both design and building phase. However, this mechanism has first to be proven.

Theoretical calculations in [2] and [10] have shown that when the horizontal stresses are high in the wall, the normal and the shear stresses in the casting joint are most probable larger than the corresponding strength levels. This indicates that slip failure in the casting occurs when there is a risk of cracking in the wall.

In Figure 3 the stress distributions are schematically described during the cooling phase for a wall cast on a slab. Before a slip failure, all stresses are larger than after eventual cracking in a casting joint. In addition, the area with high restraint effects from the slab is also significantly reduced after a slip failure in a joint.

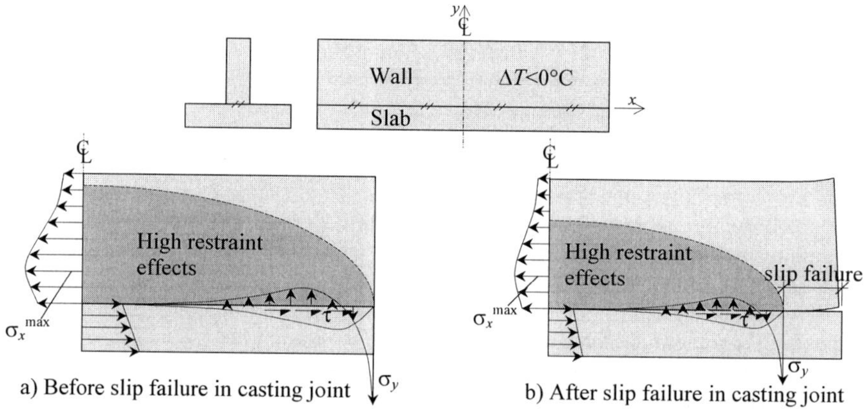

Figure 3 a) Stress distributions in a structure before slip failure in casting joint,
b) Stress distributions in a structure after slip failure in casting joint

Note the considerably reduced area of high restraint before and after the slip failure. The phenomenon of slip failure and associate effects in casting joints, may be used in both the design and building phase. A calculated slip failure in a casting joint may be regarded as a controlled form of cracking. In such joints, it is very common to use joint sealers and/or injections hoses. In those cases it is possible to benefit the effects of a slip failure, and thereby, reduce the risk of uncontrolled through cracking in a wall.

EXPERIMENTAL STUDIES

An experimental study is carried out as part of an ongoing research joint program. The project has the title "Improved Production of Advanced Concrete Structures, IPACS, and is founded by the European Union and several companies. The Division of Structural Engineering, Department of Civil and Mining Engineering at Luleå University of Technology has so far carried through one such laboratory test wall. The aim of the test is to study the effects of restraint and the effects of casting joint properties on through cracking of concrete walls cast on older slabs. In coming tests, the properties of the joints will be varied by using different amounts of through reinforcement and different surfaces and different restraint from adjacent structure. This is the first test of this kind in the world as far as is known.

Geometry and Test Arrangements for Test Wall No. 1

On a pre-cast slab, 8 m long, 1.4 m wide, 0.2 m high and prepared with 16 through holes for restraining stays, a footing was tightly bolted with 66 bolts. The footing was 6.6 m long, 0.36 m wide and 0.24 m high. On the footing a concrete wall was cast, 5.8 m long, 0.15 m thick and 1.96 m high, see Figure. In the wall, 13 steel pipes, ∅ = 25 mm, for warm and cold water were cast in. Two different amount of through reinforcement were used in the casting joint between the wall and the footing. One half contained 0 % of through reinforcement, and the other half contained approximately 1 % (2 ∅12 s150 mm).

The technique with a bolted footing has two major advantages. The first one is obvious when one test is finished and the used structure has to be removed. The 66 bolts will be loosened and the wall together with the footing can be lifted of the slab and be transported away. The second advantage is very important. When casting the footing, the form was turned upside-down. In this way, the bottom of the form will be the lower part of the future casting joint. The bottom of the form can then be used to create different and well-defined structures of casting joints. In the first test sand (fraction 0.212 - 1 mm) was glued to the bottom of the form. Hereby, other structures will easily be made and used in the future.

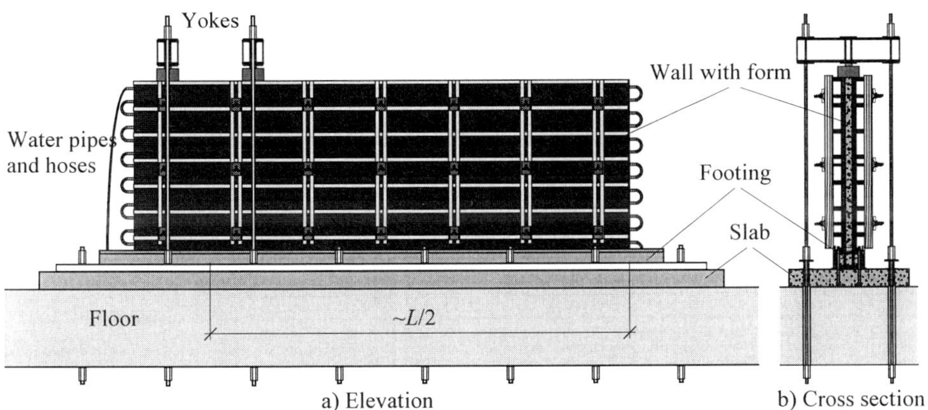

Figure 4 a) Elevation and b) cross section of test arrangement. The wall and the slab had the dimensions 6·2·0.15 m and 8·1.4·0.2 m, respectively

For test wall No. 1, the half of the joint with no through reinforcement was tested firstly, and is called test I. Secondly, the half of wall No. 1 with 1% through reinforcement in the joint is called test II. This was possible by "holding down" the half of the wall that should not be tested by two yokes between two pair of restraining stays. Each yoke was loading the top of the wall by a force of 400 kN. By using these two yokes, the centre of dilatation was moved from the geometrical centre of the wall to somewhere between the yokes. In this way the test was simulating a structure with a higher length-height ratio then had been possible without the yokes. This was done both by economical reasons and due to limited space in the laboratory. As mentioned above, the pre-cast slab contains 16 through holes for restraining stays. The holes were used for anchoring the pre-cast slab to the very well reinforced and 0.9 m thick floor in the laboratory. The restraining stays were pre-stressed with a tensional force of 350kN.

The problem with through cracking is mostly common for massive structures due to the higher internal temperature development. In spite of the slender structure used in the laboratory, it was possible to simulate a massive structure by the steel pipes in the wall. By first using warm and then cold water, the temperature-induced movements were considerably increased. Consequently, also the stresses causing through cracking were increased.

Performance

Once the fresh concrete was placed in the form of the wall, the test was started. A few very intensive hours followed, when all measuring devices and equipment were started after they had been placed on the construction. The temperature development was measured with two different systems: ConReg and ACC-2. The relative movements were measured with linear displacement transducers between the floor and the slab, the slab and the footing and between the footing and the wall along with the movements of the upper corners of the wall. The strain variation was measured at some places in the wall with so-called vibrating strain gages. In addition, the propagation of the slip failures in the casting joint was measured with crack indicators that were glued perpendicular over the joint.

Only a few hours after casting, warm water was used to increase the temperature of the wall and thereby increase the rate of hydration. During the continuation of the test, warm and cold water was used to change the concrete temperature in accordance with a pre-chosen temperature development. The water temperature was changed systematically with several degrees each time. In this way, the temperature variation within a more massive structure was simulated.

Two days after casting the two yokes were applied on the top of the wall, see Figure 4 and 5. At this time the temperature maximum was reached. Therefore, during the whole cooling phase the length/height ratio was approximately:

$$\frac{L}{H} \approx \frac{2 \cdot (6-1)}{2} = 5$$

instead of:

$$\frac{L}{H} = \frac{6}{2} = 3$$

This change of the L/H ratio is significantly increasing the risk of through cracking in the wall.

Results for the Test Wall No. 1

Due to the restraining conditions and the large temperature differences between the wall and the footing, both the casting joint and the wall cracked. The whole length of the un-reinforced casting joint did crack in test I almost immediately after the water temperature was lowered the first time. The casting joint in test II cracked piece by piece 6 times until 0.60m of the joint was cracked. Every piece of the cracking corresponded to the distance between the through reinforcing rods in the joint.

The vertical relative displacement between the footing and the wall is presented in Figure 5 together with the variation of the water temperature during the two tests. In addition, the total cracking of the un-reinforced casting joint in test I as well as the partial cracking of the reinforced casting joint in test II are shown. One obvious observation in the diagram is that after the slip failures the relative vertical displacements do correspond to the change in water temperature. The movements are rather large: ~0.4 mm at the end of the un-reinforced joint and ~0.2 mm at the end of the reinforced joint.

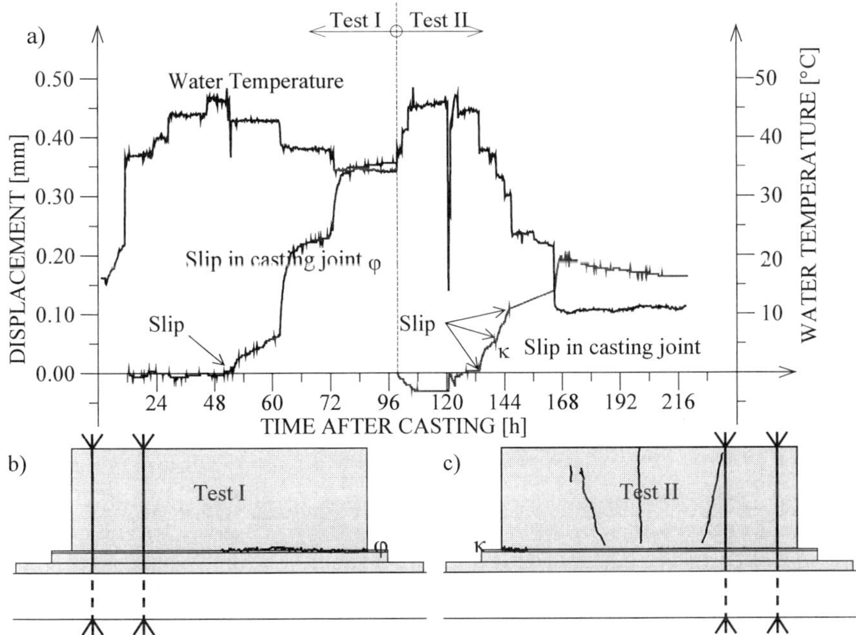

Figure 2 a) Presentation of vertical relative displacement between the footing and the wall together with the variation of the water temperature,
b) Test I; cracking of the un-reinforced casting joint,
c) Test II; partly cracking of the reinforced casting joint and through cracking of the wall

After form stripping of the wall, three through cracks were observed in the wall, see Figure 5c. It is very difficult to tell whether they arise before, during or after the slip failure in the joint. Anyhow, this result indicates that through cracks will form very likely if the restraint from adjacent structures is high and/or the restraint-transferring joint is heavily reinforced.

CONCLUSIONS

The following conclusions can be drawn about internally deforming structures founded on resilient materials and about the effects of casting joints on restraint and the risk of though cracking:

1. By use of Equation (7), it is possible to calculate the restraint in the mid span of a internally deforming structure founded on resilient material. It is also possible to calculate the length of a structure corresponding to a pre-determined amount of restraint.

2. When the wall has high horizontal stresses for the structure wall on slab, it is most probable that a slip failure at the casting joint has occurred. This is indicated by theoretical calculations. Such failures are beneficial for the risk of cracking in the wall.

3. Laboratory tests have shown that slip failure takes place at the casting joint for both no through reinforcement and for a reinforcement amount of 1 percentage.

4. As the existence of slip failure in the casting joint has been proven, the future research will in detail study different influencing parameters like amount of reinforcement and structure of the joint surface.

REFERENCES

1. BERNANDER, S. Temperature Stresses in Early Age Concrete Due to Hydration. In: Proceeding from International Conference on Concrete at Early Ages, RILEM, Paris, April 6-8 1982. Vol II. pp. 218-221, 1982

2. BERNANDER, S. Balk på elastiskt underlag åverkad av ändmoment M_l (Beam on Resilient Ground Loaded by Couples at Its Ends M_l). Göteborg: ConGeo AB. Notes and calculations with diagrams, 1993

3. BERNANDER, S. Practical Measures to Avoiding Early Age Thermal Cracking in Concrete Structures. In: RILEM Report 15, Prevention of Thermal Cracking in Concrete at Early Ages. Ed. by R. Springenschmid. London: E & FN Spon. pp. 255-314. ISBN 0-419-22310-X, 1998

4. ELFGREN ET AL. High Performance Concrete Structures - a Swedish Research Program. In: Concrete in the Service of Mankind - Radical Concrete Technology, University of Dundee, Scotland, UK on 27-28 June 1996. pp. 141-150.ISBN 0-419-21480-1, 1996

5. EMBORG, M, BERNANDER, S. Assessment of the Risk of Thermal Cracking in Hardening Concrete. Journal of Structural Engineering (ASCE), Vol 120, No 10, Oct pp. 2893-2912, 1994

6. EMBORG, M. Development of Mechanical Behaviour at Early Ages. In: RILEM Report 15, Prevention of Thermal Cracking in Concrete at Early Ages. Ed. by R. Springenschmid. London: E & FN Spon. pp. 76-148. ISBN 0-419-22310-X, 1998

7. EMBORG, M. Models and Methods for Computation of Thermal Stresses. In: RILEM Report 15, Prevention of Thermal Cracking in Concrete at Early Ages. Ed. by R. Springenschmid. London: E & FN Spon. pp. 178-230. ISBN 0-419-22310-X, 1998

8. EMBORG, M ET. AL. Temperatursprickor i betongkonstruktioner - Beräkningsmetoder för hydratationsspänningar och diagram för några vanliga typfall (Temperature Induced

Cracks in Concrete Structures - Calculation Methods for Hydration Stresses and Diagrams for some Common Typical Cases). Luleå: Luleå University of Technology. Technical Report 1997:02. 100 pp. (In Swedish), 1997

9. JONASSON, J-E. Modelling of Temperature, Moisture and Stresses in Young Concrete. Division of Structural Engineering, Luleå University of Technology. Doctoral Thesis 1994:153 D. 225 pp, 1994

10. NILSSON, M. Inverkan av tvång i gjutfogar och i betongkonstruktioner på elastiskt underlag (Influence of Restraint in Casting Joints and in Concrete Structures on Resilient Foundation). Division of Structural Engineering, Luleå University of Technology. Master Thesis 1998:090 CIV. 61 pp. (In Swedish), 1998

11. TIMOSHENKO, S. Beams on Elastic Foundation. In: Strength of Materials. Part II: Advanced Theory and Problems. New York: Van Nostrand Reinold Company. 25 pp, 1958

12. WALLIN, K, EMBORG, M, JONASSON, J-E. Värme ett alternativ till kyla (Heating an Alternative for Cooling). Division of Structural Engineering, Luleå University of Technology. Technical Report 1997

WATER PERMEABILITY AND AUTOGENOUS HEALING OF CRACKS IN CONCRETE

C Edvardsen

COWI

Denmark

ABSTRACT. The well-known practical phenomenon of autogenous healing in cracks plays a significant role in relation to the functional reliability of structures subjected to water-pressure loads. Due to the autogenous healing the water flow through the cracks gradually reduces with time, and in extreme cases the cracks seal completely. In the past there has been no deliberate technical exploitation of self-healing, because too little is known about the phenomenon itself and about the chemical/physical processes involved.

Based on theoretical and experimental research, conducted at the Technical University of Aachen, the effect of crack healing was first investigated on a larger scale. The experimental studies showed the formation of calcite in the crack to be almost the sole cause for the autogenous healing. Theoretical analyses of the chemical/physical processes concerned indicate that the crystal growth rate is dependent on crack width and water pressure, whereas concrete composition and water hardness have no influence on autogenous healing. This was confirmed by the experimental results. On the basis of the results, an algorithm that can be used to estimate the reduction in water over time as a result of autogenous healing was developed.

Keywords: Cracks, Water permeability, Autogenous healing, Service life, Testing

Dr Ing Carola Edvardsen is a Senior Engineer in the Repair and Maintenance Department of COWI, Denmark. She has experience in material testing, in the specification of concrete for special purposes and assessment of service life of new and existing concrete structures. She specialises in the prediction and modelling of concrete degradation processes, in the investigation, assessment and repair evaluation of deteriorated concrete structures, especially structures affected by reinforcement corrosion and cracking.

INTRODUCTION

Besides the durability and general stability, the watertigthness is of significant importance for the serviceability of reinforced concrete structures subjected to water pressure load, i. e. basements, water retaining structures, water service reservoirs or waste water reservoirs. Watertightness of the concrete can be ensured either using special membranes or by the concrete itself. However, a 100% watertightness can never be achieved. Cracks are unavoidable in reinforced concrete structures. The design method assumes the presence of cracks. Damages of structures show that tensile cracks resulting from the restraint of imposed deformations are normal, especially for water retaining structures. Due to the tensile cracks the concrete structure will become waterpermeable up to a certain degree depending on the crack width, the crack length, the hydraulic gradient, etc.

The practical experiences demonstrate at the same time that cracks have the ability to heal themselves, e. g. the water flow will be reduced with time. In extreme cases cracks can seal completely. The autogenous healing of cracks in concrete seems to be a complicated chemical/physical process. Although the autogenous healing has become part of the civil engineering folklore, the information about the process itself and the influencing parameters are limited. A quantification of the flow reduction degree was lacking until recently. The effect of crack healing was first investigated on a larger scale, through experimental and through theoretical studies at the Institute for Building Materials Research, University of Technology, RWTH Aachen, Germany [1-3].

WATER PERMEABILITY AND AUTOGENOUS HEALING

Water Permeability

The model used to analyse the initial flow of water (flow before the autogenous healing occurs) through cracks in concrete is derived from the parallel-plate theory and is described in fluid mechanics text-books [4]. This model assumes the flow of an incompressible fluid between parallel-sided plates where laminar flow is fully developed. The equation derived from this model to estimate the water flow through a "smooth" parallel-sided crack can be written as:

$$q_o = \Delta p \cdot b \cdot w^3 / 12 \cdot \eta \cdot d \quad (1)$$

with

- q_o = Water flow of idealised "smooth" cracks [m³/s]
- Δp = Differential water pressure between inlet and outlet of the crack [N/m²]
- b = Length of the crack (visible crack length at the surface of the concrete) [m]
- w = Crack width [m]
- d = Flow path length of a crack (thickness of the concrete structure) [m]
- η = Absolute viscosity [Ns/m²].

Equation (1), often mentioned as Poiseuille Law, shows that the crack width is the dominant factor for the water permeability as the flow rate will be proportional to the width cubed.

However, smooth parallel-sided cracks will not occur in concrete. Due to the roughness of the inner crack surfaces the water flow will be much lesser than theoretically estimated according to Equation (1). Reasons for the reduced water flow through cracks are:

- Roughness of the crack surfaces
- Micro-roughness (surface roughness of aggregates and paste)
- Macro-roughness (crack development around the coarse aggregate)
- Variation of crack width along the flow path and the visible crack length
- Local crack width reduction at the reinforcement
- Crack branching
- Physical effects (adhesion and cohesion).

Equation (1) is therefore to be modified by a reduction factor ξ:

$$q_r = \Delta p \cdot b \cdot w^3 \cdot \xi / 12 \cdot \eta \cdot d \quad \text{(Eq. 2)}$$

with q_r = Water flow of natural "rough" cracks [m^3/s]
ξ = Reduction factor comprising the roughness of cracks.

The present investigations show a large variability for ξ [5-8]. The reduction factor varied between 0.02 and 0.17 at the water permeability tests on reinforced concrete slabs [6]. The measured water flow was always less than 1/6 (17%) of the Poiseuille-flow calculated according to Equation (1). The deviation of the reduction factor was even higher with $0.04 \leq \xi \leq 0.53$ in [7].

Autogenous Healing of Cracks in Concrete

In the literature the following chemical, physical and mechanical processes are mentioned which may be the reason for the autogenous healing:

- Swelling and hydration of cement paste
- Precipitation of calcium carbonate crystals
- Blocking of flow path by water impurities
- Blocking of flow path by concrete particles broken from the crack surface due to cracking.

The most significant factor influencing the autogenous healing is the precipitation of calcium carbonate. This will be confirmed by the author's investigations.

The building practice itself gives daily evidence of this process, shown for example in Figure 1. White calcium traces at the surface of the concrete structure signify the formation of calcium carbonate at cracks.

Figure 1 Precipitation of calcium carbonate at cracks

RESEARCH STUDIES

One aim of the experimental studies was to determine the relationship between the width of a crack in concrete, as measured on the exposed surfaces, and the leakage of the water passing through the crack. The other aim was to investigate the phenomenon of autogenous healing [1-2]. During Phase I water permeability tests were conducted on small test specimens (200 x 200 x 200 (400) mm), each with a single tension crack, varying the following parameters, Table 1:

Table 1 Experimental test programme [1,2]

Aims	Dimensions (cm)	Crack charater	Test Parameters		Test amount
Leakage and Healing $\triangle$ W = 0	20 - 40; 20 x 20	Static Crack	w p I water concrete	var.	80
Crack motion $\triangle$ W $\neq$ 0	40; 20 x 20	Dynamic Crack	w $\triangle$ w	var.	10
			P I water concrete	Const.	
Artifical substances $\triangle$ W = 0	40; 20 x 20	Static Crack	Silica fume Bentonite, Micro - cement		3
			W P I concrete water	Const.	
Crack branching $\triangle$ W = 0	250; 100 x 40	Static cracks with crack branching	with skin reinforcemnet W P I concrete water	Const.	1
	250; 100 x 40	Static cracks without crack branching	without skin reinforcemnet w, p, I water, concrete	Const.	1

- Crack width (w = 0.10, 0.20, 0.30 mm)
- Crack length, i. e. thickness of the concrete element (d = 200 - 400 mm)
- Water pressure (p = 2.5 - 20 m of water)
- Hydraulic gradient (I = 6.25 - 25)
- Hardness of water (soft (4 ° dH), hard (27 °dH) and distilled water)
- Cement (Portland cement, slag cement, sulfate-resistant cement)
- Aggregate (granite, limestone, basalt)
- Filler (limestone dust, fly-ash).

For the production of a realistic tension crack and the subjection to water pressure a special testing device was developed, Figure 2. The leakage and temperature of the water were continuously measured during the exposure to water pressure. The water (before and after passing through the crack) was chemically analysed to get information about the consumption of calcium carbonate (hydrogen carbonate). After the ending of tests (approx. 5 - 20 weeks) the surfaces of the crack path were examined for blocking material by means of optical microscopy (polished and thin sections), scanning-electron-microscope-observations (SEM-EDX) and x-ray-diffraction-analyses (RSA).

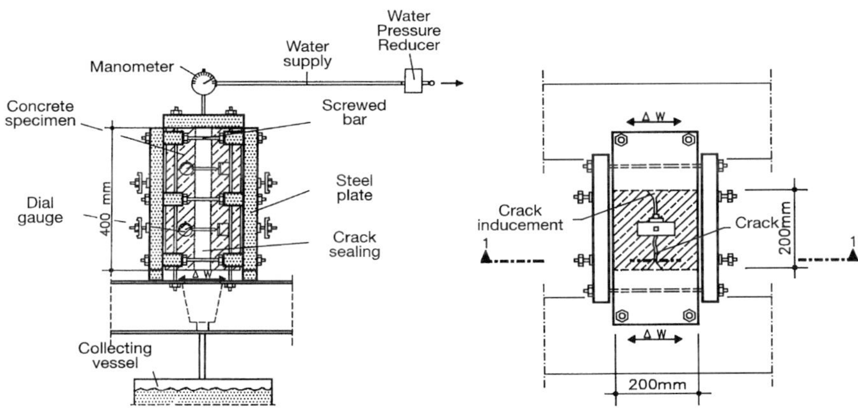

Figure 2 Testing device for the producing of realistic tensile cracks and water pressure subjection [1]

Other essential, relevant factors affecting the autogenous healing such as:

- Crack motions (mobile cracks with varying crack widths)
- Artificial substances present in the water (microcement, silica fume and bentonite)
- Crack branching near the surface of the concrete structures due to skin reinforcement

were investigated during Phase II. The influence of crack branching on the leakage of water and autogenous healing were examined by conducting water permeability tests on relevant reinforced concrete slabs (2500 x 100 x 40 mm), Table 1.

RESULTS

Stationary Cracks

Figure 3 presents some results of the investigations with single crack specimens and stationary cracks. Figure 3 shows the characteristic flow curve of concrete cracks during autogenous healing: a significantly decreasing rate at the early stages. The initially high leakages quickly drop to a low level.

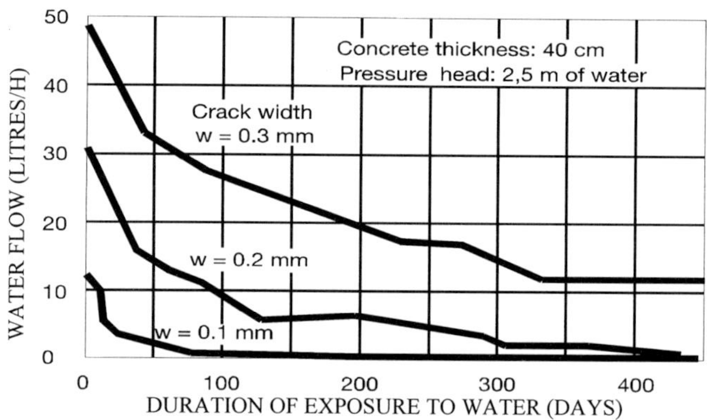

Figure 3 Relationship between water flow and time for different crack widths (per meter visible crack length) [1]

About 50% of the concrete specimens with crack width (mean value) of w = 0.20 mm and a water pressure of p = 0.25 bar (I = 6.25) healed completely during the 7 weeks of water exposure. Even for p = 1 bar (I = 25) 25% of the cracks were healed up to the extent that the flow completely stopped (w = 0.20 mm, 7 weeks of water exposure). Figure 4 shows the relationship between the initial flow and the crack width. There is a large distribution of results which can be preliminary attributed to the large variability of measured crack widths (a well-known problem at investigations with the crack width as test parameter). Despite the variability of the results the validity of Equation (1) will be confirmed which indicated that the leakage of water through cracks in concrete is proportional to the width cubed.

Mobile Cracks

Figure 5 presents the water flow of a mobile crack (crack motion corresponding to one-day cycle) in comparison to stationary crack with comparable crack widths ($w_{stat} = w_{o,mob}$) and water pressure. The figure shows that an autogenous healing also occurs at mobile cracks. The autogenous healing is comparable to stationary cracks at minimum crack width ($w_{o,mob}$), whereas it takes more time at maximum crack width ($w_{l,mob}$). Comparing to stationary cracks the time period for total healing at maximum crack width ($w_{l,mob}$) will be increased by at least 15 weeks.

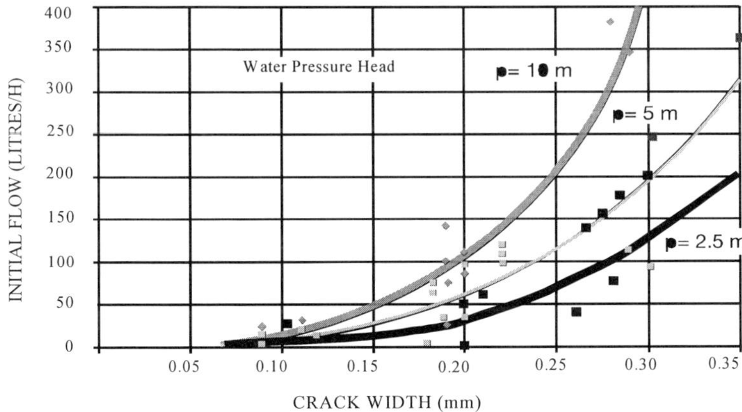

Figure 4 Relationship between initial water flow and crack width for different water pressure (per meter visible crack length) [1]

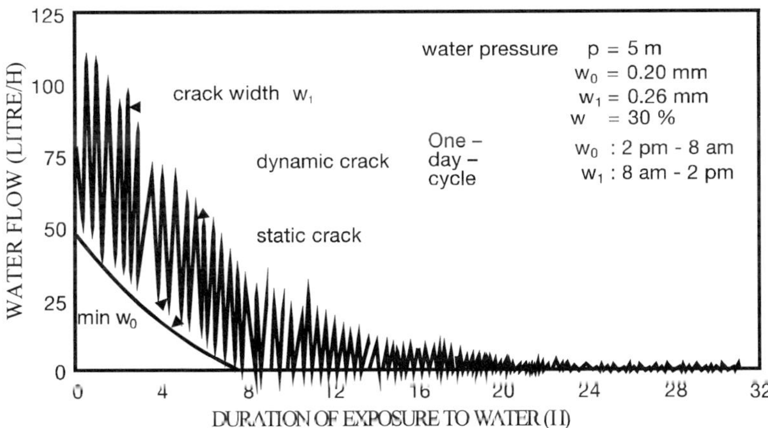

Figure 5 Relationship between flow and time for mobile cracks ($w_{o,mob}$ = 0.20 mm, $w_{1,mob}$ = 0.26 mm, Δw = 30%, p = 1.00 bar) (per meter visible crack length) [2]

Crack Branching

At comparable steel stresses (skin reinforcement considered) and comparable main reinforcement the mean surface crack width could be reduced from 0.25 mm (slab S2 without skin reinforcement) to 0.10 mm (slab S1 with skin reinforcement). Crack branching caused by the skin reinforcement produces a significant reduction of the crack widths at the surfaces of S1, whereas the crack widths in the middle of the two slabs were comparable.

Figure 6 shows that the initial flow could be reduced by 2 orders of magnitude due to the skin reinforcement and that the branched cracks of S1 healed up to negligible flow rates already after one month of water exposure. It seems that besides the crack branching visible at the surface of the S1 contractions at the inner crack path exist which further reduce the initial flow and support the autogenous healing.

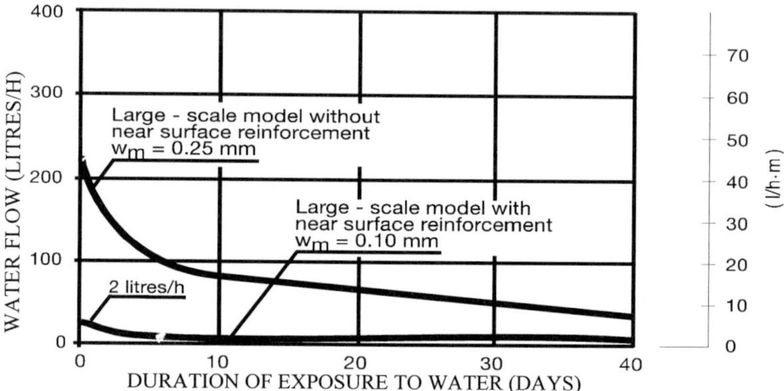

Figure 6 Relationship between flow and time for reinforced concrete slabs with and without crack branching (per m^2 concrete area under water pressure) [2]

AUTOGENOUS HEALING MECHANISM

Relevant Cause

Inspections of thin sections taken along the flow path reveal the formation of calcium carbonate on the inner surface of the crack (Figure 7) to be almost the sole cause of autogenous healing. This was confirmed by SEM-EDX-observations and RSA-investigations. The swelling of concrete and the mechanical blocking due to loose concrete particles seem to be of minor influence according to the author's observations.

Figure 7 Calcium carbonate precipitated on the inner surface of a crack (w = 0.20 mm) [3]

Autogenous Healing Processes

Equation (3a) and (3b) reveal the relevant chemical equations for the precipitation/solution of calcium carbonate at cracks in concrete:

$$Ca^{2+} + CO_3^{2-} \Leftrightarrow CaCO_3 \quad (7.5 < pH_{water} < 8) \qquad (3a)$$

$$Ca^{2+} + HCO_3^{-} \Leftrightarrow CaCO_3 + H^{+} \quad (pH_{water} > 8) \qquad (3b)$$

Calcium carbonate, CO_3^{2-} and hydrogen carbonate, HCO_3^{-}, the constituents of the carbonate hardness of the water, react with the calcium ions, Ca^{2+} exuded from the concrete. Whether precipitation or solution will occur, depends on the solubility product of $CaCO_3$ (K_{CaCo3}), which depends on the temperature, the pH-value and the CO_2-partial pressure of the water flowing through the crack. A $CaCO_3$-precipitation occurs if

$$\Omega = [Ca^{2+}] \, . \, [CO_3^{2-}] \, / \, K_{CaCO3} > 1$$

with

Ω	:	Saturation
$[Ca^{2+}]$, $[CO_3^{2-}]$	:	Concentration of Ca^{2+} and CO_3^{2-} present in the water.

A $CaCO_3$-precipitation will be promoted by:

- Increasing water temperature
- Increasing pH-value
- Decreasing CO_2-pressure.

Figure 8 shows the equilibrium concentrations at saturation of Ca^{2+}, CO_3^{2-} and HCO_3^{-} as a function of pH. The $CaCO_3$ solubility curve passes through a minimum at an intermediate pH ≈ 9.3. The minimum of $CaCO_3$ solubility occurs thus at a pH which is intermediate between that of a cement pore fluid (pH ≈ 13.5) and a typical water (pH ≈ 5.5-7.5). The $CaCO_3$ precipitation is therefore only a question of the local position along the crack path where the pH of the commingling solution of cement pore fluid and exposure water reaches the pH of the solution minimum.

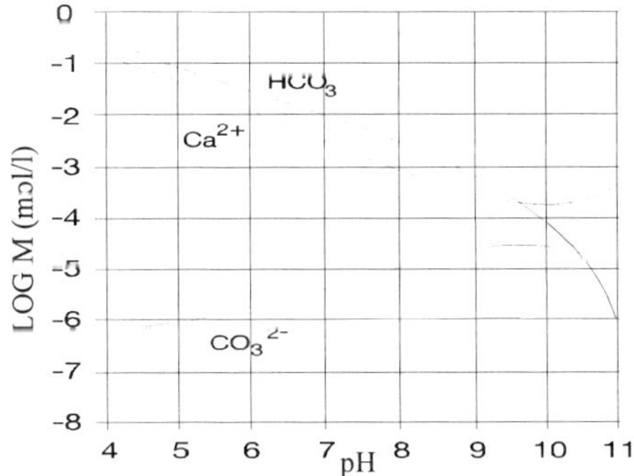

Figure 8 Equilibrium concentrations at saturation of Ca^{2+}, CO_3^{2-} and HCO_3^{-} as a function of pH [9]

Waters with pH-values between 5.5-7.5 may contain CO_3^{2-}, HCO_3^-, and Ca^{2+}, but is generally unsaturated with regard to precipation of $CaCO_3$. Most waters fulfils these basic requirements, as a result of which the autogenous healing is always able to take place by means of calcite formation. Due to the carbonic acid dissolved in the water additional Ca^{2+} will be dissolved out of the $Ca(OH)_2$ of the hardened cement. This reaction and mixing with additional alkalis (KOH and NaOH) dissolved in the pore water lead to a rise in the pH-value of the water during the water flows through the crack. The pH-value of the water and level of calcium ions Ca^{2+}, contained in the water will be lower at the centre of the crack than in the boundary area (at the crack flanks). An increase in pH-value and the content of Ca^{2+} must also be expected over the cross-sectional thickness of the concrete member (crack length). The oversaturation of the water with CO_3^{2-}, HCO_3^-, and Ca^{2+} in the boundary area of the crack as a result of the low velocity of the water leads to nucleation of the calcite crystals here. After nucleation, the crystals continue to grow, and consume the Ca^{2+} , which have been dissolved out of the boundary area of the crack. As the supply of calcium ions diminishes due to the leaching process, the growth process of the calcite crystals slows down. Figure 9 shows the distribution of calcium ions, Ca^{2+}, perpendicular to the crack flanks over the duration of the autogenous healing process. The autogenous healing process is surface-controlled as long as the calcium ions , Ca^{2+}, are available in sufficient quantity at the crack flanks. When these ions first require to be transported to the crack flanks via the pores of the cement paste, the autogenous healing process becomes diffusion-controlled.

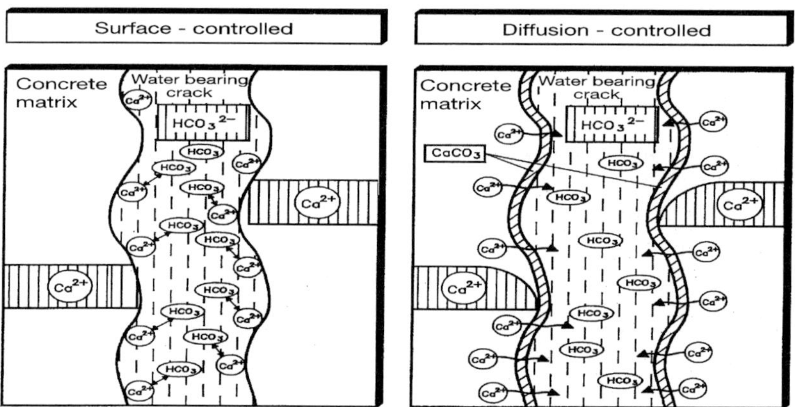

Figure 9 Surface- and diffusion-controlled process of calcite growth [3]

The transition form the surface- to the diffusion controlled process must be the reason for the characteristic form of flow curves of healing cracks. As long a sufficient calcium ions are present at the crack flanks or the near-surface region, a comparatively swift crystal growth takes place. The velocity of the crystal growth is determined by the ion content present in the near-surface region at the crack flanks. Once these calcium ions have been consumed, new ions have to diffuse from the interior of the concrete to the crack flanks. The speed of the crystal growth is then dependent on the rate at which the calcium ions diffuse to the crack flanks. Cracks through which water flows consequently reveal a strong reduction in leakage rates at the beginning of the water subjection (surface-controlled process), whereby this reduction diminishes in the course of time, as the calcium ion diffusion through the concrete to the crack flanks begins to determine the velocity of the crystal growth. Figure 10 shows the autogenous healing rate of the cracks in relation to the actual healing process. The experiments have shown that a marked slowing-down of the healing process is to be observed 50 to 100 hours after water application during the surface-controlled phase, whereby a minor slowing down occurs at the following diffusion-controlled phase.

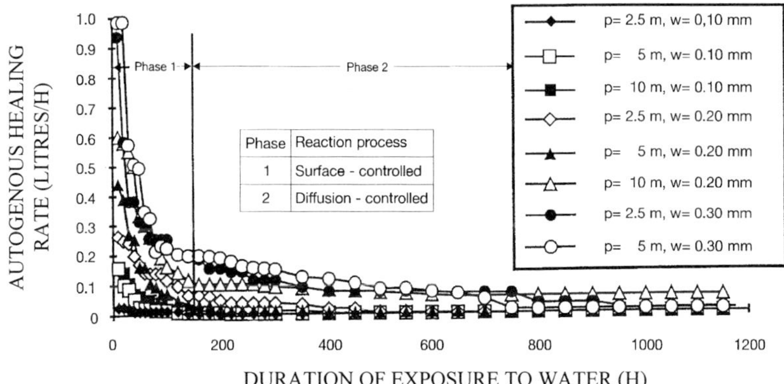

Figure 10 Healing rates (flow reduction) for different crack widths and pressure gradients (per meter visible crack length) [3]

Influencing Parameters

The autogenous healing process is influenced to a considerable degree by the crack width and the prevailing water pressure, whereas the type of cement, aggregate and filler and the hardness of water are of inferior influence.

The analyses of water revealed that not all of the CO_3^{2-} and HCO_3^-, present in the water will be consumed for the precipitation. This means that the carbonate content of the water is not the limiting factor of the $CaCO_3$-precipitation even in the case of very soft water which was confirmed by the experimental investigations

ANALYSIS OF RESULTS AND RECOMMENDATIONS

Initial Water Flow

Due to the large distribution of water leakage Equation (1) seems to be a reasonable assumption to estimate the water flow. A regression analysis of all the test results (Phase I) gives a mean value of $\xi = 0.25$ for the reduction factor in Equation (1). Assuming a water temperature of 20 °C ($\nu = \eta/\rho = 1.00$ mm²/s) the initial flow can be estimated from Equation (1) for a visible crack length of $b = 1.00$ m to:

$$q_o = 740 \cdot I \cdot w_m^{\,3} \cdot k_t \tag{4}$$

where

q_o = Initial water leakage per meter visible crack length [l/h·m]
I = Hydraulic gradient [m/m]
w_m = Crack width (mean value) at the surface [mm]
k_t = Factor comprising different water temperatures

Figure 11 shows the empirical relationship of Equation (4) for different hydraulic gradients and crack widths. Water temperatures different from 20 °C are to be corrected by the factor k_t. The graphic model is valid for stationary and mobile cracks considering the variation of crack width.

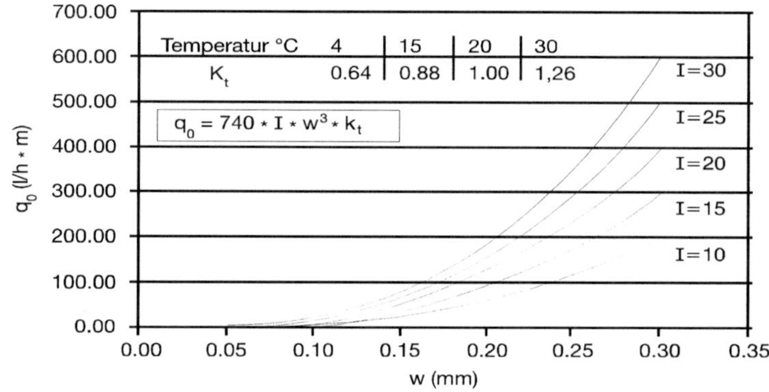

Figure 11 Graphic model to estimate the initial flow of cracks in concrete (per meter visible crack length) [3]

Rate of Autogenous Healing

The relationship between the flow (mean value) and time for all tests of Phase I are shown in Figure 12. This graphic model can be used to estimate the water flow at any time considering the autogenous healing effect (valid for stationary and mobile cracks).

A regression analysis of the results gives Equation (5) to estimate the influence of autogenous healing. As earlier demonstrated the influence of the hydraulic gradient on the water flow is less significant than that of the crack width. The differences in hydraulic gradient were therefore not considered in favour of a simplified handling.

$$q_{(t)} / q_o = 65 \cdot w_m^{-1.05}\, t^{(-1.3+4w)} - 10^5\, w_m^{\,5.8} \tag{5}$$

with

$q_{(t)}$ = Water leakage at time t [l/h·m]
q_o = Initial water flow [l/h·m] according to Eq. (4)
w_m = Crack width (mean value) at the surface [mm]
t = Water exposure time [h]

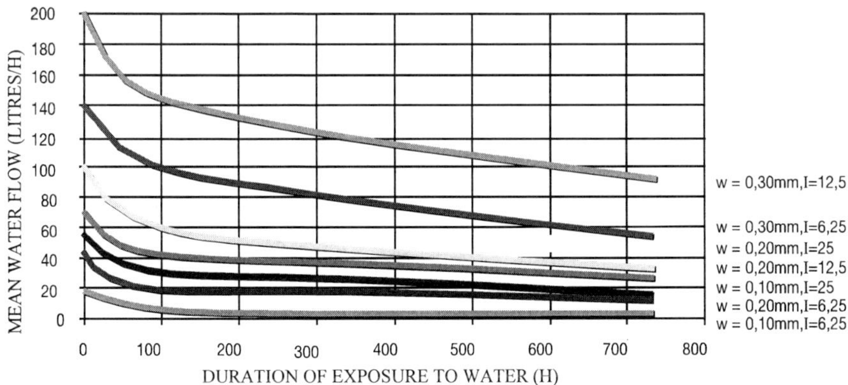

Figure 12 Graphic model to estimate the reduction of water flow with time (per meter visible crack length) [3]

Total Water Leakage

Figure 13 shows the total amount of water (mean value) for all tests of Phase I. The total water flow until the stage of total healing of the crack can be estimated (very rough estimate) at 500 -1000·q_o.

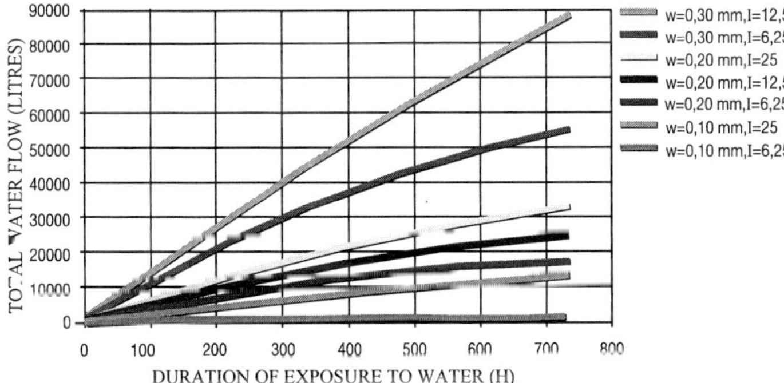

Figure 13 Relationship between total amount of water and time for different crack widths and pressure gradients (per meter visible crack length) [3]

Permissible Crack Widths

Considering the test results presented and a careful extrapolation, permissible crack widths (w_k = design crack width) for an almost total self-healing to be expected after few weeks of water pressure exposure can be derived, Table 2.

Table 2 Permissible crack widths for autogenous healing after [3]

HYDRAULIC GRADIENT [m/m]	w_k [mm] $\Delta w \leq 10\%$	w_k [mm] $10\% < \Delta w \leq 30\%$
40	0.10 - 0.15	≤ 0.10
25	0.15 - 0.20	0.10 - 0.15
15	0.20 - 0.25	0.15 - 0.20

CONCLUSIONS

1. Concrete cracks, both stationary and mobile cracks, subjected to water pressure are able to heal themselves with time.

2. The greatest autogenous healing effect occurs within the first 3 to 5 days of water exposure whereby the then still available water leakage, depending on the crack width and water pressure, makes up only 1 to 20% of the initial water leakage rate.

3. The participation of calcium carbonate crystals ($CaCO_3$) in the crack is almost the sole cause for the autogenous healing of the cracks.

4. The growth rate of the $CaCO_3$ crystals depends on crack width and water pressure, whereas concrete composition (type of cement type and aggregate, respectively) and type of water (i. e. hardness of water) has no influence on the autogenous healing rate.

5. The formation of $CaCO_3$ responds to two different crystal growth processes. In the initial phase of the water exposure the kinetics of crystal growth is surface controlled, later on it turns to a diffusion controlled crystal growth.

6. An additional skin reinforcement proves to be highly effective in supporting the autogenous healing process.

ACKNOWLEDGEMENT

The author would like to express her appreciation for the research grants made by Deutscher Beton Verein E.V. (German Concrete Association) and the Arbeitsgemeinschaft Industrieller Forschungsvereiniungen e.V. (AIF) (Working Group of Industrial Research Association).

REFERENCES

1. SCHIEβL, P AND REUTER, C. "Wasserdurchlässigkeit von Stahlbetonbauteilen im Zustand II (gerissene Betonzugzone)". Institute of Building Materials Research. University of Technology. No. F271, 1992.

2. SCHIEβL, P. AND EDVARDSEN, C. "Selbstheilung von Rissen in wasserbeaufschlagten Stahlbetonbauteilen". Institute of Building Materials Research. University of Technology. No. F361, 1993.

3. EDVARDSEN, C. "Wasserdurchlässigkeit und Selbstheilung von Trennrissen in Beton". Thesis. Institute of Building Materials Research. University of Technology, Aachen. 1994.

4. MASSEY, B S. "Mechanics of Fluids", Third edition. London. Van Nostrand Reinhold.

3. CLEAR, C A. "The effect of Autogenous Healing upon the Leakage of Water through Cracks in Concrete", *Cement and Concrete Association*. Wexham Springs. May 1985.

6. RIPPHAUSEN, B. et al. "Zur Wasserdurchlässigkeit von Stahlbetonbauteilen mit Trennrissen", *Beton- und Stahlbetonbau*. Berlin. 1989. pp. 60-63

4. MEICHSNER, H. "Über die Selbstdichtung von Trennrissen in Beton". *Beton- und Stahlbetonbau*, Berlin. 1992, pp. 95-99

8. WÖRNER, J-D. AND TSUKAMOTO, M. "Zur Durchlässigkeit von faserverstärkten Betonbauteilen mit Trennrissen. *Beton- und Stahlbetonbau*, Berlin. 1993, pp. 68-74

9. COWIE, J. AND GLASSER, F P. "The Reaction between Cement and Natural Waters Containing Dissolved Carbon Dioxide", *Advances in Cement Research 4*. No. 15, 1991/1992, pp. 119-134.

DEEP BASEMENT CONSTRUCTION IN RISING WATER TABLES

L McLennan

T Forsyth

Sika Limited

United Kingdom

ABSTRACT. This paper describes the benefits of integral watertight concrete compared to other construction methods such as external membranes that can be damaged. It also concludes that the integral watertight concrete provides the contractor with benefits in terms of time, money and risk. Jointing systems are evaluated and the refurbishable hose systems are determined to be the securest method of achieving watertight joints.

Keywords: Watertight concrete, Design parameters, Mix design, Jointing materials.

Mr L McLennan is the New Construction Manager for Sika Limited. He is an admixture specialist covering their use in readymix, precast, mortars, blocks, tunnelling and specialist applications such as shotcrete, watertight, corrosion resistant and underwater.

Mr T Forsyth is the Product Manager for Sika Limited involved in the development of new and existing admixtures along with their related systems to the new construction market.

INTRODUCTION

This paper will review the standard methods of constructing and waterproofing deep basements and will consider the use of materials to provide watertight concrete such as integral pore blockers. Reference to correct design to combat the increasing threat posed by rising ground water tables will also be taken into account.

The discussion will also include a review of materials used in ensuring watertight joints, with an insight into refurbishable construction joints using injectable hoses and more traditional methods such as integral and external waterbars along with the use of hydrophilic seals.

The design parameters of BS 8102 will be discussed and a range of solutions to comply with the requirements of the grout grades will be proposed. The presentation will also include an analysis of contracts completed within the UK.

METHOD OF CONSTRUCTON

In order to provide a watertight environment the designer has various options such as:

- Drained cavity construction
- Sheet tanking
- Internal rendering
- Integral watertight concrete

Drained cavity construction allows for water that has penetrated to the cavity to be gathered and pumped away into a drainage system. This system in effect requires the construction of an outer and inner layer which reduces the internal area of the structure which in inner city areas may be a premium requirement. This system is used on new construction.

Sheet tanking systems involve the use of bitumen and rubber/PVC to provide a barrier to protect the structure from water ingress. This procedure requires access (overdig) to install an external membrane. The risk of damage during the construction process is high and once pierced the system can no longer provide protection. This system is used on new construction.

Internal tanking relies on a render system to provide the barrier to water ingress. This system has been proven for many years, however, the main drawback is that fixings may be required to penetrate the render which could allow water ingress. This system is used in both new construction and refurbishment.

Watertight concrete is a system which allows for protection from water ingress by utilising a two component admixture combination in a basic concrete mix design. The benefit is that the structure is constructed without the need for overdig or the fixing of membranes. The contractor can speedily progress the contract to a successful completion.

THE DESIGN PARAMETERS OF BS 8102:1990

BS 8102 allows the designer a choice of protection based on variations of construction type. The selection of products individually or in combination can be used to meet the demands of a wide variety of structural designs. Listed below are some examples showing simple cost effective solutions to meet the four grades of performance listed in BS 8102.

Grade 1: Basic utility for car parks, plant rooms/non-electrical
Minor seepage or damp patches tolerable

Reinforced concrete walls and floors in accordance with BS 8110 with construction joints effectively keyed by scabbling or surface retardation and cleaning. The use of a hydrophilic strip can also be considered.

Grade 2: Better utility for workshops, plant room retail developments
No water penetration allowable

Reinforced concrete and walls in accordance with BS 8007 containing integral pore blocking and water reducing admixtures with joints sealed with refurbishable hose system.

Grade 3: Habitable for ventilated residential and work areas including offices, restaurants, leisure establishments
Essentially dry environment

Reinforced concrete and walls in accordance with BS 8007 containing integral pore blocking and water reducing admixtures with joints sealed with refurbishable hose system.

Grade 4: Archives, stores, computer rooms, hospitals
Totally dry environment

Reinforced concrete and walls in accordance with BS 8007 containing integral pore blocking and water reducing admixtures with joints sealed with refurbishable hose system. Cavity floor system with internal walls or epoxy based coatings to provide vapour free environment.

WATERTIGHT CONCRETE

Watertight concrete comprises of a basic mix design for structural concrete incorporating a dual component admixture system.

Stage 1 utilises a powerful water reducing admixture which reduces the water/cement ratio and the size and number of micro pores left in the concrete after evaporation of the free water. Permeability testing of concrete with a water/cement ratio of 0.3 shows it is approximately 1000 times less permeable than concrete with a water/cement ratio of 0.7. However, to ensure ease of placement a water cement ratio of 0.4 to 0.45 is required.

Stage 2 utilises a pore blocking admixture which lines the remaining pores and swells after contact with water. This swelling action blocks the migration paths within the concrete ensuring total impermeability of the system.

There are various types of pore blocker available for use in watertight construction with long track records, however, the market place favours environmentally friendly types for ease of production.

Watertight concrete prevents the passage of water through the concrete matrix, in doing so it also prevents the passage of chlorides and sulphates in solution. These additional benefits provide added security with regard to long term durability. The system also improves the protection to the reinforcement embedded in the structure. The additional properties of the watertight concrete are enhanced as follows.

- Increased strength both compressive and tensile
- Reduced shrinkage and bleed
- Increased workability available for ease of placement
- Ease of pumping and compaction

One requirement that is essential is the control of cracking in the structure and the designer must ensure that the structure is crack free with movement controlled by correctly positioned jointing details.

MIX DESIGN

For well designed low permeability concrete the aggregate must be continuously graded to minimise voidage and water content. Cement content should be a minimum of 350 kgs per cubic metre of concrete. The mix should have sufficient workability for complete compaction without segregation, grout loss or bleeding. The admixtures currently used by Sika Limited are dosed at the rate of:

- Sika 1 a pore blocking blend based on colloidal silicates at 7 litres per cubic metre

- Sikament 10 a powerful superplasticiser based on polymers at 3.5 litres per cubic metre

Independent testing has demonstrated that this system can withstand pressures up to 100m head (10 bar).

The use of cementitious replacements will improve the performance of the system and should be encouraged if available.

The selection of other superplasticisers provides the ability to vary the properties of the concrete to achieve any of the following.

1. Accelerated strength development for early release of form work

2. High strength to afford the designer more flexibility

3. Flowing concrete for ease of placement

4. Self compacting by utilising Sika's Viscocrete technology

5. Retarded setting time for long haul supply situation

JOINTING MATERIALS

Watertight systems require the ability to provide a range of jointing solutions in order to ensure a watertight structure.

Water bar has been used in the UK for some considerable time and comes in all shapes and sizes for both internal and external applications. Whilst they provide a solution in the majority of cases it is widely recognised that around 10% fail due to a number of factors such as movement. Whilst placing concrete, difficulty to install correctly due to congested reinforcement and lack of compaction due to presence of the water bar.

Hydrophilic strips have also been established in the market place as a simple solution, with the strips swelling capability ensuring the blockage of the water's path. The use of these types of system tends to be in less critical situations.

An unique joint sealing system based on hypalon and an epoxy adhesive can be used for expansion joint and over bandage joint sealing. The advantage of this system is that it can be installed on damp surfaces and can accommodate large movements. It can also deal with extrusions through the concrete such as pipes.

In recent years the use of refurbishable hose systems has become more commonplace with the ability to reseal the joint in the future if the structure was damaged or had moved greater than anticipated. These systems are easily installed and afford the client and contractor the ability to test the joint and if necessary seal with a cementitious or resin based product. The latest innovation is to combine the hose system with hydrophilic strips to further enhance the contractors options in terms of success.

SUMMARY

Today's concrete structures should be built with tomorrow's needs in mind. However, changing environmental legislation can mean that current designs are inadequate for future requirements.

Containment structures and basements often require the basic concrete design and joint details to be protected with external sacrificial membranes or coatings to prevent structural deterioration or seepage.

Should these coatings or membranes break down before the service life of the structure is complete, costly repairs and inconvenience to the client are incurred.

As a worldwide leader in both concrete production and the repair and maintenance of underground and containment structures Sika Limited recognise that the future lies in the technology of today providing the client, specifier and the contractor with the solution to deep basement construction. This has enabled us to simplify the process and achieve the requirement of watertight construction with Sika's watertight system.

Table 1a, b and c show UK contracts which have been successfully supplied by Sika Limited as at April 1999.

Table 1a Watertight concrete references - UK Contracts

PROJECT	CLIENT	SPECIFIER/ CONSULTANTS	MAIN CONTRACTOR	SUPPLIER
GPO tunnels & survival chambers, Birmingham, Manchester, Oban Circa 1958-62	NATO			
St Nicholas Centre, Sutton, Surrey Circa 1985	P & O / Norwich Union	Bowden, Sillet & Partners, London	Balfour Beatty Building Limited	Ready Mixed Concrete (SE) Ltd
Hinde Court, Kingston, Surrey Circa 1987	London & Paris Property plc	Bowden, Sillett & Partners, Architects - Elsworth, Sykes & Partners, London	Pearce Construction Ltd	Pioneer Concrete Ltd
1995				
Brook House and Underground Car Park, Park Lane, London	Brook Developments	Sir Alexander Gibb	Higgs & Hill, Surrey	
Victoria Street, London Eight storey office block		Scott Wilson Kirkpatrick	Wiltshier Limited	
1996				
Imperial House, Temple Row, Birmingham Underground basement	Barberry House Properties	Fairbrother & Partners	Bryant Construction	
Barr Foods, Nine Mile Point, Newport Underground access tunnel	Barr Foods	John S Clegg & Associates	Dinas Constrcution	
1997				
Astra Charnwood, Loughborough Underground lift pits and service tunnel	Astra Charnwood	Amec Design	John Doyle	
Peebles Hydro Leisure Pool Peebles, Scotland	Hotel		Bardon Group	
Faggs Road Bridge		Hounslow County Council	Henry Boot	Tarmac Topmix S/Counties
Safeway, Temple End, High Wycombe		Paul Owen Associates	P J Carey Limited	Tarmac Topmix S/Counties

Table 1b Watertight concrete references - UK contracts

PROJECT	CLIENT	SPECIFIER/ CONSULTANTS	MAIN CONTRACTOR	SUPPLIER
1998				
St John's Centre, Queens Arcade, Liverpool		Curtins Consulting Engineers	Balfour Beatty	
Rolls Royce, Bilston, West Midlands			Roger Bullivant Sub: Framework Contractors	
Oracle Building, Blythe Valley, Solihull, West Midlands		Cameron Taylor Bedford	Birse Construction	
Manweb Sub-Station, Speke Business Park, Liverpool		Manweb	David McLean Contractors Ltd	
Toyota Casting Facility, Deeside, North Wales		Mouchelle International	Snimizu (UK)	
Fazakerley Waste Water Treatment Works, Liverpool		Bechtel	Pierse Constrcution	
Doxford International Business Park - Phase V11, Building 3		Mott MacDonald	Sillars (NE) Limited	
New Theatre, Keswick, Cumbria		Dewhurst McFarlane	Thomas Armstrong	
Abbey Panels Ltd, Coventry		Elliott Project Development Ltd	Beignton Construction	
The Weald Sports Centre		Douglas Rose & Partners	R Durtnell & Sons Ltd	Pioneer Willment Conc.
Canary Wharf Building D56			Swift Construction	RMC London
Senate, Nene College, Northampton			Stepnell Limited	RMC Transite
Downlands Industrial Estate		Greatorex Consulting Engineers	Oliver & Moody	Pioneer Willment SE
Canning Town Sub-Station			C F Harrington	London Concrete, Bow
Pfizer, Sandwich, Kent			John Doyle Ltd	RMC SE, Margate
HSA Hotel, Heathrow		Bowden Sillet & Partners	Laing	Tarmac Topmix S/Counties

Table 1c Watertight concrete references - UK contracts

PROJECT	CLIENT	SPECIFIER/ CONSULTANTS	MAIN CONTRACTOR	SUPPLIER
Aeltc Phase 2, Wimbledon		BDP Consulting Engineers	Byrne Bros (Formwork) Ltd	Pioneer Willment, Wimbledon
30-31 Golden Square, London W1		Bowden Sillett & Partners	Ballast Wiltshier	
Cannon (UK) HQ, Reigate, Surrey		Curtins Consulting Engineers	Crispin Borst Limited	
4 Spital Square, London D & B contract			Elmer Construction	

REFERENCES

1. BRITISH STANDARDS INSTITUTION. BS 8102: 1990 Protection of structures against water from the ground, 40pp

2. BRITISH STANDARDS INSTITUTION. BS 8007: 1987 Design of concrete structures for retaining aqueous liquids, 32pp

THEME FIVE:

PRECAST CONCRETE CONSTRUCTION

Keynote Paper

MODERN PREFABRICATION TECHNIQUES FOR BUILDING STRUCTURES

A Van Acker

Addtek International

Belgium

ABSTRACT. The current developments in precast concrete building techniques are inspired by a number of new market demands such as structural efficiency, optimum use of materials, speed of construction, quality, flexibility in use, adaptability and environmental issues. Prefabrication possesses a large potential to respond to these requirements. The possible solutions which are at our disposal lie not only within the classical advantages related to working conditions, technology and speed of execution, but also in new developments in the domain of materials such as high performance concrete, building systems such as hybrid structures, manufacturing technology e.g. automation, service integrated products and others.

Keywords: Prefabrication, Concrete, Building structures, Products, Systems, Applications

Arnold Van Acker completed his Masters degree in Civil Engineering at the University Ghent Belgium in 1961. Currently, he is a Director of Research and Development in Addtek International, a new group of companies for prefabricated concrete with 40 plants in Europe. He has been involved in the precast concrete business for 35 years in the areas design, execution and development. Arnold Van Acker is member of the Belgium Standardization Commission for Concrete, member of the CEN/TC 250/SC2, Eurocode2, and convenor of the working group Structural Elements of the CEN/TC229 "Precast Concrete".

He is chairman of the FIP Commission on Prefabrication since 1986 - now FIB since the merger between CEB and FIP in May 1998. He was awarded the FIP Medal for outstanding contribution to the development of precast concrete at the FIP Congress, Washington in 1994.

INTRODUCTION

The ongoing developments in the building activity are to a large extend inspired by a number of social and economic factors which marked our society during the last decades:

- A murderous competition between different construction materials and techniques: concrete versus steel, precast concrete versus cast in situ concrete or steel.

- Successive economical crises during the last decades

- A growing environmental consciousness

- Higher demands with regard to labour circumstances and comfort in general

The key words which are governing the evolution of the construction activity in general and precast concrete in particular are efficiency, durability and respect for the environment. Prefabrication possesses a large potential in this field to become a leading actor in the building activity of the future.

MARKET DEMANDS

The above mentioned trends started already during the eighties. In the mean time they have gradually been taking shape. Before getting into technical details, it might be interesting to describe more in detail the meaning of the most prominent factors that are at the base of the evolutions.

Structural Efficiency

The idea behind this is to design structures and to develop systems in such a way that the building offers maximum efficiency to the user. For example create maximum exploitation capacity of the available building space by using more slender building components like in slim floor structures etc.

Competition between different construction materials and systems is more and more judged in terms of performances and costs. Systems offering more m^2 inside the building volume are showing greater competition.

Flexibility in Use

Certain types of buildings are to be frequently adapted to the user needs. This is especially the case for office spaces, but also housing might require more adaptability in the future. The most suitable solution to this effect is to create a large free internal space without any restriction to possible subdivisions.

Optimum Use of Materials

Each construction material possesses specific properties and optimum applications. Until recently, the structure of a building was mostly build in the same material. The present evolution tends to use in the same construction, those materials which are best suited for different functions.

Speed of Construction

Because of the slowness of traditional in-situ construction methods, people are getting accustomed to long construction delays. However, during the last decades, we see that the request for quick return on invested capital is getting more and more importance: the decision to start the work is postponed until the last moment, but the initially agreed construction delay has to be met. In addition, projects are getting more complex, which is not in favour of short construction delays. One of the solutions taken by the general contractor, is to put more responsibilities on subcontractors.

Quality Consciousness

Quality has a broad meaning. Not only the quality of materials and execution has to respond to higher standards than before, but also the quality in the domain of user friendliness, comfort and aesthetics has become a dominant exigency in modern buildings. Several precasting companies have already obtained the ISO-9000 label, and a lot of work is being done in the field of European Standardization.

Adaptability

In the future, design of buildings shall have to take into account not only the direct costs for construction and exploitation, but also the deferred cost for adaptation or demolition, in other words we will have to design for the whole building's life span from cradle to grave. In this context, there will be much less demolition of entire buildings, but owners will rather choose for renovation. The initial concept of the entire building will therefore have to take account of the life span of the different components of the building: load bearing structure: 50 to 100 years and more; external envelope: 20 to 60 years; services: 10 to 20 years; finishes: 5 to 15 years.

Consequently everything apart of the main structure should be designed replaceable and renewable within the normal building life to avoid early termination; thus we have to design our buildings for sufficient inherent flexibility and adaptability to permit full reconfiguration of spaces and services. Periodic refurbishment, major modifications, replacements and improvements throughout the building life shall be possible. An inherent condition for reuse will be the complete inspectability of all building components.

Protection of the environment

In many countries the design of buildings and structures continues to be determined solely by the factors of "capital" and "labour" employed by ecologically unregulated market economies. This will become expensive when "nature" is added as the third market-determining factor and the full price also shall be paid for it.

The thought consciousness to preserve nature is getting increasing importance on world scale. In a number of European countries, governments are already imposing regulations with respect to environmental burdening by some materials like plastic, recuperation of packing, recycling of waste, soiling of ground by chemicals, etc. It is expected that also in the domain of construction, more severe exigencies will be imposed, e.g. with respect to emissivity of materials, shortage of raw materials, waste dumping, energy consumption etc.

All this fits into a strategy of sustainable development, which was put forward at the Rio conference in 1991. Sustainable development means adopting business strategies that meet the needs of the enterprise today while protecting, sustaining and enhancing the human and natural resources that will be needed in the future: balancing of the economic, the social and the ecological interests. For the construction industry, this means that the concept and design will undergo a drastic change of mentality.

SOLUTIONS OFFERED BY PRECAST CONCRETE

Prefabrication has always been a fore runner with respect to modernization in many fields: working conditions, advanced manufacturing technology, speed of construction and environmental friendliness. The last developments to respond the market needs are described here after.

New Materials

High performance concrete

High performance concrete, with compressive stresses exceeding 80 N/mm^2, is now being used every day in precast concrete manufacture. The applications are in the first place for heavy loaded columns, where an important reduction of the cross section can be obtained. Other applications are heavy bridge beams. roof beams with long spans, products with high requirements on durability, etc.

A recent development is the so-called ultra high strength concrete, up to 300 N/mm^2 and more. These materials are characterised by extremely low water/powder ratios (typically 0.14 - 0.20) high contents of micro silica or an equivalent microfiller (typically 20 - 25 % by weight of cement) and high contents of superplasticisers. Among the properties that are obtained are low permeability, high water resistance, high chemical resistance and compressive strengths up to 300 N/mm^2. Equally important is the fact that these materials can be produced and cast easily by the use of conventional mixes and vibration equipment.

The ultra high strength concrete has one drawback, common to all types of high strength concrete, namely a rather brittle behaviour when compared to conventional concretes. This can be modified by the introduction of rather high ratios of steel or carbon fibres to reinforce the matrix. New types of steel fibres are being introduced on the market recently. It concerns very thin high carbon steel fibres with strengths up to 2.000 N/mm^2.

The applications are still at the experimental stage, for example in containments for nuclear waste, products for extreme severe corrosive environments, products that were traditionally in steel such as manhole covers for sewerage, etc.

Steel fibre concrete

Concrete reinforced with steel fibers is taking increasingly importance. The reason lies in either the economy of materials, for example in tunnel linings, garage boxes etc, or in the possibility for automated production. Indeed, it has been proven that steel fibres are able to replace conventional stirrups, especially in prestressed concrete structural components. This opens new perspectives for automation of the production process for a number of precast structural products.

Products

Hollow core units

They are now the most widely used type of precast flooring: in Europe the annual production is about 20 million mR. This success is largely due to efficient design and production methods, choice of unit depth and capacity, surface finish and last but not least structural efficiency.

For many years, the maximum depth of prestressed hollow core floor units was 300 mm. However, recent developments in Northern Europe have led to a 500 mm deep unit.

In the context of sustainable construction, precast hollow core scores better than other types of floors, because of the rational use of the materials. The presence of longitudinal voids in the crosssection leads not only to a saving of half of the concrete needed for a plain slab, but reduces at the same time the amount of prestressing steel by 30 % because of the lower self weight.

Hollow core products suite also very well the purpose of flexibility and adaptability because of the large span/performance capacities. Presently, it is common practice to realize with a hollow core floor of 400 mm thickness, a span of 17 m for a live load of 5 kN/mR.

In Finland, the 500 mm thick unit permits already to span 21 m with a live load of 5 kN/mR. In Sweden, the concept to span from one facade to the other without intermediate support is currently applied in office buildings. This evolution is still progressing.

Light roof units

One of the inconveniences of concrete is its large self weight. On the other hand, mass in a building does control heat flow in such a way as to conserve energy. However, the latter property is still ignored by many designers. For roofs of industrial buildings, the energy conservation effect of mass is not really preponderant compared to the limiting span-load capacity by the large self weight. The latest developments for roof units are slender beams with large openings and complex wafer-form cross sections, with wall thicknesses of 20 mm. The realization of such thin concrete sections is now possible with the help of fibre reinforced concrete. As a result, roof units with a span from 6 to 30 m can now be made with a self weight of only 100 to 200 ka/mR.

New Systems: Hybrid Structures

The term "hybrid" is used to describe the combination of prefabricated concrete with one, or more than one other type(s) of structural component(s) in the same structure or building. It should not be confused with composite construction, which also uses both precast and another material, but where the structural performance relies on the interaction between the two. Prefabricated concrete is most commonly mixed with structural steelwork, cast in-situ concrete and masonry, and this form of construction is getting more popularity in new building design. The major beneficial advantages of hybrid precast construction lies in the possibility to maximise the overall building performance at both the conceptual level and the detailed structural level, where the large number of options available to designers will widen the scope for savings. One of the great virtues of hybrid construction is versatility.

The initial choice of the frame material for a project is influenced by many interrelated factors. Individually, each structural material has merits, yet there is greater benefit in combining materials: the advantages of one compensate for the drawbacks of another. For example, the combination of hollow core floors supported on steel girders enables to realize a flat ceiling. Such a system is not yet possible with concrete beams. In their turn, hollow core floors contribute substantially to the fire resistance of the system, where steel floor decking has no other solution than to apply an expensive fire insulating protection.

Production Processes

Automation is the challenge for the precast concrete industry in the coming decade. There are already good examples in the domain of pipes and small products with large series, such as terrazzo tiles, railway sleepers etc. In the area of building components, most has still to be done. Among the precursors are the hollow core slab, lattice girders and some small products. Also for internal wall units there exist fully automated production systems. The most suited solution for the main reinforcement of beams and floor units, is prestressing on long beds. The advantage of prestressing compared to mild steel reinforcement is that it gives high capacity for low labour input. The only remaining obstacles are the stirrups. Automatic incorporation of stirrups of different shapes and sizes seems to be excluded. The solution lies probably in the use of fibre concrete: steel fibres or other types of fibres to be developed.

In several countries, waste dumping is discouraged by high taxation. For precast concrete plants, as a consequence, rejected elements, concrete debris, waste concrete, slurry as a residue of the cleaning of concrete mixers or sawing the concrete, grinding and other processes, all this material will have to be processed. Recuperation and recycling of waste concrete and slurry is already done in several precasting factories. They are ideal for recycling because of the fact that the origin of the waste concrete is known. Research has been carried out on the evolution of the technological properties of concrete made with recycled aggregates. The study reveals that the percentage of coarse recycled material in structural concrete can be raised without any risk to 20% of the total amount of aggregates in the concrete mix.

EXAMPLES OF APPLICATION

Industrial and Commercial Buildings, Parking Garages etc

Systems, using frames composed of columns and beams, are still widely applied. The most prominent current changes are larger spans, lighter structures, new types of connections. A new system for industrial buildings is using load bearing sandwich walls in architectural concrete and long span light weight TT roof units. The latter have a self weight of 180 to 200 kg/mR and are spanning from facade to facade over up to 30 m. The system offers a more rational, economical and aesthetic construction.

Housing

Large wall panel systems combined with hollow core slab floors are still currently used for housing projects. However, the design has become much more flexible. The formerly rigid bloc shaped realizations hare now replaced by much more lively architectural designs.

Office buildings

Current trends in office buildings point towards more prefabrication, more efficiency e.g. through a flat under surface for floors without underneath beams and corbels, reduced site activity by incorporating ducts and conduits in the floor elements, and safer and faster construction.

a) *Slim floor structures:* Slim floor structures offer an effective solution to reduce the total floor construction height and to support the floor elements on the bottom flange of a steel beam. This enables us to realize a shallow floor in which beams and slab elements are integrated within the same depth. The combined units form a composite structure. The FIP Commission on Prefabrication has just published an Guide to Good Practice with design guidelines for "Composite floor structures".

b) *Hybrid structures:* There are numerous good examples of hybrid precast systems are available and some are already widely used. This is illustrated by the following examples.

- Timber roof beams can be combined with precast concrete columns and facades, mainly for architectural variety, e.g. in sport complexes, super markets, and others.

- Building frames are composed of combinations of precast concrete, in situ concrete and structural steel. There are for example cast in-situ columns with precast beams, or precast columns with steel beams and precast hollow core floors, etc.

- Floors can also be made with precast planks in combination with cast in-situ concrete. Another example of hybrid floors is the beam-bloc floor. The system is still widely used in Europe.

- Precast concrete is extensively used to provide structural and non-structural cladding. It offers the opportunity for almost unlimited architectural expression. Besides the self-finishes, the cladding units can also be covered by thin veneer stone, clay brick masonry or other materials.

- Stone is highly regarded as a cladding material or finish. Its expense can be mitigated by using thin veneers in precast concrete facade units. The system provides not only economy, but also better quality, speed of construction and higher safety than traditional in situ solutions.

c) *Split structure façade:* It concerns a facade construction in which the two leaves (inside leaf and outside leaf) are fabricated separately and erected separately. The load-bearing leaf of the facade consists of simple framed panels placed with the smooth moulded side towards the interior of the building.

   The precast floor units are supported on these elements. Afterwards an air tight joint sealing is applied and an insulation layer is attached to the exterior face of the wall panels and finally the exterior cladding panels are erected. The exterior cladding can be made in precast concrete or in other materials. Very often the window frames are placed over several storeys giving a large architectural freedom to the design

d) *New types of hidden corbels:* Column corbels often are inconvenient, especially in residential and administrative buildings. Recent developments go towards hidden steel corbels. The advantage of the solution is that the intersection between beam and column is neat, without an underlying corbel. The connection is also attractive from the aesthetic point of view. Various solutions are available on the market, of which some are seven below.

Other types of buildings

Stadia and grandstands are often constructed in precast concrete because of the extra short construction delays. There are numerous good examples in many countries. Last year, a large stadium for ice hockey was constructed in Helsinki in a time lapse of about 6 months.

SUMMARY

Prefabrication has always been a fore runner with respect to modernization of construction in many fields: working conditions, advanced manufacturing technology, speed of construction and environmental friendliness. The paper describes the different innovations offered by precast concrete in the domain of materials, products, new systems and production processes. They are for a great deal inspired by new market demands such as structural efficiency, flexibility in use, optimum use of materials, speed of construction, quality consciousness, adaptability and protection of the environment. In the domain of materials we see a increasing use of high strength concrete. Also steel fibre concrete is taking increasingly importance. For products, the tendency goes towards lighter products especially for floors and roofs. As far as construction system are concerned, the innovation is often based on a combination of precast units together with other types of structural components such as structural steelwork, masonry, wood, etc. Finally, in the domain of production processes, the emphasis is put on automation and environmental friendly processes.

THE BEHAVIOUR OF HOLLOWCORE FLOOR SLABS

E Fallon

Dublin Institute of Technology

J C Robinson

Trinity College

Ireland

ABSTRACT. The project forms one of a number of strategic and research development programmes with the Dublin Institute of Technology. It deals with the layout, design and production of pre-stressed Hollowcore slabs. Software modules are being developed which will integrate this type of pre-cast floor design into one package. The project began with a comparison of various codes of practice currently in use together with the new eurocode[6] EC2. It has now been extended to interact with CAD files of floor layouts and database files storing information on the floor geometry.

Keywords: Hollowcore slabs, Pre-cast flooring, Pre-stressing, Floor layout, Pre-cast slabs.

Edward Fallon MSc CEng MIStructE is a lecturer in the Dept. of Engineering Technology at the Dublin Institute of Technology, Dublin 1. He lectures in Computer applications/modelling and has a special interest in the development of software packages for the pre-cast industry.

James C Robinson PhD lectures in the Dept. of Civil Engineering in Trinity College, Dublin 2, Ireland. He specialises in numerical methods and, in particular, the Finite element method.

INTRODUCTION

In essence, pre-cast pre-stressed hollowcore slabs are simple pre-stressed concrete components. As such the general theory regarding their behaviour and design should be no different to that governing other pre-stressed components such as T-beams. A hollowcore slab could simply be regarded as a wide pre-stressed beam with longitudinal voids. In a more engineering sense, one would see a hollowcore slab as a series of pre-stressed concrete I-sections.

Whilst the theory and design approach governing pre-stressed concrete generally also applies to hollowcore slabs, the geometry and nature of the cross section also gives rise to other behavioural and design criteria that have to be considered. Such criteria are more likely to be contained in specialist literature or specific design recommendations regarding hollowcore slabs[1-4] than in general codes of practice and regulations for pre-stressed concrete design[5-9]. The manufacturing process for hollowcore slabs is such that it is generally not possible to include shear reinforcement or transverse reinforcement (to act as distribution steel) in the slabs.

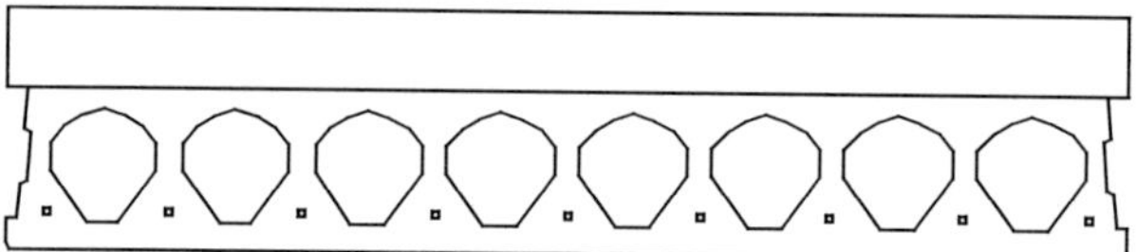

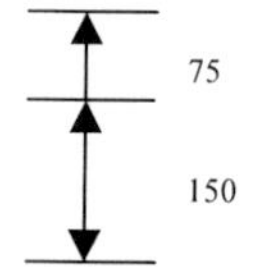

Figure 1 150 thick Hollowcore slab with 75 topping.

DESIGN OF HOLLOWCORE SLABS

There are up to four distinct stages in the design of hollowcore slabs. These are:

1. Stress Distribution for Serviceability Limit State.
2. Moment and vertical shear capacity for the Ultimate Limit State.
3. Deflection checks for Serviceability Limit State.
4. Horizontal shear checks at the slab topping interface.

STRESS DISTRIBUTION

The principle design check is carried out at the Serviceability Limit State and the edge fibre stresses are compared with allowable limits. In the discussion that follows, the stresses are all calculated and checked according to the rules of BS8110. Figures 2 and 3 give a breakdown of the stress components in a typical simply supported Hollowcore slab. The stresses have all been calculated at the point of maximum bending moment.

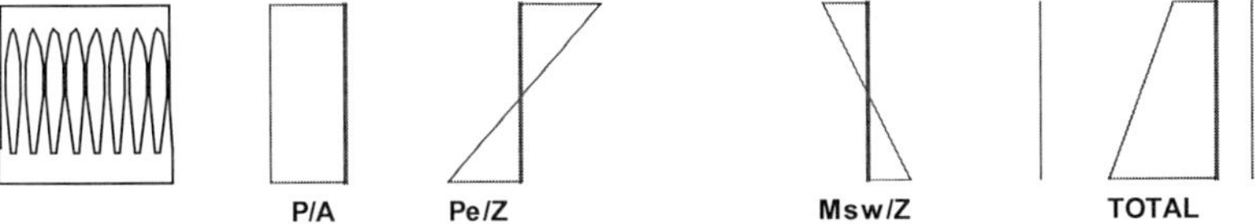

Figure 2 Distribution of transfer stresses

P/A: Effective pre-stressing force after losses up to that point in time divided by the cross-sectional area of the section. The slight additional stiffness due to the strands is ignored.

Pe/Z: The effective pre-stressing moment (P x eccentricity) over the section modulus of the section.

Msw /Z: The stress in the section due to self weight.
In the above example the total of these stresses shows a slight over-stress at transfer in the bottom fibres.

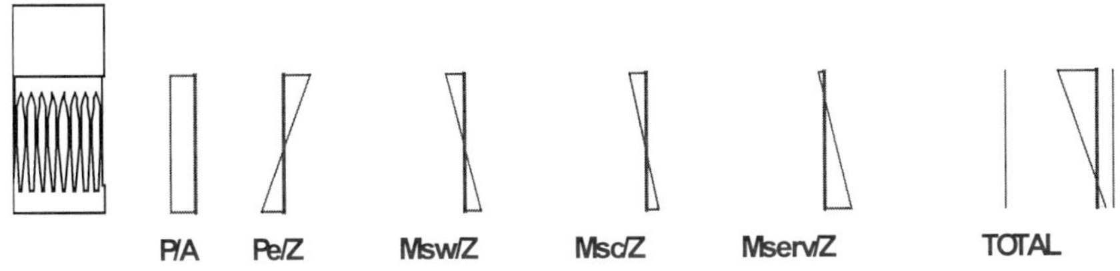

Figure 3 Distribution of service stresses

Msc/Z: The stress in the section due to topping.

Mserv/Z: The stress in the section due to applied Dead and Live loads.
The section modulus (Z) will vary from the time of transfer to the time of service if there is a topping applied. This means that loads, applied after the topping has hardened and become integrated into the profile, will result in lower stresses as the section will now distribute stress in proportion to its composite modulus.

The permissible stress limits allowed by BS8110 have been superimposed onto the TOTAL stress diagram and appear as two vertical red lines.

MOMENT CAPACITY FOR THE ULTIMATE LIMIT STATE

Moment capacities are calculated by equating the horizontal tension and compression forces. In so doing, the lever arm between these forces is obtained and thus the moment capacity (MR). As there up to four layers of strand possible in any cross-section, it is important that the centre of strand force be calculated for determining the moment capacity. If instead, the centre of strand area is used then an error of up to 20% underestimate of the moment capacity can result. The problem does not arise where only one layer of strand is used.

MR = Tension x Lever arm = Compression x Lever arm.

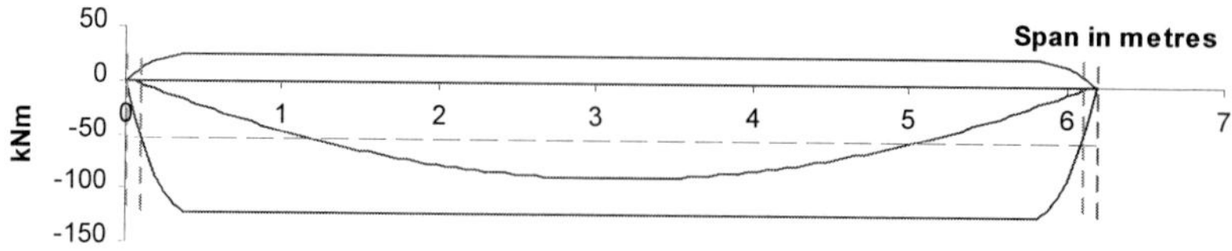

Figure 4 Moment in kNm due to applied Loads at ULS

The red lines show the positive and negative moment capacities. The horizontal dotted line shows the cracking moment capacity. The negative moment capacity is nominal, usually between 20 - 50 kNm for a typical slab. However, it can be enhanced for cantilever situations by including top strands in the profile.

VERTICAL SHEAR CAPACITY FOR THE ULTIMATE LIMIT STATE

The diagram in Figure 5 shows a capacity envelope superimposed onto the applied shear force due to factored gravity loads. The capacity envelope consists of two regions. The first, near the supports, is a function of the un-cracked capacity and is shown as a straight line. It is not truly straight as it will taper off to zero within the transmission length of the tendons. The transmission length is the distance within which the pre-stressing force is developed and is zero at the end faces of the unit.

The un-cracked shear capacity (V_{co}) is given by:

$$V_{co} = b_v \,.\, (I/S) \,.\, (\text{average shear stress})$$

Where b_v = Effective width of concrete available to resist shear and is the overall width of section less the total width of all cores.

(I/S) is the ratio of the Second moment of Area to the First moment of Area and can be considered to be the effective thickness of concrete resisting shear. Both I and S refer to the section capacities of the section without any composite action that would be available if a topping is present. This is because the average shear stress is primarily a function of the pre-stressing force acting on the original section. No mechanism exists which would suggest that the pre-stressing force could ever effect the level of stress within the thickness of the topping.

Within the top and bottom curved regions, away from the supports, the shear capacity is determined by cracked strength of the profile. The BS8110 equation is given below and is expressed in its two parts for convenience:

$$V_{cr}\,(1^{st}) = (1-0.55*Loss\_F\_x*(Jacking/100))*vc*bv*d\_ave$$
$$V_{cr}\,(2^{st}) = Mo*|SF\_x\_u|/|M\_x\_u|$$

Where :
Loss_F_x = fraction loss of pre-stress at the section considered.

Jacking = The level of jacking expressed as a percentage, usually 70%.

v_c = Shear stress resistance of the cracked section and includes the contribution of both the concrete and steel.

b_v as defined above.

d_ave = Distance from the compression fibres to the centroid of the strands.

M_o = Cracking moment capacity. This is the applied moment which would cause zero tension in the bottom fibres using an effective prestress of 80% of its actual value.

SF_x_u = Applied ultimate shear at section x.

M_x_u = Applied ultimate Moment at section x.

The second cracking shear term represents the residual shear resistance after the section has cracked.

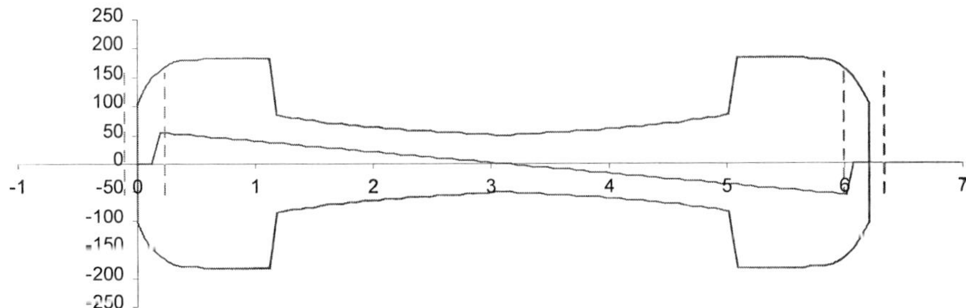

Figure 5 Shear in kN due to applied Loads at ULS

DEFLECTION CHECKS FOR SERVICEABILITY LIMIT STATE

The calculation of deflection involves the double integration of the slab curvature (M/EI) along the span[9]. Such calculations are usually undertaken by computer because of the number of variables involved. Deflections may be broken down into short term and long term. Within each, further sub-divisions are made giving a total of four types of deflection that need to be checked.

1. **Upward camber**: This is the deflection due to both the self weight and the pre-stressing force at transfer measured relative to a horizontal soffit. In the case of long spans the resulting deflection might even go downwards. The camber should not exceed 1/350 of the span.

2. **Deflection when finishes are applied**: The level of pre-stress at the time the finishes are applied is taken into account. This upward deflection is added to the downward deflection arising from all the permanent gravity loads.

3. **Long Term Deflection:** The E value for the section includes the effect of creep and is given by E long term = E short term / (1 + ϕ) , where ϕ = creep coefficient, usually taken as 2. This deflection includes both the service pre-stress and all gravity loads. A limit of 1/350 of the span is applied.

4. **Deflection after finishes:** The main interest here is to determine the relative deflection, which will occur in the long term, arising from finishes. Here a limit of 1/350 of the span is used for flexible finishes and 1/500 of the span in the case where the finishes are considered brittle.

HORIZONTAL SHEAR CHECKS AT THE SLAB TOPPING INTERFACE

If a topping is present then a check needs to be done , at the Ultimate Limit State, to ensure that the shear stress at the concrete/topping interface is not exceeded.

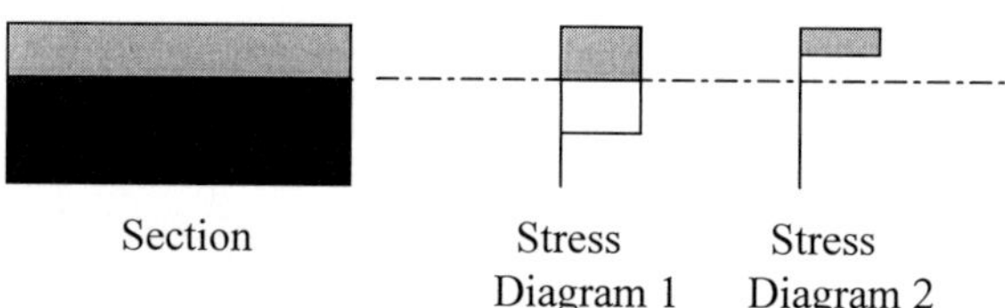

Figure 6 Stress Distribution at ULS

The compressive stress to be transmitted across the interface is shown by the grey areas in Figure 6. Its magnitude is found from the integration of the ultimate stress diagram at the point of maximum bending moment. The average horizontal shear stress is then calculated by taking this maximum topping force and dividing it by that area of slab/topping interface between the point of maximum bending moment and the point of zero bending moment. The resulting average shear stress is assumed to vary in direct proportion to the applied shear force diagram. In the first diagram , the compression depth exceeds the topping depth and the compression force at the interface is limited to the depth of the topping. In the second stress diagram the depth doesn't exceed the topping depth and the force is not restricted.

A hollowcore slab may have service openings which will reduce the contact area. This in turn, makes the resulting calculations rather tedious.

BS8110 allows 0.6 N/mm² stress a t the interface using a topping of 35N/mm² cube strength, which is typical for most floors. At this level of stress , only those floors that have short spans and high applied loads will experience difficulty in regard to horizontal shear.

DRAWING LAYOUT

When it has been decided that hollowcore slabs are to be used in a particular building project, the pre-cast supplier requests the project drawings. Increasingly these days, such drawings are in AutoCAD™ format. Typically the AutoCAD drawings show general details such as floor plans and sections. On receipt of these details, the pre-cast supplier extracts his particular information by re-tracing over the drawing using the built-in 'Polyline' function. A polyline describes on plan the boundaries of those areas to be covered with hollowcore slabs.

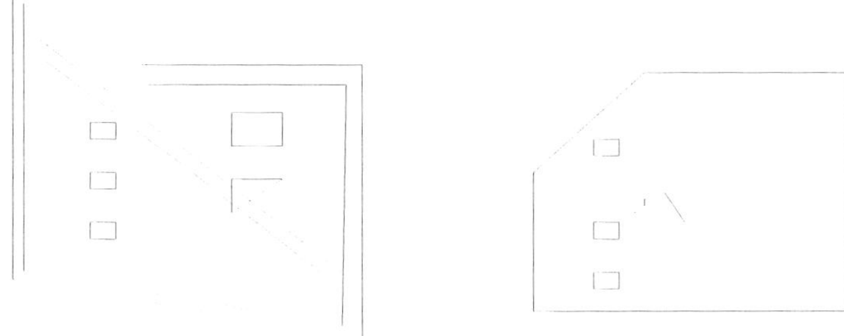

Figure 7 Floor Plans using Line and Polyline entities

Figure 7 shows a simple example of a floor plan. It contains several floor openings which are drawn as polylines. The plan on the left has an inner perimeter drawn as a polyline and an outer perimeter drawn as a series of line entities. In addition, there is a pair of parallel lines, also line entities, which will act as a reference guide for imposing line loads due to walling. If a polyline, which is a closed entity, has an area of less than 0.5 m² then it is treated as an opening in the floor, otherwise it is assumed to be a polyline containing hollowcore slabs. Software tools are used to perform various operations on the drawing. These are:

1. Slicing: The areas are cut into a series of parallel slices. The thickness of the slice is determined by the width of the Hollow Core unit, usually 1200.
2. Trimming: Slabs need to be trimmed at all openings and edges. Invariably this will result in slabs which are not rectangular in shape. Such slabs are avoided where possible, as they result in expensive cutting and handling.
3. Floor Loading.

TYPES OF HOLLOWCORE PROFILES

1. **Hand cast**. Here the concrete is placed in the conventional way, using a concrete gang and assisted with tampers and vibrators. It is normal practice to introduce rectangular void formers in polystyrene.

2. **Machine cast**. The machine, which is either of a "slip former" or "extruder" type, moves slowly along a pre-cast bed. The bed can be up to 150m in length. A large hopper on top of the machine is kept continuously filled and the profile is formed either by horizontal cylinders moving in and out of the concrete – slip forming, or by being squeezed out under its own weight – extrusion.

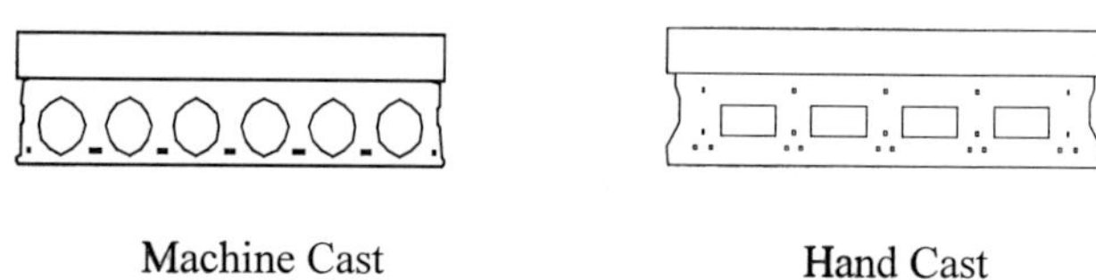

Machine Cast Hand Cast

Figure 8 Profile Types

WORKED EXAMPLE

A design example has been chosen which illustrates the complexities involved with wall loads and openings having to be taken into account. The profile used is 200mm thick and contains 11 cores. It is made using an ECHO Engineering slip former machine.

One feature of this design example is the use of propping mid-span during construction. Propping has the effect of reducing the final tensile stresses by locking in a counter compression force. This is shown in Figure 9. In the first diagram then bending moment of the wet screed looks like that of a two span slab. Thus the bottom fibre stress at mid-span is compressive. At this point the section stiffness is obtained only from the slab. However when the topping has set the stiffness increases and the prop can be removed. The effect of removing the prop is equivalent to applying a point load mid-span. An overall benefit is gained as the total tensile stress in the bottom fibres at service has now been reduced and propping will therefore extend the span range of the slab.

Figure 9 Propped and un-propped Moment diagram due to Topping

This example was deliberately chosen to fail. This is shown in Figure 12 of the results where the Moment Strength envelope doesn't fully encompass the applied Moment diagram.

Creating with ConcreteDundee, 6-10 September 1999		
Profile Analysis: Design to BS8110-200mm Slab	Tel +44(1382) 344 347	
Client: **Creating with Concrete**	Date: 31/08/98	JOB REF: Sample

Basic Design Data

Self Weight(unjointed)	4.78 kN/m²	Dead Load Factor(ULS)	1.4
Self Weight(jointed)	4.91 kN/m²	Live Load Factor(ULS)	1.6
Screed Strength	35 N/mm²	Ixx:	672.0E6 mm4
Slab Strength @ Transfer	35 N/mm²	Ybtm:	98.27 mm
Slab Strength @ Service	50 N/mm²	I Comp:	1.7E9 mm4
Creep Factor	2.0	Y Comp:	146.42 mm

Slab Details

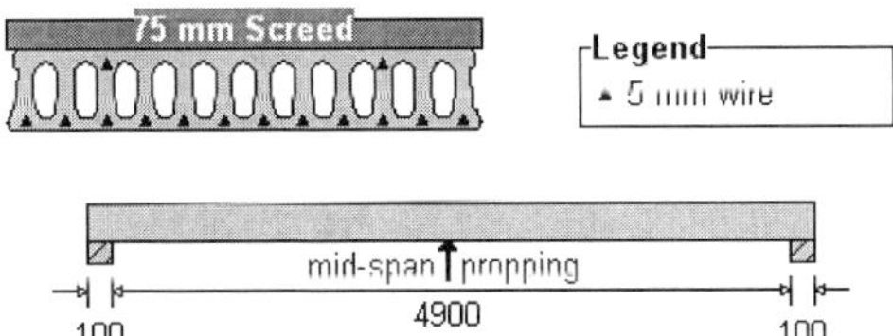

Applied Loads

UDL No: (kN/mNo: 1
Dead Load: 3.0
Live Load: 3.5

Wall No: (kN/nNo: 1
Dead Load: 10.0
Live Load: 0.0

Openings

Opening No:	**No:1**	**No:2**	**No:3**
***Left (mm):**	2000	2000	2000
Width (mm):	1000	1000	500
Height (mm):	100	100	100
Inset (mm):	950	150	550

*Left: distance from the left face of the opening to the left end of the slab

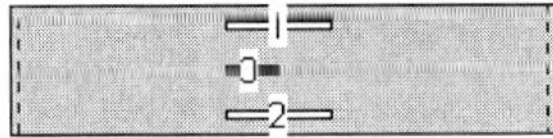

Results generated using HC-DESIGN 1.0 PAGE 1 Copyright FLOORCAD LTD. 1998

Figure 10

Creating with ConcreteDundee, 6-10 September 1999
Profile Analysis: Design to BS8110-200mm Slab
Tel +44(1382) 344 347
Client: Creating with Concrete
Date: 31/08/98
JOB REF: Sample
14.00
1.55
-0.05
-2.66
Top Fibre Stress (N/mm²) @ Transfer
PASS
14.00
3.50
0.00
-2.66
Bottom Fibre Stress (N/mm²) @ Transfer
PASS
16.50
3.22
-0.02
-3.18
Top Fibre Stress (N/mm²) @ Service
PASS
16.50
2.12
-3.18
-3.06
Bottom Fibre Stress (N/mm²) @ Service
PASS
Results generated using HC-DESIGN 1.0
PAGE 2
Copyright FLOORCAD LTD. 1998

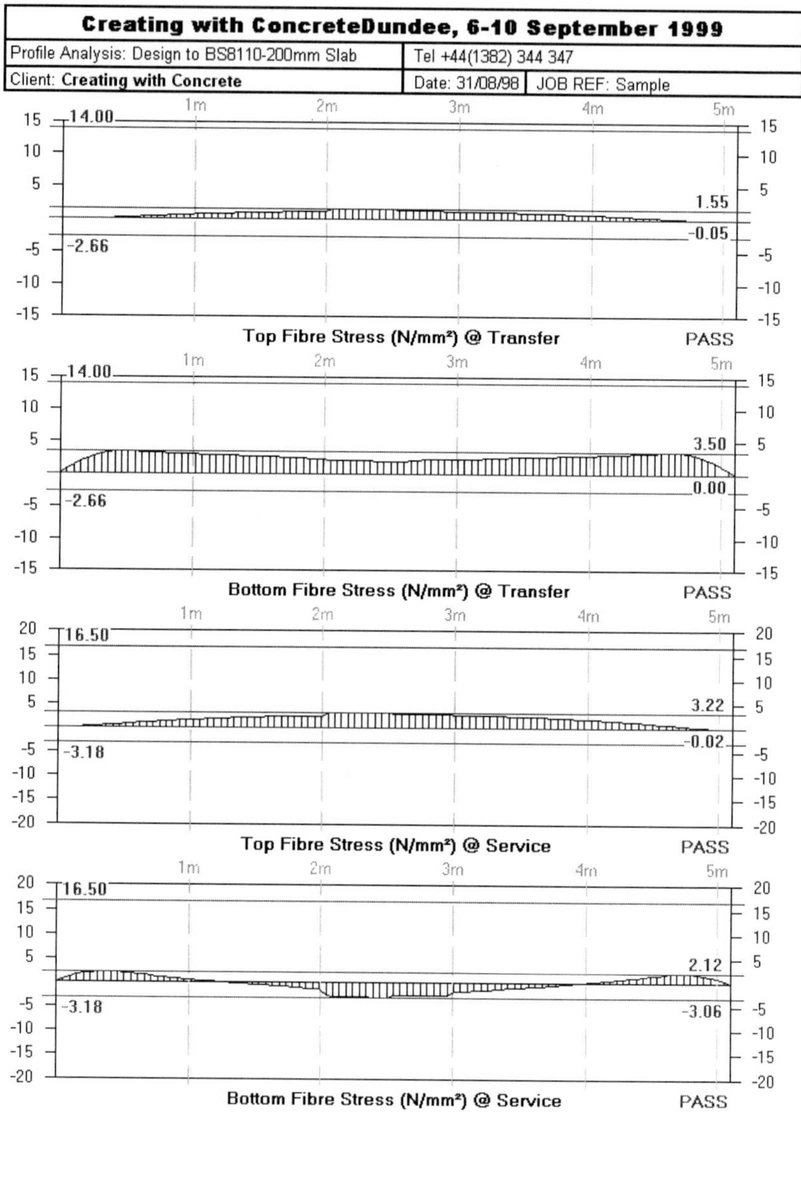

Figure 11

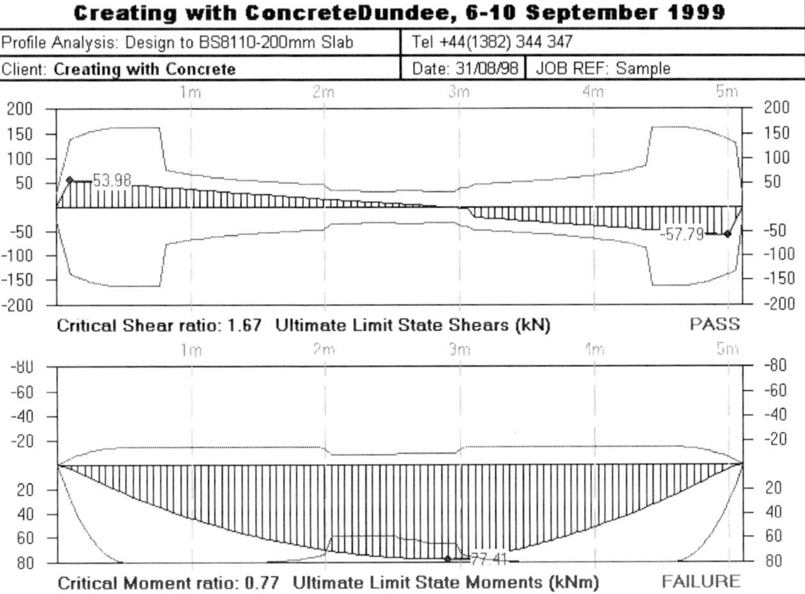

Results generated using HC-DESIGN 1.0 PAGE 3 Copyright FLOORCAD LTD. 1998

Figure 12

CONCLUSIONS

1. Pre-stressed hollowcore slabs are extremely versatile structural floor elements.

2. A propped composite topping can be used to advantage in limiting both deflections and service tensile stresses.

3. Holes and Notches can be accommodated easily in Hollowcore slabs. Unfortunately such openings often occur directly at strand positions and has the effect of making the resulting calculations long and tedious. Appropriate computer software is now available to tackle such situations.

4. Further work needs to be carried out in such areas as diaphragm action, continuous supports, mechanism for transferring prestressing force at the ends of slabs and the effect of large openings.

REFERENCES

1. FÉDÉRATION INTERNATIONALE DE LA PRÉCONTRAINTE: FIP Recommendations – Precast prestressed hollow core floors, Thomas Telford, London, 1988.

2. TECHNICAL COMMITTEE CEN/TC 229: prEN 1168 Floor of precast prestressed hollow core elements. European Committee for Standardisation (CEN), 1993.

3. PCI HOLLOW CORE SLAB PRODUCERS COMITTEE: PCI Manual for the design of Hollow Core Slabs, Precast/Prestressed Concrete Institute, Chicago, 1985.

4. GROUPE SPÉCIALISÉ N° 3 DE LA COMMISION TECHNIQUES: Cahiers des prescriptions techniques communes aux procedes de planchers Titre III (1ere Partie) Planchers confectionnes à partir de Dalles Alvéolées en Beton Précontrainte, Centre Scientifique et Technique du Bâtiment, Paris, 1992.

5. BRITISH STANDARDS INSTITUTION CIVIL ENGINEERING AND BUILDING STRUCTURES COMMITTEE: BS 8110: Part 1: 1985 Structural use of concrete - Code of practice for design and construction. British Standards Institution, 1985.

6. EUROPEAN COMMITTEE FOR STANDARDISATION: DD ENV 1992-1-1 : 1991 Eurocode 2 Draft for Development: Design of concrete structures - Part 1: General rules and rules for buildings. European Committee for Standardisation (CEN), 1991.

7. ACI COMMITTEE 318: ACI 318M-95 & ACI 318RM/95 - Building Code Requirements for Structural Concrete & Commentary. American Concrete Institute, Farmington Hills, 1995.

8. MOSLEY, W, H, AND BUNGEY, J, H. Reinforced Concrete Design 4th Edition, Macmillan, London, 1990.

FEASIBILITY OF CONSTRUCTING SHELL ROOFS FOR LARGE-SPAN PUBLIC BUILDINGS

A S Zhiv

V A Matveyev

S V Nikolayeva

Vladimir State University

Russia

ABSTRACT. This paper considers the approach to the design and construction of two public buildings in the cities of Vladimir and Ivanovo (Russia). Both buildings situated in the historic centres of the cities provide accomodation for public gatherings during sport events and concerts. The building of a swimming pool in the city of Vladimir is spanned by double curvature cast-in-place and precast shell measuring 78.0 x 78.0 m on plan. The building in the city of Ivanovo replacing the old circus building is intended for giving concert and circus performances. The building is spanned by 43 m diameter cast-in-place and precast dome. The construction of such large-span buildings demonstrates new possibilites for using modern construction techniques as it involves industrialized methods of construction, makes efficient use of materials in structures as well as of innovative techniques for controlling production sequences.

Keywords: Shell of double curvature, Dome, Subdiagonal bracing, Arches, Erection tower.

Professor A S Zhiv Vladimir State University. He specializes in the design and construction of reinforced concrete shell roofs and spatial structures.

Mr V A Matveyev is Pro-rector for Construction at Vladimir State University, Vladimir, Russia. He is a post-graduate student at the Department of Construction Technology and specializes in the design and construction of reinforced concrete shell roofs and spatial structures.

Mrs S V Nikolayeva is a senior lecturer at the Faculty of Civil Engineering and Architecture of Vladimir State University, Vladimir, Russia. She specializes in the construction of buildings and structures.

INTRODUCTION

The experience in urban development and dialectics of public relations favour the view that types of buildings and the way they interact show how physical environment may influence the tenor of life. That's why problems connected with the design and construction of large-span buildings and structures, the choice of efficient and functional structural designs, and siting of buildings in a city come to the forefront. Sport facilities are good for creating the most favourable atmosphere during sporting festivals, public holidays and the like and their spatial volume should be optimal. In sport facilities hall is the main composition core which gives possibilities for creating architectural forms.

Basically the architectural design is governed by the configuration of the plan taking into account the composition of the structure and the functional, technological, acoustical and mechanical requirements to the hall, as well as the shell roof design. The above-mentioned requirements determine the aesthetically satisfying design and when taken together dictate the horizontal and vertical dimensions of the hall, i.e. its volume and shape.

SWIMMING POOL

The design objective for the swimming pool in the city of Vladimir with 80 m length of the swimming bath was to accomodate the large-span structure on the relatively small plot of land measuring 90 x 90 m. on plan. This called for placing the bath not parallel to the sides of the building but at the angle of 45^0 in the diagonal direction.

Correspondingly, natural lighting of the building had to be reconsidered with regard to the direction of the swimming track (Figure 1).

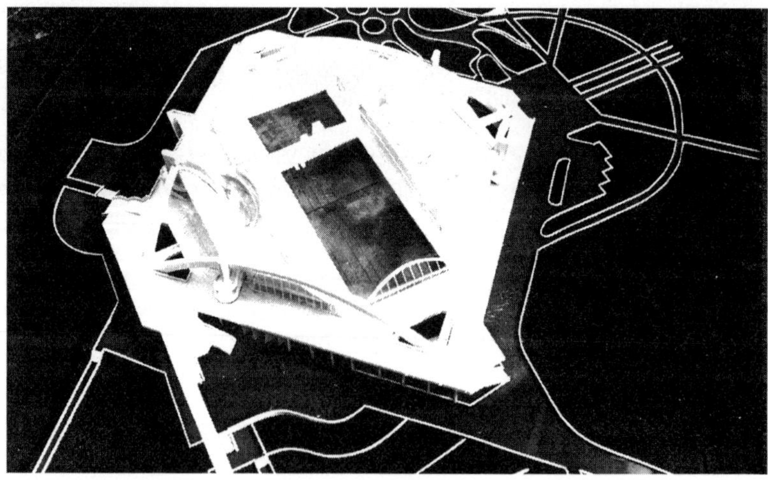

Figure 1 Plan view of swimming pool (scale model)

During the preliminary design stage, a number of architectural configurations and structural schemes were investigated. Many aspects of the design were dictated by handling and erection capabilities and the construction machinery available.

The feasibility study confirmed the final decision which was to use cast-in-place and precast concrete shell of positive Gaussian curvature measuring 78 × 78 m on plan (about axes).
The original design of shell roof called for a convex polyhedron formed of a system of folded vaults. The parameters of the shell roof (rise, supporting arc curvature) have been taken to satisfy the sloping criteria.

The shell structure consists of 6 m precast flat ribbed slabs and cast-in-place units adjoining the supporting contours. Cast-in-place braced arches spaced 18 and 24 m from the main axes are inscribed into the shell at the angles of 45 and $30^0 30'$ (parallel to the supporting contours. The cast-in-place supporting contour of the shell in the form of a curvilinear arch is supported by precast columns with a spacing of 6 m. The shell slabs are 60-200 mm thick and the ribs are 300 mm high.

To provide the top lighting some of the slabs in the middle have 1800 ◈ 2400 mm openings. The elements making up the shell were made of B30 grade heavy-weight concrete. Arch bracing was provided by prestressed reinforcing bars (Figure 2).

Figure 2 Plan of swimming pool shell roof

The shell roof has been calculated for the uniformly distributed loading of 5.5 kN/m^2 in accordance with the Moment Theory by V.Z.Vlasov [1] and for the concentrated loads of 30 kN arising from hanged up equipment and advertizing boards.

The construction of the large-span shell roof has presented a challenging task mainly due to the shell roof large dimensions on plan.

According to the structural engineering concept the shape of the shell roof is created by flushing sand matrix accompanied by layer-by-layer soil compaction with the subsequent construction of 50 mm thick plain concrete base for the erection of precast roof slabs. The erection of the precast slabs was done with the of MI-8 helicopters. After joining together protruding bars of the slabs the construction joints were grouted with B30 grade concrete.

The shell roof form removal is done in the course of the earthwork when excavators move the flushed soil away for the site planning.

CIRCUS DOME

The cast-in-place and precast dome in the city of Ivanovo with the diameter of 43,860 mm is assembled of 96 units of two standard sizes made to the shape of a trapezium (Figure 3).

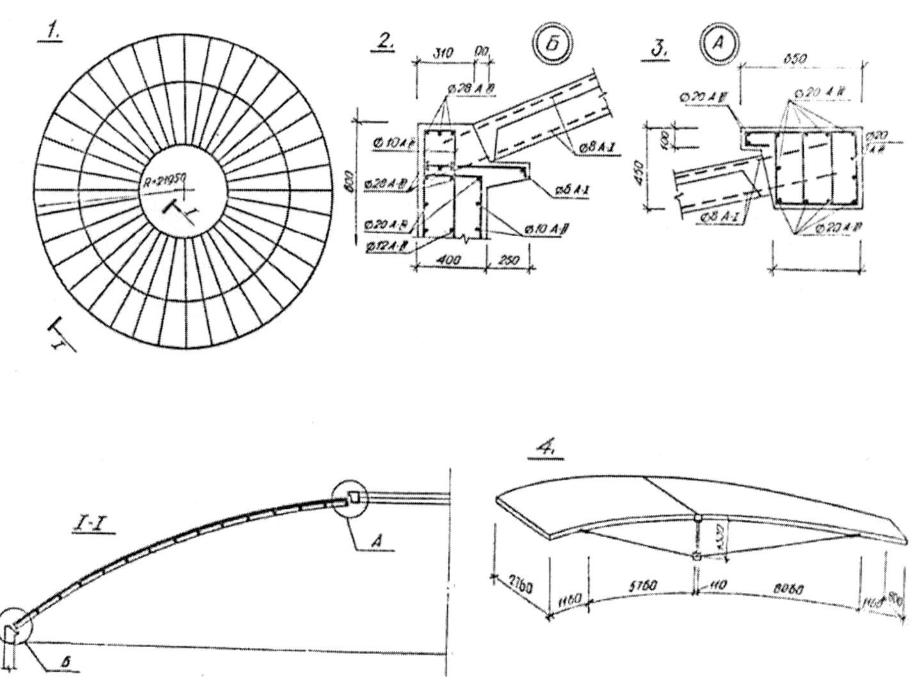

Figure 3 Circus dome structure,
Unit 1: shell plan, Units 2 and 3: cross - sectional view, 4: Erection modular unit

The dome camber makes 1/6 of the diameter. There is a 1,340 mm cutout in the central part of the dome, cutting out being done in the meridian and circumferential directions.

The lower and upper supporting rings having the sections of 400 x 800 mm and 450 x 600 mm accordingly were made of cast-in-place reinforced concrete. The precast units of the dome circling the perimeter of the roof were designed ribbed. The meridian ribs were 200 mm high and the height of the ring ribs was 180 mm. The flange thickness was assumed constant (50 mm) for all the units.

The lower and upper supporting rings were reinforced using 20-28 mm diameter nonprestressed A-III grade reinforcement. Reinforcing of the meridian and ring ribs was done with reinforcing cages, the diameter of the rods being 14 and 16 mm accordingly; the flange units were reinforced using 150/150/3/3 single reinforcing fabric.

Joining cast-in-place units together was done by welding protruding rods and connecting them with precast supporting rings was done by embedding in concrete the corresponding protruding rods of the meridian ribs of the slabs. B25 grade heavy-weight concrete was taken for all precast units of the dome, cast-in-place supporting rings and for grouting field joints. The dome has been calculated for the uniformly distributed load of 5.0 kN/m^2 and for the concentrated load of 15 kN. The concentrated load was applied to the shell roof in eight points.

The dome was erected using larger sectors to produce which top and bottom slabs were combined using embedded fittings on the principle of subdiagonal bracing.
According to the building construction program the large dome sectors were mounted onto the erection modular unit installed in the centre of the shell roof, one side of the sectors being supported by the formwork of the lower ring. The sectors were jointed together by arc welding, the upper and lower rings were concreted and the butt joints between the sectors were grouted with heavy-weight conrete.

CONCLUSIONS

In conclusion it may be stated that the construction of these two cast-in-place and precast large-span buildings in relatively small Russian cities is a significant achievement of the Russian science of building.

REFERENCES

1. VLASOV, Y Z. Cylindrical shells and new ways of developing thin- walled spatial systems in structural mechanics, Concrete shell roof construction 1-3 july 1957, Teknisk ukeblad, Oslo, Norway, P. P. 106-130.

BEHAVIOUR OF CONCRETE AND REINFORCED CONCRETE AT RANDOM LOADING

T L Chyrva

Krivoy Rog Technical University

V N Chyrva

Pridneprovye

Ukraine

ABSTRACT. This article describes experimental and theoretical researches of stressed-deformed state of concrete and reinforced concrete elements under random loadings on the basis of statistical test method. Analytical dependences for deformations and strength of concrete as a function of loading cycle value have been obtained. The results of the concrete investigation have been taken as a principle of suggested determination methods of eccentrically compressed element strength at random loadings. Trustworthiness of methods has been confirmed by experimental researches. The present methods of approach to the concrete research can be used at investigation of construction strength and reliability.

Keywords: Prop, Concrete, Reinforced concrete, Random loading, Strength, Deformation, Number of cycle, Relative eccentricity value, Persent of reinforcing.

Professor Tetyana L Chyrva is Associate Professor at Department of Architectural constructions, Krivoy Rog Technical University, Ukraine. Her main research interest includes the behaviour of reinforced concrete elements under conditions of non-multiple fluctuating loads and at random loading.

Professor Volodimir N Chyrva is Head Director of Building Firm "Pridneprovye", Krivoy Rog, Ukraine. He is engaged in researching and working out buildings and construction strengthening methods under conditions of industrial seismic and also in creating dynamic models of constructions at random loading.

INTRODUCTION

Probability approarch to studies of external actions and characteristics of using materials is taken as the basis of strength and safety of building constructions.

In the process of exploitation most of constructions are under conditions of non-multiple fluctuating loads. Their value in some of cases can exceed exploitation levels and occur at random. That sort of actions includes earthquake, technological, emergency, temperature and humidity loads. At the construction design the building regulations of today do not take into account special features of reinforced concrete work in natural conditions, exposed to random loads. Meanwhile, such loading under the action of growth and accumulation of residual deformations, crack number increase, faults of bond between concrete and steel, can lead to exhaustion of load-carrying capacity and to untimely failure.It is necessary, therefore, to carry out experimental and theoretical research of reinforced concrete constructions under random loadings.

METHODS OF EXPERIMENTAL RESEARCH

The estimation of experimental research volume was carried out on the basis of an active experiment.While making up of a plan of a full factorial experiment the following variable values were taken: class of concrete R_b, the number of cycles n, the relative eccentricity value e/h_0 , persent of reinforcing μ_S.

The props with the section of 150x150 mm and the length of 600 mm were constructed. The given length of the props excluded the influence of flexibility on the work of the work of the element. The samples were made of heavy concrete R_b=12MPa and R_b=30MPa. The bar deformed reinforcement of class A-III, the diameters of which are 10 mm and 12 mm were taken as effective reinforcement in the samples.

The value of the given excentricity was within the limits of the section. The number of the cycles varied from 10 to 100.

By nature of action of the longitudinal load the samples were divided into two groups, the first one included the samples tested at static single load, and the second one tested at fluctuating load with the further transient load. According to the results of the tests of the first group samples the levels of repeated loadings were established. Such methods of experiment allowed to bring out the action of random loadings on strength and deformation characteristics of constructions. Strength value at a single loading had been taken as a standard one, referring to which the strength at random loadings was measured. The character of concrete damage at eccentrically pressed samples from compressive stress of the concrete at the action of random load depended on the value of assumed eccentricity and practically did not depend on the loading level and cycle number. The results of the investigation showed that the random loading for tested samples had led to the increase of the load-carrying capacity on 18-30% in comparison with the single loading, depending on the value of eccentricity and loading level. The random loading with the increase of cycle number ($n \leq$ 30) promotes the realization of strength abilities owing to concrete structure compaction.

Therefore, in order to work out the methods of construction strength calculation, it is necessary to research the concrete strength at random loadings.

To describe such loadings the methods of random values and processes are used. In the present paper such an approach was implemented on the basis of the Monte-Carlo method. The technique of its application is as follows: simulating a set of pseudorandom numbers on a computer, an optional realization of random loading is being simulated as a function of cycle number according to the specified distribution law. After testing of concrete prisms in accordance with specified programme the realizations of stressed-deformed concrete parameters are obtained. Carrying out a statistical analysis of gained realization, the probability characteristics of investigated parameters are being calculated.

Simulation methods of random actions use the sequence of uniformly distributed random numbers in interval (0, 1). To obtain them certain tables or programme transmitters are used. On the basis of the sequence of uniformly distributed numbers, the sequences with other laws of distribution are obtained.

RESULTS OF RESEARCH

In the submitted paper an upper level of external action has been considered as a uniformlydistributed one, and has been modelled as follows:

$$\eta = a + b \cdot RND(x), \qquad (1)$$

where,

η	: an upper level of the external action;
a	: the lowest value of the action level;
b	: the interval of varying;
$RND(x)$	: the random number in interval (0, 1).

It has been specified, that a = 0.4; b = 0.5, i.e. the possible upper level of loading is equal to 0.9.

The testings were held on the concrete prisms, worked out of heavy concrete R_b=30MPa. Dependences $\sigma \sim \varepsilon$ permit carrying out an analysis of loading history influence on the parameters of stressed deformed state of a running cycle.The total number of cycles was specified as 30. If the loading level on the present cycle is $\eta_n > \eta^{max}$ in comparison with those of the previous cycles, the hysteresis loop width is considerable, and a high increase of residual deformations takes place. The following cycle with $\eta^{n+1} < \eta_n$ has insignificant width of the loop indicating of concrete plastic flow release and of its elasticity work, i.e. concrete adaptation for loading regime is observed. At $\eta^{n+1} > \eta_n$ a new increasing of loop growth of residual deformation occurs.

Statistical processing of testing results permits achieving approximate expressions to carry out a plot construction of dependences $\sigma \sim \varepsilon$ at each loading cycle (n = 30):

$$\varepsilon^{*n}_{r} = 49 + 58\, log(n), \qquad (r=0.93)\ (2)$$

$$\sigma_{n\uparrow} = 0.4236\, n^{1.9626*10\text{-}2} (\varepsilon^* - \varepsilon^{*n}{}_{r})^{0.7987}\, e^{-2.017*10\text{-}3(\varepsilon^* - \varepsilon^{*}nr)}, \qquad (r = 0.92) \quad (3)$$

$$\sigma_{n\downarrow} = 0.4768\, n^{3.7955*10\text{-}3} (\varepsilon^* - \varepsilon^{*n+1}{}_{r})^{0.6716}\, e^{5.5843*10\text{-}3(\varepsilon^* - \varepsilon^{*}n+1\, r)} \qquad (r = 0.85) \quad (4)$$

where, $\varepsilon^* = \varepsilon * 10^5$;
$[\sigma]$: stress in, MPa;
$\sigma_{n\uparrow}$: stress on *n*-semicycle of the loading;
$\sigma_{n\downarrow}$: stress on n-semicycle of load relief.

The investigates of the second derivative from functions 3 and 4 allow to analyse the presence of points of contraflexure and accordingly of convexity of these functions:

$$2c\beta \cdot 1/\varepsilon + c^2 + \beta(\beta - 1) \cdot 1/\varepsilon^2 = 0 \qquad (5)$$

where, $c = 2.0170\ 10^{-3}$, $\beta = 0.7987$ for $\sigma_{n\uparrow}$;
$c = 5.5843\ 10^{-3}$, $\beta = 0.6718$ for $\sigma_{n\downarrow}$.

Function 3 before rearching its maximum meaning, when $\varepsilon = 39.5$, does not have the points contraflexure and is turned with convexity to axis.

Analysis of equation 3 and 4 shows that with the growth of cycle number the elasticity modulus value increases. That is subsequently, one of preconditions of a load-carrying capacity increasing of eccentrically pressed elements at cyclical loadings.

Correlation function of present random process, constructed for residual deformations, is as follows:

$$K(\tau) = 11.16 - 9.33*10^{-3}\tau, \qquad (6)$$

where, $\tau \in [0; 30]$.

The analysis of given function shows that it is lineal and subsides slightly, depending on distance, which presents the cycle difference, i.e. the residual deformation value of given cycle depends considerably on loading history of concrete.

Taking $\varepsilon^{*n}{}_{r} = 0$ and $\varepsilon^{*n+1}{}_{r} = 0$, the equation 3 and 4 make it possible to determine the energy dissipation at each cycle of a concrete work:

$$E_n = \int \sigma_{n\uparrow} d\varepsilon - \int \sigma_{n\downarrow} d\varepsilon. \qquad (7)$$

In order to confirm the correctness of energy computation 6 on the basis of equation 3 and 4 numerical integration of dependences $\sigma \sim \varepsilon$ has also been used in this paper.

The given method of approach is realized on a computer. Energy dissipation values at random loadings have been obtained (Figure 1)

Let $\eta^{max}{}_{1}, \eta^{max}{}_{2}, \ldots, \eta^{max}{}_{k}$ denote the maximal levels of loadings. Then, we can conclude, that the energy dissipation splash will take place when $\eta^{max}{}_{i} > (\eta^{max}{}_{i-1}, \ldots, \eta^{max}{}_{1})$. Thus, on the basis of precondition that the failure can take place in case when total energy of destruction at single loadings before destruction is equal to the critical value of summary energy dissipation at random loadings, follows, that the operation with the increasing level of loading is the most dangerous to the concrete. With the increasing of loading cycle number the energy dissipation decreases. The accumulation of summary dissipation energy is being stabilized and one can consider the given concrete under random loading as a cyclically stable material.

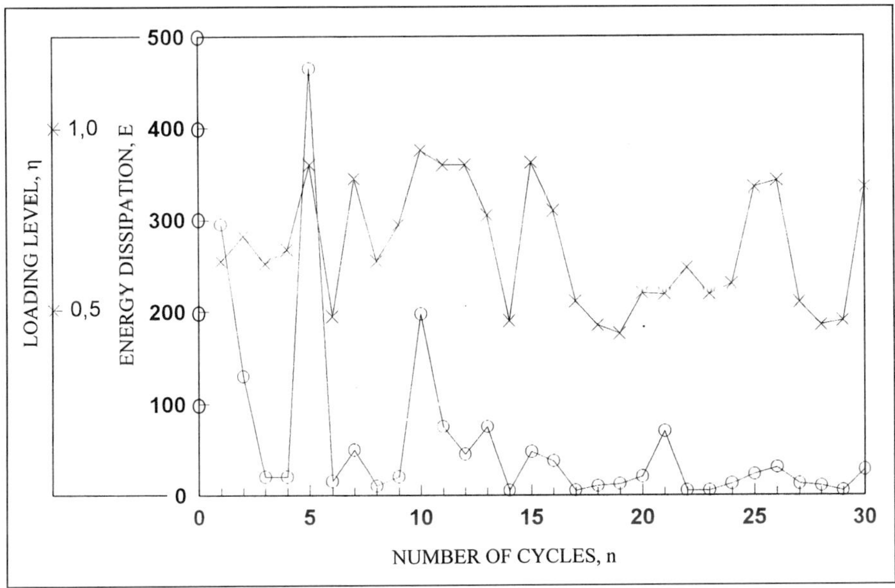

Figure 1 The dissipation of energy

PRACTICAL APPLICATION OF RESULTS

To confirm possibility to use the prism test results at the research of stressed and deformed condition of eccentrically stressed elements, their load-carrying capacity is determined with the formula:

$$N = b \int \sigma_b \, dx + N'_s \pm N_s, \tag{8}$$

where, N'_s : the force in reinforcement at the most compressed side;
N_s : the force in reinforcement at the least compressed side;
b : the width of section;
the value $b \int \sigma_b \, dx$: presents itself a force, taken by the concrete.

Statistical analysis of deformation diagram, $\sigma \sim \varepsilon$ obtained at the experimental investigate of concrete, shows, that the deformations under single compression approximate satisfactorily with the fifth-degree polynomial

$$\sigma_b = R_b\,(6.9436\cdot10^{-12}\,\varepsilon^{*5} - 3.3059\cdot10^{-9}\varepsilon^{*4} + 556.10\cdot10^{-9}\,\varepsilon^{*3} - 60.021\cdot10^{-6}\varepsilon^{*2} + 9.3866\cdot10^{-3}\varepsilon^{*} + 8.5309\cdot10^{-3}), \qquad (9)$$

where, $\varepsilon^* = \varepsilon * 10^5$;
$[R_b]$: strength of concrete in, MPa;
$[\sigma_b]$: stress of concrete in, MPa.

From equation 9 one can determine the compression value in each point of the section. Statistical processing of testing results gives possibility to construct the dependence $\sigma = f(x)$ for various eccentricities of external action application.

According to the results of prism random load tests it was determined, that the increase of prism strength at cyclical loadings approximate satisfactorily by the following dependence:

$$R_{b,c} = R_b \cdot n^{2.8*10-2}, \qquad (10)$$

where, $R_{b,c} = f(n)$: the prism strength of concrete when loaded with the n-cycles.

It is recommended to fulfill the calculation of eccentrically compressed elements with small eccentricities according to the following formulae:

$$N(n) \le n^{2.8*10-2}\, R_b\, bx + R_{sc}\, A'_s - \sigma_s\, A_s, \qquad (11)$$

$$N(n)\cdot e \le n^{2.8*10-2}\, R_b\, bx(h_0 - 0{,}38x) + R_{sc}\, A'_s(h_0 - a'). \qquad (12)$$

CONCLUSIONS

Determination of the load-carrying capacity by given procedure is in accordance with the values of load-carrying capacity, obtained with the help of regression equations, set up on the basis of a full factorial experiment. Discordance of experimental values of eccentrically compressed element load-carring capacity in comparison with the theoretical ones calculated by the equation 8 does not excess 2%. Given method of the approach to the concrete research can be used in strength and reability investigation of constructions that work under various external actions using the apparatus of random variables and processes.

REFERENCES

1. CHIRVA, T, CHIRVA, V, PERLOVA, H. Calculation of strength of eccentric compression reinforced concrete elements at random loading.XLIII Konferencja Naukowa KIL i WPAN i KNPZITB "KRYNICA '97" ,1997, 19-24, tom IV, Poland.

CREEP AND SHRINKAGE EFFECT ON REINFORCED CONCRETE SLAB-AND-BEAM STRUCTURES

E J Sapountzakis

J T Katsikadelis

National Technical University of Athens

Greece

ABSTRACT. In this paper a solution to the problem of reinforced concrete slab-and-beam structures including creep and shrinkage effect is presented. The adopted model takes into account the resulting inplane forces and deformations of the plate as well as the axial forces and deformations of the beam, due to combined response of the system. The analysis consists in isolating the beams from the plate by sections parallel to the lower outer surface of the plate. The interface forces are established using continuity conditions. The influence of creep and shrinkage effect relative with the time of the casting and the time of the loading of the plate and the beams is taken into account. The solution of the arising plate and beam problems which are nonlinearly coupled, is achieved using the Analog Equation Method (AEM). The adopted model, compared with those ignoring the inplane forces and deformations, describes better the actual response of the plate - beams system and permits the evaluation of the shear forces at the interfaces, the knowledge of which is very important in the design of prefabricated ribbed plates.

Keywords: Elastic stiffened plate, Reinforced plate with beams, Bending, Ribbed plate, Creep, Shrinkage, Slab-and-beam structure.

Dr Evangelos J Sapountzakis is a Lecturer in Civil Engineering Department, National Technical University of Athens. He specialises in static, dynamic analysis and elastic stability of structures using analytical and numerical methods. He has also great experience in bridge engineering.

Professor John T Katsikadelis is Director of the Laboratory of Structural Analysis and Aseismic Research of National Technical University of Athens. His main research interests include the fields of mechanics of deformable bodies and of analysis and behaviour of structures subjected to static and dynamic loads (elasticity, stability of structures, theory of beams, plates and shells) by developing and using boundary elements. He is a past Director of the Earthquake Planning and Protection Organization of Greece and of the European Center on Prevention and Forecasting of Earthquake.

INTRODUCTION

The interest in reinforced concrete slab-and-beam structures has been widespread in recent years due to the economic and structural advantages of such systems. The extensive use of the aforementioned plate structures necessitate a rigorous analysis. In the extensive literature on static analysis of slab-and-beam structures, their behaviour was initially approximated by converting this system to an equivalent homogeneous slab of constant thickness using the stiffness properties of the beams and applying the orthotropic plate theory [1]. This approximation may be applicable only when the stiffened plate satisfies two limitations. The first one is that ratios of spacing between two consecutive stiffeners to slab boundary dimensions are small enough to ensure approximate homogeneity of stiffness. The second limitation is that the ratio of stiffener rigidity to the slab rigidity must not become so large that the beam action is predominant. Subsequently, in more refined approximations the adopted models for the analysis of the plate - beams system isolated the beams from the plate and neglected the shear forces at the interfaces [2-3]. This assumption results in discrepancies from the actual response of the stiffened plate. Moreover it does not allow the establishment of these forces, which are necessary for the design of composite or prefabricated structures.

In this paper the analysis of reinforced concrete slab-and-beam structures is presented. The adopted structural model is that employed by Sapountzakis and Katsikadelis [4], which contrary to the models used previously takes into account the resulting inplane forces and deformations of the plate as well as the axial forces and deformations of the beam, due to combined response of the system. Using this model, the study of the behaviour of a stiffened plate subjected to a lateral load and to the effects of creep and shrinkage, either for simultaneous or for separate casting of the plate and the beams is attempted. The analysis consists in isolating the beams from the plate by sections parallel to the lower outer surface of the plate. The forces at the interface, which produce lateral deflection and inplane deformation to the plate and lateral deflection and axial deformation to the beam, are established using continuity conditions at the interface. The solution of the arising plate and beam problems which are nonlinearly coupled, is achieved using the Analog Equation Method (A.E.M.) following a procedure similar to that presented in [5-6]. The adopted model describes better the actual response of the plate beams system and permits the evaluation of the shear forces at the interface, the knowledge of which is very important in the design of reinforced concrete prefabricated ribbed plates. The evaluated lateral deflections of the plate - beams system are found to exhibit considerable discrepancy from those of other models, which neglect inplane and axial forces and deformations. Finally, the influence of the time interval between the casting of the plate and the beams to the behaviour of the stiffened plate is examined.

STATEMENT OF THE PROBLEM

Consider a thin reinforced concrete plate having constant thickness h_p, occupying the domain Ω of the x, y plane and stiffened by a set of parallel reinforced concrete beams. The plate may have J holes and is supported on its boundary $\Gamma = \cup_{j=0}^{J} \Gamma_j$, which may be piecewise smooth (Figure 1), while the beams may have point supports. For the sake of convenience the x axis is taken parallel to the beams. Let the time of the casting of the beams

t_{bc} be the beginning of the time considered t, t_{bl} be the time at initial loading of the beams, t_{pc} be the time of the casting of the plate and t_{pl} be the time at which the plate is initially subjected to the lateral load $g = g(\mathbf{x})$, $\mathbf{x}:\{x, y\}$. The solution of the problem at hand is approached by isolating the beams from the plate by sections in the lower outer surface of the plate and taking into account the tractions at the fictitious interfaces (Figure 2). These tractions result in the loading of the beam as well as the additional loading of the plate. Their distribution is unknown and can be established by imposing displacement continuity conditions at the interfaces and using the procedure developed in this investigation.

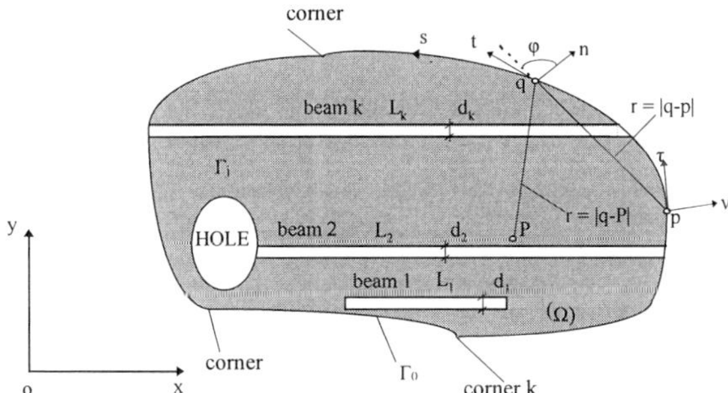

Figure 1 Two dimensional region Ω occupied by the plate

The integration of the tractions along the width of the beam result in line forces per unit length which are denoted by q_x, q_y and q_z. Taking into account that the torsional stiffness of the beam is small, the traction component q_y, in the direction normal to the beam axis, may be ignored. However, in a more refined model the influence of this component may also be considered. The other two components q_x and q_z produce the following loadings along the trace of each beam.

A. In the plate

(i) A lateral line load $-q_z$ at the interface.
(ii)A lateral line load $-\partial M_p / \partial x$ due to the eccentricity of the component q_x from the middle surface of the plate. $M_p = q_x h_p / 2$ is the bending moment.
(iii) An inplane line body force q_x at the middle surface of the plate.

B. In each beam

(i) A transverse load q_z.
(ii)A transverse load $\partial M_b / \partial x$ due to the eccentricity of q_x from the neutral axis of the beam cross section.
(iii)An inplane axial force q_x.

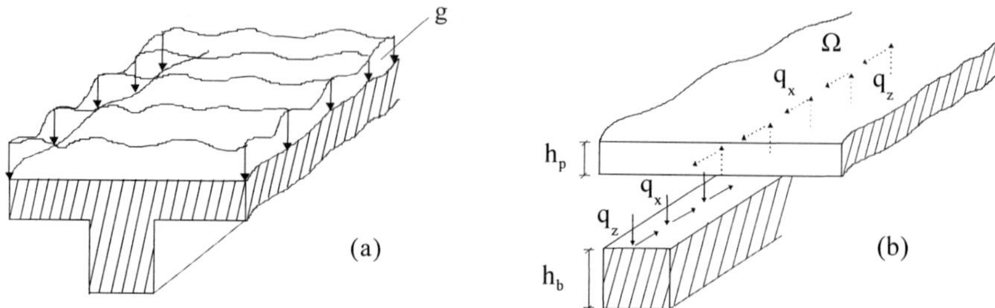

Figure 2 Thin elastic plate stiffened by (a) beams and (b) isolation of the beams from the plate

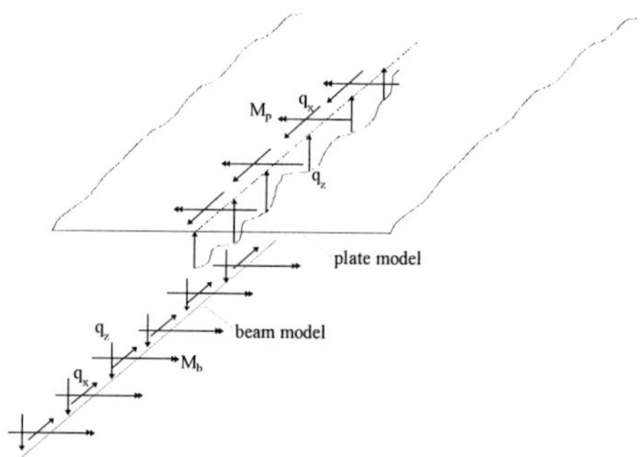

Figure 3 Structural model of the plate and the beams

The structural models of the plate and the beams are shown in Fig.3. On the basis of the above considerations the response of the plate and of the beams may be described by the following boundary value problems.

A. For the plate

The plate undergoes transverse deflection and inplane deformation. Thus, for the transverse deflection we have

$$D\nabla^4 w_p - \left(N_x \frac{\partial^2 w_p}{\partial x^2} + 2N_{xy} \frac{\partial^2 w_p}{\partial x \partial y} + N_y \frac{\partial^2 w}{\partial y^2} \right) = g - \sum_{k=1}^{K} \left(q_z^{(k)} + \frac{\partial M_p^{(k)}}{\partial x} \right) \delta(y - y_k) \text{ in } \Omega \quad (1)$$

$$\alpha_1 w_p + \alpha_2 V_n = \alpha_3 \qquad \beta_1 \frac{\partial w_p}{\partial n} + \beta_2 M_n = \beta_3 \qquad \text{on } \Gamma \qquad (2a,b)$$

where $w_p = w_p(\mathbf{x},t)$ is the time dependent transverse deflection of the plate; $D(t) = E_p(t)h_p{}^3 / 12(1-v^2)$ is its time dependent flexural rigidity with E_p being the elastic modulus and v the Poisson ratio; $N_x = N_x(\mathbf{x},t)$, $N_y = N_y(\mathbf{x},t)$, $N_{xy} = N_{xy}(\mathbf{x},t)$ are the membrane forces per unit length of the plate cross section at time *t*; $\delta(y-y_k)$ is the Dirac's delta function in the y direction; M_n and V_n are the bending moment normal to the boundary and the effective reaction along it, respectively. Finally, a_i, β_i $(i=1,2,3)$ are functions specified on the boundary Γ. The boundary conditions (2a,b) are the most general linear boundary conditions for the plate problem including also the elastic support.

If the stresses are kept within the limits corresponding to the normal service conditions, assuming linear relationship between creep and the stress causing the creep and denoting by $t_p = t_{pl} - t_{pc}$, Trost's theory [7] gives the tangent modulus of elasticity as

$$E_p(t) = E_{pl} / \left[1 + \chi\varphi(t,t_p)\right] \tag{3}$$

where E_{pl} is the tangent modulus of elasticity of the plate at time t_p; χ is an ageing coefficient depending on strain development with time; $\varphi(t,t_p)$ is the creep coefficient related to the elastic deformation at t_p days which is defined as [8]

$$\varphi(t,t_p) = \phi_{RH}\beta(f_{cm})\beta(t_p)\beta_{cp}(t-t_p) \tag{4}$$

where φ_{RH}, $\beta(f_{cm})$ and $\beta(t_p)$ are factors depending on the relative humidity, the concrete strength and the concrete age at loading, respectively, which are defined as

$$\varphi_{RH} = 1 + (1 - RH/100)/\left(0.10\sqrt[3]{h_0}\right) \quad \beta\left(f_{cm}\right) = 16.8/\sqrt{f_{cm}} \quad \beta\left(t_p\right) = 1/\left(0.1 + t_p^{\,0.20}\right) \tag{5a,b,c}$$

where RH is the relative humidity of the ambient environment in %; $h_0 = 2A_p / u_p$ is the notional size of the plate in mm; A_p is the area of the plate cross section; u_p is the plate perimeter in contact with the atmosphere; f_{cm} is the mean compressive strength of concrete in N/mm^2 at the age of 28 days. Moreover, $\beta_{cp}(t-t_p)$ in eqn.(4) is a coefficient for the development of creep with time, which is estimated from

$$\beta_{cp}\left(t-t_p\right) = \left[\left(t-t_p\right)/\left(\beta_H + t - t_p\right)\right]^{0.3} \tag{6}$$

where β_H is a coefficient depending on the relative humidity RH, given as

$$\beta_H = 1.5\left[1 + (0.012RH)^{18}\right]h_0 + 250 \leq 1500 \tag{7}$$

Since linear plate bending theory is considered, the components of the membrane forces N_x, N_y, N_{xy} do not depend on the deflection w_p. They are evaluated from the displacement components $u_p = u_p(\mathbf{x},t)$ and $v_p = v_p(\mathbf{x},t)$ of the middle surface of the plate arised from both the line body force q_x and the temperature distribution $T_p(\mathbf{x},t)$ due to the plate shrinkage. The latter are established by solving independently the plane stress problem (Navier's equations of equilibrium)

$$\nabla^2 u_p + \frac{1+\nu}{1-\nu}\frac{\partial}{\partial x}\left[\frac{\partial u_p}{\partial x} + \frac{\partial v_p}{\partial y}\right] + \frac{1}{G_p}q_x\delta\left(y-y_k\right) - 2\alpha\frac{1+\nu}{1-\nu}\frac{\partial T_p}{\partial x} = 0 \tag{8a}$$

$$\nabla^2 v_p + \frac{1+\nu}{1-\nu}\frac{\partial}{\partial y}\left[\frac{\partial u_p}{\partial x} + \frac{\partial v_p}{\partial y}\right] - 2\alpha\frac{1+\nu}{1-\nu}\frac{\partial T_p}{\partial y} = 0 \quad \text{in } \Omega \tag{8b}$$

$$\gamma_1 u_n + \gamma_2 N_n = \gamma_3 \qquad \delta_1 u_t + \delta_2 N_t = \delta_3 \qquad \text{on } \Gamma \quad (9a,b)$$

in which $G_p(t) = E_p(t)/2(1+v)$ is the shear modulus of the plate; α is the linear coefficient of thermal expansion; N_n, N_t and u_n, u_t are the boundary membrane forces and displacements in the normal and tangential directions to the boundary, respectively; γ_i, δ_i $(i = 1,2,3)$ are functions specified on Γ. Assuming that creep and shrinkage are independent, the temperature distribution $T_p(\mathbf{x},t)$ is given as [8]

$$T_p(\mathbf{x},t) = \varepsilon_{sp}(t - t_{pc})/\alpha \tag{10}$$

where $\varepsilon_{sp}(t - t_{pc})$ is the shrinkage strain calculated from

$$\varepsilon_{sp}(t - t_{pc}) = \varepsilon_{sp}(f_{cm})\beta_{RH}\beta_{sp}(t - t_{pc}) \tag{11}$$

where $\varepsilon_{sp}(f_{cm})$, β_{RH} are factors depending on the concrete strength and the relative humidity, respectively, which are defined as

$$\varepsilon_{sp}(f_{cm}) = \left[160 + \beta_{sc}(90 - f_{cm})\right]10^{-6} \tag{12a}$$

$$\beta_{RH} = \begin{cases} -1.55\left(1 - (RH/100)^3\right) & \text{,for } 40\% \leq RH \leq 99\% \text{ (stored in air)} \\ +0.25\left(1 - (RH/100)^3\right) & \text{,for } RH \geq 99\% \text{ (immersed in water)} \end{cases}$$

where β_{sc} is a coefficient depending on type of cement. Moreover, $\beta_{sp}(t - t_{pc})$ in Eqn.(11) is a coefficient for the development of shrinkage with time, which is estimated from

$$\beta_{sp}(t - t_{pc}) = \left[(t - t_{pc})/\left(0.035h_0^2 + t - t_{pc}\right)\right]^{0.5} \tag{13}$$

B. For each beam

Each beam undergoes transverse deflection and axial deformation. Thus, for the transverse deflection we have

$$E_b I_b \frac{d^4 w_b}{dx^4} - N_b \frac{\partial^2 w_b}{\partial x^2} = q_z - \frac{\partial M_b}{\partial x} \qquad \text{in } L_k,\ k = 1,2,\ldots,K \tag{14}$$

$$a_1 w_b + a_2 V = a_3 \qquad b_1 \frac{\partial w_b}{\partial x} + b_2 M = b_3 \qquad \text{at the beam ends } x = 0,\ l \quad (15a,b)$$

where $w_b = w_b(x,t)$ is the time dependent transverse deflection of the beam; I_b is its moment of inertia; $N_b = N_b(x,t)$ is the axial force at the neutral axis; V, M are the reaction and the bending moment at the beam ends, respectively; a_i, b_i $(i = 1,2,3)$ are coefficients specified at the boundary of the beam; $E_b = E_b(t)$ is the time dependent tangent modulus of elasticity of the beam given as [7]

$$E_b(t) = \frac{E_{bl}}{1 + \chi\varphi(t, t_b)} \tag{16}$$

where E_{bl} is the tangent modulus of elasticity of the beam at time t_{bl}; χ is an ageing coefficient depending on strain development with time; $t_b = t_{bl} - t_{bc}$; $\varphi(t, t_b)$ is the creep coefficient related to the elastic deformation at t_b days, which is defined as [8]

$$\varphi(t, t_b) = \phi_{RH}\beta(f_{cm})\beta(t_b)\beta_{cb}(t - t_b) \tag{17}$$

where φ_{RH}, $\beta(f_{cm})$, $\beta(t_b)$ and $\beta_{cb}(t - t_b)$ are creep coefficients for the beam similar to those for the plate.

Since linear beam bending theory is considered, the axial force of the beam does not depend on the deflection w_b. The axial deformation of the beam arised from both the inplane axial force q_X and the temperature distribution $T_b(x, t)$ due to the beam shrinkage is described by solving independently the boundary value problem i.e.

$$E_b A_b \frac{\partial^2 u_b}{\partial x^2} = -q_x + \alpha \frac{\partial T_b}{\partial x} \qquad \text{in } L_k,\ k = 1, 2, \ldots, K \tag{18}$$

$$c_1 u_b + c_2 N = c_3 \qquad \text{at the beam ends } x = 0,\ l \tag{19}$$

where N is the axial reaction at the beam ends. Similarly to the plate, the temperature distribution T_b for the beam is given as

$$T_b = \varepsilon_{sb}(t - t_{bc}) / \alpha \tag{20}$$

where

$$\varepsilon_{sb}(t - t_{bc}) = \varepsilon_{sb}(f_{cm})\beta_{RH}\beta_{sb}(t - t_{bc}) \tag{21}$$

where $\varepsilon_{sp}(f_{cm})$, β_{RH}, $\beta_{sb}(t - t_{bc})$ are shrinkage coefficients for the beam similar to those for the plate.

Equations (1), (8a), (8b), (14), (18) constitute a set of five coupled partial differential equations including seven unknowns, namely w_p, u_p, v_p, w_b, u_b, q_x, q_z.

Two additional equations are required, which result from the continuity conditions of the displacements in the direction of the z and x axes at the interfaces between the plate and the beams. These conditions can be expressed as

$$w_p = w_b \tag{22}$$

$$u_p - \frac{h_p}{2}\frac{\partial w_p}{\partial x} = u_b + \frac{h_b}{2}\frac{\partial w_b}{\partial x} \tag{23}$$

It must be noted that the coupling of eqns (1) and (8a,b), as well as of eqns (14) and (18) is nonlinear due to the terms including the unknown membrane and axial forces, respectively.

NUMERICAL SOLUTION

The numerical solution of the described plate bending problem, plane stress problem, beam bending problem and axial beam problem is achieved using the Analog Equation Method (AEM) as this is presented in detail by Sapountzakis and Katsikadelis in [4].

NUMERICAL EXAMPLES

On the basis of the analytical and numerical procedures presented in the previous sections, a computer program has been written and representative examples have been studied to demonstrate the efficiency and the range of applications of the developed model. In all the examples treated, the numerical results have been obtained using *54* constant boundary elements and *162* constant domain rectangular cells. The following data have been used for the numerical results : Concrete *C25/30*, $f_{cm}=25N/mm^2$, $RH=40\%$, $t_p=t_b=28days$, $E_{pl}=E_{bl}=E_{c28}=32.55kN/mm^2$, $h_p=0.20m$, $v=0.154$, $\beta_{sc}=5$ (normal or rapid hardening cement), $a=10^{-5}\ {}^oC$.

Example 1

A rectangular plate $a \times b = 18.0 \times 9.0\ m$ subjected to a uniform load $g = 10kN/m^2$ and stiffened by a beam with width *1.0 m* through the center line of the plate has been studied. The plate is simply supported along its small edges, while the other two edges are free. In this example both the plate and the beam are simultaneously casted ($t_{bc}=t_{pc}=0$).

According to the obtained deflections of the stiffened plate, in Table 1 the time development of the deflection w at its center for different values of the height h_b of the beam is presented. It is worth noting that the obtained deflections for the stiffened plates with low height beams, though increased in the early age of concrete exhibit a local maximum and they are afterwards decreased. This is due to the predominant action of creep compared to shrinkage in the early age of concrete which is later opposed. For higher values of the height of the stiffening beam the action of creep is at all times predominant. In Figure 4 the distributions of the deflection w at the interface for the height of the beam $h_b=0.60m$ and for various instants are shown. Moreover, in Table 2 the deflections w at the center of the stiffened plate for different values of the height h_b and for various instants are shown as compared with those obtained from a FEM solution [9], which cannot include the inplane forces and deformations, using *162* square elements. The discrepancy of the results is obvious, while the obtained deflections from the FEM solution are monotonically increased for all values of the height of the stiffening beam. According to the obtained interface forces q_x of the stiffened plate, in Table 3 the time development of these forces at point $x=0.5m$ of the interface for different values of the height h_b of the beam is presented.

Example 2

The casting of the beam of the stiffened plate of example 1 preceded the casting of the plate at a time interval of $T=t_{pc}-t_{bc}$ days. In Fig.7 the time development of the deflection w at the Center of the stiffened plate for the height of the stiffening beam $h_b=0.60m$ and for different values of the time interval T is presented.

Table 1 Time development of the deflections w (m) at the plate center of Example 1

AGE OF CONCRETE T (days)	STIFFENING BEAM				
	1.00x0.20m	1.00x0.40m	1.00x0.60m	1.00x1.25m	1.00x2.00m
30	0.0600	0.0218	0.0091	0.0014	0.00041
60	0.0750	0.0288	0.0122	0.0019	0.00055
90	0.0766	0.0308	0.0133	0.0021	0.00061
150	0.0749	0.0317	0.0140	0.0022	0.00064
200	0.0734	0.0321	0.0143	0.0023	0.00067
300	0.0707	0.0323	0.0148	0.0024	0.00070
500	0.0666	0.0322	0.0151	0.0025	0.00073
1000	0.0610	0.0314	0.0153	0.0027	0.00077

Table 2 Deflections w (m) at the center of the plate of Example 1

AGE OF CONCRETE T (days)	STIFFENING BEAM					
	1.00x0.20m		1.00x0.60m		1.00x1.25m	
	AEM	FEM	AEM	FEM	AEM	FEM
30	0.0600	0.3987	0.0091	0.0478	0.0014	0.0076
150	0.0749	0.7123	0.0140	0.0796	0.0022	0.0119
500	0.0666	0.8467	0.0151	0.0914	0.0025	0.0125
1000	0.0610	0.8990	0.0153	0.0944	0.0027	0.0128
3000	0.0544	0.0152	0.0946		0.0027	0.0128

Table 3 Time development of the interface forces q_x (kN) of the stiffened plate of Example 1

AGE OF CONCRETE T (days)	STIFFENING BEAM				
	1.00x0.20m	1.00x0.40m	1.00x0.60m	1.00x1.25m	1.00x2.00m
30	3432	2508	1972	1165	794
60	3087	2384	1913	1144	782
90	2851	2289	1862	1124	769
150	2668	2208	1818	1107	758
200	2544	2150	1785	1093	749
300	2379	2067	1736	1073	736
500	2192	1966	1675	1047	720
1000	1987	1845	1598	1014	698

Table 4 Time development of the interface forces q_x (kN) of the stiffened plate of Example 2

AGE OF CONCRETE T (days)	STIFFENING BEAM 1.00 x 0.60m			1.00 x 1.25m		
	T=60days	T=120days	T=300days	T=60 days	T=120days	T=300days
30	1956	1952	1946	1154	1152	1149
60	1903	1897	1889	1139	1137	1133
90	1854	1850	1842	1121	1119	1116
150	1812	1808	1800	1104	1103	1100
200	1780	1777	1770	1091	1090	1088
300	1733	1730	1745	1072	1071	1069
500	1673	1671	1667	1046	1046	1044
1000	1598	1597	1595	1013	1013	1012

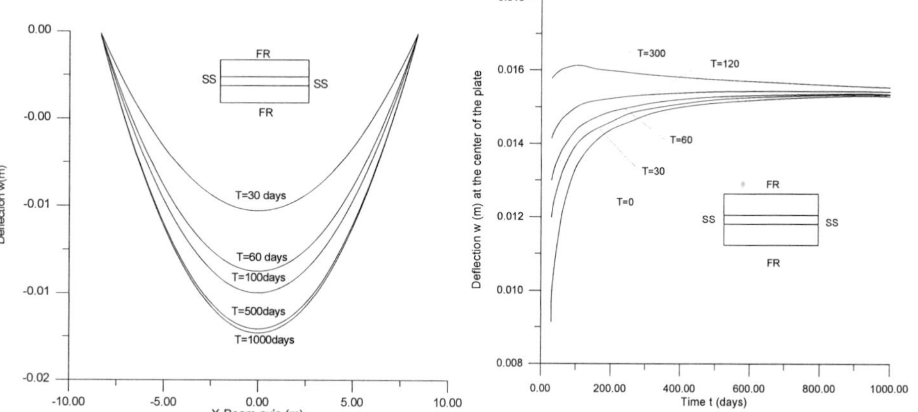

Figure 4 Deflections w(m) along the axis of the beam

Figure 5 Deflections w(m) at the center of the plate

Moreover, in Tables 4 and 5 the time development of the interface forces q_X at point $x=0.5m$ and q_Z at point $x=8.5m$ of the interface, for different values of the time interval T are presented. From the obtained results it is easily concluded that the reinforced concrete stiffened plate is not significantly influenced from the time interval between the casting of the plate and the beams.

CONCLUSIONS

1. The proposed model permits the study of the behaviour of a stiffened plate due to the opposed effects of creep and shrinkage either for simultaneous or for separate casting of the plate and the beams.

2. The evaluated lateral deflections of the plate - beams system are found to exhibit considerable discrepancy from those of other models, which neglect inplane and axial forces and deformations.

3. The adopted model permits the evaluation of the inplane shear forces at the interface between the plate and the beams, the knowledge of which is very important in the design of prefabricated plate beams structures (estimation of shear connectors).

4. Reinforced concrete slab-and-beam structures are not significantly influenced from the time interval between the casting of the plate and the beams.

REFERENCES

1. PAMA, R P AND CUSENS, A R. Edge Beam Stiffening of Multibeam Bridges. Journal of the Structural Division, ASCE, Vol.93, ST2, April 1967, pp 141-161.

2. TANAKA, M AND BERCIN, A N. A Boundary Element Method Applied to the Elastic Bending Problem of Stiffened Plates. Boundary Element Method XIX, 1997, pp 203-212.

3. NG, S F, CHEUNG, M S AND XU, T. A Combined Boundary Element and Finite Element Solution of Slab and Slab-on-Girder Bridges. Comp. and Struct., Vol.37, 1990, pp 1069-1075.

4. SAPOUNTZAKIS, E J AND KATSIKADELIS, J T. Analysis of Plates Reinforced with Beams. Proc. 5th National Congress of H.S.T.A.M., Ioannina, 1998, 27-30 August.

5. KATSIKADELIS, J T. The Analog Equation Method-a Powerful BEM-Based Solution Technique for Solving Linear and Non-linear Engineering Problems. Boundary Element Method XVI, 1994, pp 167-182.

6. NERANTZAKI, M S. Solving Plate Bending Problems by the Analog Equation Method. Boundary Element Method XVI, Computational Mechanics Publications, 1994, pp 283-291.

7. TROST, H AND WOLFF, J. Zur Wirklichkeitsnahen Ermittlung der Beanspruchungen in Abschnittsweise Hergestellten Spannbetontragwerken. Der Bauingenieur, Vol.45, 1970, pp 155-169.

8. EUROCODE NO.2. Design of Concrete Structures. Part 1 : General Rules and Rules for Buildings. Eurocode 2 Editorial Group, 1991.

9. CUBUS, A G. Software. Cedrus-3 User's References Manual, Zürich, Switzerland, version 1.56, June, 1995.

A NEW BUILDING SYSTEM USING JOINTS OF ULTRA HIGH-STRENGTH FIBRE REINFORCED CONCRETE

L P Hansen

Aalborg University

B C Jensen

Carl Bro

Denmark

ABSTRACT. Precast elements have been used for decades in the building industry. The main reasons for this are reduction in price, reduction in erection time and increase of the quality as production in factories often reduces the possibility of faults. Since the beginning of the 1970's there have been several attempts in Denmark to change from large panel buildings into building systems which allow much more flexibility in the use of the building including changes during the lifetime of the building. Column/beam/slab building systems have been introduced but the flexibility is still limited. Column/slab buildings are preferred but due to transportation and lift capacities on the site, the size of the slab elements is limited. A new building system has been developed. The system consists of a column/slab system with 6 m x 6 m distance between the columns. The slabs are precast elements (2.9 m x 5.9 m) connected through joints of ultra high strength fibre reinforced concrete (Densit ®). Also the connection between columns and slabs are made of this material. Using this material very short anchorage lengths for the reinforcement can be applied and thus a very simple connection can be used. The column distance can of course be changed and the slab elements can also be cantilevered. Several laboratory tests as anchorage tests, slab joint tests and column/slab joint tests have been carried out. The building system has been used for some new buildings at Aalborg University and a hospital using the system is in preparation.

Keywords: Building system, Ultra high strength fibre reinforced concrete, Joints, Slab elements, Ultimate load, Fire resistance

Lars Pilegaard Hansen is a reader Head of the Structural Research Laboratory, Aalborg University, Denmark. He specializes in static testing as well as fatigue and vibration testing for building materials and civil engineering structures.

Professor Bjarne Chr Jensen is Chief Consultant, working in the consulting firm Carl Bro as, where he is dealing with R&D in Danish and international projects. He has experience in design of buildings and structures, especially in reinforced concrete.

INTRODUCTION

A new building system has been developed by the Danish architect firm Dall and Lindhartsen and the Danish consulting firm Carl Bro as. An essential part of this new system are the joints between prefabricated concrete slabs and these joints are made of ultra high strength fibre reinforced concrete - abbreviated UHS FRC - developed at the Cement and Concrete Laboratory at Aalborg Portland. The Structural Laboratory at the Department of Building Technology and Structural Engineering has tested parts of this new building system. The new building system has been applied for new buildings at Aalborg University. Part 1 was built in 1995 / 96 and was ready for use in the summer of 1996. Part 2 is built 1998 / 99 and will be ready for use in the summer of 1999. An extension of a hospital in Denmark is also in progress. The principle for the new building system is shown in Figure 1.

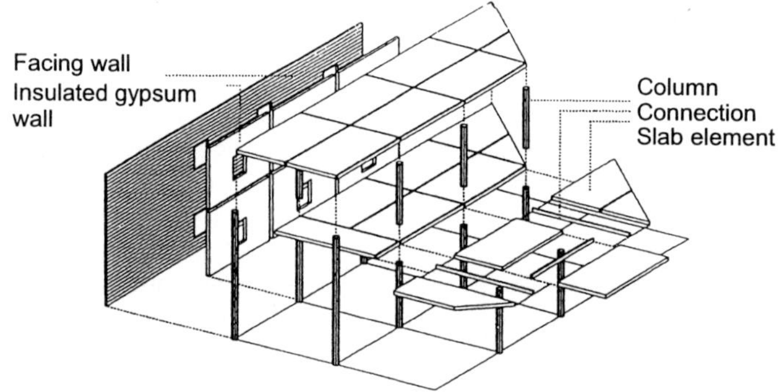

Figure 1 Sketch of principle for the new building system

DESCRIPTION OF THE NEW BUILDING SYSTEM

The load bearing system for the new building system is prefabricated concrete slabs at a thickness of 200 mm and concrete columns (circular at a diameter of 350 mm) in a square net of 6 m x 6 m. The slabs are cast as 2.9 m x 5.9 m slab elements and are made of normal strength concrete. At the building site the slabs are connected with UHS FRC thus forming a continuous slab with the columns as supports. No beams are used in this building system.

The new building system gives the architects more freedom to create the plan for the different rooms in the building. It is also possible to allow different solutions at the bounderies As no beams are present in the system there are good possibilities to arrange the technical installations between the slab elements and the non-bearing ceilings.

The columns are erected at a distance of 6 m and on the top of the columns slab elements are placed. As the slab elements are 2.9 m x 5.9 m there is an opening of 100 mm between the slabs. The slabs are reinforced in both sides with a net of 8 mm reinforcement - Ks 550 - and 80 mm of the reinforcement jutting out the slab. Ks550 are ribbed bars with yield stress higher than 550 MPa.

After placing the slab elements on the columns, longitunal reinforcement bars are placed in the joints and the joints are cast with UHS FRC. This means that slab elements now act as a continuous slab as the joints are able to transfer moments and forces. The extremely short anchorage length can be used, because the UHS FRC gives this joint a sufficient load bearing capacity.

With the joints "welding" the slab elements together to a monolithic slab, it is possible to

- change position of the column
- change size of the slab element
- produce cantilevered parts of the slab

The vertical forces are transferred through the slabs to the columns. For the first building (1995 / 96) at Aalborg University the horizontal forces were transferred through the slabs to some stabilizing concrete wall elements. For the second building (1998 / 99) at Aalborg University the horizontal forces are transferred only through the columns and the stabilizing concrete walls are thus not needed.

Tests have been carried out on different types of joints and the final version of the joints between two slab elements is shown in Figure 2.

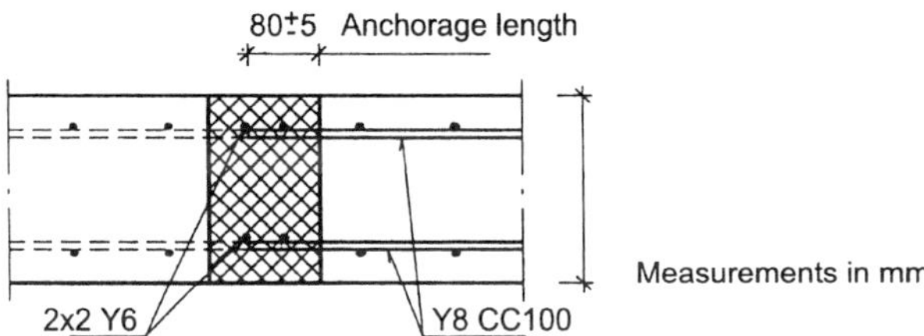

Figure 2 Typical joint between concrete slabs

In the following chapters a short description of the UHS FRC and some of the test results with parts of the new building system are given. More information can be found in the references. All the test results are given in the reports [1] and [2], where [1] contains the test results for the 1995 / 96 building and [2] contains the results for the 1998 / 99 building. The two reports are in Danish.

THE ULTRA HIGH STRENGTH FIBRE REINFORCED CONCRETE

In ultra high strength concrete materials the binder is often composed of Portland cement powder and other fine particles such as fly ash, steel powder, Al_2O_3 etc. Often silica fume with ball shaped particles is added too and the binder is very dense and strong. Such binders can be used to produce ultra high strength concrete but such concretes are brittle materials. This problem can be solved by adding steel fibres to the concrete. The material used in this investigation is based on the binder Densit® which is silica fume based and developed at Aalborg Portland, see [3].

In the concrete the water / powder ratio is 0.15 - 0.18, the silica fume content is 20 - 25 % and the compressive strength is at the interval of 100 - 200 MPa. The steel fibre content is 6 % by volume and the length of the fibres is 12 mm and the diameter is 0.4 mm. This matrix is called Compresit. Further description of this material can for example be found in [4].

In this dense Compresit matrix the water absorption and the air permeability are very low. The chemical resistance is remarkably good. Freeze / thaw damages were not detected after the test according to RILEM Recommendation No. 117 - FDC. The matrix is practical immune to carbonation and the chloride diffusion is extremely low compared to ordinary concrete. The resistance to fire is described in [1] and [10].

The failure criterion for Compresit is investigated in [5]. One of the results is that the theory of plasticity can be used in the ultimate limit state. Therefore, the design can be done according to modern theories of concrete design as for example described in Eurocode 2 for shear in beams and in construction joints.

The price for Compresit is three to four times the price of ordinary concrete which means that a general substitution of concrete with Compresit will not take place but an evident use of Compresit is in joints carrying forces between structural elements. This is due to the eminent anchorage of reinforcement bars, see the next section.

The properties of the UHS FRC are typical:

- Uniaxial compressive strength 100 - 200 MPa
- Initial modulus in compression 50 GPa
- Uniaxial tensile strength 10 - 15 MPa
- Initial modulus in tension 50 GPa
- Fracture energy 10 - 15 kN / m

STATIC TESTS WITH DIFFERENT PARTS OF THE BUILDING SYSTEM

In this section a short description of some of the tests carried out in connection with the two buildings at Aalborg University is given.

Anchorage Tests

Previous to the testing at Aalborg University of different connections some tests were carried out at the Cement and Concrete Laboratory at Aalborg Portland with small specimens subjected to pure tension. The diameters for the reinforcement were 8 mm, 12 mm and 16 mm and these tests are reported in [6] and [7]. On the basis of these tests an expression for the anchorage length as function of the concrete cover, the tensile strength of the reinforcement, the compressive strength of the UHS FRC and the reinforcement perpendicular to the main reinforcement has been established, see [4]. The results from these tests form the basis for the design of the connections reported in [1].

As greater diameters for the reinforcement also are of interest anchorage tests for 20 mm reinforcement (and also 16 mm) have been applied in connection with the tests of joints reported in [2]. Five specimens for both diameters have been tested.

As an example some test results for the specimens with 16 mm reinforcement are given in the following. The average yielding strength for the reinforcement was 598 MPa and the average ultimate strength was 683 MPa. The average compressive strength for the UHS FRC used in these tests was 128 MPa. The anchorage length for the 16 mm reinforcement bar is 170 mm corresponding to approximately 11 times the bar diameter. A typical load - displacement (slip) curve is shown in Figure 3 and it is seen that there is a clear yield plateau. An average value for the yielding force was 112.8 kN corresponding to a stress of 561 MPa and an average for the ultimate force was 132.8 kN corresponding to a stress of 661 MPa.

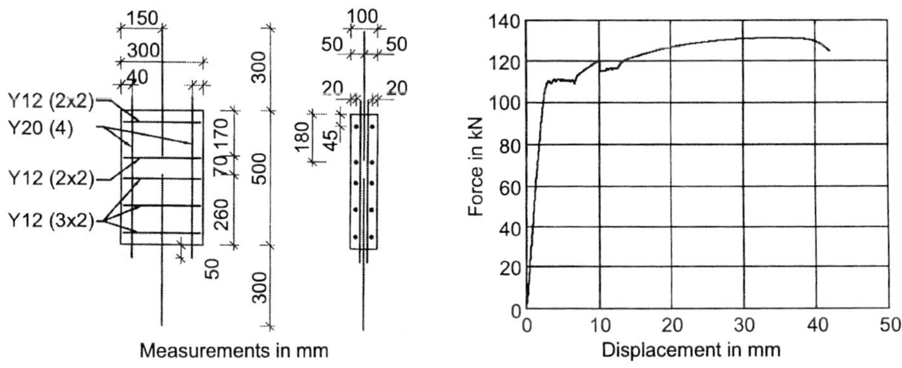

Figure 3 Test specimen and load - displacement curve for a 16 mm reinforcement bar

The anchorage yielding stress is thus only 37 MPa less than the yielding stress for the reinforcement bar and the ultimate anchorage strength is 22 MPa less than the ultimate strength for the reinforcement bar. Most important however is that the anchorage length makes the failure to be a failure in the reinforcement bars.

Beam and Slab Tests

In connection with the 1994 / 95 building at Aalborg University 18 beams were tested for static load. In addition 8 beams exposed to fire were tested. It is important to note that no concrete beams are used in this building system as explained in the introduction. The words beam tests are only used in this context for tests with slabs with a width of 300 mm instead of the width for the slabs used in the slab tests where the width was more than 1 m. So, beam tests stand for slab tests with less width than for the normal slab tests.

A sketch of a beam with loading and cross sections is shown in Figure 4.

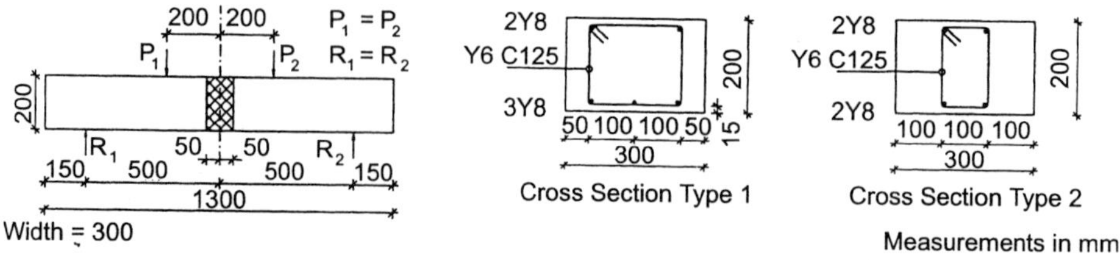

Figure 4 Beam with loading and cross sections

These series of beam tests consist of six different test series and the main purpose was to find the best joint to connect the two beam parts. Joints are in the figures shown as hatching. In more detail the purpose was:

- to develope a connection where the reinforcement is not pulled out of the UHS FRC. Beams with such a connection will then act as normal continuous beams with respect to strength and stiffness

- to estimate the sensitivity to the anchoring length and the placing of the reinforcement in the horizontal direction

- to estimate the influence of horizontal reinforcement perpendicular to the principal reinforcement in the length direction

In these tests the two forces in Figure 4 were of the same magnitude thus the joint was in a state of pure bending and no shear in the region between the two forces.

During the load tests different forces, displacements and strains were measured together with the crack development. A typical load-deflection curve is shown in Figure 5 and it is seen that the beam acts in a plastic and ductile manner.

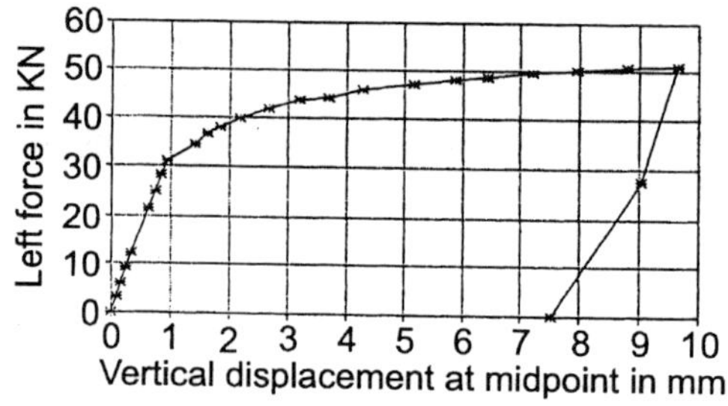

Figure 5 Load - deflection curve for beam with UHS FRC joint in the middle

The conclusion for these beam tests was that a connection as shown in Figure 2 at an 80 mm anchorage length for the 8 mm reinforcement bars and a total length of the joint of 100 mm was a practical, reasonable and in static sense good connection.The measured value for the ultimate moment capacity was in close agreement with the calculated value.

In connection with the 1998 / 99 building at Aalborg University further nine beams were tested. Three of the beams were identical to three of the beams tested in connection with the 1994 / 95 building and six of the beams were in two new beams series with three beams in each series. These new series were specially focused on the combination of bending and shear in joints. A sketch of the beams in the first test series is shown in Figure 6 to the left and the surface between the UHS FRC and the beam (made of normal strength concrete) is rugged. The beams in the second test series are shown in Figure 6 to the right and the surface between the beam parts and the UHS FRC is smooth.

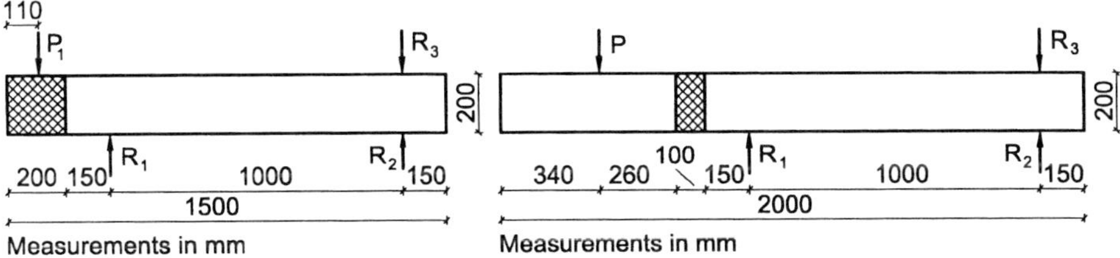

Figure 6 Beams with rugged and smooth surfaces

It is seen that in these two test series the connections with UHS FRC are acted by both moments and shear forces. Also in these tests the beams in the final stage act in a plastic and ductile way with yielding in the reinforcement and a state of beginning crushing in the compression zone of the concrete. Different upper and lower bound solutions according to the theory of plasticity have been calculated and the measured values from the tests were found to lie in the interval between these bounds both for the yielding state and for the ultimate state. Further details can be found in [2].

In connection with the 1994 / 95 building at Aalborg University four different types of slabs were tested. The slabs were constructed as parts of greater slabs and the load on the slabs simulate some typical load cases and support conditions. The slabs were only loaded in vertical direction.

A typical load - deflection curve is shown in Figure 7 and the conclusion for these load tests was that the connection are so strong and ductile that the slabs constructed with these connections will act as normal concrete slabs and the standard calculation methods used for reinforced concrete can be applied.

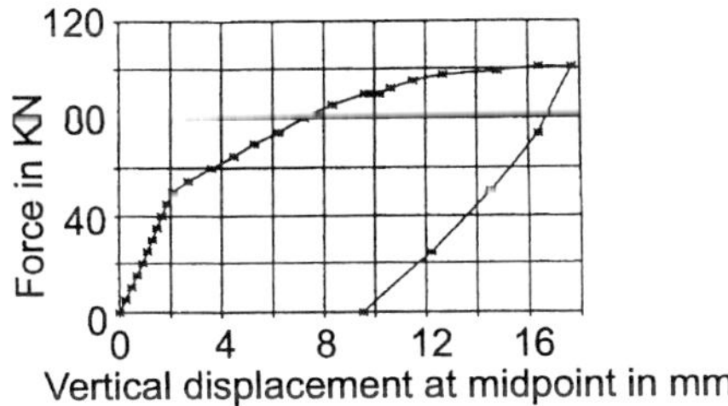

Figure 7 Load - deflection curve for a slab

Also in connection with the 1998 / 99 building some additional slab tests have been carried out. These tests correspond to the beam tests with focus on combination of bending and shear in the joints mentioned earlier and thus three slabs with rugged surfaces and three slabs with smooth surfaces between the parts with normal concrete and the UHS FRC were tested.

The conclusion for these tests was also the same as described for the beam tests. The load capacity was for all beams and slabs greater than the lower bound solution. The lower bound solution corresponds to the exact load capacity if the beams and slabs were not strengthened in the regions outside the connections. The introduction of a joint of this kind cannot be estimated to give any reduction in the load capacity. This means that the usual theory of plasticity can be used even if these vertical joints with UHS FRC are present in the slabs.

Column - Slab Tests

As mentioned in the introduction the horizontal forces on the building are transferred only through the columns in the 1998 / 99 building and stabilizing concrete walls are thus not necessary.

Before the construction of this building some tests with column / slab connections have been carried out to be sure that this important connection has the necessary load bearing capacity. Two types of tests were made. Type 1 with an inner circular column at a diameter of 350 mm and acted by both vertical and horizontal forces and Type 2 with a boundary column with a cross section corresponding to half of a circle and also with at a diameter of 350 mm.

Only the region of the slab near the column was examined in the tests and the slab dimensions were 1.3 m times 1.4 m at a thickness of 0.2 m. A photo of the test arrangement for a Type 1 column is shown in Figure 8 to the left and a sketch of the system is shown in Figure 8 to the right.

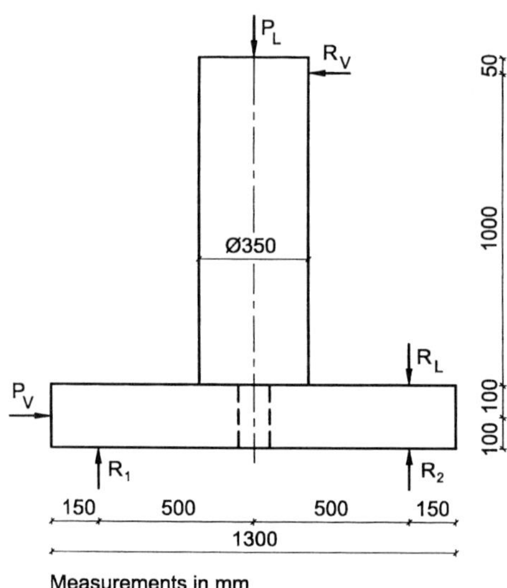

Figure 8 Test rig for inner column / slab connection

Four main reinforcement bars at a diameter of 20 mm from the column were anchoraged in the UHS FRC joints forming a cross near the column. As seen in Figure 8 the column / slab connection is turned 180 ° in relation to the situation in practice where the slabs transfer the vertical loads to the column. This was only to simplify the testing system. Several forces, displacements and strains were measured during the tests.

The vertical load starting with 0 was increased to 550 kN and simultaneous the horizontal load was increased to 90 kN. At this stage the vertical load was kept at a constant value and the horizontal load was then increased to its maximum value. The horizontal displacement difference between a point at the top of the column and a point at the slab was approx. 30 mm. At this stage the concrete near the region where the column met the slab was crushed and there was yielding in the reinforcement. All the tests showed a ductile rupture.

For the Type 2 tests with the column / slab connection a more complicated test setup was constructed with the possibility to force the column with horizontal loads in opposite directions together with a vertical load. For this type the maximum vertical load was 275 kN (half the value for Type 1) and a horizontal load of about 50 kN. Also for these tests the behaviour in the final stage was ductile with crushing in the concrete and yielding in the reinforcement.

FIRE TESTS

The very dense UHS FRC could have a high risk of explosive spalling when exposed to fire. Due to the low water content and steel fibres the risk do not exists after a resonable drying period. Due to the mix of UHS FRC the decomposing processes are also different to the processes in ordinary concrete, see [4]. Eight beams with joints exposed to fire were tested to verify the fire resistance and they are reported in [1].

CONCLUSIONS

From the test series and the experience both during the casting of the test specimens and during casting of the joints in practice at the construction site the following can be concluded:

1. Casting with the UHS FRC turns out to be easy to do in this type of joints

2. The joints make the slab acting as a ductile structure with strength which can be calculated according to the traditional methods

3. The joints and the UHS FRC make it possible to anchorage deformed bars for example at a diameter of 8 mm and yield strength of 550 MPa at a length of only 60 mm. In practice an anchorage length of 80 mm was used in this project

4. It is possible to use the building system without stabilizing walls. The horizontal forces can be transferred through the slab / column connections to the foundations

5. The joints can be classified as fire resistant in at least 60 minutes. After a fire exposure in 60 minutes and cooling the residual load bearing capacity is 75% of the original capacity.

ACKNOWLEDGEMENTS

The authors would like to express their appreciation for the research grants made by The Building Department, Ministry of Education (Byggedirektoratet, Undervisningsministeriet) to carry out the reported work, which forms part of a major research programme on ultra high strength fibre reinforced concrete.

REFERENCES

1. BYGGEDIREKTORATET, UNDERVISNINGSMINISTERIET. Udvikling af et nyt konstruktivt system for betonelementbyggeri. (Development of a new structural system for prefabricated concrete elements). In Danish. Undervisningsministeriet, 1994, 89 pages + 164 pages

2. BYGGEDIREKTORATET, UNDERVISNINGSMINISTERIET. Forsøg med et nyt konstruktivt system for betonbyggeri. (Tests with a new structural system for a concrete building system). In Danish. Undervisningsministeriet, 1998, 166 pages + appendices A-O.

3. BACHE, H H. Compact Reinforced Composite, Basic Principles. CBL Report 41, Aalborg Portland, 1987, 87 pages

4 JENSEN, BJARNE CHR. An Example of the Structural Use of Steel-Fibre Reinforced Ultra High Strength Concrete. Accepted for publication in Structural Engineering International

5. NIELSEN, C V. Ultra High-Strength Steel Fibre Reinforced Concrete, Part I and Part II. Department of Structural Engineering, Technical University of Denmark. Report No. 323 and 324, 1995, 188 pages and 154 pages

6. EUREKA project: EU264 - COMPRESIT. Technical Report, Subtask 1.7 - Anchorage of Reinforced Bars, 1992, 29 pages

7. CARL BRO as. MINISTRUCT Task 3. Laboratory Experiments on Structural Properties. Sub-task 3.1: Effect of Different Types of Reinforcement. Anchorage of Ribbed Steel Bars and Prestressing Strands, February 1996, 18 pages

8. OLESEN, JOHN FORBES: The Concrete Weld: Monolithic Slabs from Precast Elements. Structural Engineering International, Vol. 4, 1995, pp. 230

9. NIELSEN, C V, OLESEN, J F, and AARUP, B K. Effects of Fibres on Bond Strength of High-Strength Concrete. 4.th International Symposium on Utilization of High-Strength/High Performance Concrete. Paris, 1996, pp. 1209 - 1218

10. JENSEN, B C, and AARUP, B. Fire resistance of fibre reinforced silica fume based concrete. 4th International Symposium on Utilization of High strength - High performance Concrete. Paris, 1996, pp. 551 - 560

VARIOUS USES OF FLOURESCENT RESIN CONCRETE

A Fujita

Meijo University

H Nakanishi

Taiyu Kenetsu Co Ltd

T Maruyama

Nagaoka University of Technology

Japan

ABSTRACT. We have developed an inorganic fluorescent material using fluorescent ceramics, fluorescent acryle which radiated various color bright light during night time through Ultoraviolet light lamps irradialion, to be used for road brightness, warning etc. A number of uses can be considered, this time we consider strength of emitted light from fluorescent mixed resin concrete. We measured compressive strength and bending strength of the light emitted from fluorescent mixed Resin Concrete.

As a result we found that measured factors are identical to the results as in the case of natural concrete aggregate. On the other hand characteristic of irradiated light was in direct proportion to the radiation of UV strength fluorescent aggregate. We have discussed all these observed results.

Keywords: Resin concrete, Fluorescent ceramics, Fluorescent acryle, Ultoraviolet rays lamps, UV strength, Degree of strength of the light, Scene material, Light color, Careful arouse material

Dr Akihiro Fujita is Associate Professor at Meijo University, Nagoya, Japan, Faculty of Science and Technology Department of Civil Engineering, his research field is Pavement Engineering. Recently his work is concentrated especially to develop material to be used for inorganic fluorescent pavement and to study optical characteristics such as brightness and warning signs at curves.

Mr Hiromitsu Nakanishi is working as a General Manager of the Central Research Institute of m/s Taiyu Kensetsu Co., Ltd. Nagoya, Japan, and he is a keen Researcher in the field of Asphalt, Resin Mix Concrete.

Professor Teruhiko Maruyama is Doctor of Engineering and he is Professor at Nagaoka University of Technology, Japan, his main research field is Environment Construction in the Road Engineering Department.

INTRODUCTION

The purpose of this research is to develop new fluorescent ceramic material. Acryl material ceramic material used is strong in adverse weather conditions for creating beautiful land scapes through various colors of bright light. This fluorescent material has a property that it emits variety of light colors, when illuminated by UV light.

This can be used as a luminescent pavement material for brightness in the tunnel, cross-section and dangerous points to provide safety on dark roads. With a purpose of using this material as a pavement material for Resin concrete. This paper reports the study of various dynamic characteristics and Luminescence.

EXPERIMENTAL DETAILS

This newly developed material fluorescent ceramic is crushed Glass type ceramic & inorganic fluorescent material (moist mixed) and kept in an alumna pot heated at a high temperature in an electric furnace. Then cooled down and crushed in a crushing machine and it finally become aggregated material, this is called a fluorescent ceramic material. On the other hand fluorescent acryl is obtained through acrylic resin mixed with inorganic fluorescent pigment and crushed (grinded) and aggregated. Fluorescent Acrylic resin's specific gravity is lower than the natural aggregate. Table 1 and 2 show the specifications of various materials. Luminescent material showed green color. Out of luminance green color only was used in this experiment

Table 1 Fluorescent ceramics

Tension strength	(kg/m^2)	600
Bending strength	(kg/m^2)	900
Diamond strength	(%)	< 30
Specific strength		1.772
Absorb rate	(%)	0.3
L. A. wear lose weight	(%)	3.06

Table 2 Fluorescent acryl

Tension strength	(kg/m^2)	450
Bending strength	(kg/m^2)	800
Diamond strength	(%)	<45
Reaction to Climate change	(Ea^*b^*)	3.12
Reaction to Temperature		
180...30min		<0.01
200...30min		<0.01
Specific gravity		1.80
Absorb rate	(%)	0.04
L. A. wear lose weight	(%)	3.00

SPECIMEN

Resin concrete aggregate (particle size 13.15mm) natural aggregate only, fluorescent ceramic aggregate, fluorescent Acryl aggregate. Table 3 shows 10%, 20%, 30%, 40% of those aggregate as contents of the specimen mix proportion of specimen is shown.

Table 3 Proportion of specimen, g

w/c = 40%		W	C	S	G_1	G_2	P
Crushed Rock only		54	250	500	500	0	83
Percentage mixed Ceramics aggregate (%)	10	54	250	500	500	47	83
	20	98	450	900	720	170	150
	30	54	250	500	350	142	83
	40	98	450	900	540	340	150
Percentage mixed Acryl aggregate (%)	10	54	250	500	500	34	83
	20	98	450	900	720	122	150
	30	54	250	500	350	102	83
	40	98	450	900	540	345	150

W = water, c = cement s = sand , G1 = natural aggregate G2 = fluorescent aggregate,
P = polymer (RBS)

RESULT OF EXPERIMENT

Strength of Resin Concrete

Compressive strength and bending strength after one day age limit is reported in Figure 1 and Figure 2. Figure 3 and Figure 4 shows the 3 and 7 days age limit.

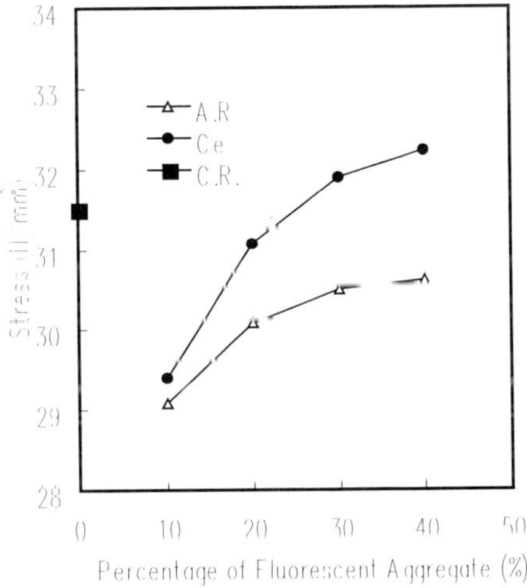

Figure 1 Percentage mixed flourescent aggregate and compressive strength of relation

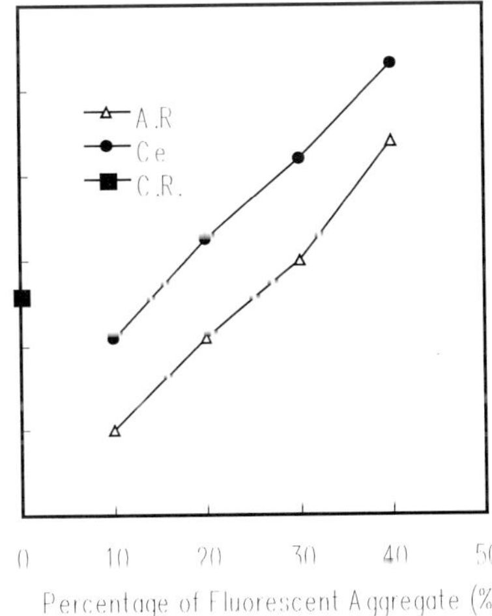

Figure 2 Percentage mixed flourescent compressive strength of relation

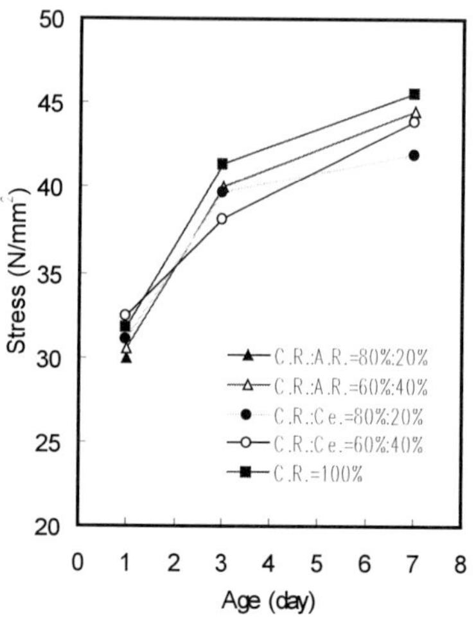

Figure 3 Age and compressive strength of relation

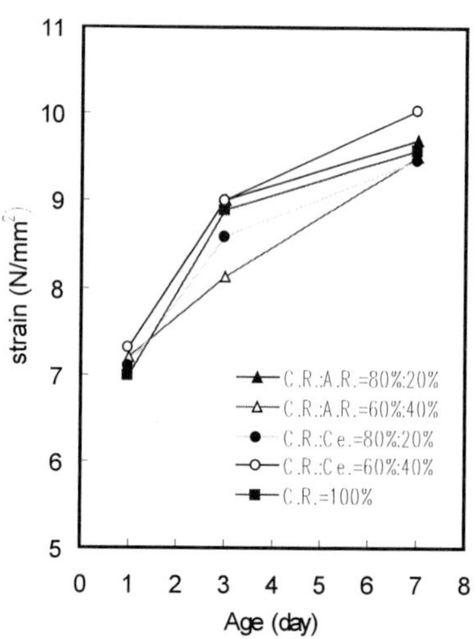

Figure 4 Age and bending strength of relation

Compressive strength for one day shows higher in proportion to their mixed quantity for both type of aggregates. Fluorescent aggregate, if mixed 20%, the specimen's strength is similar to the natural aggregate 31 N/mm^2.

Compressive strength and bending strength showed similar characteristics shows 7.1 N when amount of fluorescent aggregate is 20%. On the other hand relationship between compressed strength directly proportional, but 3 days fluorescent aggregate regardless of the quantity mixed, only natural aggregate strength of specimen is similar. Again bending strength shows similar result. all the fluorescent resin materials improved about 36% while strength for 7 days age compared with one day age. but bending strength showed similar result.

Measurement of Degree of Strength of the Light

Concrete slab(300*300*50),10mm thick Resin concrete pasted and polished by shot pleats. Measurement was carried out in a dark room by UV Lamp light from top of the specimen and degree of strength of light is measured at 45.using Minolta CS-100 these results are shown in Figure.5

The degree of the strength of light of Acryl and ceramic specimen measured simultaneously fluorescent ceramic (mixed Quantity 40%) 12.4 cg/m^2 while fluorescent Acryl Resin concrete.(mixed Quantity 40%) 1.68cg/mm^2 This represents that Acryl is lower than ceramic. In general fluorescent material degree of strength of the light is affected largely by strength of is affected largely by strength of UV light and the other is total are a of fluorescent material lightened, but both show lower response irrespective of the areas lightened. The reason is the amount of fluorescent material mixed that was rather small in quantity fluorescent ceramics 30% and fluorescent Acryle 1% only.

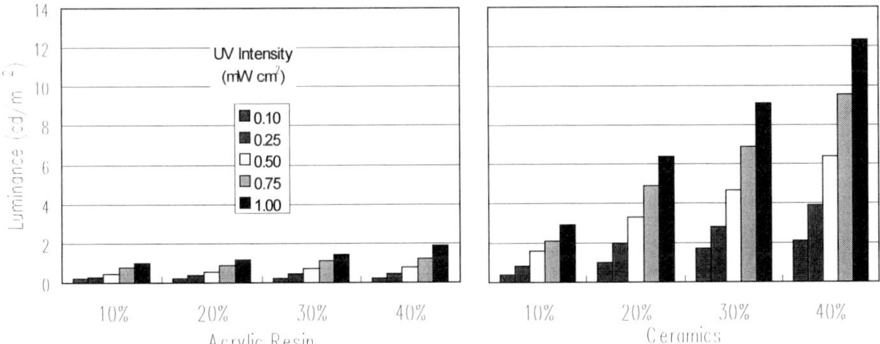

Figure 5 Percentage of fluorescent aggregate mixed and degree of strength of light.

Table 4 Percentage mixed area fluorescent aggregate

FLUORESCENT AGGREGATE	PERCENTAGE OF FLUORESCENT AGGREGATE (%)			
	10	20	30	40
Acrylic Resin	5.9	8.1	9.7	14.2
Ceramics	2.6	7.0	9.7	13.3

CONCLUSIONS

1. Compressive strength of R.C and bending strength of fluorescent aggregate increases in proportion to the quantity is mixed 20%, it shows similar results as for natural aggregate only.

2. The relation between strength, compressive strength and bending strength is proportional to the days but after 3 days shows similar results as for the natural aggregates only. Fluorescent ceramic Resin material and fluorescent Acrylic Resin material didn't show any difference with the natural fluorescent material. Fluorescent ceramics aggregate fluorescent acryl aggregate Resin concrete no big difference was observed.

3. These both Resin concrete UV, percentage of fluorescent aggregate mixing degree of strength of the light.

4. Difference in fluorescent material mixing percentage is effected strongly to degree of strength of the light

5. Fluorescent Resin material can easily be used for out side pavement, color pavement, warning sign pavement.

REFERENCES

1. AKIHIRO FUJITA, TERUHIKO MARUYAMA ET. AL. Optical Properties of Fluorescent Material Illuminated by Ultraviolet(UV) light, Proceedings LUX PACIFICA'97(3rd Pacifica Basin Lighting Congress), 1997.pp.F-7.F-11.

2. AKIHIRO FUJITA, TERUHIKO MARUYAMA, TETSUYA SHIMODA, HIDETO TAKEISHI The Technical characteristics of Fluorescent material especially used for Road use, Pavement, Vol.32.No.6,1997,pp.15.20.

3. AKIHIRO FUJITA ET. AL. The proposal of a new kind of sign system by Ultraviolet light, Traffic Engineering, Vol.32,No.6,1997,pp.43.49.

4. AKIHIRO FUJITA. Current situation and model cases of Artificial aggregate and color aggregate. Pavement, Vol.30,No.11,1995,pp.29.32.

FIBRE REINFORCED CONCRETE FOR TELECOMMUNICATION NETS PRECAST SYSTEM

P Spinelli

L Galano

University of Florence

D Migliori

SEIEMAC SpA

Italy

ABSTRACT. This paper presents the application of steel fibre reinforced concrete in precast pits for telecommunication nets. Two different prototypes are considered. Outer dimensions of the first prototype (referred to as 125 × 80 pit) are 145 cm in length and 100 cm in width (the second prototype measures 108 cm in length and 88 cm in width and is referred to as 90 × 70 pit). The prototypes are composed by: a base element in steel fibre reinforced concrete and two or more precast reinforced concrete rings to be placed upon the base. Thin layers of cement mortar are used to joint two different precast elements. A suitable stiff steel plate is used to close the upper part of the pits so to permit the road transit. The main technical characteristics of these elements are furnished in the paper. To verify the reliability of the project for serviceability and ultimate limit states some experimental tests on the prototypes were carried out. The relevant results of these tests together with the behaviour of the pits up to collapse are discussed. The behaviour of the precast elements under road loading was also investigated by numerical modelling using the finite element approach. The results showed that the precast system using fibre reinforced concrete is valid alternative to the ordinary reinforced concrete systems. However, to avoid brittle failures of the upper ring elements, representing the critical zone, steel bars and stirrups should be also provided.

Keywords: Precast system, Telecommunication nets, Fibre reinforced concrete (FRC), Experimental tests, Numerical modelling.

Professor P Spinelli, is Full Professor in Design of Structures at Department of Civil Engineering, Faculty of Engineering, University of Florence, Italy. His main interests are wind effects on structures and prestressed and precast concrete structures.

Dr L Galano, is a Doctor of Philosophy in Structural Engineering from 1993, Faculty of Engineering, University of Florence, Italy. His main research interests are the seismic behaviour of reinforced concrete structures, with particular attention to experimental tests on prototypes and to theoretical aspects on this topic.

Dr D Migliori, is a Civil Engineer, General Manager of SEIEMAC S.p.A., and industry in Florence of precast concrete element, which produces reinforced concrete pits for telecommunication networks.

INTRODUCTION

In the so called SOCRATES program of Telecom Italia the cabling of all major Italian cities with optical fibres is programmed. A high production of controlled quality of concrete pits is necessary to satisfy the needs. Usually these pits are of reinforced concrete. The firm SEIEMAC has adopted a type of pit reinforced with steel fibres. This production method is very suitable as it permits a quality controlled production and gives the pits a very durable performance. Also the static performance of the pits is encouraged as demonstrated in the following note. In fact, due to the very compact shape and the help of the ground fill, the values of tensile stresses in the pits are lower than the allowable stresses. Experimental tests confirm substantially the structural behaviour.

MATERIALS AND DESIGN SPECIFICATIONS

Materials

Fibre reinforced concrete (FRC) is a composite material in which steel fibres are added to the cementitious matrix. Previous experimental works have demonstrated that steel fibres improve the flexural response of the plain concrete because of increased toughness and limited spreading of cracks; so, more favourable durability properties are obtained [1, 2]. However, with relatively low content of fibres the increase of the tensile strength is negligible.

The improved characteristics of FRC well adapt to the design of these precast pits; in the present case mild steel fibres with length variable from 20 to 30 mm and aspect ratio equal to 50 were used (aspect ratio is defined as the ratio between length and diameter of the steel fibres). An average value of 25 kg/m^3 of fibres was assumed for design purposes. Following the ACI recommendations [1] tensile strengths in flexural conditions are given by:

$$\sigma_{cf} = 0.843\, f_r\, V_m + 425\, V_f\, l/d_f; \qquad \sigma_{cu} = 0.970\, f_r\, V_m + 494\, V_f\, l/d_f \quad (psi)$$

where,

σ_{cf} = first cracking strength, and
σ_{cu} = ultimate flexural strength (modulus of rupture)

V_m and V_f indicate the volumetric fraction of matrix and fibres, l/d_f is the aspect ratio and f_r the modulus of rupture of the cement matrix without fibres. More caution values are proposed by the actual Italian code on reinforced concrete structures [3]. The design of the pits was based on a concrete C40 (f_{ck} = 40 N/mm^2) adding the steel fibres as previously said and using deformed steel bars (FeB 44 k with mechanical characteristics as specified in the Italian national standards [3]).

The design approach was based on the evaluation of the maximum principal tensile stresses in the different parts of the pits and checking that allowable values were not overcome. Allowable values of stresses were assumed with reference to σ_{cu} using a suitable safety coefficient (γ_c = 1.6, [3]).

Main Dimensions of the Two Pit Prototypes (see Figure 1 for the 90 × 70 type)

	Pit 1	Pit 2
- Inner base dimensions:	125 × 80 cm	90 × 70 cm
- Outer base length:	145 cm	108 cm
- Outer base width:	100 cm	88 cm
- Height (with two ring elements):	91.5 cm	80 cm
- Thickness of base slab:	10 cm	8 cm
- Thickness of the lateral walls:	10 or 4.5 cm	9 or 4.5 cm.

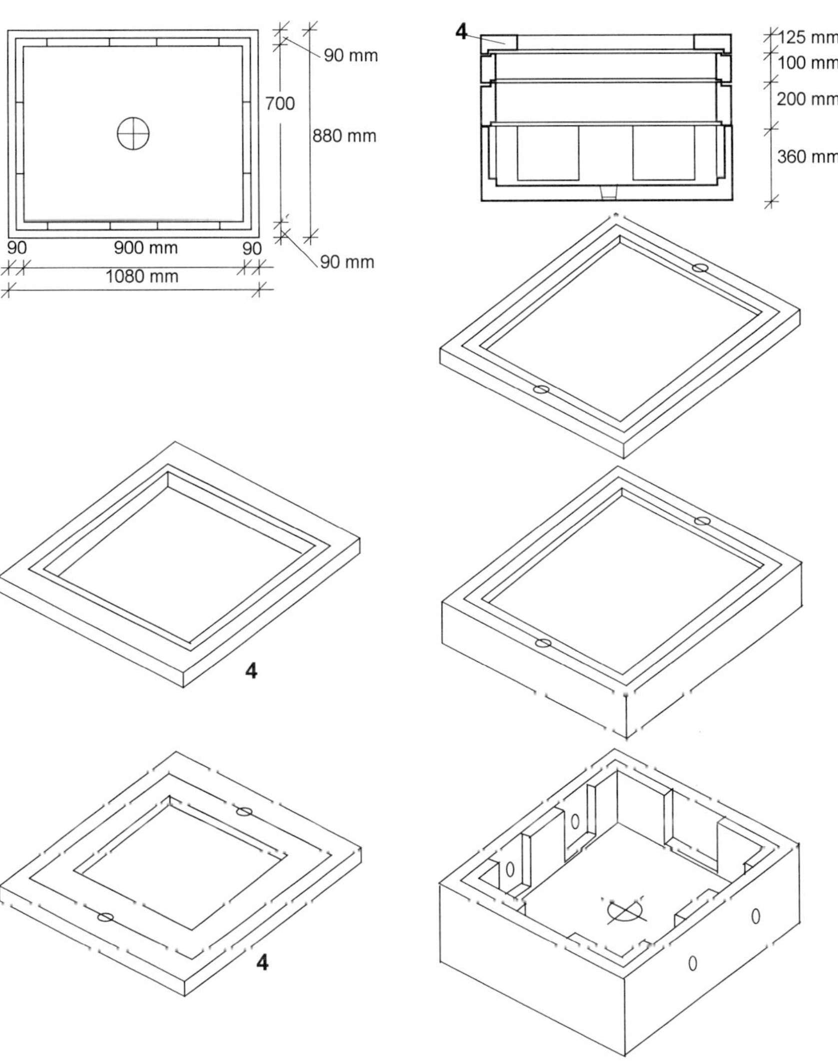

Figure 1 Dimensions and possible assemblage for the 90 × 70 pit

Design Ground Properties

- Specific weight: $\gamma_t = 19$ kN/m^3
- Friction angle: $\varphi = 30°$
- Friction angle earth-concrete: $\delta = 0°$
- Active pushing coefficient: $k_a = 0.271$.

Loading Conditions

With reference to the 90 × 70 pit, Figure 2 shows the two basic loading conditions used for design. The dead weight of the pit (G_1), the dead weight of the steel closing element (G_2), the dead weight of road-bed (G_3) and the horizontal push of the earth (S_t) were considered in both the cases. Furthermore, in the first condition (a) the weight Q of road wheels (10 kN, [4]) was applied on the centre of the closing element. In the second condition (b) the lateral forces S_q due to the road wheel of 10 kN applied on the lateral zone of the pits were assumed. Loading combinations were:

- $(G_1 + G_2 + G_3) + S_t + Q$ (serviceability limit state, S.L.S., condition a)
- $1.4\ (G_1 + G_2 + G_3) + S_t + 1.5\ Q$ (ultimate limit state, U.L.S., condition a)
- $(G_1 + G_2 + G_3) + S_t + S_q$ (serviceability limit state, S.L.S., condition b)
- $1.4\ (G_1 + G_2 + G_3) + 1.4\ S_t + 1.5\ S_q$ (ultimate limit state, U.L.S., condition b).

To simulate the interaction between the concrete and the ground elastic springs both vertical and horizontal were considered (Winkler model) as explained in Figure 2 and assuming: $k_{v1} = 0.05$ N/mm^3, $k_{v2} = 0.0125$ N/mm^3 and $k_h = 0.015$ N/mm^3.

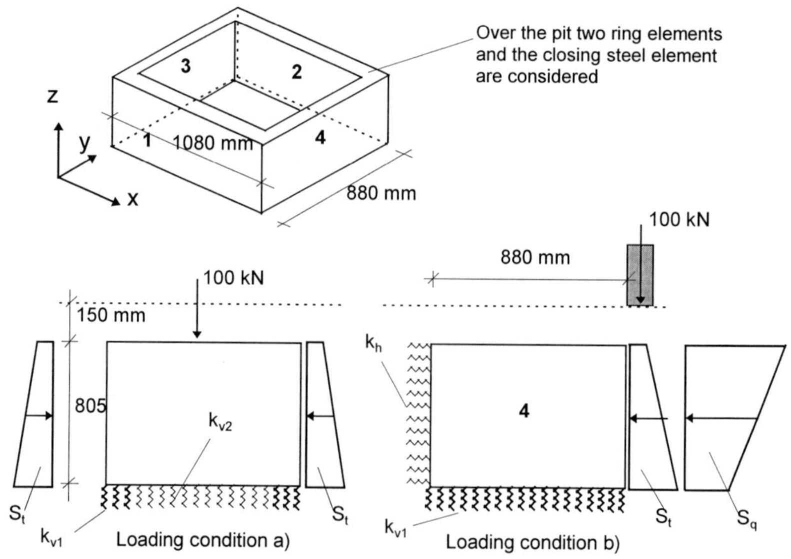

Figure 2 Loading conditions a) and b)

EXPERIMENTAL DETAILS

Loading Setup

Experimental tests were performed on the two different pit prototypes, namely 125 × 80 pit and 90 × 70 pit. Figure 3 shows the loading setup used for the tests. The prototypes were composed by: the base element, a first ring element, 200 mm height and the upper ring element, 100 mm height. A stiff steel plate (element 4 in Figure 3), to simulate the true closing element used on the road, was placed above and a vertical load was applied on the centre of the plate by an hydraulic jack capable of 500 kN. Thin layers of cement mortar were used to combine the different parts of the pits. Below the base a layer of sand, about 50 mm thick, was used to assure an uniform distribution of vertical stresses. Three different 125 × 80 pits were tested: P1, P2 and P3. All the elements of these prototypes were in fibre reinforced concrete without steel bars. Other three pits (90 × 70) were tested: P4, P5 and P6. P4 prototype was in fibre reinforced concrete whereas P5 and P6 had the upper ring element reinforced by 4 ϕ 8 longitudinal steel bars and transversal stirrups ϕ 6, step 10 cm.

Two different tests were performed on the same prototype (with the exception of P4); in the first test the load was increased with velocity of application of 2 kN/s up to about 150 kN (test a). This level of load corresponded to 1.5 times the maximum weight of a wheel as provided by Italian rules on road bridges [4]. Therefore, the pits behave satisfactorily if elastic behaviour is observed in these tests. A second test was performed increasing the load with the same velocity until the collapse, to evaluate the safety coefficient (test b). During the tests the displacements S_1 and S_2 on two opposite sides of the pits (Figure 3) were measured together with the applied load and other displacement transducers were used to reveal the formation of the first cracks. A 15 channel acquisition system with frequency of 5 Hz was used. Figure 4 shows a pit during the loading test.

Results

The main results are presented in Figures 6 and 7 in which the load-displacement diagrams are given (displacement was evaluate as the average value of S_1 and S_2) and in Table 1, in which the maximum load values are reported. Similar behaviours were observed for all the pits with the only exception of P4. In the following, P1a and P1b tests are explained.

Table 1 Results of experimental tests on precast pits

FIRST ELASTIC TESTS			COLLAPSE TESTS		
PIT	TEST	MAX. LOAD kN	PIT	TEST	MAX. LOAD kN
P1 (125 × 80)	P1a	151.3	P1 (125 × 80)	P1b	345.2
P2 (125 × 80)	P2a	151.2	P2 (125 × 80)	P2b	336.3
P3 (125 × 80)	P3a	151.1	P3 (125 × 80)	P3b	327.3
P4 (90 × 70)	/	/	P4 (90 × 70)	P4b	151.5
P5 (90 × 70)	P5a	151.2	P5 (90 × 70)	P5b	265.3
P6 (90 × 70)	P6a	151.1	P6 (90 × 70)	P6b	267.9

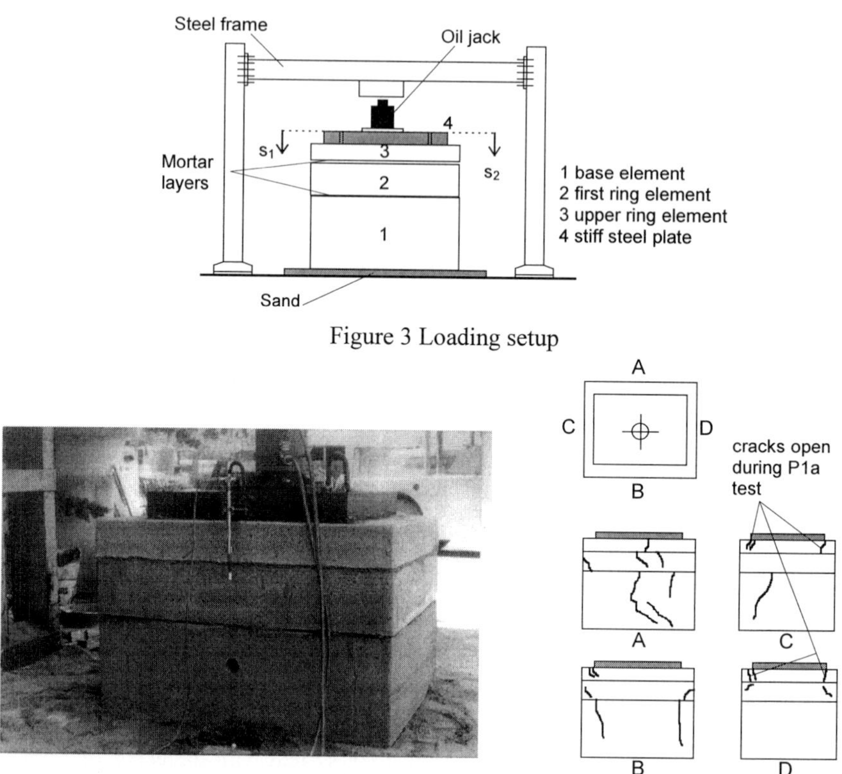

Figure 3 Loading setup

Figure 4 A prototype during the loading test

Figure 5 Cracking pattern after P1b test

Figure 6 shows that an elastic response of the materials was achieved during experiment P1a; however, five minor cracks developed near the corners of the upper ring element (Figure 5). These cracks disappeared when the load was removed. Hence, the prototype behaved satisfactorily. During P1b test, increasing the load from 150 to 250 kN several cracks appeared and spread near the corners of the ring elements; when the load was increased up to 300 kN four main cracks appeared in the vertical sides of the base element and propagated up to the base. Further increased load (350 kN) caused a speedy increase of vertical displacements and the test was stopped (collapse state). Figure 5 shows the cracking pattern in the final state. Load-displacement curve of Figure 7 indicate an elastic behaviour under 200 kN of loading followed by a plastic phase in which maximum displacement was greater than 9 mm.

No relevant differences were observed in the other tests; the 90 × 70 pits (P5 and P6) showed a lower collapse load level than 125 × 80 pits but with similar cracking patterns. The main difference in experimental behaviour resulted for the P4 prototype. During the first test, at the load of 150 kN and before unloading an unexpected and sudden failure of the pit occurred, with brittle character. This was probably due to the use of an upper ring element without steel bars and reinforced only with steel fibre. On the whole the experiments (with P4 exception) showed good performance of the precast pit prototypes.

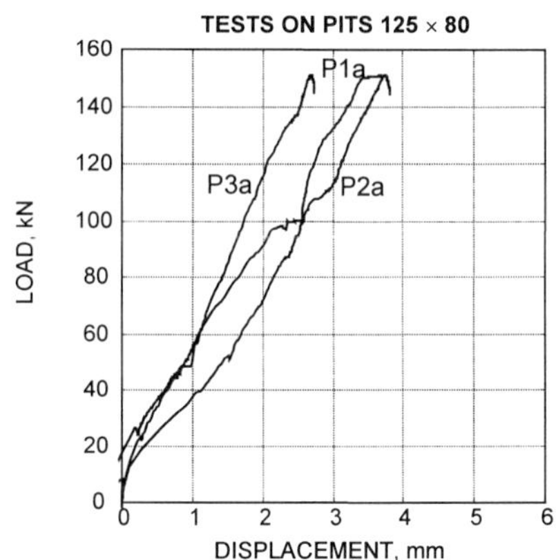

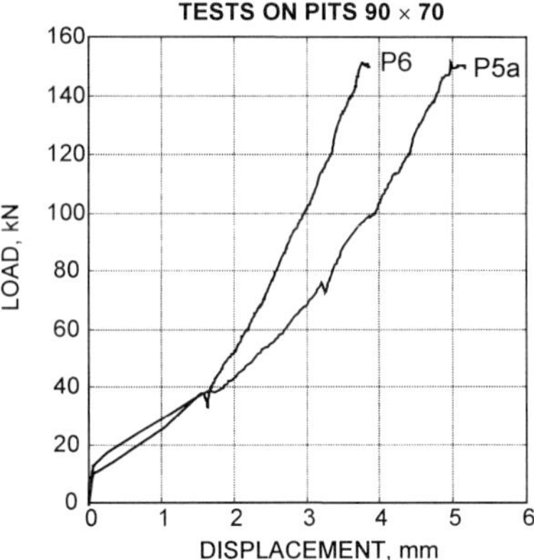

Figure 6 Elastic tests on precast pits

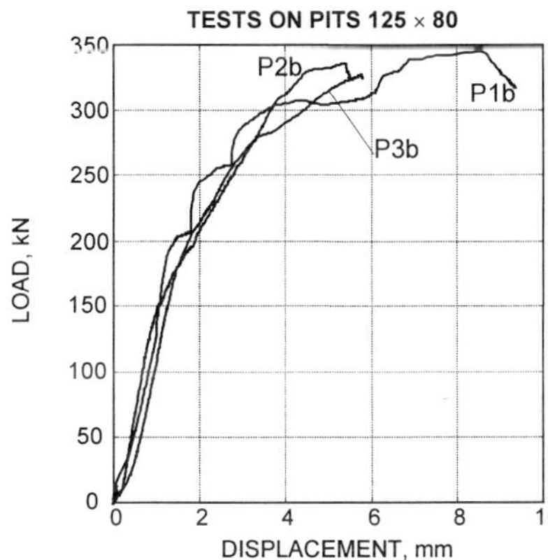

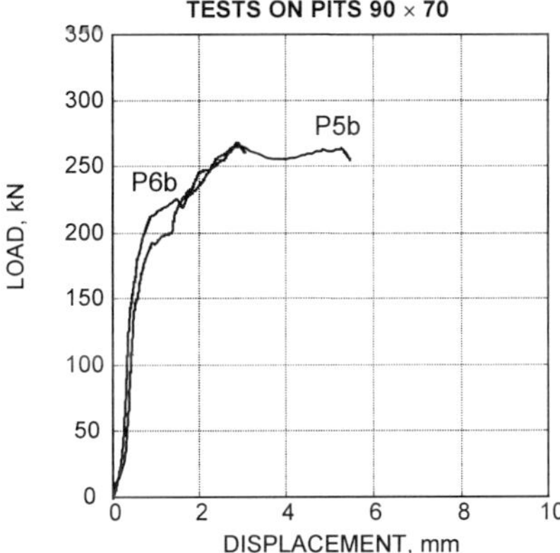

Figure 7 Collapse tests on precast pits

NUMERICAL MODELLING

To evaluate the state of stress in the pit prototypes under loading conditions a) and b), finite element models were developed. A standard computer program capable of simulating the geometrical characteristics of the pits was used; tridimensional models, 8 node brick isoparametric elements were used to the aim. Figure 8a shows a view of the finite element mesh of the 90 × 70 prototype (base element). Shell elements were used to model the closing steel plate in the upper part. Thin brick elements were used to simulate the mortar layers interposed between different part of the pit. The most relevant values of the input parameters are listed in the following hypothesis of linear elastic behaviour of all the materials.

- Young modulus of the concrete: $E_c = 38237$ N/mm^2
- Young modulus of the thin mortar layers: $E_m = 5000$ N/mm^2
- Poisson ratio: $\nu = 0.20$
- Thickness of shell elements: t = 40 mm.

To simulate the possibility of breaking off the lateral walls to accomplish the optical fibres net cables, modified models with five holes were also developed. Finally, a sensibility analysis varying the Winkler coefficient k_{v2}, the E_m modulus and the k_a coefficient was also performed. In the following the various solutions and models for the 90 × 70 prototype are indicated as (a similar group of analyses were performed for the 125 × 80 pit prototype):

- P1a1: basic model without holes, loading a), S.L.S.
- P1a2: basic model without holes, loading a), U.L.S.
- P1c1: model with holes, loading a), U.L.S.
- P1b1: basic model without holes, loading b), S.L.S.
- P1b2: basic model without holes, loading b), U.L.S.
- P1d1: model with holes, loading b), U.L.S.
- P1a3: as P1a2 but with k_{v2}= 0.05 N/mm^3, i.e. equal to k_{v1}
- P1a4: as P1a2 but with E_m = 30000 N/mm^2
- P1b3: as P1b2 but with k_a = 0.5 instead of 0.271.

Numerical Results

More representative results for the 90 × 70 pit prototype are given in Tables 2 and 3, in which the maximum values of stresses in the base slab and in the wall 1 are reported. σ_I and σ_{III} indicate the principal tensile and compressive stresses. The whole examination of the results showed that the load condition a) was more severe for the base slab (σ_I = 1.57 N/mm^2 for U.L.S. that increased to 1.62 N/mm^2 in the model with the holes); the same load condition was the more severe for the wall 1 (σ_I = 1.71 N/mm^2 for the model with holes and for U.L.S.). However the effects of the five holes were not globally relevant. Sensibility analyses allowed to track other general conclusions about the stress state in the concrete. The parameters k_{v1} and k_{v2} strongly affected the σ_I values in the base slab (σ_I = 3.38 N/mm^2 with analysis P1a3). The influence of the E_m parameter was negligible, whereas k_a coefficient strongly affected maximum tensile stresses in the wall 1 (σ_I = 2.52 N/mm^2 with P1b3).

Although the general results obtained with the models of the 125 × 80 pit prototype were the same previously examined, it was noted that higher values of tensile stresses were computed (as an example, maximum value of σ_I in the wall 1 was 1.89 N/mm^2). Figure 8b shows the stress distribution of σ_{xx} for wall 1 with P1b2 analysis and Figure 9 the stress distribution in the base slab both for σ_{xx} and σ_{yy} components with P1a2 analysis. The distribution and the values of the stresses evaluated by the finite element analyses showed that zones subjected to tension stress were limited; if a value of σ_{all}=2.01 N/mm^2 is assumed as design tensile stress (σ_{cu}/1.6) in the ultimate limit state conditions, basis analyses furnished stress values always lower than σ_{all}. Furthermore, flexural tests on fibre reinforced concrete used for making the pits, furnished a characteristic flexural strength f_{cfk} equal to 3.73 N/mm^2, greater than theoretical σ_{cu}.

However, failure mode observed during the experimental tests on the pit prototypes and the sensibility analyses, suggested that only the base element had a safe and suitable behaviour when realised in FRC without main steel bars. The ring elements and, particularly, the upper ring elements which directly withstand the weight of the road loading, should be further reinforced with standard steel deformed bars and stirrups.

Table 2 Maximum stress values in the base slab (N/mm^2)

	σ_{xx}		σ_{yy}		σ_{xz}		σ_{yz}		σ_I	σ_{III}
P1a1	0.68	-0.79	1.04	-1.04	0.33	-0.33	0.21	-021	1.04	-1.04
P1a2	1.02	-1.18	1.57	-1.54	0.50	-0.50	0.31	-0.31	1.57	-1.54
P1a3(*)	2.33	-2.73	3.38	-3.53	0.96	-0.96	0.67	-0.67	3.38	-0.31
P1a4(*)	1.01	-1.16	1.56	-1.53	0.50	-0.50	0.32	-0.32	1.56	-1.53
P1b1	0.44	-0.30	0.64	-1.02	0.18	-0.18	0.11	-0.12	0.65	-1.03
P1b2	0.64	-0.43	0.92	-1.48	0.27	-0.27	0.15	-0.17	0.95	-1.49
P1b3(*)	1.20	-0.78	1.65	-2.67	0.54	-0.54	0.26	-0.30	1.69	-2.69
P1c1	0.98	-1.13	1.62	-1.57	0.61	-0.61	0.29	-0.29	1.62	-1.57
P1d1	0.70	-0.45	0.99	-1.51	0.28	-0.28	0.16	-0.19	1.02	-1.52

(*) Sensibility analysis

Table 3 Maximum stress values in the wall 1 (N/mm^2)

	σ_{xx}		σ_{yy}		σ_{xz}		σ_{yz}		σ_I	σ_{III}
P1a1	0.76	-0.96	0.38	-3.50	0.46	-0.46	0.18	-0.33	0.76	-3.54
P1a2	1.16	-1.43	0.59	-5.24	0.69	-0.69	0.26	-0.49	1.16	-5.30
P1a3(*)	1.60	-1.41	1.43	-5.22	1.47	-1.47	0.51	-0.48	1.66	-5.28
P1a4(*)	1.08	-1.38	1.20	-5.31	0.67	-0.67	0.26	-0.57	1.47	-5.38
P1b1	0.79	-0.79	0.85	-1.01	0.28	-0.28	0.29	-0.20	0.90	-1.22
P1b2	1.16	-1.17	1.25	-1.44	0.41	-0.41	0.42	-0.29	1.33	-1.76
P1b3(*)	2.17	-2.21	2.38	-2.44	0.77	-0.77	0.74	-0.55	2.52	-3.03
P1c1	1.70	-1.44	0.59	-5.20	0.71	-0.71	0.21	-0.49	1.71	-5.25
P1d1	1.19	-1.26	1.28	-1.62	0.48	-0.48	0.45	-0.30	1.37	-1.95

(*) Sensibility analysis

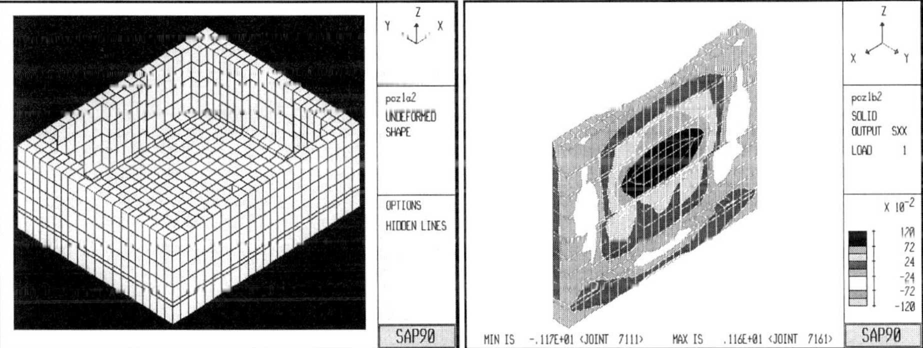

Figure 8a Modelling of the base element of 90 × 70 prototype

Figure 8b Stress σ_{xx} in wall 1 (P1b2 analysis)

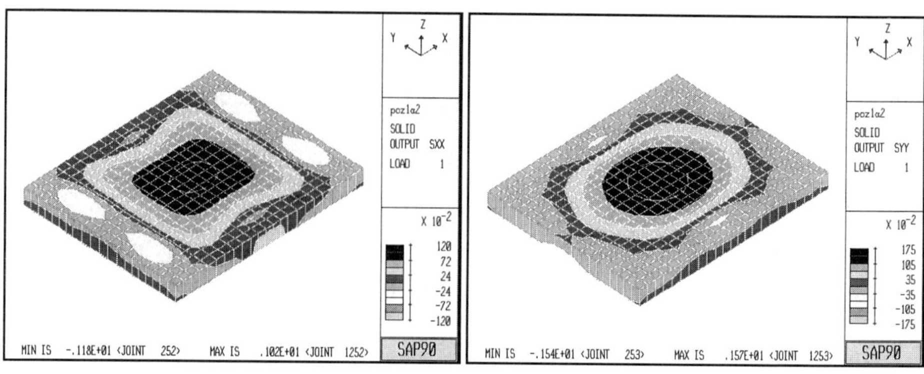

Figure 9 Stresses σ_{xx} and σ_{yy} in the base slab (P1a2 analysis)

CONCLUSIONS

1. Given the advantages of adding reinforcing fibres in the cement matrix, the usage of FRC in the design project of the pit prototypes is possible and recommended.

2. Finite element analyses showed that a criterion based on limiting maximum values of the tensile stresses was a good approach for design purposes.

3. Experimental tests on flexural strength of the FRC materials furnished values of modulus of rupture notably higher than that assumed in the design. Further experimental tests on the pit prototypes showed that: an elastic response with none or minor cracks was observed until the applied load was lower than 150 kN. With the exception of a pit the collapse load was always at least 1.5 times the previous value.

4. The failure mode of all the pits indicated that the upper ring element is the more critical part of the pit and suggested that further reinforcing steel bars should be added to the fibres to achieve a safe design.

REFERENCES

1. ACI COMMITTEE 544-4R-1988. Design Considerations for Steel Fiber Reinforced Concrete, 1988.

2. ASSOCIAZIONE ITALIANA DEL CEMENTO ARMATO NORMALE E PRECOMPRESSO, A.I.C.A.P. Raccomandazioni tecniche A.I.C.A.P. per l'impiego del conglomerato cementizio rinforzato con fibre metalliche, 1990 (in Italian).

3. D.M. 09/01/1996. Norme tecniche per il calcolo, l'esecuzione ed il collaudo delle strutture in cemento armato, normale e precompresso e per le strutture metalliche, 1996, (in Italian).

4. D.M. 4/5/1990. Aggiornamento delle norme tecniche per la progettazione, esecuzione e collaudo dei ponti stradali

DRY CAST, CAST STONE: A REVIEW OF ACCELERATED AND NATURAL WEATHERING PERFORMANCE

M R Jones

R K Dhir

Y Wang

University of Dundee

United Kingdom

ABSTRACT. This paper describes a study carried out to determine the weathering performance of dry cast, cast stone and the effect of binder type and waterproofer concentration. A novel test method to accelerate weathering is described. This is based on the rate of particulate accumulation on the cast stone surface and the effect on colour tone and chromicity. A large number of cast stone specimens were tested, which were representative of the production of manufacturing members of the United Kingdom Cast Stone Association (UKCSA). Four natural stones and 3 non-waterproofed cast stone samples were used as reference materials. In addition, unique data is presented of concurrent natural weathering tests on duplicate samples stored at the UK Building Research Establishment's Myers Hill exposure site in Strathclyde, Scotland. Results show that all normal cast stone products, which were waterproofed, behaved similarly to natural stone, even though they were from different sources and had a wide range of characteristics. The waterproofer is shown to be the most important characteristic affecting weathering. Based on the results given in this paper manufactures can review the amount of waterproofer added to a mix for different exposure environments, depending on whether rapid or resistant weathering performance is required. The results also indicate that the accelerated weathering test method can reliably be used for differentiating the natural weathering performance of cast stone.

Keywords: Dry cast cast stone, Waterproofer, Surface particulate retention, Accelerated weathering, Natural exposure tests, Colorimeter analysis, GGBS, PFA and metakaolin, Weathering test method development.

M R Jones is a chartered civil engineer and senior lecturer in the Concrete Technology Unit at the Department of Civil Engineering at the University of Dundee. His research focuses mainly on binder technology, concrete durability and repair and maintenance.

R K Dhir is Director of the Concrete Technology Unit and Professor of Concrete Technology at the University of Dundee. He is a member of numerous national and international technical committees and has published extensively on many aspects of concrete technology, binder science, durability and construction methods.

Y Wang is currently undertaking research into the development of performance specifications for cast stone in Concrete Technology Unit at the University of Dundee.

INTRODUCTION

Dry cast, cast stone is pigmented, precast concrete intended to resemble and be used in a similar way to natural stone [1, 2]. As well as a primary building facade, cast stone products can be used as ornamentation and architectural trim, sills, lintels, copings, balustrades, and door and window trimming [3]. The appearance of cast stone will change with time with the effect of rain, sunlight and pollution, ie weathering and is commonly manifest as discolouration of the exposed faces due to the accumulation of particulate matter, both of organic and inorganic origin [4]. As a result many manufacturers add an integral waterproofer to the concrete mix, which is designed to reduce the degree of surface water absorption and improve weathering performance.

In most cases, the retention of a pristine finish is an important requirement [5]. However, there is no direct test method which can determine the rate of weathering of cast stone and indeed there are little data on the subject. As a result, an accelerated weathering test method was developed at the University of Dundee and a representative sample of UK produced cast stone products were studied. In this paper the effect of the waterproofer dosage and various binders on the weathering performance of cast stone is reported. The accelerated tests have been validated against measurements taken on a duplicate series of test samples stored at an external exposure site for 12 months.

DEVELOPMENT OF AN ACCELERATED WEATHERING TEST

Rainfall plays an important role in the weathering process, since it may carry in suspension grit, dust, smoke and diesel particles and support chemical reactions in the surface zone. Moisture also allows the development of organic growths and a thin layer of water may also encourage the attachment of dirt. On the other hand, rainfall can be considered a cleansing agent [4]. These factors were taken into account in designing the weathering test method.

Brief Description of Accelerated Weathering Test Method

A schematic layout of the weathering system is shown in Figure 1. The chamber accommodates up to 40 test specimens of approximately 100 mm^2 facing mix. Test samples are set slightly tilted with exposure (test) surface facing to the main stream of particle-loaded air. Samples are dried at room temperature and all faces except exposure face are sealed before test. The samples in the weathering chamber can be cyclically wetted, by introducing fog (fine water drop), and dried, by dry air blown in with fans at any required interval. Several types of pigment, both dry and wet, were considered but it was found that PFA particles were the easiest to use and could be introduced into either the drying or wetting cycles [6]. Having experimented with various cycles of wetting and drying, it was found that the following was found to be the most effective:

- 1 hour wetting of test samples (water spray drawn directly from mains).
- 2 hours 'dirty' air (air drawn from lab and PFA of mean particle size 30μm injected at a rate of approximately 15g/hour).
- 1 hour 'clean' air drying (drawn directly from lab).

This gave a total of 4 hours per cycle and hence 6 cycles per day. Colour change on the surface was measured by the use of a standard colorimeter (Minolta Chroma Meter Model No CR-210).

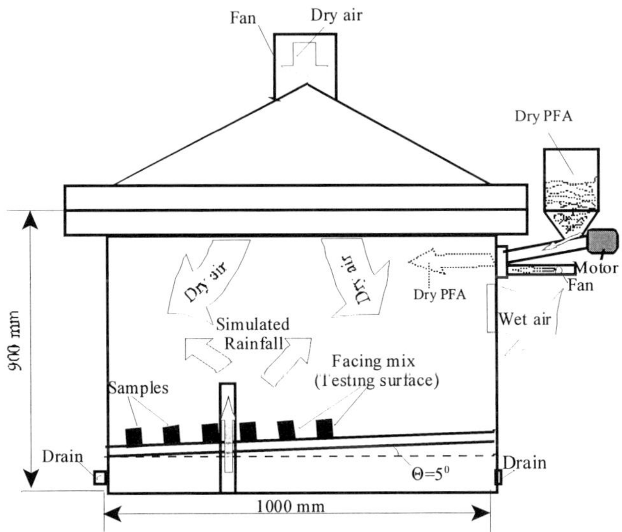

Figure 1 Schematic layout of accelerated weathering chamber

Determination of Surface Colour Change due to Weathering

The initial colour, L and a and b, of test samples are measured by the colorimeter and then remeasured after a prescribed number of accelerated weathering cycles. *Lightness* (L), *chroma* (C) and *colour size* (E) can then be calculated. The changes in these characteristics, ie ΔL, ΔC and ΔE provide an indication of the weathering rate of the samples. For ΔL and ΔC, a positive value means that sample colour has become lighter and paler, for example due to efflorescence, and negative value means that colour gets darker and deeper, due to surface particulate retention. On the other hand, ΔE, which is always positive, represents the difference in colour.

EXPERIMENTAL STUDY

Test Specimens

A total of 9 cast stone products from 7 members of United Kingdom Cast Stone Association (UKCSA) were tested. Four natural stone control samples, viz sandstone (yellow and red) Bathstone (dark yellow) and Portland stone (light grey), were also tested. In addition, three cast stone samples without waterproofer were tested for comparison. The characteristics of the samples are given in Table 1 and it can be seen that there was a wide variation in strength (25.0 to 48.0 N/mm^2) as well as initial colour group.

A further series of test samples were considered with varying waterproofer dosage and binder type as given in Tables 2 and 3. It should be noted that the specimens with varied dosages of waterproofer had the same mix proportions and curing regime. For specimens containing different binders, the same total binder and water/binder ratio were fixed but part of Portland cement content was replaced on a direct weight for weight basis with other binders, as given in Table 3. As a result, strength varied in a great range (13.0 to 38.0 N/mm^2).

Table 1 Characteristics of test samples of cast stone products and control natural stone

SAMPLE	CODE	F_{cu}*	COLOUR GROUP
Control Natural Stone	NS1 (Sandstone)		Yellow
	NS2 (Sandstone)		Red
	NP (Portland Stone)		Light Grey
	NB (Bath Stone)		Dark Yellow
Normal Cast Stone Products (Waterproofed)	C1	33.5	Yellow
	C2	42.0	Yellow
	C3	25.0	Yellow
	C4	35.0	Yellow
	C5	48.0	Yellow
	C6	40.0	Yellow
	C7	40.0	White
	C8	44.0	White
	C9	35.0	Red
Cast Stone (Non-waterproofed)	CN1	No Data	Yellow
	CN2	40.0	Light Grey
	CN3	44.0	White

*Fcu - 28 day cube compressive strength, N/mm^2

Table 2 Characteristics of bespoke cast stone samples with varied dosages of waterproofer

CODE	DOW*	BINDER	COLOUR GROUP	CODE	DOW*	BINDER	COLOUR GROUP
PC0	0	White PC: 420kg/m^3	Yellow	G0	0	White PC 80% + GGBS 20% : Total Binder : 420kg/m^3	Light Yellow
PC25	0.25			G25	0.25		
PC50	0.50			G50	0.50		
PC75	0.75			G75	0.75		
PC100	1.00			G100	1.00		

* DOW : Dosage of Waterproofer, % Weight of Binder

Table 3 Characteristics of bespoke cast stone samples with different binders

CODE	BINDER	Fcu*	COLOUR GROUP
SO	PC only	38.0	White
G2	PC80%+GGBS20%	33.0	Yellow
G4	PC60%+GGBS40%	26.0	Yellow
M15	PC85%+MK15%	34.0	White
SF	PC90%+SF10%	29.0	Dark Grey
PFA	PC70%+PFA30%	13.0	
L10	PC90%+LS10%	23.0	
L20	PC80%+LS20%	20.0	

Results

Weathering performance

The colour change (ΔE) of the test specimens are compared in Figure 2a with the control natural stone. Figure 2(b) gives the concurrent changes in lightness (ΔL) and chroma (ΔC) after 800 weathering cycles. The weathering behaviour of the test samples can be divided into 3 groups, ie the non-waterproofed series (CN1 to 3), which as expected underwent the largest colour change, the normal cast stone series and the natural stone controls and finally C1 and C4 which did not retain particulates and remained essentially pristine even after 800 weathering cycles.

The results also show that the initial colour group has a significant effect on weathering performance, particularly, for samples with light colour. For example, the samples, C7 and C8 which were from the white initial colour group underwent a larger colour change than those form the yellow initial colour group. This may be due to a larger accumulation of particulates or that the colorimeter test was more sensitive to the presence of the particulate on a lighter background.

The results suggest, however, that in general the weathering performance of normal cast stone products was broadly similar to that of natural stone but that the waterproofer was vital in attaining this property.

Effect of waterproofer dosage

A further series of test samples were manufactured with waterproofer dosages varied from 0% to 1.0 % weight of binder (note the typical dosage is 0.5and 0.75%) using both white PC and white PC with 20% GGBS. These samples were again exposed to 800 accelerated weathering cycles and the resulting changes in lightness, chroma and colour are shown in Figure 3.

The samples without waterproofer, PC0 and G0, were found to accumulate surface particulates significantly faster than the waterproofed samples containing the same type of binder, ie PC0 weathered faster than PC25, PC50, PC75 and PC100 and G0 weathered faster than G25, G50, G75 and G100.

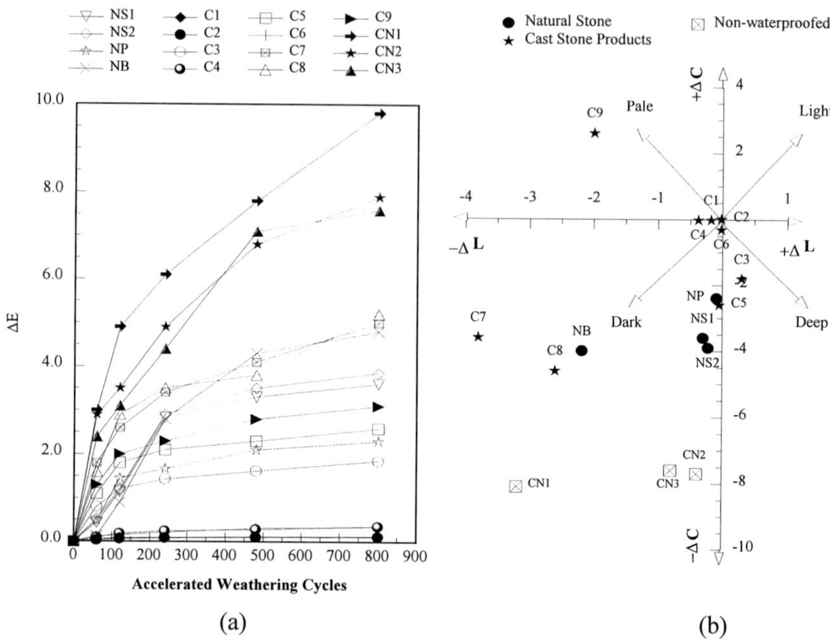

(a) (b)

Figure 2 Accelerated weathering results of (a) colour change with weathering cycles and (b) lightness and chroma changes after 800 cycles for cast and natural stone samples

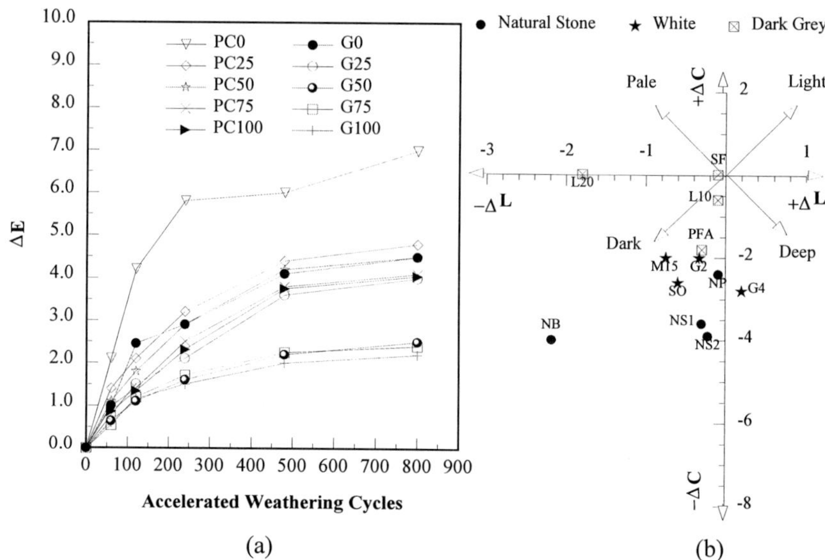

(a) (b)

Figure 3 Accelerated weathering results of (a) colour change with weathering cycles and (b) lightness and chroma change after 800 cycles for cast stone samples with varied dosage of waterproofer

Figure 4 gives the relationship between waterproofer dosage and colour size change. As can be seen there was a continuous improvement in weathering resistance from 0.25% to about 1% waterproofer however above certain concentrations there appears to be little advantage of adding further waterproofer.

There are different effective dosages of waterproofer for practical weathering performance for different binder. For the PC only cast stone samples 0.75% appeared to be the most efficient waterproofer dosage, compared with 0.50% for PC+20%GGBS cast stone. This may be related to the amount of waterproofer required to cover the surface of pores and induce a hydrophobic surface. In turn, this may be reflected in more refined capillary pore system produced by the GGBS.

The results do, however, suggest that the cast stone industry is using the correct dosage of waterproofer, which is unsurprising given its high unit cost (≈1200/tonne) and the production refinement that manufacturers have undertaken over many years.

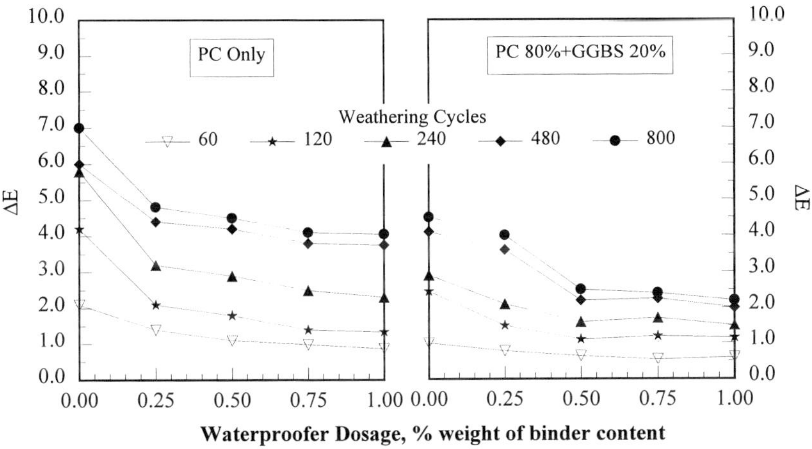

Figure 4 Effect of waterproofer dosage on weathering resistance

Effect of binder type

Above results show that the type of binder used influences the weathering performance of cast stone. To study the effect of binder type on weathering further, additional samples were manufactured with white PC and 40% GGBS (G4), 30% PFA, 15% metakaolin (M15), 10% and 20% limestone (L10 and L20) and 10% silica fume (SF). The waterproofer dosage was fixed at 0.75% weight of binder and the results are shown in Figure 5.

In these tests there was a poor overall relationship between binder type and weathering performance. Initially, it was considered that this was due to the manufacturing process, which adopted a direct replacement mix method and, therefore the cast stone strength was inevitably different (Table 3). However, analysing the results further showed for example

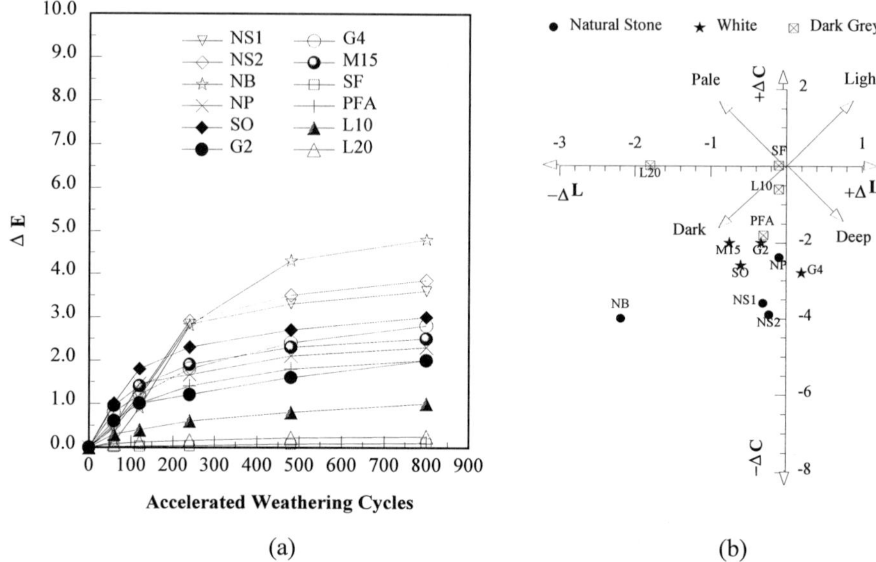

Figure 5 (a) change in colour during accelerated weathering and (b) lightness and chroma change after 800 cycles for bespoke cast stone samples with different binders

that G4 and SF had a similar strength but very different weathering behaviour. On the other hand, SF, L10 and PFA which had large differences in strength yet weathered similarly.

Classifying the samples into initial colour groups, it can be seen that the group weathering performance is similar irrespective of binder type or strength. It should also be recognised that the PFA, SF and LS produced a dark grey initial colour group which was not greatly affected by the dark particulate build-up. However, with the lighter initial colour groups it was found that there were differences in the weathering performance of test specimens with different binders. In this case the largest colour change was found with the MK specimen. Given the results noted above with the GGBS mix, it was expected that the MK should have produced a smaller colour change. Although, this cannot be shown from the present results, it is felt that the initial colour differences between the PC, GGBS and MK mixes effected the overall weathering performance, even though the ΔE value is supposed to take this into account. This suggests that if the initial specimen colour is lighter then a higher dosage of waterproofer should be used to provide equal performance to darker units.

Natural Weathering Tests

One sample of each specimen was placed at the UK Building Research Establishment exposure site at Myers Hill, see Table 4.

Table 4 Topographic information for Myers Hill exposure site

Location	18 km south of Glasgow, ~50 km from coast (see map)
Altitude	300 metres above mean sea level
Rainfall	Approximate 1000 mm/annum;
Driving Rain	3000 mm/yr (SE), 300 mm/yr (NE)
MDT^{+}	10.0^{0}C Max./ 4.5^{0}C Min.
Air Frosts	Typically 70 times/annum
F/T cycles*	Typically 33 times/annum

$^{+}$ MDT : Mean Daily Temp. * Temp goes from >=+1^{0}C to <=-1^{0}C and back to >=1^{0}C as 1 F/T cycles.

The initial colorimeter values of these specimens and after 6 and 12 months exposure were correlated with the accelerated weathering data, as shown in Figure 6. The results for natural exposure have been plotted at the same scale as those for accelerated weathering using a multiplication factor as shown in Figure 6. It can be seen that there is a great deal of symmetry between the plots for accelerated weathering and those for natural weathering. This is indeed encouraging and suggests the accelerated test method could be used to rapidly differentiate the weathering performance of cast stone, although much further work will be needed to refine this.

Based on these results a predictive equation has been developed to allow the equivalent period of natural weathering to be estimated from the accelerated test results. Using 6 accelerated cycles per day results in 1 year of accelerated weathering testing being approximately equivalent to 15 years of natural exposure for *lightness* change (ΔL), as calculated as below :

$$1\,Year\ Accelerated \cong Factor \times 365\ days \times \frac{6\,Cycles}{800\,Cycles} \approx 15\,Year\ Natural\ Exposure$$

Where the factor in this case is 6 for ΔL, as shown in Figure 6.

Similarly, 1 year of accelerated weathering is equivalent to about 20 years for *chroma* change (ΔC) and about 15 years for *colour size* change (ΔE). The results show that chroma change was a little faster in accelerated weathering compared to lightness and colour size change. The reason may be that the particulates used in the accelerated test are darker than those typically encountered in the natural environment. As a result, chroma change may be over exaggerated in the accelerated tests, particularly as the Myers Hill exposure site is located well away from a city/industrial area and is subject to a high degree of rainfall.

CONCLUSIONS

The results show that the weathering performance of normal (waterproofed) cast stone was similar to natural stone, even where their characteristics varied significantly. The waterproofer had the most significant influence on the weathering performance of cast stone.

However, increasing the dosage of waterproofer beyond 0.75% by weight of binder did not significantly improve weathering performance. On the other hand, it was found that the binder type used had only a minor effect on weathering, although lower waterproofer dosages

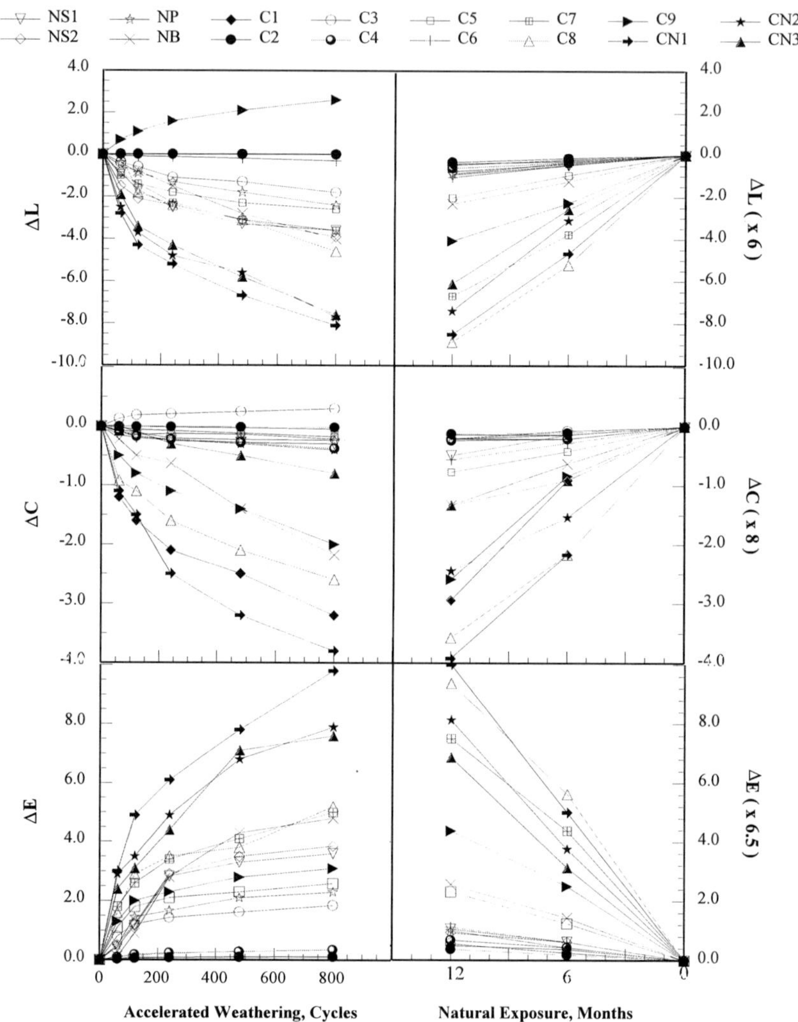

Figure 6 Comparisons of results of natural exposure with accelerated weathering

could be used to give a similar weathering performance if 20% GGBS was used in the mix. Comparing the natural exposure and accelerated test data showed that the long-term weathering performance of cast stone could be determined rapidly. Although further work will be needed to confirm the weathering data with existing structures it was estimated that 1 year accelerated testing was approximately equal to 15 to 20 years natural exposure.

It should be emphasised, however, that in city/industrial areas with a large concentration of particulate air pollutants, such as diesel smoke, this may be underestimation of the weathering rate.

ACKNOWLEDGMENTS

The authors would like to acknowledge the support provided for the project by the UK Department of the Environment, Transport and the Regions and the UK Cast Stone Association, Castle Cement Ltd, ECC International Ltd and Scottish Power - Ash Sales Ltd. The authors also would like to thank Dr T Yates, Dr A Stupart and I Murray of the BRE Scottish Laboratory for their assistance in the natural exposure tests carried out at Myers Hill.

REFERENCES

1. BRITISH STANDARDS INSTITUTION, BS 1217 : Specification for cast stone, 1997

2. UNITED KINGDOM CAST STONE ASSOCIATION, UKCSA Specification for cast stone, UKCSA, Crowthorne, June 1994 pp1-4

3. TAYLOR, H.P.J. Precast Concrete Cladding, Edward Arnold, London, 1992, pp23-57

4. HAWES, F, The Weathering of Concrete Buildings, Cement and Concrete Association, Wexham Springs, Slough, 1986, pp1-18

5. DAWSON, S, Cast in Concrete, Construction Research Communications Ltd, London, 1995, pp 6-13

6. BRITISH STANDARDS INSTITUTION, BS 3892 : Part 2 : Specification for pulverized-fuel ash to be used as a Type 1 addition.

PRECAST CONCRETE CONSTRUCTION: THE SCOTTISH WIDOWS HEADQUARTERS BUILDING IN EDINBURGH

M Peden

W A Fairhurst and Partners

United Kingdom

ABSTRACT. The recently completed, £70 million, Scottish Widows Headquarters building in Edinburgh, is one of the finest examples of modern building within Scotland. This paper describes the design and construction of the building, with particular emphasis being placed on the extensive use of precast concrete construction.

The building has two levels of basement, which were constructed utilising Omnia precast concrete construction. Above ground level the construction splits into three main buildings, one of which is curved and elegantly rises up to eight storeys high above ground level, with an interesting parabolic shaped roof. Alongside, are two lower level buildings each of five storeys which set the overall development into the existing street-scape.

The three buildings above ground level are constructed using exposed precast concrete structures. Virtually all of the interesting curved shapes of the building, which architecturally express the structure both internally and externally, are achieved using precast concrete. Nearly two thirds of the structural frame is built in honey coloured concrete, as part of the Architect's design statement.

Keywords: Precast concrete, Architectural concrete, Cladding panels, Construction speed.

Murray Peden, BSc, CEng, M.I.C.E., M.I.Struct.Eng., is a Technical Director of W A Fairhurst & Partners, Consulting Structural and Civil Engineers. He joined the consulting engineers in 1980, following graduation from Dundee University and has worked with them, primarily in their Edinburgh office in Scotland, since then. In 1991 he was promoted to the post of Technical Director and in the intervening period has been responsible for the structural aspects of numerous building and civil engineering projects, constructed in both concrete and structural steelwork. He is a past winner of both the Institution of Structural Engineers Graham Wood Prize and their Scottish Branch Prize.

INTRODUCTION

The new £70m Scottish Widows Headquarters office development at Port Hamilton, is located in the heart of the Exchange, Edinburgh's Financial District. The construction of the building, in Scotland's capital city, commenced in the Spring of 1995 and was completed towards the end of 1997. Construction of the 35,000 square metres of office accommodation, for Scottish Widows, one of Europe's leading life assurance companies and a key financial player within Scotland, was undertaken by Edinburgh Development Group on behalf of Scottish Widows. Architect for the project was Building Design Partnership and civil and structural engineer was W A Fairhurst & Partners. A list of the full Project Team is shown below.

The building was procured utilising a traditional management form of contract, with the construction being split into different work packages by management contractor, Laing Management (Scotland) Ltd. The Management Contractor, in keeping with the Design Team, were keen to use precast concrete for as many aspects as possible, to both improve quality and aid the construction of what was a large building in a city centre site. The Architect was also very keen to expose and express the reinforced and precast concrete frame as part of his design statement for the building. As such almost two thirds of the structural frame was exposed and built in honey coloured concrete.

The 6.5 acre site was the former location of Port Hamilton, a terminal of the Union Canal. As well as high quality office accommodation above ground level the building also provides car parking for more than 500 cars, located in two levels of basement which occupy 24,000 sq.m. of additional space below ground level. Above ground level the construction splits into three main buildings. The Crescent building, curved to reflect the City's influential crescents architecture, rises up to eight storeys high above ground level and is topped with an interesting parobolic shaped roof. Alongside are two lower level buildings, each of five storeys above ground level, which set the overall development into the existing street-scape. The basement levels also provide main vehicular servicing routes, which include facilities for articulated lorries, which impose substantial loads on the suspended construction. By design, no servicing of the building is required to be carried out from street level. Figure 1 shows an illustration of the building.

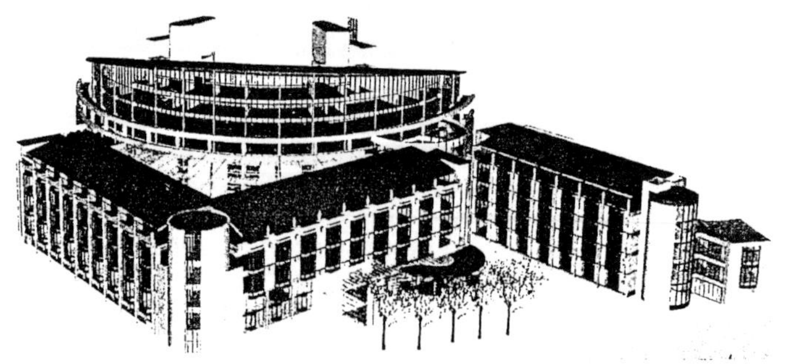

Figure 1 The Scottish Widows Headquarters Building

PROJECT TEAM

The development was undertaken by the Edinburgh Development Group Ltd for Scottish Widows. The principal members of the Project team were;

Architect:	Building Design Partnership
Civil, Structural & Environment Engineers:	W A Fairhurst & Partners
Services Engineers:	Cundall Johnston & Partners
Quantity Surveyors:	Turner & Townsend
Project Administrators:	Turner & Townsend Project Management
Management Contractors:	Laing Management (Scotland) Ltd

BASEMENT & INSITU CONCRETE CONSTRUCTION

The foundations, insitu reinforced concrete basements and the stair and lift cores for the precast concrete superstructures, were constructed by John Doyle Construction. This work was carried out under two separate work packages, valued at approximately £10 million.

The foundations generally comprised simple pad foundations, resting directly on moderately weathered mudstones or sandstone. In the north west corner of the site, bored concrete contiguous piles which were constructed to stabilise excavations, in the close proximity of existing tenement buildings, were also used in conjunction with isolated bored piles to support vertical loads from the frame.

The structural grid in the basement generally reflects the structural grid in the office areas over them, which utilise a precast concrete frame. The 6 metre wide bay width of the office blocks have columns on an orthogonal grid spacing of either 6 metres, 9 metres or 12 metres depending on the location. The slabs, usually 300mm thick, therefore typically span 6 metres and the insitu reinforced concrete beams span either 6 metres, 9 metres or 12 metres. This less prominent construction is all constructed in ordinary 'grey' concrete. The insitu concrete beams are typically 1000mm wide with an overall thickness of 600mm, although in places they are up to 750mm deep. The beams are constructed with horizontal slots through their webs, at regular widely spaced centres, to facilitate the distribution of services, see Figure 3. This approach was adopted to minimise overall construction depth and cost.

Figure 2 Solid Omnia slab and Omnia slab with void formers

The slab of the floor plates within the basement, which are exposed to view as the basement houses either car parking or plantrooms, are formed using an Omnia wide plank precast system, supplied by Birchwood Omnia. The Omnia wide plank system uses a thin precast concrete slab construction as part of a thicker composite unit. On this project the planks were normally supplied as 2.4m wide by approximately 5 metres long (the distance between the edges of beams) and were 75mm thick. This precast slab contained the bottom 'sagging moment' reinforcement in the continuous construction. The Omnia planks which were propped, normally at their 1/3 points, during construction of the remainder of the composite unit were supplied with polystyrene void formers fixed to them. These were simply glued to the precast unit. The Omnia slabs as well as carrying the main bottom steel also had a projecting 'A frame' reinforced system, which stiffens the unit during construction, projecting upwards between the polystyrene void formers. Following temporary propping, insitu concrete was poured over the units to form a composite and continuous unit. The specified strength of all concrete was 40N/mm².

Where the Omnia construction were subject to concentrated wheel loads, from articulated vehicles, of 100 kN (Nominal) the precast units were specified as solid, the polystyrene void formers simply being omitted, to provide the higher shear capacities required with this construction, see Figure 2.

The stair towers and lift cores, which rise to the full height of the building and provide stability to the three separate buildings which rise above the basement, were all constructed in reinforced concrete. Precast concrete was considered for these elements but was discounted for both technical and financial reasons. The construction of these cores and the safety of the operatives during construction of these high and slender elements, was aided by the use of climbing formwork supplier by Peri.

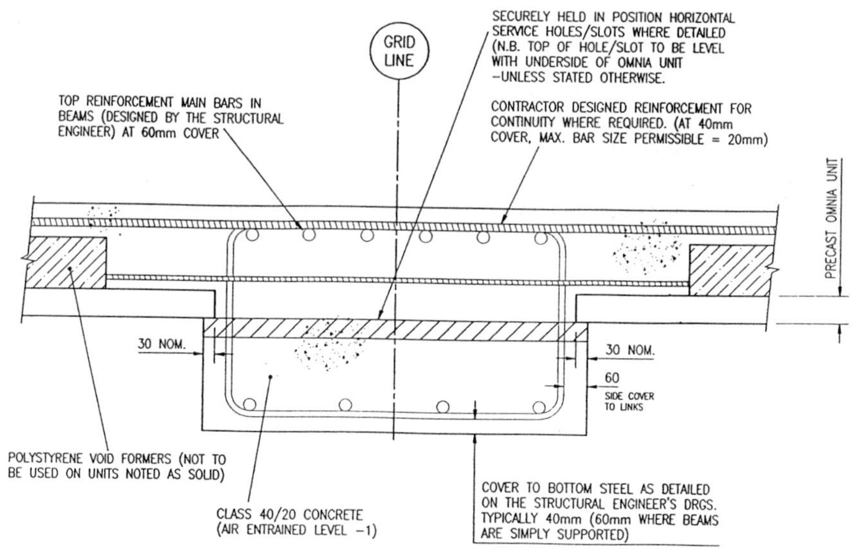

Figure 3 Typical Beam/Omnia Unit Intersection

PRECAST CONCRETE FRAME

As is common practice with most precast concrete frames, the precast concrete frame was analysed as a pin jointed braced frame. The insitu reinforced concrete stair and lift enclosures, forming the cores to the buildings, were used as shear walls. These shear walls were designed to carry the horizontal wind loading by means of a simple cantilever action from the foundations. The precast concrete floor system formed horizontal diaphragms to transfer the horizontal loading back to the cores.

The contract to supply and erect the precast concrete frame was won by Costain Dowmac at a value of approximately £8m (which included the precast cladding panels). Costain Dowmac, were very soon after acquired by Tarmac Precast Concrete[1] who carried out the detailed design and manufacture of the components.

The specification for the precast concrete was extremely demanding, as most of the structural frame had to be produced in honey coloured architectural concrete. The specification ruled out the use of a colour pigment in the concrete, to ensure a consistent colour was achieved throughout. The concrete mix used in the 50 N/sq.mm frame combined Blue Circle's Antique white cement and Dowlaw limestone coarse aggregate from Derbyshire with Hyndford sand. All of these materials had to be transported to Tarmac's Coltness factory, in Scotland, for the manufacture of the concrete. Experiments took place to get the desired surface texture, both acid etching and dry abrasive blasting were rejected in favour of wet abrasive blasting techniques that permitted close observation of texture by the operative.

Most of the precast concrete comprised complex shapes and architectural features. All the exposed columns were elliptical shaped with corbels at their ends for connections. Those columns which were not exposed were rectangular and cast in normal grey concrete. Many of the beams were also curved in plan. Figure 4, shows a photograph of the precast concrete during erection.

The specified use of pinned connections by the structural engineers offered the advantage that they can be fairly easily constructed on site and in a number of ways. The precise detail of these connections is often, for economy of construction, left to the specialist manufacturer. On this project outline details, which achieved the overall design criteria of the building, were prepared by the structural engineers who approved the final details developed by Tarmac Precast Concrete. The design team prepared full details of the shapes and layouts of all the members in the precast frame.

During detailed design by Tarmac Precast Concrete, they suggested an alternative four storey column with corbelled connection to the beams. This was suggested as a cost saving exercise. Multistorey columns are in fact a common approach in precast concrete frames. This approach was adopted in the non exposed columns but following trials for reasons of quality, single storey columns had to be adopted where the columns were exposed. Other columns dictated by the architectural design required to be double height.

Most of the floors are constructed in precast prestressed hollow-core slabs. These are generally 150mm or 200mm deep and do not utilise a structural screed. A structural screed is not normally required, when using this form of construction, to achieve the diaphragm action required to transmit wind loads to the shear walls [2].

Figure 4 Photograph of Precast Concrete During Erection

ROBUSTNESS & PRECAST CONCRETE FRAME CONNECTIONS

In the basement areas where construction is predominantly insitu concrete the robustness requirements, to reduce the potential for disproportionate collapse, are achieved through normal detailing. No special provisions were required to meet the statutory requirements for horizontal and vertical ties.

The precast frame layout, was detailed in a manner so as to avoid the use of key elements. That is, no columns bear directly on beams. Normal vertical and horizontal ties did, of course, need to be incorporated. The interface between the insitu basement construction and the precast frame was achieved by means of projecting reinforcement from the columns being cast into 75mm diameter well void formers (plastic tubes forming an oversize hole in the insitu concrete). The arrangement is shown in Figure 5. This interface, following normal alignment and plumbing of the columns was then grouted. At higher levels, within the precast frame, vertical continuity was provided in a similar manner, although threaded reinforcement into cast in couplers was also used, as was stud/plate connections, depending on location. Curved and balcony beams incorporate cast in plates and site welded connections to provide torsional resistance.

Horizontal perimeter ties were provided by the reinforcement within the edge beams, by welding billet connections and cast in plates, by couplers cast into precast columns and insitu cores, threaded reinforcement into couplers, and by unstressed prestressing wire strand set into the insitu infill between hollow-core and edge beams. Internal ties between slabs and beams etc was achieved in a similar manner.

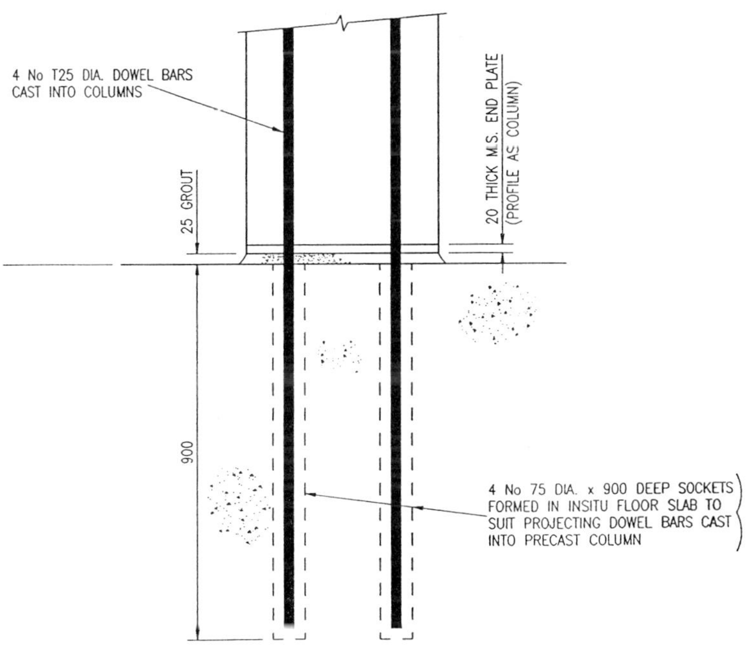

Figure 5 Typical Precast Column/Insitu Beam Detail

CLADDING

The elevation to all cores, some of which are curved, are all clad in precast concrete panels of varying size, up to 6 metres by 4 metres. These precast concrete units are faced in panels 600mm wide by 400mm high, in either Clashach sandstone or honey coloured concrete. Both types of construction were constructed and erected by Tarmac Precast Concrete.

Clashach sandstone facing is used on the more prominent elevations, using 50mm thick panels mounted on the concrete.

The Clashach sandstone, which is quarried in Scotland, has rich veins of colour running through it, which has been used to provide a deliberate and random feel to the cladding. As a more cost effective solution at the rear of the building, precast reconstituted stone cladding panels, which are intended to closely resemble the Clashach stone panels, have been used as an economical alternative for these less prominent elevations. The overall effect of quality is marginally reduced, as there is clearly less variation, but is not compromised to any great extent.

The composite panels, which weigh up to 14 tonnes, incorporate integrally cast stainless steel brackets. Figure 6 illustrates the typical cladding support detail. On delivery to site the panels were set on temporary A-frame supports while foil-backed foam insulation slabs was fixed to the inner face. Erection of the storey-height panels then involved craning them into place so that the stainless steel brackets bear onto pockets cast into the core walls, or onto perimeter beams. Bottom restraint was provided by bolting through the brackets into cast-in channels, while top restraint involved fixing cleats attached to the back of each panel into further pockets with channels. Once erected, the joints between panels were sealed with brown mastic to match the pointing between the sandstone slabs and the precast concrete backing panel.

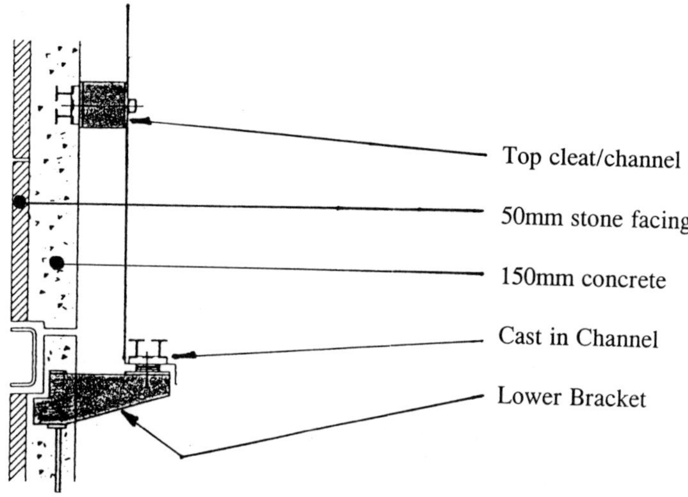

Figure 6 Typical Cladding Support Detail

MOVEMENT JOINTS

As with all buildings of this magnitude but principally due to the complex shapes present in this building, the location of all movement joints required careful consideration. As the three buildings rose from a common basement, which extended well beyond their plan shape, it was considered appropriate to provide movement joints within the basement around the perimeter of all buildings in the same plan location.

On the Crescent block and the Courtyard building a further joint was introduced on each, at mid point, to divide both buildings into two equal segments. Again this was reflected in the basement construction. The buildings, either side of each joint were then designed to be stable in isolation.

Due to the complex shapes which the joints had to follow, in particular in the basement areas, it was considered a requirement to permit joints to both open or close across the joint but also to permit small 'shearing' movements along the length of the joint. To facilitate this requirement, without the use of a double column arrangement (one either side of the joint), Staifix dowels by Ancon Clark were specified. The high strength stainless steel dowels, were capped at their free end with elongated caps. This allowed the slabs relative lateral movement but restricted any differential vertical movement. Figure 7 illustrates the staifix dowel arrangement. In virtually all cases the Staifix dowel arrangement were carrying significant loads of up to 400 kN and the surrounding concrete was therefore provided with local reinforcement to accommodate these loads.

In the precast superstructure the direction of movement was much easier to predict. The movement joints allow horizontal movement across the joint and restrict vertical movement. Lateral 'shearing' movement is also restricted. The joints at the centre of the Crescent building, is provided with a conventional halving joint arrangement. A small rubber bearing, which permits sliding movement, being placed between the interface of beam and slab. Movement of beam seatings at the movement joint is accommodated with a similar principle. Unlike the basement construction, it was not considered necessary to utilise the staifix dowels in the precast concrete frame construction.

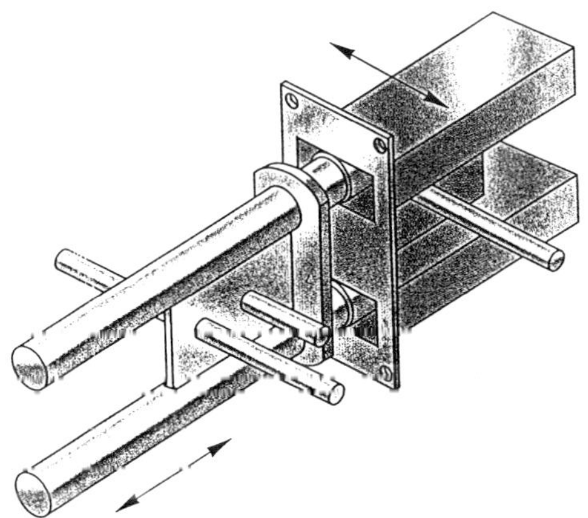

Figure 7 Staifix Dowel Arrangement in Substructure

QUALITY AND TOLERANCES

At all stages of the project, from inception to completion, quality played a key role in the selection of both components and works package contractors. At no time was quality ever sacrificed for the sake of speed of construction. Quality and consistency of product was a key factor in the Design Team's decision to use a precast frame and this became of fundamental importance, with the architects approach of exposing and expressing the structural frame as part of his design statement. Undoubtedly, the quality and consistency achieved in the construction of this large development would, given the tight timescales available for construction, been almost impossible to achieve using insitu concrete.

The management of the interface between the insitu reinforced concrete frame and the precast concrete frame, the design of which was only being developed during the construction of the reinforced concrete superstructure, left the Management Contractor and the Design Team with a considerable challenge. Despite the tight tolerances specified there were inevitably difficulties of compatibility at the interface between different contractors. It is a credit to the Design Team and suppliers that these were successfully resolved. The successful realisation of this project demonstrates that with proper consideration of tolerances and issues of buildability, that precast concrete can be used as a fast track form of construction alongside the use of other building materials.

CONCLUSIONS

The new £70 million Scottish Widows Corporate headquarters building, which overlooks Edinburgh's famous Castle, has become one of the finest examples of modern building within Scotland. The imaginative use of architectural precast concrete has undoubtedly improved the buildings quality, not least by permitting much of the construction to be completed in factory conditions off site.

The Scottish Widows staff at the Port Hamilton site now enjoy high levels of ambient light from a large atrium space, which is formed with a glass roof spanning between the Crescent Building and the lower level Courtyard Building. They also enjoy the bold and interesting building architecture achieved through the use of the honey coloured architecturally expressed precast concrete frame, which helps impose the design of the building on the capitals skyline. The building benefits from an intelligent building management system, allowing the building management system to expand and contract in direct response to the future changes in world wide technology, a point which serves to emphasise the forward thinking design approach of the overall building.

REFERENCES

1. TAYLOR, H P J. The new Scottish Widows HQ, Edinburgh. The Structural Engineer,Volume 74 Nos 23 & 24 10 December 1996, pp 423-425.

2. ELLIOT, K S, DAVIES, C, OMAR W. Experimental and theoretical investigations of precast hollow cored slabs used as horizontal diaphragms. The Structural Engineer, Vol 70 No. 10 May 1992.

ASSESSMENT OF LARGE PANEL SYSTEM BUILT DWELLINGS FOR ACCIDENTAL LOADING

S L Matthews

B R Reeves

T G D Canisius

Building Research Establishment

United Kingdom

ABSTRACT. This paper presents a brief historical perspective on the appraisal of large panel buildings and discusses issues relating to a programme of full-scale static load tests which were conducted upon a Bison and a Reema large panel block. All the elements tested were able to sustain a load equivalent to at least 17 kN/m^2 without significant distress, deflection or failure. Some elements were able to sustain appreciably larger loadings. The testing is considered to have shown that the buildings concerned would have been able to resist the specified notional loading of 17kN/m^2 for a building without a piped-gas supply. These findings have significant implications for the assessment of other Bison and Reema LPS blocks in which a reasonable standard of workmanship was achieved.

Keywords: Load testing, Large panel system construction, Bison Wallframe, Reema Conclad, Static pressures, Ronan Point, Piped gas explosion, Workmanship, Structural assessment.

Dr S L Matthews is the Director of the Centre for Construction Repair and Refurbishment, Building Research Establishment. He has gained a wide variety of experience in the investigation and appraisal of existing buildings using a variety of investigation techniques. This includes the assessment of deteriorating structures and the specification and design of remedial measures for repair and restoration. Dr Matthews has published widely and serves on a number of technical committees.

Mr B R Reeves is a senior researcher in the Centre for Construction Repair and Refurbishment, Building Research Establishment. He has gained wide experience in the field of structural investigations and appraisal of a wide range of concrete structures employing a range of non-destructive and partially destructive procedures.

Dr T G D Canisius is a senior engineer in the Centre for Structural Performance, Building Research Establishment. His main research interests are in the use and application of static and dynamic finite element analysis as a tool in the assessment of the behaviour of a wide range of buildings and structures. Dr Canisius serves on a number of technical committees.

INTRODUCTION

There was a huge push in the late 1950's to provide increasing numbers of housing units within a very short space of time and making the maximum use of restricted site space. These constraints and the associated cost limits led to the introduction of large panel construction in the United Kingdom (UK) using existing technology first developed in Denmark in 1948. One such system was the Larsen Neilsen system which was erected in the UK under license by Taylor Woodrow – Anglian Ltd.

The first structure of this type erected in the UK was for the London County Council in 1963. When it came into existence in 1965 the London Borough of Newham commissioned the construction of nine, 22-storey Larsen Neilsen blocks. The construction of the now infamous Ronan Point block was started on 25 July 1966 and was the second to be completed and handed over to the London Borough of Newham on the 11 March 1968. At the time of the gas explosion, which occurred at 5.45am on Thursday 16 May 1968, only eight of the flats remained vacant. Fortunately four of these were in the south-east corner which collapsed.

A significant factor in the incident was that the piped gas explosion occurred five floors from the top of the building where the pre-load due to the dead weight of the floor and wall components was relatively low. The explosion generated a dynamic pressure which acted simultaneously on the external flank wall and floor and ceiling panels of the flat. The upward forces lifted the top four floors momentarily while the, now unrestrained, flank wall was blown outwards. When the dead load from the upper four floors returned the supporting walls were no longer present to offer any resistance. Hence the dead weight of some of the upper construction descended through a storey height before impacting on the next lower floor to the south east corner of the block. Once initiated, the kinetic energy contained within the falling portion of the structure grew rapidly allowing the collapse front to progress at almost free fall speed.

Estimates of the pressures arising from the gas explosion were made by assessing the magnitude of the forces which were required to deform a number of items in the flat. The best estimate which could be made at the time was that the explosion had induced an overpressure of 5 psi (pounds per square inch) acting on all surfaces bounded by the explosion. This pressure was subsequently adopted as the criteria for checking the robustness of LPS structures with piped gas.

Following the collapse, the then Minister of Housing and Local Government instructed local authorities in August 1968 to appraise the structural design of existing and proposed large panel buildings (LPS) in order to reduce the probability of progressive collapse in the event of the loss of 'key' load bearing elements [6]. This resulted in a nationwide program of assessment and the strengthening of many exiting LPS blocks, together with a review/modification of the design of the many LPS blocks then under construction.

By the mid 1980's some blocks were beginning to experience problems with weather-tightness and questions were being raised about their long-term performance and durability. In October 1984, the Minister of Housing and Local Government announced a BRE programme for investigating the existing condition of such dwellings and measures for their appraisal and repair, where appropriate. A BRE report *"The structural adequacy and durability of large panel system dwellings"* was published in 1987 [4]. Part 1 of this report concluded that, based on the condition of buildings inspected by BRE at the time, the LPS buildings generally showed no signs of distress sufficient to give cause for the safety of the occupants. However the report did highlight that there may be a worsening of the condition of the external envelope of LPS buildings with time.

The second part of the report recommended that appraisals should be carried out of buildings required to exceed 25 years service life, providing guidance on the nature and scope of such appraisals. This part contained a number of measures for continued durability and safety under normal loading and some options for assessing the sensitivity of a building to damage in the event of an accident. These options were concerned with the requirement to avoid disproportionate collapse, introduced as a result of investigations of the partial collapse of Ronan Point in May 1968.

Although LPS buildings constructed before this time did not have any designed provisions against disproportionate collapse, it was considered essential to ensure that such buildings with five or more storeys should be safe during their future service life under normal as well as accidental loading conditions. The assessment recommended in the BRE report was analytical, including checking of local strength against a static overpressure of 34 kN/m^2 for buildings with a piped gas supply and 17 kN/m^2 for buildings without piped gas. The BRE report did not include any method for load-testing buildings for this purpose.

On the basis of this report many of the existing 1960's LPS blocks have been re-appraised in recent years, or are due to be appraised. Direct enquiries and information passed to BRE suggest that, in order to satisfy the accidental loading requirements, appraising engineers have recommended strengthening measures for a significant number of these blocks; some which have been in service for over 30 years. In some cases strengthening has been recommended in previously strengthened buildings.

Although the structural adequacy of a building under normal loading conditions can be assessed with the help of guidance provided in the 1987 BRE report [4] and the Institution of Structural Engineers' report *"Appraisal of existing structures"* [5] a satisfactory method of dealing with accidental loading is not currently available.

The form of construction used in LPS buildings gives them inherent strength which is difficult to quantify by simple calculation procedures. BRE considered that this form of construction would make LPS structures more able to resist the forces associated with a gas explosion than the results of simple hand calculation procedures would seem to predict. This view is supported by previous laboratory studies undertaken by BRE and by the results of experimental testing carried out upon LPS components and assemblies carried out by others and made available to BRE. Unless there are particular problems with the workmanship in a block, the current view held by the BRE is that further extensive strengthening is unlikely to be necessary for many types of blocks without a piped gas supply.

However, up until now there has been no way of establishing this definitively because of the lack of information on the performance of LPS buildings under accidental loads. The satisfactory in-service structural performance of the general population of UK LPS buildings over the past 30 or so years also adds weight to the view that widespread additional strengthening is unlikely to be required.

MEANS OF AVOIDING DISPROPORTIONATE COLLAPSE IN AN EXISTING HIGH-RISE LPS BLOCK

The principal concern in respect of disproportionate collapse of an existing high-rise (in this context high-rise is defined as five or more storeys, including basements) LPS block is a gaseous explosion. Those originating from a piped gas supply are different from other explosions, such as those from liquefied petroleum gas (LPG) or aerosols, because of their potentially much greater severity. Research has shown[3] that the pressures generated during explosions involving LPG accord fairly well with the $17kN/m^2$ overpressure used to appraise buildings without a piped gas supply. Theoretically the pressures generated during a LPG explosion could be much larger. However, venting via windows and other relatively weak elements of the construction such as stud partitions, typically reduce the peak pressure generated to less than the recommended overpressure of $17kN/m^2$.

All vertical load bearing elements in LPS blocks are considered to be key elements. Therefore they either have to be able to resist the forces which would be applied during an event such as an explosion, or the structure has to be able to mobilise alternative load paths in the event of loss of a key member. In such cases the structure should be able to bridge over the missing element; albeit in a severely deformed condition.

The latter approach is achieved by the provision of effective tying within the structure. The nature of the construction of some LPS blocks built prior to May 1968 is such that the provision of effective tying reinforcement cannot be demonstrated practically. Accordingly, it is often not possible to show that the structure would be able to mobilise alternative load paths in the event of the loss of a load-bearing wall element.

BRE RESEARCH INTO THE STRUCTURAL BEHAVIOUR OF TWO LPS SYSTEMS UNDER ACCIDENTAL LOADS

BRE considered that there was a need to obtain experimental data for the appraisal of the principal types of LPS construction which would establish definitively their performance in relation to the currently employed overpressure loading criterion of $17\ kN/m^2$. The opportunity arose to conduct load test upon a Bison Wallframe block and a Reema Conclad LPS block. Typical construction details for these large panel systems are given in references [1] and [2].

There are two principal structural issues of concern here. Firstly there is the ability of the vertical load bearing wall elements and the associated joints to resist the lateral forces arising during the explosion. In many instances the form of failure which is of greatest concern arises from an explosion located in a room adjacent the flank wall. In this scenario, the bottom (or occasionally the top) of the flank wall adjacent the explosion site could be pushed outwards by the blast.

Logic would suggest that panels in the upper stories of a building are generally taken to at greatest risk because of the reduced dead load pre-compression in the wall elements towards the top of the building.

Secondly there is flexural failure (or possibly total displacement) of the floor slabs immediately above and below the explosion site. Theoretically, floors at all levels in a building are exposed to nominal similar levels of risk, depending on whether they are subjected to an upward or downward acting overpressure.

However, the consequences of a floor failure adjacent a flank wall of an LPS block would generally be considered to be more significant because of the associated reduction in the lateral support to the external flank wall.

BRE PROGRAMME OF FULL-SCALE LOAD TESTING OF A BISON WALLFRAME AND A REEMA CONCLAD BLOCK[7]

The programme of load testing involved two phases of work. These being :

Phase 1: A desk study and exploratory site investigations and measurements to establish details of onstruction (and compliance with the design intentions) and the standard of workmanship achieved in the construction of the two blocks. The blocks had been built to a standard form employed prior to the collapse of Ronan Point.

The studies found that although a reasonably good standard of workmanship had been achieved in both blocks, the blocks were not totally free from defects or variations in construction.

Phase 2 : Static load testing of load-bearing walls and floors in selected rooms to establish the load capacity and load-deflection characteristics of the floors and/or walls. The load tests were carried upon short and long span floors at different levels within the blocks. The test rooms were located both adjacent the flank walls and within the central region of each block.

The locations chosen corresponded to the high and medium risk zones identified in Figure 1. The most critical tests exploring the combined floor and wall behaviour were carried out on the floor below the top storey of the buildings, thereby minimising the pre-compression effects in the walls of the dead load.

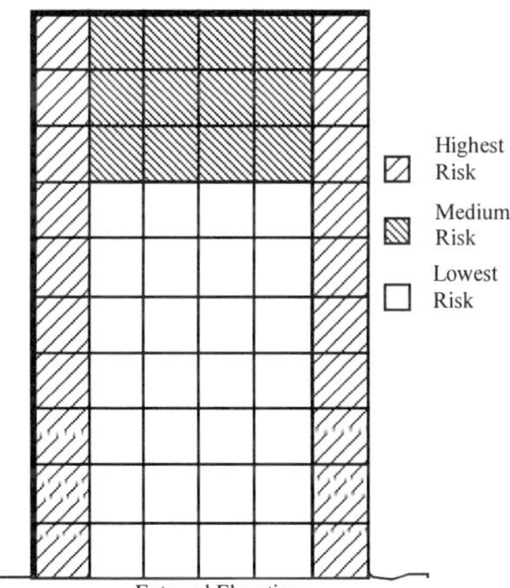

Figure 1 Accidental loading - Risk of disproportional collapse

Figure 2 illustrates the nature of the in-plane accidental loading generated during a gas explosion and the test arrangement used during the tests to simulate this loading using static forces applied via a hydraulic loading system.

Figure 3 presents an isometric view showing a typical arrangement of flank/cross wall load distribution beams and safety steelwork.

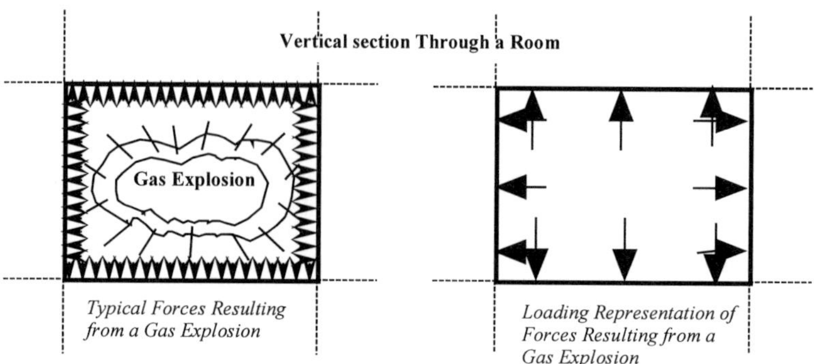

Figure 2 Nature of the in-plane loading generated during a gas explosion and selected load cases used to simulate this loading using forces applied via a hydraulic loading system

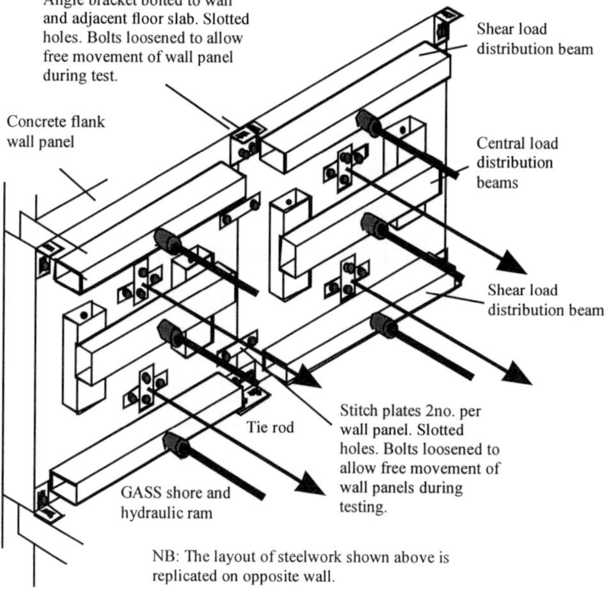

Figure 3 Isometric view showing typical arrangement of flank/cross wall load distribution beams and safety steelwork

Previous research [8] has indicated that it is acceptable to employ an equivalent static pressure to simulate gas explosion loading. Static loading is considered to be a more severe test than an equivalent gas explosion loading because it does not take advantage of the strain rate effects associated with the faster rate of loading which occurs during a gas explosion.

Since the static loads were applied at several discrete locations for practical reasons (refer Figure 3), rather than as a uniformly distributed pressure loading (UDL) which would be associated with a gaseous explosion, it was necessary to derive an estimate of the equivalent UDL. This was undertaken after the test were completed using linear elastic finite element analysis using ANSYS™ finite element program. In the analyses, a number of simplifying assumptions were made concerning factors such as the boundary support conditions of the floor slabs and wall panels, the concrete strength, the form of the panels and the distribution and strength of the main and tie reinforcement.

Equivalent UDL's were calculated for two conditions in the measured load-response curves. A lower bound equivalent UDL was determined where the measured response indicated that there was a significant change in the behaviour of the elements. This was taken at the point where the response changed from approximately linear to significantly non-linear or, in the case of the floor slabs, a lower flexural stiffness as a result of cracking developing. A higher value equivalent UDL was also calculated for the maximum test load applied.

Thus the linear elastic analyses produced a range of equivalent UDL pressures for the static load tests, depending upon the analysis parameter considered.

Whilst the limitations of the simplified linear elastic approach must be recognised, the approach is considered to be conservative. A more sophisticated analysis taking into account non-linearities due to factors such as material behaviour, the geometric form of the hollow floors and the changing interactions between the wall and floor components etc., would be expected to have produced somewhat higher equivalence values.

CONCLUSIONS

Static load testing was performed on two high-rise Bison Wallframe and Reema Conclad blocks built prior to the partial collapse of Ronan Point. The results are considered to have conclusively shown that the blocks would be able to resist the specified 17 kN/m^2 overpressure loading for a building without a piped gas supply.

As such, no key vertical load-bearing elements were removed, or damaged sufficiently, for there to be a need to mobilise alternative load paths. Accordingly the buildings tested by BRE were judged sufficiently strong to resist the accidental loads associated with typical non-piped gas explosions.

The findings of the load tests conducted by BRE have significant implications for the structural assessment of other high-rise Reema Conclad and Bison Wallframe LPS blocks which were of a similar design and built prior to the partial collapse of Ronan Point, are without a piped gas supply and in which investigations are judged to have demonstrated that a reasonable standard of workmanship has been achieved.

The results confirmed what many appraising engineers have long suspected, that significant lateral frictional / adhesive forces exist at the joint at the bottom of individual load bearing wall panels. Some elements of the current guidance for appraisal indicate that such forces should be disregarded, leading to difficulties in justifying the lateral stability of loadbearing wall panels under the specified accidental loading overpressures.

These forces appear to be present even in situations where the uplift forces applied to the soffit of the upper floor of the test room (located two floors below roof level) were actually seen to be lifting the part of the building above the test room.

The work has also demonstrated, in suitable circumstances, static load testing could potentially be used to assess the safety of particular designs of LPS buildings where analytical methods are not able to demonstrate an adequate margin of safety for the appropriate accidental loading criterion.

It should, however, be recognised that in the statistically unlikely event of a very severe explosion [3] removing one or more load-bearing wall elements; the post-damaged behaviour of the building would be controlled primarily by its ability to mobilise alternative load paths. The above programme of work has made no investigation of this aspect of LPS building behaviour. In such circumstances progressive collapse might occur. The risks of such an occurrence are, on a statistical basis, extremely small. In these circumstances such damage may not be disproportionate to the cause.

ACKNOWLEDGEMENTS

The work presented in this paper draws upon valuable contributions made by the BRE staff, its contractors and upon the findings of a number of previous research programs. The BRE wishes to acknowledge the co-operation and assistance provided by the Building Research Housing Group, Leeds City Council, Sandwell Metropolitan Borough Council and its contractors. Funding for this work was provided by the Housing Directorate of Department of the Environment, Transport and the Regions..

REFERENCES

1. BRE report BR116 : 'Reema large panel system dwellings : construction detail', 1987.

2. BRE report BR118 : 'Bison large panel system dwellings : construction detail', 1988.

3. ELLIS B R AND CURRIE D M, 'Gas explosions in buildings in the UK - Regulation and risk', The Structural Engineer, Vol 76, No 19, 6 October 1998, p373-380.

4. BRE Report BR107: 'The structural adequacy and durability of large panel system dwellings', 1987.

5. INSTITUTION OF STRUCTURAL ENGINEERS REPORT : 'Appraisal of existing structures', October 1996.6. Ministry of Housing and Local Government Circular 62/68 : 'Flats constructed with precast concrete panels. Appraisal and strengthening of existing high blocks. Design of new blocks'.

6. BRE Seminar : Assessment of Large Panel High-Rise Residential Buildings, 19 November 1998.

7. ELLIS B R AND TSUI F, 'Testing and analysis of reinforced concrete panels subject to explosive and static loading', Proceedings of Institution of Civil Engineers, Structures and Buildings, 1997, V 122, August, p233-304.

THEME SIX:
HOUSING

Keynote Paper

HOUSING THE HOMELESS HORDES

J Morris

Unversity of the Witwatersrand

South Africa

ABSTRACT. Chronic and acute housing shortages exist in many parts of the world. The problem is most acute in the developing world where burgeoning birth rates and mass unemployment dominate the social scene.The problem is seldom purely technical since there are many ways of creating housing but often the situation is bedevilled by poverty, politics and a lack of skills. In this paper the author considers the problem using Southern Africa as an example.The development of the shortage, various attempts at alleviating the problem, the current situation and the contribution that innovative technical developments can make are described. In conclusion the author speculates about the role that cement and concrete can play in addressing the problem of housing the poor.

Keywords: Housing, Roofing, Low cost, Innovative technology, Poverty, Afforability

John Morris, Dr ès Sc, Leuven, Belgium, Professor of Building Science, University of the Witwatersrand, Johannesburg. Formerly Chief Director of the National Building Research Institute (now BOUTEK), Pretoria, South Africa. Honorary Member of the Institute for Housing of Southern Africa .

INTRODUCTION

In 1964 the eminent American chemist, Carl Djerassi, told a congress of South African chemists that *'South Africa in common with the rest of the developing world is making the mistake of introducing death-control through better health care services before introducing birth-control.'* This he claimed would lead to a population explosion that the country was ill-equipped to manage.

He has been proved to be correct as Southern Africa has been riven by political strife and civil wars that have limited the economic growth but done little to reduce the birthrate - and the increase in the survival rate. In South Africa the birthrate is currently at about 2,4% while the economic growth over the last decades has hovered around zero.

Unemployment is claimed to be anywhere between 25 % and 55%, depending on the source of the statistics and the particular area examined. The turmoil in the world's money markets of the past year has contributed to the slowing of the already fragile economies of the these states and the political ineptitude of some of the leaders has done nothing to improve the situation.

In South Africa the rapid urbanisation initiated by the second World War and the accompanying industrial development led to the growth of squatter settlements (informal settlements) around the major cities and when the demands of war eased, there was a concerted effort to upgrade the accommodation of the informal settlers by building thousands of houses in the so-called townships of Soweto, Mamelodi and others.

In 1948 came the political change that led to the formalisation of the Apartheid policies of the then South African government. In very simplistic terms this policy propounded the theory that the black settlers in the townships were temporary sojourners who would move to 'their own areas' as these were developed and became economically self-sufficient. Therefore there was no need for more housing around the 'white' cities. During the following thirty years virtually no houses were built and people were not encouraged to build their own houses - except in the so-called homelands. These homelands were not developed as the theoreticians had anticipated - 'white' capital was excluded and there was very little 'black' capital available to kick-start such development.

During the 1980's the fallacy of this theory became ever clearer and changes in legislation were introduced and changes in political will were evident. Then came the release of Nelson Mandela and the pace of change accelerated.

In October of 1994 a nation-wide housing conference was held at which the 'stakeholders' from Government, the private sector, NGO's and all shades of society, agreed that housing should be a priority and that the many obstacles that had arisen during the previous forty years were to be swept aside.

The development of a new policy was beset by problems of interpretation, personal ambition, the death of the Minister and the 'education' of the new Minister.

Finally the state offered a subsidy in the form of an outright cash grant of Rands 15 000 for the purchase of a serviced stand and the erection of as much 'top-structure' as could be built for what was left of the grant. Depending on the soil conditions and the associated cost of the infrastructure there was often very little money left to erect any shelter at all. 'Houses' of 10 m^2 have been built as 'starter' houses and have been criticised as being 'unfit for human habitation'. It is clear that the problem has not been solved by this approach.

It may be that the wrong problem is being addressed since it is evident that if someone is unemployed and has no income, he or she cannot maintain a house nor can they pay for the services.[1] Should the question not rather be how to expand the economy, create jobs for people who have been given access to skills training so that they are able to sustain themselves and hence provide their own housing?

The author has dwelt at some length on the development of the South African version of the housing problem but although details may differ I expect that the central problem of 'affordability' is crucial in all developing countries that suffer from a housing shortage. In 1992 I was told that India [2] needs 31 million more houses - that might mean 150 million people that are homeless and although the problem in other countries may be smaller in total numbers, the relative scale would probably be of the same order. In South Africa it is claimed that we need an extra 1 to 2 million houses to shelter an eighth to a quarter of our current population without providing for the expected increase.

DISCUSSION

How should we approach the problem?

The two extremes that might be considered are those of mass housing in which large contractors, probably using highly mechanised techniques, are required to build hundreds (perhaps thousands) of houses in the shortest time possible. The other extreme is to provide serviced land and let the homeless fend for themselves. In terms of the provision of large numbers of houses the former process offers the most efficient production of houses if sheer numbers is the only criterion. Self-build offers the cheapest solution but the lack of skills and the time it takes for an unskilled individual to construct his or her own house, suggest that it will not solve the problem either.[3]

It can be argued that there is no simple solution and that different approaches would be appropriate in different situations.

It seems that there must be a compromise between large scale industrial production of houses and self-build efforts. A compromise that is appropriate to particular situations and individual needs and abilities. As already noted that for me job-creation is fundamental to solving the problem.

In this context it is tempting to equate the role of building materials with the role of materials in the manufacturing industries but there are some fundamental differences that must be borne in mind. In most manufacturing industries materials are selected on the basis of their properties and their processed costs. To that extent there is a similarity with the selection of building materials but the durability demanded of building materials is of a different order to that expected of most consumer durables. This is reflected by the provisions that are made for the purchase of homes, since even the humblest dwelling is expected to outlast the duration of the mortgage term by a factor of at least two. What other product is expected to have a commercially useful life of more than six decades?[4]

The properties of particular materials interact with those of other materials. The properties of the materials also influence the environment inside and outside the house. To achieve the desired levels of comfort in, durability and cost of the complete building, the fitness for purpose of the materials, individually and in combination, is an essential but complex component of the construction process.

'Any investigations into the history of building construction are complicated by its close interrelation with the other two practical aspects of building; namely, the materials and the structural systems. A given structural system is fundamentally affected - even completely determined in most cases by the materials employed in its realization' [5]

If one grafts onto this consideration that of ease of construction by relatively unskilled individuals and acceptability in the cultural climate of the community, the prospect of self-build processes providing a serious contribution to the alleviation of the housing need appears remote. Yet as noted, the need for job creation and for keeping whatever money there is, in circulation among the people of the community is sufficiently important to encourage attempts at making self-build a viable alternative.

Traditional construction techniques in Africa ranged from thatched beehive shaped huts, through wattle and daub, and sun-dried clay blocks to stone shelters. *'Probably the most remarkable variety of such house forms is found in Africa, where each tribe, each clan, even each sub-clan has had its own distinctive type of habitation. The forms are as varied as the languages, as numerous as are the dialects. A given house type, like the speech of the natives who inhabit it, is a persistent reality, highly resistant to change . In fact some anthropologists have noted that, in most cultures , house forms are undoubtedly less subject to foreign influence or contamination by adjacent cultures than any other artifacts, crafts or practices including language itself.'* [6]

Some of these construction techniques were well suited to the environmental demands of the climate but lacked durability. The thatched huts from sub-tropical Zululand were thermally ideal but had to be rebuilt every so often while the clay brick or wattle and daub constructions of the high inland plateaus were thermally attuned to the large diurnal temperature fluctuations but dissolved in the heavy summer rains and required frequent maintenance.

If, therefore, some form of traditional construction were contemplated as the basis for self-built alleviation of the housing shortage there had to be changes made to accommodate the decreasing availability of thatching, for example, and the unacceptability of such ongoing maintenance. Such changes would however, have to accommodate people's cultural needs as well.

Despite Fitchen's assurance that fashion , in the form of influence by other cultures does not play a major role, there is an increasing demand for houses such as the other population sectors have. Most South Africans now require houses of brick and mortar or something that simulates brick and mortar closely.

During the spate of house building, by the State, in the immediate post-war period, the houses were made of brick or concrete block and mortar and the building process was 'industrialised' in the sense that there was a factory-like assembly line with teams of foundation diggers, concrete placers, masonry corner builders and masonry wall builders moving from stand to stand.[7]

The European approach to industrialised building of medium to high rise blocks of apartments was never seriously considered and that was probably fortunate in the light of the Ronan Points and dehumanised high rise council flats in London, Glasgow and other cities.

Many industrialised systems were proposed and evaluated by the South African Board of Agrément but few have met with any success in providing accommodation of appropriate cost and acceptability on any worthwhile scale. Panel-type construction systems using fibre-cement boards have been approved by the South African Board of Agrément. Innovative approaches such as spray applied concrete over wire mesh, expanded polystyrene permanent formwork for filling and rendering with cement mortar and cement mortar used in conjunction with wooden formwork have all been accepted as potential contributors to the alleviation of the housing problem.

We have been involved in the examination of cement stabilized earth blocks as there are a number of systems available in South Africa.[8] The claimed advantages include the ready availability of the raw material, the possibility of manufacturing the blocks on site and thereby involving the community in their production. Since the blocks interlock it is also claimed that the training of workers to lay the blocks is quick and simple. That one of these systems is successful is evidenced by the sale of 34 block-making machines to Rwanda for the construction of rural schools and the claims that 500 houses have been built in Argentina (August 1998).

The possibility of stabilizing earth blocks through the use of proprietary chemical preparations and bituminous emulsions has also been examined. The proprietary chemical system was fairly successful but more expensive than stabilization with cement and therefore offered no advantage.

The modification of traditional sun-dried earth block construction to ensure greater durability has been examined by Ngowi.[9] In this he describes the development of the indigenous house construction process in Botswana and his attempts at modifying the process to improve the durability of the moulded blocks by the addition of cement and hence of the structures thereby reducing the need for ongoing maintenance while remaining sufficiently true to the traditional to ensure acceptability by the community.

It has been my contention[10] that the consumption of cement is a measure of the development in a country. Compare for example, the per capita consumption of cement in some countries of Africa: those countries that consume less than, say, 150 kg of cement per person per year appear to be those that have internal political conflicts or major economic problems and suffer from arrested development. See Table 1.

THE ROLE OF CEMENT

Since this is a conference about cement and concrete, let us examine some of the contributions that cement and concrete can make to the provision of housing and possibly also development.

I have already mentioned the cement stabilisation of earth. The problem with earth construction is that of durability and the on-going maintenance that is required.

Table 1 Cement consumption per head of capita per year

COUNTRY	kg/CAPITA/YEAR
Libya	653
Tunisia	446
Egypt	306
Morocco	238
South Africa	220
Namibia	186
Swaziland	168
Gabon	150
Ghana	91
Zimbabwe	89
Senegal	89
Congo (Brazzaville)	66
Benin	62
Nigeria	56
Tanzania	41
Kenya	34
Sudan	32
Zambia	29
Mozambique	28
Cameroon	25
Angola	16

The Namibian Clay House Project has built some attractive houses of unstabilised clay which are also very pleasant to live in but it has become clear that the maintenance required to ensure not only long-term durability but in fact the safety of the inhabitants might well be beyond what the home owner is able or prepared to invest. The vaulted roofs can be structurally dangerous if the maintenance is inadequate. Ngowi[9] has explored the possibilities of stabilising hand-moulded 'sun-dried' earth blocks for home building in Botswana. Morris and Blight[8] have examined the amount of cement needed to stabilize a range of earth types to ensure adequate strength and durability in hydraulically compressed dry-stacked earth blocks and found that at relatively low concentrations of cement, namely 5-6%, strengths of >4 MPa can be achieved and that the resistance to water induced erosion is excellent. This system of construction is apparently gaining ground since the machines used in our experiments are now being assembled in Argentina and manufactured in India as well as South Africa. In addition there has been extensive use of these machines to produce blocks for school construction in Uganda, Rwanda and Malawi.

In South Africa the most commonly accepted construction system for houses for low income families is still that based on concrete blocks which may be factory produced or made on site by so-called 'egg-laying' hand operated block-making machines. One of the difficulties experienced with these site-made blocks is that their dimensional stability leaves something to be desired and very often cracks appear along the mortar lines. In areas where driving rain occurs the weather tightness of the structures suffers.

Numerous variations on the traditional wattle-and-daub construction have been proposed; most variations being intended to counter the lack of durability of the original clay plaster. One of the earliest systems to gain a measure of acceptance in South Africa was the Zenzele [11] system in which structural timber framing in the form of creosote treated 'gum-poles' (Eucalyptus) is linked by steel mesh on either side of the poles. The space between the two layers of mesh is filled with rubble or balls of clay. The whole is then plastered with a cement based rendering to provide a smooth finish that is weather resistant. Tens of thousands of houses of this type have been built in the Eastern Cape Province of South Africa.

Similar in concept is the use of wire mesh that is coated with cement mortar by using a 'Gunite'-type application to produce a relatively thin skin of wire reinforced concrete. Such a system has been granted Agrément recognition under the name of Wirewall. It does require the use of sophisticated equipment and considerable skill on the part of the operator.

During the late 1970s the National Building Research Institute (NBRI) of South Africa, now known as the Division of Building Technology (BOUTEK), developed a construction system using shuttering that could be manhandled and once erected, was filled with a weak cement/sand mortar to create the outer and inner walls of simple houses. This system, known as BRISC (Building Research In-situ Concrete), was used extensively in northern Namibia and was modified for use in an experimental housing project in Kabokweni [12] in the then Northern Transvaal Province of South Africa. It also formed the basis for the development of a system of house construction developed for the Comores. In these islands, particularly on Grand Comore, there is an acute shortage of building materials and to avoid having to import material at great expense, it was proposed that a modified shutter system be filled with lumps of volcanic rock and consolidated with a weak mortar of coarse volcanic gravel and cement. The surface was given a skim coat of mortar to provide a smooth finish for painting. It was found that this system used approximately half the quantity of imported cement that would have been needed had normal concrete blocks been used. [13]

It is interesting to note that the same approach was adopted independently by the Namibian Clay House Project in Windhoek where there is no suitable clay. They used stones gathered by the people from within the township of Katutura, placed in the formwork and consolidated with cement mortar. The first house has been completed with apparent success and acclaimed by the local populace.

Many heavy-weight concrete elements have been proposed for what might be termed industrialised construction of houses and schools.

Factory produced building elements such as walls have been produced and brought to site. In other cases the floor slab has been cast and then used as a casting floor on which to cast the walls which, when cured, can be lifted up to complete the house or school. These all require considerable engineering input and sophisticated equipment such as cranes. They have the disadvantage in areas such as the winter rainfall area of the Western Cape Province that the concrete is too dense and that leads to condensation problems and mould growth during winter.

So far only to walls have been discussed but an expensive component of houses is the roof and particularly where most houses are single storeyed the contribution of the roof to the overall cost of the house can be high.

The use of thatch is still popular among the more affluent both for its appearance and for its thermal properties and among the rural poor because of its availability. However, it is not a solution to the housing of the urban poor.

Arrigone[14] suggested the use of pre-cast concrete troughs made on site as a possible solution. However, they have the limitation that extending the roof when extending the house would be difficult. They are heavy to handle and would require considerable skill in casting.

Commercial producers of concrete roof tiles claim that they can produce good quality tiles which with suitable roof designs can be as cost effective as any other system. However, there have also been some innovative attempts at making it possible for individuals to produce their own concrete roof tiles using small hand operated machines or machines driven by 12 volt batteries. A Cuban built version of this type of machine [15] is claimed to have been used to produce several million squares metres of roofing successfully and several versions of such machines are available in Africa. [16]

During the 1970s the oil crises focussed attention sharply on the need to conserve energy. Currently the threat of global warming due to the production of greenhouse gases is doing the same thing. Either way, for the poor, the need to ensure acceptable living conditions in their homes is a very real practical problem. The South African Board of Agrément has shown that in summer in Johannesburg (altitude 1700m) the maximum indoor temperature in a 'standard' brick house is likely to reach 30 ^{0}C whereas in a galvanised steel shack the temperature would reach 38 ^{0}C. What is more, to heat the standard house in winter would require 50,36 kWh.m^{-2} per year to maintain temperatures within the comfort range. In the shack the energy demand would be 202,22 kWh.m^{-2} per year which the poor could hardly afford. The insulation of informal, and formal housing is important for the environment and for the inhabitant of the house.

Because the houses are generally single storeyed the roof plays a bigger role in the energy balance of these houses than it would in multistoried houses. It has been suggested that as much as 60% of the heat lost is radiated through the roof to the cold night sky on a typically cloudless winter's night in Johannesburg.

The insulation of such roofs is a matter of some concern to researchers in the field and we too have examined the possibility of producing light weight insulation at reasonable cost from micro-cellular concrete. Some success has been achieved in the production of foamed cement boards of around 400 kg.m^{-2} of adequate strength and suitably low thermal conductivity.[17] There is however and vast gap between the laboratory production of an insulating ceiling board and the general application of such boards in houses that are yet to be built or to retrofit existing shacks with insulation.

An aspect of the provision of housing that is the subject of ongoing debate is that of standards. How much control should there be on the type of houses built for or by the homeless? The very real danger exists that if the standards, intended to protect the new home owner or occupier, are set too high the likelihood of that prospective occupier being able to afford the house becomes even more remote. Yet if there are no standards set, the problem for the occupier and the authorities, mortgage lenders and everyone else will be the need for maintenance, the associated costs and the likelihood that the houses will deteriorate into slum dwellings.

The South African Board of Agrément have taken the bold step of setting criteria which they describe as MANTAG (Minimum Agrément Norms, and Technical Advisory Guides) which are intended to ensure that house building techniques covered by these criteria will provide shelter, safety and durability. The Certificates cover single storey free-standing homes only, and consider aspects such as structural integrity, health risks, fire risks and durability. The TAG section advises on how the homes can be improved but these recommendations are not mandatory.[18]

CONCLUSIONS

Ultimately, despite all the technological innovations and the effort that has been put into the physical provision of housing for the newly urbanised hordes throughout the world the solution lies in the fields of sociology, economics and politics. However, it behoves us as scientists, engineers and entrepreneurs to ensure that we have the skills and technology available to provide those houses in association with the prospective occupiers when the other requirements are met.

It is the authors belief that cement and concrete in one form or another will be crucial to the provision of housing and that the increased consumption of cement that any large scale housing development will entail will be reflected by an improved level of economic and social development in the country concerned.

REFERENCES

1. MORRIS, J. Editorial Comment, Housing in Southern Africa, August 1994, pp 2

2. GOVERNMENT OF INDIA, National Housing Policy, 1992, pp 1-22

3. MORRIS, J, VILJOEN, C A, FINLAYSON, K A and BOSMAN, R E. Housing Delivery Systems: a comparative analysis, NBRI R?BOU 1521, Proceedings of the XIV International Association of Housing Science World Congress on Housing, Berlin 1987, pp 1-10

4. MORRIS, J. Building Materials : Fitness for purpose, NBRI R/BOU 1439, Symposium on engineering Materials in South Africa, 1986, pp 1-7

5. FITCHEN, J. Building Construction before Mechanization, The MIT Press Cambridge Massachusetts, 1986, p 21

6. FITCHEN, J. Building Construction before Mechanization, The MIT Press Cambridge Massachusetts, 1986, p 214

7. NATIONAL BUILDING RESEARCH INSTITUTE. Low Cost Housing, BRR642, 1987, pp i – xii

8. MORRIS, J and BLIGHT, G F. Cement Stabilisation of Soil for the Production of Building Blocks, Concrete in the Service of Mankind, Appropriate Concrete Technology, Dundee, E & F N SPON, 1996, pp 181-188

9. NGOWI, A B. The Potential for Alleviating Housing Needs through Modification of Indigenous Technology - a case study of Botswana, PhD Thesis, Univeristy of the Witwatersrand, 1998

10. MORRIS, J and SEPHTON, S S. The Role of the Cement and Concrete Industry in Developing Countries, Concrete in the Service of Mankind, Dundee, E & F N SPON, 1996 pp 28 – 41

11. AGRÉMENT BOARD of SOUTH AFRICA. Zenzele Building System, MANTAG Certificate, 1986 M1, Current Agrément Certificates and MANTAGs, March 1986, p 18

12. SCOTT, T W. Low Cost Housing with Formwark, NBRI C/BOU 1073, 1983, pp 1- 50

13. NBRI ANNUAL REVIEW. Houses from Volcanic Lava, 1986, pp 6 – 7

14. ARRIGONE, J L. Low Cost Roof Building Technology - three case studies using locally manufactured building components, NBRI, R/BOU 1161, 1983, pp 1 – 6

15. LATIN AMERICAN MCR NETWORK. History of the MCR Tile, published by the Grupo Sofonias, May 1997 pp 1 – 41

16. WRATTEN, W. Production of a Viable Cement Mortar Roof Tile for Low Cost Housing Schemes, Undergraduate Research Report, University of the Witwatersrand, 1996

17. DE ROSE, L. The Influence of Mix Design on the Properties of Micro-cellular Concrete Boards, MSc Thesis, University of the Witwatersrand, 1998, pp 1 – 200

18. MORRIS, J and SCHLOTFELDT, C J. Affordable Quality in Low Cost Housing, CIB Congress Proceedings, Theme III, Volume II, Paris 1989, pp 442 - 446

DEVELOPMENT OF HOUSING KIT USING AEROCRETE

A Swamidurai

P D Manoharan

Anna University

India

ABSTRACT. Economy in construction can be achieved by cost effective construction methods and use of local building materials. A housing kit has been developed using Aerocrete, a light weight material developed in Anna University using flyash as the major constituent. The suggested system consists of precast column elements, wall elements, joists, roof elements and prefabricated window frames with or without sunshade. In this paper the method of manufacture of precast Aerocrete elements and construction of a model house of size 3.15 m x 3.15 m is presented. The model house is tested for its strength and serviceability. The model house could be assembled in 12 days using only unskilled labourers.

Keywords: Housing kit, Aerocrete, Column element, Wall element, Lintel element, Reinforced roof panel, Joists, Window frame, Door frame.

Dr A Swamidurai is Professor of Structural Engineering, Anna University, Chennai, India. He specialises in construction technique using local materials and industrial wastes with special emphasis on prefabricated structures. Professor Swamidurai has published widely and serves in many Technical Committees.

Dr P Devadas Manoharan is Assistant Professor of Structural Engineering, Anna University, Chennai, India. He specialises in cost effective construction technique using local materials and industrial wastes.

INTRODUCTION

Considering the scattered rural population, dearth of conventional materials, skilled labour and high initial investment for industrialised building construction, semilarge panel prefabrication techniques are suitable to cater to the housing demands of a country like India. Economy in construction can be achieved by the construction methods also over and above the use of local building materials. The objective is to manufacture the components of a house in a centralized factory and will be made available to the users like any other factory made products. This can lead to:

i. Speed of construction
ii. Elimination of hoisting equipment
iii. Ease in the method of assembly
iv. Reduction in the labour cost
v. Increase in durability
vi. Improved thermal comfort
vii. Improved weather resistivity
viii. Promote self housing aptitude
ix. Achieve cost reduction
x. Effective disposal of industrial wastes such as fly ash

A housing kit was developed using Aerocrete.

AEROCRETE

Aerocrete is a lightweight material developed in the structural Engineering Department, Anna University, Chennai, India using fly ash as the major constituent (2). The light weight is obtained by adding extra water than that is required for the chemical reaction. The water solid ratio is so adjusted that the fine solid particles will be in floating equilibrium and as the green mortar hardens it entraps the microscopic discontinuous air voids and induces light weight.

Manufacture of Aerocrete

The ingredients of Aerocrete are identical to aerated concrete. Cement and fly ash are weigh batched and the required quantity of water is added and the matrix is mixed well in a tilting type of concrete mixer to get uniformity. By thorough mixing, the micro fine solid particles form a slurry of density 16N/litre. The slurry hardens leaving discontinuous spherical voids inside the matrix due to excess water.

HOUSING KIT

The suggested system consists of precast column elements, wall elements, joists, roof elements and prefabricated window frames with or without sunshade. The size and other details of these elements are optimized so as to achieve reduction in transportation and hoisting cost. The suggested housing kit is intended for mass housing schemes. To generate much greater consumer acceptance a model housing unit of size 3.15m x 3.15m was constructured and tested for its strength and serviceability.

The housing kit consists of precast components which can be manufactured easily and assembled at a faster rate with lesser labour cost. The plan and sectional elevation of the model house for the suggested technique is shown in Figure 1.

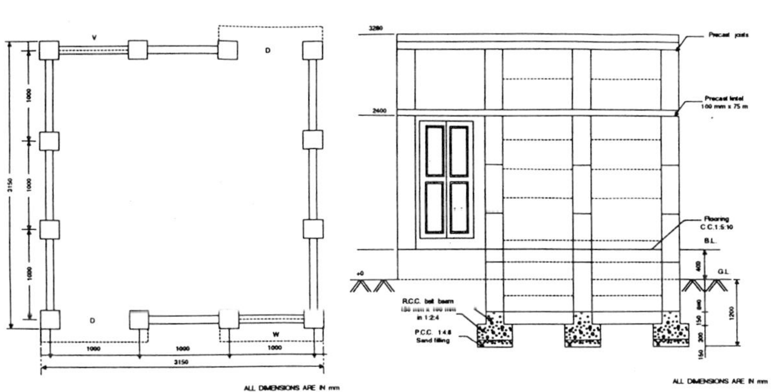

Figure 1 Plan and sectional elevation of housing kit

The components of suggested lowcost housing kit are shown in Figure 2.

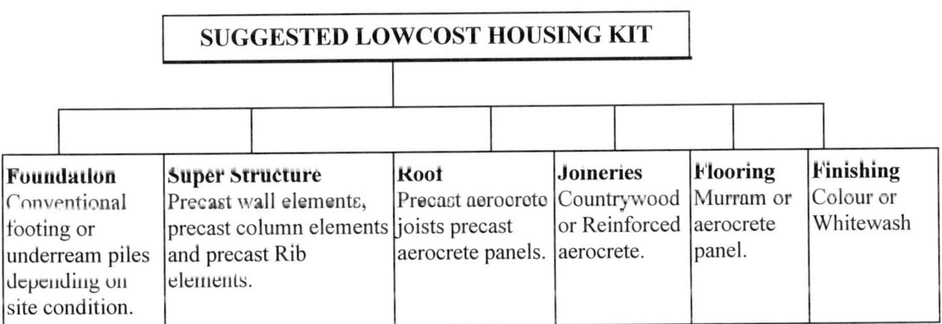

Figure 2 Suggested lowcost housing kit components

Wall Elements

Details of wall elements are shown in Figure 2. Overall size of each panel is 900 mm x 300 mm x 100 mm. Each panel weighs less than 34kg, therefore these panels can be manually lifted and can be assembled at site. The grooves provided at the top and bottom provide water tight horizontal joints (Figure 3). Vertical tongues provided at the ends seat the slab in the grooves provided in column elements.

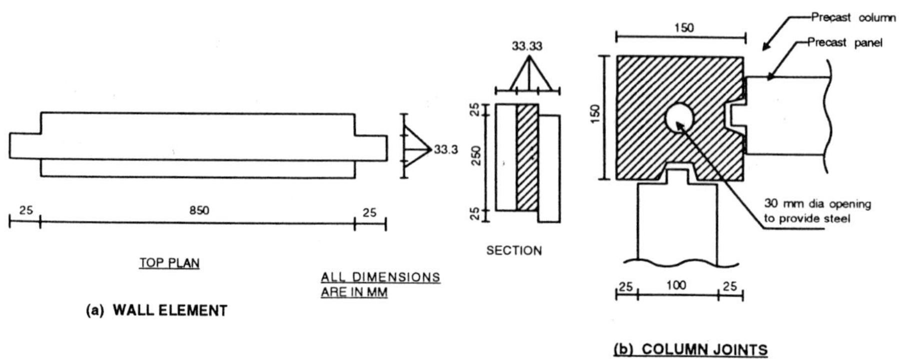

Figure 3 Details of wall element and column joints

Column Elements

The height of the column elements up to lintel levels are 750mm and above lintel level are 600 mm. Central hole of 30mm diameter is provided for all type of columns. Reinforcing steel extended from the foundation passes through this hole. After positioning the column elements these holes can be easily grouted. Grooves are provided in necessary faces to place wall elements in position (Figure 3).

Lintel Elements

Lintel elements shown in Figure 4 are used at the door and window level where the wall and column element ends. The width of the lintels are fixed as 150 mm so as to provide a horizontal continuous rib and to avoid uneven settlement. If necessary, precast sunshade cum lintels can also be used over the door and window opening.

Roof

Consists of precast reinforced Aerocrete panel elements and partially precast reinforced Aerocrete joists. Construction technique is similar to brick panel roof (1). Dimensions of precast slab element are of size 960mmx 470mm x 75mm reinforced with two 6mm diameter mild steel bars and five 6mm diameter transverse reinforcement.

Precast joists

Joists are of sections 130mm x 130mm x 3750mm designed as per limit state method and are provided with triangular stirrups extending outside the joists to a height of 75mm. Details and construction sequences are shown in Figure 4. Spacing of the joists are 1000mm c/c.

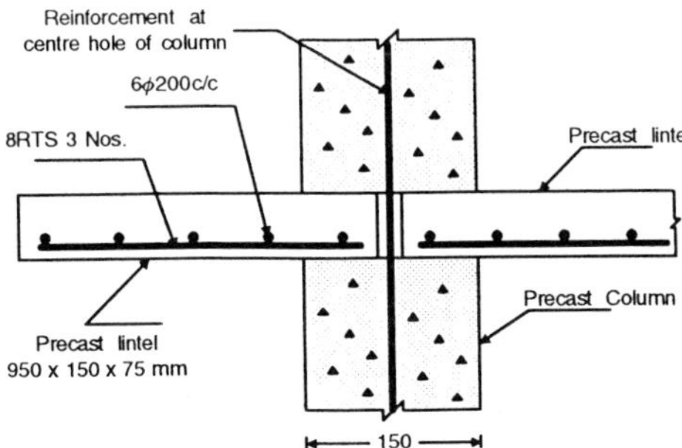

Figure 4 Details of lintel element

Screed Concrete and Weathering Course

The panels are spread in between the joists. Mild steel rods of diameter 6mm are provided at every panel joint perpendicular to the span of joists. M15 grade concrete is of thickness 30mm, over which 75mm thick brick jelly concrete in lime mortar is provided as weathering course. Over this weathering tiles are laid

Joineries

Frameless doors with G.I. shutters of size 800mm x 2100mm. Precast Aerocrete windows with lintel and sunshade are used. Use of Aerocrete jolly ventilator of size 860mmx600mm is suggested. The spacing of columns and the width of panels are so provided that the positions of doors and windows can be suitably changed by eliminating wall panels.

Flooring and Finishing

Filling upto the basement can be done in the conventional way and precast Aerocrete slabs of size 500mm x 500mm x 50mm can be paved for flooring. Suitable colour or white wash can be provided taking into account the cost.

MANUFACTURE OF PRECAST AEROCRETE ELEMENTS

Wall panels

Using reusable wooden mould 124 numbers of panels of size 900mm x 300mm x 100mm were cast on level ground. Aerocrete of strength 10 N/mm^2 was used.

Column elements

48 columns of size 150mm x 150mm x 750 mm and 12 numbers of 150mm x 150mm x 600mm were cast using wooden moulds. A central hole was provided while casting for inserting reinforcement during construction. Strength of Aerocrete used was 15N/mm^2.

Casting of Window Frames with Sunshade

The lintel cum sunshade was cast using special mould.

Reinforced Roof Panels

Panels were cast on level ground using steel side forms. The size of the panels were 470mm x 960mm x 75mm.

Aerocrete Joists

The cross sectional details of the joists can be seen from Figure 5. The length of the joists were 3.75m so as to provide an over hang of 300mm from the faces of the wall. Holes were provided at predetermined places to allow the rod from the column to pass through. The reinforcement grill was provided with chicken mesh caging and the joists were cast at level ground using steel channels and side forms.

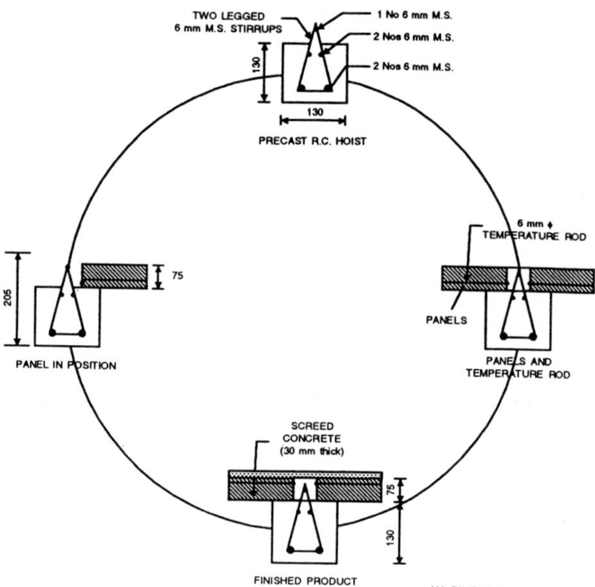

Figure 5 Sequence of construction of roof

CONSTRUCTION OF MODEL UNIT

The model unit was constructed at the rear side of the Structural Engineering Department, Anna University, Chennai-25, Tamilnadu.

Foundation

Since the strata was hard, isolated column footing of size 600mm x 600mm to a depth of 1200 mm were provided. Sand filling was done to a depth of 150mm over which 1:5:10 PCC was laid to a depth of 200mm from which the 10mm diameter rod for column was extended.

Sequence of Construction of Side Walls

Column units of height 750mm were positioned at the predetermined points by inserting the rod through the holes. In between the column units one row of the wall panel units were inserted atplinth beam level and joints were grouted. The second and third layers were inserted and grouted by cement mortar. The sill for the door was fixed and the construction process repeated for other sides till they reached sill level for the windows and ventilators. Necessary openings were left for window, doors and ventilator and the construction continued for 3.0m height to reach the roof level. At this level a continuous lintel of size 1000mm x 150mm x 75mm with two 8mm RTS projecting from both ends were fixed to get a level surface for keeping the roof.

Doors, Windows and Ventilators

Frameless doors with G.I. shutters were provided. Reinforced Aerocrete window was used and the shutters were of Aerocrete planks reinforced with weld mesh and chicken mesh.

Laying of Roof and Floor

Precast joists were manually carried to the site and kept over the column head inserting the rods through the holes provided during casting. Totally four joists were placed at the roof level. Sufficient struts were provided for the joists so as to avoid cracking while laying the panels. These struts were retained till the roof attained the composite action. Reinforced Aerocrete panels were carried manually and placed in between the joists. 24 panels were spaced at the roof level and 6mm diameter mild steel rods were tied perpendicular to the direction of the joists as temperature rods and 30mm thick screed concrete using 4.75mm jelly was laid. After 14 days the roof was tested for serviceability by sand bag loading. From the load deflection hysterisis for the test roof (Figure 6) it is evident that the roof satisfies the serviceability requirement. Precast Aerocrete panels were used for flooring.

Finishing

Suitable colour wash was provided for the side walls and for the joineries. The constructed model house can be seen from Figure 7.

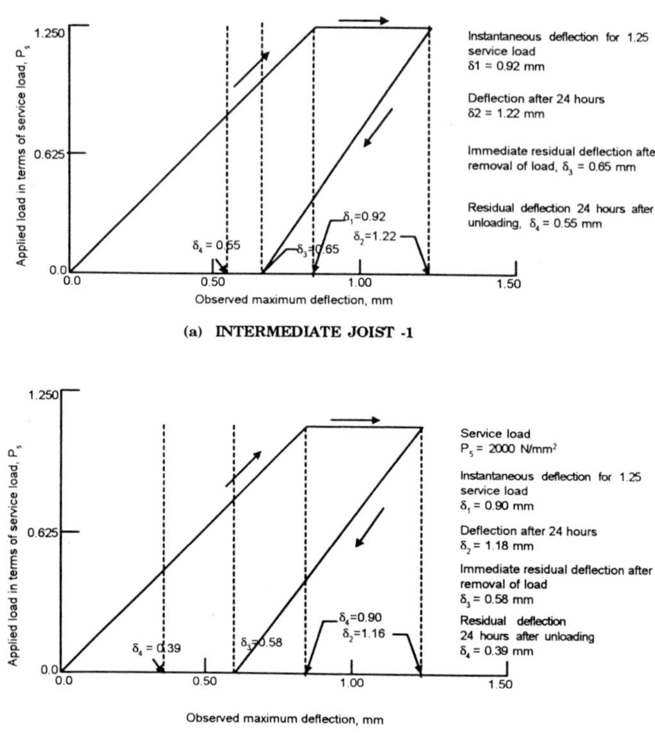

Figure 6 Load-deflection hysterisis

Figure 7 Model house

Manpower Required

Since the use of construction equipment was eliminated fully, the manpower requirement at each stage of work was closely observed and refinement was attempted to reduce labour cost. The model house could be assembled within 12 days. Only semi skilled and unskilled labour were used.

CONCLUSIONS

The total requirement of fly ash per unit was 60kN, which envisages the feasibility of bulk disposal of fly ash through building industry, contributing to environmental pollution control.

REFERENCES

1. CHITHARANJAN, N. Long span brick roofs for mass housing. The Indian Concrete Journal, Vol.60, No.1, 1986, pp.306-314.

2. CHITHARANJAN, N. SUNDARARAJAN R. and DEVADAS MANOHARAN P. Development of Aerocrete, A new lightweight High strength material, International Journal of Cement Composite and Light weight concrete, U.K., Vol.10, No.1, Feb 1988, pp.219-232.

3. CHITHARANJAN, N. and KALIAPPAN T.P. Use of Industrial Waste for precast wall elements, proceedings of the 8th International Brick/Masonry Conference, Ireland, 1988, pp.177-185.

MASS HOUSING AND REHABILITATION OF SLUMS: INDIAN EXPERIENCE

A K Jain

Grasim Industries (Cement Division)

India

ABSTRACT. The state of housing in any country is a direct measure of its level of social and economic development. Housing on one hand affects the quality of life of the people and on the other hand has great influence on the ecology and environment of the place. In most of the urban areas, housing is posing increasingly difficult problem due to lack of resources, infrastructure and imbalanced growth. In India there is a sharp division of have's and have not's due to the legacy of social, cultural and political systems followed over the centuries. Only 20 to 30 % of population, in metropolitan cities, have access to proper housing and support facilities, while majority of people are living in highly inadequate dwelling systems. The housing problem in most of the developing countries can be tackled through mass movement involving government, non-government, social and religious organisations and individuals in updating the existing housing systems and making up the deficiencies. The problem of housing in India is two faced; on one side large number of dwelling units have to be constructed to meet the net shortage of housing and on the other side, the slums and other similar living areas have to be improved to make them habitable. The problem is further aggravated by the unprecedented growth of population and increasing pressure of migration from rural to urban areas, coupled with the lack of resources. Housing in general has been accorded low priority in most of the developmental plans. In the prevailing scenario, the optimization of available resources and implementation of bold and innovative schemes is the key to make a dent in this formidable problem.

Keywords: Housing, Technology, Funds, Financial institutions, Basic amenities, Slums, Urban, Rural, Low cost housing.

Mr A K Jain, is Asst. Vice President (Technical Services), Grasim Industries (Cement Division), a Unit of Aditya Birla Group, India. He has 30 years of experience in planning, design and construction of various civil engineering works. He is now associated with quality, customer services, market and business development, related to cement based applications. He has published and traveled widely and serves on many Technical Committees.

INTRODUCTION

Shelter is one of the three most basic needs of any human being. The other two are food and clothing. The problem of housing has assumed gigantic proportions in India. The very large population base, low per capita income and severe deficiencies of infrastructure facilities are mainly responsible in forcing a large segment of population to live in slums and under sub-human conditions. The lack of employment opportunities in rural India and excessive pressure on agriculture are further aggravating the problem due to large scale migration of population from rural to urban areas. The minimal civic amenities in urban areas are choking up due to population pressure. Large pockets in cities are converting in to slums. The acute shortage of housing is forcing people to use pavements and public places. The problem has engaged the attention of national and state governments, but no satisfactory solution appears to be in sight in the foreseable future due to shortage of resources and inadequate purchasing power of large sections of population.

MACRO ECONOMIC REVIEW AND OUTLOOK

India is a country of striking contrasts and enormous ethnic, linguistic and cultural diversity. India is a federation of 25 states, 13 states have population in excess of 20 million people, 6 have more than 60 million, 3 have above 80 million and 1 has a population of above 140 million. These states have vastly different natural resources, languages, ethnic composition, religious orientation and economic and social performance. The major indicators on India are given in Table 1.

Table 1 India at a glance [4]

S.No.	INDICATOR	PERFORMANCE
1	Land area (1,000 Sq Km.)	3,287.6
2	GNP percapita (US $)	350.0
3	Population mid - 1997 (million)	979.6
4	Population growth rate (1990 - 95)	1.8
5	Total fertility rate (Births per woman)	3.7
6	Life expentancy at birth (Years)	62
7	Infant mortality (Per 1000 live births)	68
8	Illiteracy rate (Percentage of population age15 +)	48

India's population is expected to reach 1016 million by Dec'99, making India the world's second largest populous country and its per capita income is expected to be US $ 365. India has the world's largest concentration of poor people with about 38 % living below the official poverty line. The country also faces high level of illiteracy and low learning achievements. On an average, the adult labour force has only 2.4 years of schooling.

India otherwise has a large pool of highly qualified business professionals, engineers and scientists of international calibre. It is estimated that India's construction industry employs about 32 million personnel [5], the rough break-up of which is as follows:

Engineers and Executives	:	3 million
Supervisors and Qualified Personnel	:	5 million
Workers	:	24 million
		---
		32 million

There is a tremendous strain on infrastructure. Though substantial urbanization coupled with a burgeoning economy, is accelerating the growth of housing, the gap between supply and demand is continuously increasing due to high population growth rate and low level of investment. Low per capita income of nearly 40 % of Indian population, which is living below the poverty line, is the main cause of poor availability of housing. The houses which are built also lack basic infrastructural facilities due to low level of investment by the Government.

The strengths of housing sector in India come from:

- Substantial pool of highly qualified and experienced technical and managerial manpower.
- Abundant labour force.
- Well established engineering practices.
- Abundant supply of basic construction materials.
- Social and personal priority accorded for owing a house.

And the weaknesses of the housing sector come from:

- Little access to latest technology.
- Lack of mechanization.
- Absence of practical and suitable systems of financing.
- Lack of suitable infrastructure.
- Lack of unified and consistent training facilities for Construction workmen.
- Inadequate gross purchasing power of a large section of population.
- Very low expectations and attitude towards modern living.

URBAN HOUSING PROFILE IN INDIA

Housing is the main urban challenge facing the developing world. The rate of housing construction lags behind the galloping population growth in a majority of the developing countries. A United Nations survey (1990) shows that while 42 houses were built for every 100 new household formed in 1970-74, it has decreased to 38 new houses for 100 new households in 1985-89. As compared to this depressing trend in the developing world, housing construction growth has out-paced the growth in household formations in the developed world. The developing world scenario on housing is not very optimistic.

The United Nations Centre for Human Settlement (HABITAT) has estimated that world wide 2.2 billion people will be living in slums and poor housing by the turn of this century.

The Indian scenario is not much different from this international picture. According to 1991 census [5], 162 million household (urban plus rural) were living in 131 million houses indicating a housing backlog of about 31 million. It is estimated that about 31 % of this shortfall has occurred in the urban areas which accommodate about 29.25 % of total households in the country. The national housing backlog for the year 2001 is estimated at 41 million.

Even though the urban households in India have grown at a higher rate than the rural households (3.7 % as compared to 2.12 %)(3), the urban housing stock has kept pace with this increasing demand for housing, however, in absolute terms the national housing shortage is alarming. The backlog in housing stock is surmountable, provided housing grows at 6 % per year as compared to the present rate of about 3 %. The country should mobilize about Rs.1500 billion (US $ 35 billion) during the current five year plan period, but the government lending Institutions have been asked to raise only Rs.350 billion, about 25 % of the required investment.

In the 12 Urban agglomerations, which share 26.4 % of the total urban population in India, there has been an increase in the number of households living in slums. Table 2 [4] gives estimated slum population in 12 major metropolitan cities in India.

Table 2 Estimated slum population in 12 major cities in India [4]
(Population in 100,000)

S.No	CITY	1991			2001		
		Total population	Slum population	%	Total population	Slum population	%
1	Greater Mumbai	125.962	43.205	34.3	170.701	58.55	34.30
2	Calcutta	110.219	36.262	32.9	131.147	43.241	32.97
3	Delhi	84.191	22.480	26.7	122.204	32.628	26.70
4	Chennai	54.220	15.251	28.1	69.823	19.620	28.10
5	Hyderabad	43.444	8.593	19.8	62.964	12.466	19.80
6	Bangalore	41.303	5.162	12.5	63.597	7.494	11.78
7	Ahmedabad	33.122	6.724	20.3	43.629	8.859	20.31
8	Pune	24.940	4.065	16.3	35.299	5.753	16.30
9	Kanpur	20.299	4.172	20.6	24.875	5.124	20.60
10	Lucknow	16.692	2.778	16.6	22.581	3.748	16.60
11	Nagpur	16.640	5.308	31.9	23.212	7.405	31.90
12	Jaipur	15.182	4.418	29.1	22.108	6.433	29.10
	TOTAL	586.214	160.418	27.37	792.140	211.321	26.68

Though urbanization could be cited as one of the reasons for housing shortage, it does not seem to be the primary cause. The major factor influencing the shortage is the decreasing level of investment in the housing sector. From the decreasing level of investment pattern, it is observed that the government is entrusting the private sector with the responsibility to house the growing national population.

The private sector could face up to this challenge only if proper investment environment is provided through speedy and sound legal, economic and administrative reforms. Figure: 1 gives the allocation on housing in India for each five year plan.

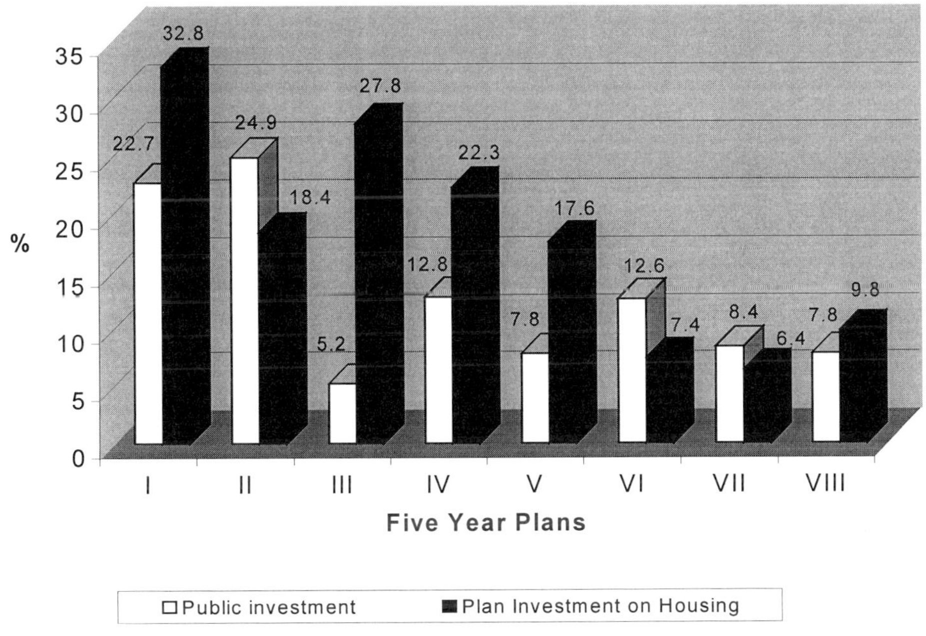

Figure 1 Plan allocation on housing in India [3]

The role of the private sector in the provision of housing needs of the country draws significance. During the past five year plans, the public sector, inspite of its public housing supply development pradigms, has been able to contribute only 10.2 % of the total sector investment. The share of public investment has fallen from 34 % during the first five year plan to 8 % during the eighth plan[3]. The institutional funding by banks, the Life Insurance Corporation (LIC), General Insurance Corporation (GIC), Housing Urban Development Corporation (HUDCO) and other housing finance institutions such as National Housing Bank, Public Provident Fund and other loan schemes are expected to add up to Rs.155,650 million (US $ 3700 million). This development perspective indicates the extent of challenge placed on the private sector to meet the housing shortage in India, both in urban and rural areas.

POSITION OF HOUSING IN RURAL INDIA

As per the Census of 1991 [5], the total households in rural India, which comprised, Pucca, semi Pucca, Kutcha, huts etc. stood at 92.98 million dwelling units.

The net shortfall was estimated 22.4 million dwelling units. In rural India, apart from the problem of net shortage, the upgradation of dwelling units and provision of basic amenities such as drinking water, lighting and sanitation assume great significance. These facilities are greatly lacking in rural houses. Table 3 gives the position based on 1991 census report.

Table 3 Position of basic amenities in rural and urban houses in India [3]

S.No.	CHARACTERSTIC	PERCENTAGE COVERED AS ON 31ST MARCH			
		1985	1990	1993	1995
1	Drinking water supply (Tap Water)				
	Rural	15.47	19.21	23.27	27.54
	Urban	64.19	72.11	82.84	91.34
2	Sanitation facilities				
	Rural	0.7	2.4	3.3	3.59
	Urban	28.4	45.9	47.9	47.9
3	Lighting (Electricity)				
	Rural	23.4	27.04	34.4	39.7
	Urban	62.48	74.38	81.27	89.45

METHODS USED FOR REHABILITATION OF SLUMS

To tackle the acute shortage of housing for poor sections of the society and for rehabilitation of slums in urban areas, number of options have been tried in India. The details of some of the options and their effectiveness is as follows:

Deep Subsidies

The government and local municipal authorities extended assistance to the urban poor in owning a shelter. It was believed that home ownership offered the urban poor new reasoning to work and save. The urban poor can not capitalise housing cost out of their current income and a policy of deep subsidies to limited urban poor was therefore conceived. The policy worked to a limited extent in the initial phase. The privatisation process of the 1980s cut down the subsidies and considerably reduced the scope of the scheme. However, the policy did not succeed as there were no long term controls on the resale of such properties. After couple of years, the houses instead of being owned by the urban poor were purchased by the wealthier buyers. The urban poor again occupied the public land and created new slums. The policy failed and lost public support.

Land as a Finance Source

The urban poor in most of the cities are occupying unauthorized public land. The slums in Mumbai, like at any other place, are located at premium sites – close to the mass transit routes or close to the major job centers.

The land under slums is valuable, and a squatter due to many years of occupancy have got a right on it, however illegal it may be from strictly legal point of view.

The government of Maharashtra is 1995 initiated a very bold and innovative scheme "Shivshahi Punarvasan Prakalpa" to rehabilitate 4 millions slum dwellers in Mumbai[1]. The main focus of this scheme was put on using `land as a resource' due to the un-affordability of the slum dwellers to pay initial capital cost. By participating in the scheme, the squatter is foregoing his right in lieu of a 225 sq.ft. tenement.

The Government or the local authority's contribution is by way of allowing extra FSI (Floor Space Index) to the developers. The developer contributes his entrepreneurial skills and initial capital and all the three parties are adequately compensated for their inputs. Squatter by a house; local authority by way of fee for extra FSI and developer by the profit he makes on the scheme. Therefore to call it a `free housing' may be more appropriate.

Under the scheme, the private developer is allowed a FSI of 2.5 instead of 1.5. For this concession, the developer is required to provide tenement to squatter as per original permissible FSI and a fee of Rs.840 per sq.m. to the Government for strengthening the civic structures. In return, he is authorised to sell extra tenements in the open market which would become available due to higher FSI. The scheme does not strain already over stretched civic infrastructure such as water and electric supply, storm and sewerage drainage, transport network etc. These facilities will be created out of the fee paid by the developer to the Government.

The scheme when initiated in 1995 was considered practical and workable as the property rates in Mumbai during that period were very high and ranged between Rs.8,000 to Rs.10,000 per Sq.ft. for residential area even in suburbs[1]. The steep fall in property prices from 1996 onwards has created unsurmountable barriers in the implementation of the scheme. Till date the construction of only 82,500 tenements is under progress, against a target of 2,50,000 tenements per year. The scheme at present is under jeopardy due to recession in real estate market.

Land as a finance source has limitations, it is highly dependent on the real estate market. Real estate market is cyclic, therefore the impact or utility of land as finance source is varied. In the boom period it is highly effective but as the boom ends, the value of land goes down and with that its effectiveness. Second limitation is that it is highly location specific. Lands in the business districts or prime residential locations are scarce and highly valuable. On the periphery the lands are cheaper.

To be effective, the policy should provide for these differentials in land prices. Therefore to base housing policy on the "land as resource" without any alternative during the recession phase will become unsustainable. Rehabilitation scheme in Mumbai is the case in point. It has suffered heavily post April, 1996 period.

Co-operative Societies

The low income housing co-operative societies have also been tried in some of the cities. The development of low income co-operative society is a complicated, time consuming and costly process. It demands long term commitment on by its key members and the politicians.

The absence of regular contributions by the co-operative members and lack of continuous support by the government to such societies and frequent changes in the establishment and policies created hindrance in working of co-operative societies. The experiment has not been found very successful except in few cases.

Low Cost Technology

Development of appropriate low cost housing technology to make shelter affordable to the urban poor has also been one of the options. Urban poor can pay for the housing cost at the maximum of 30 percent of his total emoluments. Urban poor has a priority of `sheltering space' over the `level of service'. For him level of service becomes secondary. But for the community, the level of service commands a priority over affordable `sheltering space'. It is thus a conflict between the preference of an individual and the community which has given rise to concept of appropriate technology.

The components of the infrastructure for basic services to prevent low cost housing going down to slums would be;

- Preparation of site.
- Providing marginal access to houses.
- Water supply and sanitation.
- Management of waste.
- Providing electricity.

The low cost housing technology has to adopt the approach of minimum cost of construction; maximum utilization of space, minimum comfort and maximum durability. A number of innovative methods of construction to reduce cost have been experimented. The notable techniques are; cavity walls, filler slabs, RCC door frames, flyash bricks, lime sand mortar, compressed earth blocks, ferrocement roofing channels, micro concrete roofing tiles and precast components. The cost of construction by use of these methods has been brought down by approximately 30 %[2]. Even the cost of construction of low cost house in India is about Rs.2150 per Sqm. (US $ 50 per Sqm.). A unit of 24 Sqm. including cost of land is approximately Rs.2,00,000 (US $ 4800) and many urban poor find it unaffordable.

SUPPORT APPROACH FOR SHELTER AND SERVICES

The attempts to provide housing through government projects have never been cost effective, hence not affordable. The government can not subsidise so called `low-cost' housing any more. The support approach to shelters may provide an economic and affordable solution. In this approach, the housing construction and maintenance would be under the control of the individual household.

Housing agencies would provide only the technical, financial and social support for the individual and the community. This will boost up the cost saving. Active participation of individual and community in construction, development of services and subsequent maintenance is necessary for cost reduction and sustaining the scheme.

The community and the individual are expected to fulfil their respective roles. The responsibility will be manifested through following stages.

- Involvement in planning and selecting technology.
- Determining the rate of improving services.
- Organizing and management.

The capital cost of the infrastructure may be paid by the sponsoring authority and the operation and maintenance costs by the households. This scheme has promise as `free housing' without physical and financial involvement of individual have proved futile in the past.

CONCLUSIONS

The slums in Indian cities are a physical reality. They have a life and economy, which is integrated with rest of the city. They are in utter state of squalor and there is crying need to address the situation. The resettlement of slums will provide new lease of life to the cities. Large development of slums will release valuable land, which can be used for housing and other purposes such as hospitals, schools, community centres etc. Integration of the slum dwellers in the main stream will release tremendous social and economic pressure.

The problem is, however, formidable. So far either we have avoided the problem wishing it away or have done meaning less little about it. The scheme was launched with great fanfare in Mumbai in 1995 but it has yielded no tangible results.

Provision of shelter to each citizen has been on top of the list in the agenda of various National and State Governments, but it remains an unfulfilled dream even after 50 years of Independence. There is only solace that at least we are speaking, looking at and focusing on this burning problem.

REFERENCES

1. GILL, G S. Free houses: Concrete solution or a hollow promise. Proceedings, National Seminar on low cost housing, Mumbai, India, October 1998.

2. KUMAR, A. Sustainable building technologies for low cost housing. Development Alternatives, New Delhi, India, 1999.

3. GOVERNMENT OF INDIA Economic Survey, 1995-96. Ministry of Finance, India. Published in Gazette of India, 1997, published by Ministry of Urban Development in 1998.

4. GOVERNMENT OF INDIA. Statistical data. Town and Country Planning Organisation.

5. GOVERNMENT OF INDIA. Census report 1991, India, 1993.

PRINCIPLES OF COMPOSITIONAL BUILD-UP OF HEAT RESISTANT MATERIALS MADE WITH ALKALINE CEMENTITIOUS MATERIALS

E K Pushkaryeva

G S Stanetskii

Scientific Research Institute on Binders and Materials

N M Mkhitaryan

Aerospace Academy of Ukraine

Ukraine

ABSTRACT. The study deals with the peculiarities of compositional build-up of the heat resistant concretes based on alkaline cements. It is shown that the regulation of heat resistance, and hence, the increase in durability of the heat resistant stone may be reached due to formation of a fragmentary microstructure of the stone when aluminosilicate additives, thermoactivated at various temperatures, are introduced into the cement composition to provide a target synthesis within the dehydration products of needle-shaped substances similar in crystal structure and chemistry which possess the increased density, heat conductivity and the lower coefficient of linear thermal expansion (C.L.T.E.) as compared with initial cement constituents. It is demonstrated that the highest thermal resistance after firing at T=1000°C (86-130 MPa) is characteristic of the chamotte-slag mixes with an additive of thermoactivated kaolinite at T=600-700°C in a quantity of 10-20 % by mass. The heat resistance of concretes made with the developed cements exceeds by 10-15 water heat cycles that of the reference concretes made with the same cements without additives.

Keywords: Alkaline Cement, Chamotte, Coefficient of linear thermal expansion (C.L.T.E.), Dehydration, Heat resistance, Hydration, Kaolinite, Slag, Thermo-resistance.

DrSc E K Pushkaryeva is Head of the Specialty Cements and Concretes Division, Scientific Research Institute on Binders and Materials, Kiev, Ukraine. She specialises in the synthesis of heat-resistant, thermo-resistant and other special cements and concretes and the durability of such concretes when affected by high temperature environments.

DrSc N M Mkhitaryan is Professor in Building Materials and Articles, the Aerospace Academy of Ukraine, Kiev. He specialises in the heat treatment, in particular, solar energy treatment of different types of concretes.

Doctorate G S Stanetskii is a Researcher, Scientific Research Institute on Binders and Materials, Kiev, Ukraine. His scientific interests cover the development of thermo-resistant materials made with alkaline cements.

INTRODUCTION

An analysis of contemporary theories in the field of producing heat-and thermo-resistant artificial stone ("maximum/extreme stresses" theory [1]," two phases" theory [2], "structural or fragmentary" theory [3], statistical or theory of "weak linkage"[4]) shows that in the background of phenomena initiating thermal failure of concretes lay the processes which may be caused by occurrence of stresses in the material. The thermal stresses may be initiated by a temperature gradient (stresses of the first kind) and the presence of structure re-arrangement processes (stresses of the second kind), which are connected with the flow of chemical reactions and anisotropy of crystals growth and change in the coefficient of linear thermal expansion (C.L.T.E.) as well [1].

According to the known criteria for assessment of thermal resistance, the last may be presented as property of a material dependent upon strength, C.L.T.E., elasticity modulus, Poisson's ratio, heat conductivity [3].

Taking into account that heat resistant concretes are, in their structure, heterogeneous materials with considerable porosity and developed crack resistance, their thermal resistance, measured as resistance to crack propagation during heat impacts, is of sense to assess with consideration of the fragmentary theory [3, 5].

As it follows from [6, 7], the increase in thermal resistance of heat resistant concretes may be reached by regulating the degree of fragmentation of the structure by:

1. using mixes with various C.L.T.E. [1];
2. mixing refractory substances with different baking ability [3];
3. using aggregates with various grain size [5];
4. manufacturing articles combining high density and thermal resistance, for example, owing to the flow of baking reactions [1];
5. producing microstructures of eutectics [3];
6. using damping additives (perlite, vermiculite, zeolites) [8];
7. a directed regulation of structure due to the effect of crystals growth anisotropy and extension of pores to the crystals' boundaries;
8. changing a character of pores being formed (a material with cracks developed from the spherical pores is known to be more resistant to main cracks propagation as compared with that with elliptical pores [9]);
9. a directed formation of certain composition of hydration products in the fired heat resistant materials, f.e. with synthesis of crystals of thread- shaped and needle-shaped structure (calcium dialuminate, mullite), which do not create a close contact with the aggregate grains, since the axis of needle-shaped crystals are directed perpendicularly to the surface of the aggregate grain. Such build-up of the contact surface promotes relaxation of the thermal stresses and the increase in thermo-resistance of the material [3].

Thus, the solution to the problem of producing an artificial stone with target thermo-mechanical characteristics requires not only meeting the known regulations of formation of the heat-resistant conglomerates, but taking into account the fundamental principles in order to improve thermo-resistance of the materials due to the directed changes of their composition and structure at different hierarchical levels.

An analysis of world trends on the use of cementitious materials of different types of hardening [10] aimed at production of fire- and thermo-resistant concretes testifies to prospects of the approach of using the alkaline cementitious materials, the peculiarities of chemical and mineralogical composition of which open up wide possibilities to synthesise the composite materials with target properties.

By the investigations earlier held [11], it was established that strengthening of structure of the materials during hydration of the alkaline cements is reached due to the realisation of topotaxial mechanism of recrystallisation of the hydrate phases (magnesium hydrosilicates, hydrogarnets, zeolite-like hydroaluminosilicates) into anhydrous compounds without developing considerable destructive stresses in the material structure, the increase in criterion of the similarity of new formations and decline of the coefficient of destruction owing to increasing a symmetry of new formations crystallised as well as increasing the similarity degree of the crystals parameters capable to intergrow being observed.

The regulation of thermo-resistance, hence, the increase in durability of the heat-resistant stone made with alkaline cements, may be reached by the formation of a fragmentary microstructure due to the introduction into the cement of correcting additives which are similar in composition but which were thermoactivated at various temperatures as well as due to a directed synthesis within the dehydration products of needle- shaped and thread- shaped substances similar in crystals and chemistry and which have the increased density, heat conductivity, and the decreased C.L.T.E.

TESTING TECHNIQUE

In order to reveal major regularities of formation of the thermo-resistant materials made with alkaline cements, there was conducted the modelling of processes of synthesis of heat-resistant stone of increased thermo- resistance with the use of blast furnace granulated slag (BFGS), finely dispersed chamotte and kaolinite additives, thermoactivated at T=500°C, 600°C, and 700°C. The chemical composition of the initial constituents is given in Table 1.

According to the data obtained with the use of X-ray microscope, the blast furnace granulated slag is represented by amorphosized substance (Figure 1a, curve 1) and the chamotte includes, together with mullite (d=0.342; 0.339; 0.269; 0.254; 0.228; 0.220 nm), cristobalite (d= 0.410; 0.252; 0.205; 0.182 nm) (Figure 1a, curve 2).

In the X- ray grams of the samples of thermoactivated at T=500°C kaolinite the main diffraction reflections of the mineral (d=0.714; 0.436; 0.417; 0.385; 0.357; 0.332; 0.256; 0.249; 0.238 nm) are detected (Figure 1b, curve 1), and at T=600°C there is observed the failure of the structure with formation of the amorphosized aluminosilicate substance and insignificant quantity of $Al_2O_3SiO_2$ (d=0.336; 0.341 nm) (Figure 1b, curve 2).

The firing temperature increase up to 700°C results in a full decomposition of kaolinite with formation of the amorphosized substance (Figure 1b, curve 3).

Table 1 Chemical composition of initial constituents

NAME	OXIDE CONTENT, % by mass								
	SiO_2	Al_2O_3	Fe_2O_3	CaO	MgO	SO_3	MnO	TiO_2	R_2O
BFGS	39.5	6.49	0.12	47.1	3.1	1.7	1.15	-	-
Chamotte	51.5	42.0	3.4	0.4	0.7	-	-	1.2	0.6

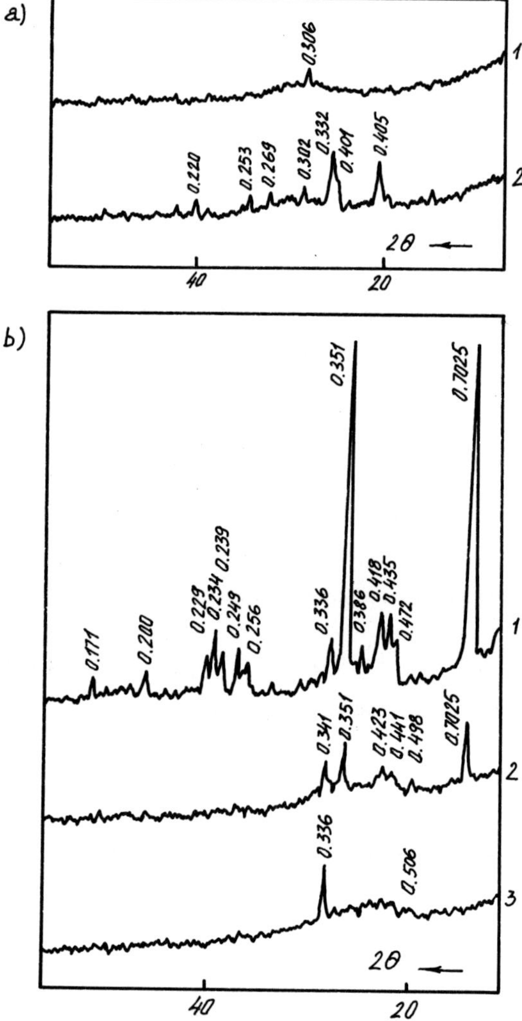

Figure 1 X-ray grams of initial constituents blast furnace granulated slag (1) and chamotte (2); kaolinite, thermoactivated at T=500°C (1), 600°C(2), and 700°C(3)

Used as an alkaline constituent were sodium soluble glasses with a silicate modulus (Ms)= 1; 1.75; 2.5 and density equal to 1250 kg/ m^3.

The cement mixes were prepared from a finely ground blend of blast furnace granulated slag and chamotte taken in the ratio of 1:1, a correcting additive of thermo-activated kaolinite was introduced in a quantity of 10- 30 % by mass in excess of 100 %. The cement mix design was made using calculations and experimentally with the use of a mathematical planning of experiment [12].

The specimens, prepared using vibration and made from the mortar of normal plasticity, were stored under normal conditions for 7 days, then were subjected to drying at T=100°C for 32 hours and firing at T=800°C and 1000°C for 4 hours.The physical and mechanical characteristics of the specimens, inclusive of thermal resistance, were determined according to the USSR standards 310.4-80 and 20910-90.

The optimal cement mixes were used in the production of heat-resistant concrete specimens (aggregate-chamotte) to determine their thermoresistance, as specified by the USSR standard 20910-90.

The new formations compositions of the mixes under investigation in different conditions of thermal treatment within a temperature range of 100-1000°C were detected using X-ray phase and differential-thermal analyses. The data obtained were used in the determination of conditions of synthesis of thermo-resistant and heat-resistant stone with taking into account the degree of crystallo-chemical similarity of new formations and their thermo-mechanical properties (the C.L.T.E. and coefficient of heat conductivity).

EXPERIMENTATION DISCUSSION

To put into practice the developed principles of producing thermo- and heat-resistant materials with the use of a three-factor three-level method of planning an experiment the use was made of the mixes made with blast furnace granulated slag and chamotte taken in the ratio of 1:1 and the quantity of thermoactivated kaolinite was 10- 30 % by mass.

In accordance with the results obtained from physical and mechanical tests, there have been calculated the equations of regression and the isoparameter diagrams of variations of strength of the specimens after hardening in normal conditions, which was followed by drying at T=100°C and firing at T=800°C and 1000°C, have been plotted (Figure 2).

Taken for reference were the specimens made from a slag-chamotte blend without additives in combination with sodium soluble glass solution with Ms=1; 1.75; 2.5.

Analysing the isoparameter diagrams allowed to conclude that the highest strength, when allowed to harden in normal conditions (as well as after consequent drying and firing) demonstrate the mixes comprising additives of kaolinite, thermoactivated at T=600°C and 700°C.

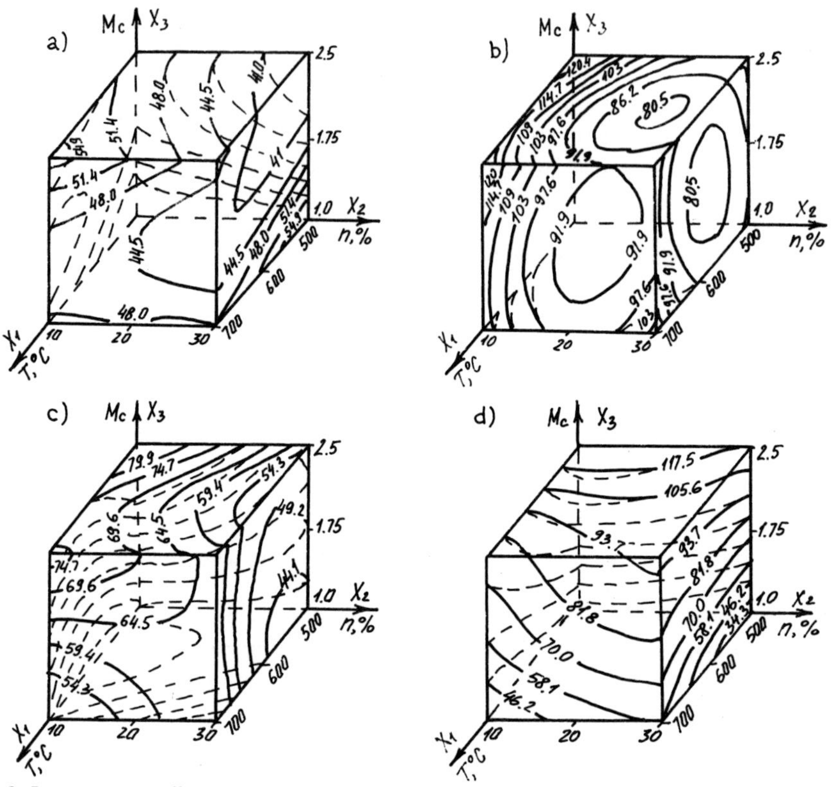

Figure 2 Isoparameter diagrams of variations of strength of the stone after 7 days storage (a); drying at T=100°C (b); firing at T=800°C (c), and 1000°C (d).

The use of thermoactivated at T=500°C kaolinite additive allows to produce a stone with rather high strength after drying at T=100°C at the expense of the presence within the hydration products of the X-ray detected as amorphous zeolite-like new formations NaA and low- basic hydrosilicates CSH(B) along with the kaolinite.

However, after firing, there may be observed the strength decline which is caused by the dehydration of kaolinite with the formation of mullite which is accompanied by a sharp re-arrangement of structure and development of considerable deformations.

At the same time, the use of kaolinite, thermoactivated at T=600°C and 700°C and represented by a substance which is X-ray detected as amorphous, promotes the directed formation of zeolite-like phases of the NaA composition, the hydration process of which at T=800°C takes place more smoothly (without development of considerable stresses in the material's structure) with the formation of distensillimanite Al_2O_3, SiO_2 (d=0.377; 0.318; 0.302; 0.299; 0.252; 0.221; 0.197; 0.182 nm) and insignificant quantities of mullite. After the mixes are fired at T=1000°C, a β-nepheline (d=0.37; 0.3; 0.256; 0.233; 0.184 nm) is detected within the dehydration products (Figure 3, curve 3).

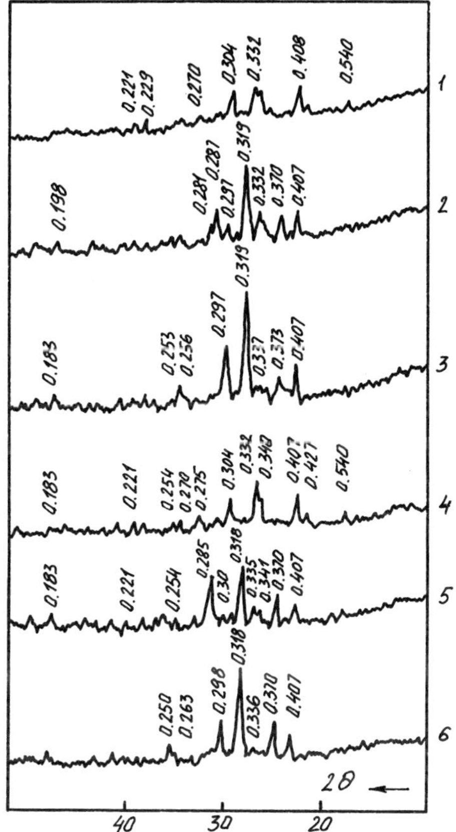

Figure 3 X-Ray grams of the cements made from a slag- chamotte blend with 20 % kaolinite, thermoactivated at T=700°C, which was hydrated in the presence of sodium soluble glass with Ms=2.5 (curves 1- 3) and Ms=1 (curves 4- 6) after 7 days hardening (curves 1,4) and firing at T=800°C ((curves 2,5) and T=1000°C (curves 3,6)

The kinetics of strength variations of the specimens after drying and firing depends, to a great extent, upon a silicate modulus of sodium soluble glass. When sodium soluble glass with Ms=1 is used, the hydration process of the slag-chamotte mixes accelerates, however, the dehydration process is accompanied with development of considerable destructive stresses and leads to considerable strength decline (after firing at T=1000°C)- the value of residual strength of mixes with sodium soluble glass with Ms=1 is 46- 48 %; with Ms=1.75- 74- 100 %, with Ms=2.5- 120- 130 %.

The mixes comprising kaolinite additives, thermoactivated at T=600°C and 700°C, and which were introduced in a quantity of 10- 20 % feature not only high thermal strength but the increased thermo-resistance as well.

Thermo-resistance of the concretes made from additive-free chamotte-slag mixes was 30-40 water-heat cycles, whereas when using the optimal mixes with the kaolinite additive- 40-50 cycles, what may be attributed to the formation, during firing the concrete, of fragmentary structures due to a directed synthesis, within the firing products, of crystallo-chemically similar compounds: mullite, β-nepheline and distensillimanite of thread-shaped and needle-shaped form (Table 2), which, with the temperature rise up to 1200-1300°C undergo re-crystallisation into mullite without development of considerable thermal stresses, thus not affecting negatively the strength of the fired stone. The dehydration products formed are characterised by increased density, increased heat conductivity and the decreased C.L.T.E. as compared with the initial phases, that is they possess a set of target properties which affect favourably the formation of a thermo-resistant structure of the material.

Table 2 Thermo- physical and crystallo-chemical characteristics of the dehydration products of the mixes investigated

NAME	SYNGONIA	PARAMETERS OF CRYSTAL LATTICE			DENSITY, kg/ m^3	C.H.C., W/m K	C. L.T. E., 10^{-6}, 1/K
		a	b	c			
Mullite	Rhombic	7.50	7.65	5.75	3.03-3.16	0.4-0.9	0.4-0.6
Disten-sillimanite	Rhombic	7.44	7.59	5.75	3.23-3.27	0.7-1.0	0.3-0.6
β-nepheline	Cubic	7.37	-	-	2.61	0.5-1.0	0.4-0.6
Kaolinite	Cubic	5.16	8.94	7.38	2.58-2.60	-	0.5-0.6

C.H.C.- coefficient of heat conductivity

Thus, there has been proved experimentally the possibility to enhance thermo-resistance of the stone due to a directed formation in the dehydration products of the crystallo-chemically similar substances which exhibit the increased density, heat conductivity, the lower C.L.T.E., promoting the formation of a stable fragmentary structure with the porosity varying inconsiderably during following heating and cooling of the material, predetermining the production of an artificial stone with stable thermo- mechanical characteristics.

CONCLUSIONS

1. With the use of a three-factor method of planning an experiment the formulations of alkaline cements with a kaolinite additive have been developed and optimised to produce thermo- and heat-resistant concretes. It was shown that the highest thermo- resistance (86-130 MPa) after being fired at T=1000°C (regardless of curing conditions) demonstrate the chamotte- slag mixes with thermoactivated at T= 600-700°C kaolinite additive introduced in a quantity of 10- 20 % by mass.

2. It was established that the increase in thermo-resistance of heat-resistant concretes may be reached by the formation, after firing, of a fragmentary microstructure due to a directed synthesis of needle-shaped crystallo-chemically similar substances (distensillimanite, mullite, β-nepheline) which are known to have the increased density, heat conductivity and the decreased C.L.T.E. as compared with initial cement constituents.

REFERENCES

1. KINGERY, U D. Transformations at High Temperatures, Metallurgizdat, Moscow, 1963, pp 237.

2. HASSELMAN, D P M. Thermal stress resistance parameter for brittle refractory ceramics: a compendium. Amer.Ceram.Soc.Bull., 1970, Vol.4, No.49, pp 22-28.
3. STRELOV, K K. Theoretical Bases of Refractory Materials Technology, Metallurgia, Moscow, 1985, pp 480.

4. PISARENKO, G S, RUDENKO, V P, TRETJACHENKO, G N. Strength of Materials at High Temperatures, Naukova Dumka, Kiev, 1966, pp 245-256.

5. STRELOV, K K. Structure and Properties of Refractory Materials, Metallurgia, Moscow, 1982, pp 208.

6. STRELOV, K K AND GOGOTSI, G A. Current state of the theories of thermo-resistance and prospects of their development. Journal Ogneupory, Moscow, 1974, No.9, pp 39- 47.

7. GOBERIS, S YU. Physico- Chemical and Technological Aspects...: The investigation of thermo-resistance of heat-resistant concretes made with water glass, Nauka, Moscow, 1986, pp 123- 136.

8. GORLOV, YU P, ERJEMIN N F, SEDUNOV B U. Refractory and Heat- Insulating Materials, Stroyizdat, Moscow, 1976, pp 192.

9. KINGERY, U D. Introduction into Ceramics, Stroyizdat, Moscow, 1967, pp 499.

10. KRIVENKO, P V, PUSHKARJEVA, E K. Durability of Slag Alkaline Concrete, Budivel'nik, Kiev, 1993, pp 223.

11. PUSHKARJEVA, E K. Heat-resistant alkaline binders. Proc.First Int.Conf.Alkaline Cements and Concretes, Ed. P V Krivenko, VIPOL, Kiev, 1994, pp 245- 256.

12. VOZNESENSKIY, V A, LJASHENKO, T V, OGARKOV, B L. Numerical methods of solving building- technological tasks using ECM, Vyscha Shkola, Kiev, 1989, pp 326.

ADAPTING CONCRETE FOR INDIGENOUS HOUSE CONSTRUCTION

A B Ngowi

J Morris

University of the Witwatersrand

South Africa

ABSTRACT. The desired characteristics of concrete are influenced by the quality of the constituent materials and their mix ratios as well as the procedures followed in its production. Strength has often been the design target in concrete construction and to achieve the desired strength within reasonable cost, the proportion of large particles in the concrete has to be as high as possible. However, in most low cost houses, especially the ones constructed using indigenous technologies, durability of the walling material, not its strength is the critical parameter that needs proper attention.

A study carried out in Botswana found that the walling could be improved substantially by using cement-stabilised bricks. To improve the foundation, standard test cubes were produced using the locally available stones, soil and "pit sand". On testing the cubes, it was found that concrete produced using "pit sand" could be used for both the foundation and floors of low-cost housing.

Keywords: Indigenous house, Indigenous technology, Low-strength concrete, Durability and Pit sand.

Dr Alfred Ngowi is a Senior Lecturer in the Department of Building and Quantity Surveying, University of the Witwatersrand, South Africa. His main research interests include the transfer of various construction technologies between communities of different culture and technological backgrounds.

Professor John Morris is the PPC Professor of Building Science in the Department of Building and Quantity Surveying of the University of the Witwatersrand, Johannesburg. Formerly Chief Director of the South African National Building Research Institute (NBRI) of the Council for Scientific and Industrial Research, he has been involved in research into building materials for many years. In 1987 he was elected an honorary member of the Institute of Housing of South Africa in recognition of his contribution to the NBRI book on Low-cost Housing.

INTRODUCTION

The history of civilization shows that, in the prehistoric times man depended entirely on natural resources of energy and available local material to form his habitat. Early efforts to build permanent structures were limited to the materials at hand. Trees, clay, mud and rocks provided only primitive structures, and few historic examples survive. However, the technical abilities to produce copper and bronze tools for making and preparing stone made it possible to cut stone for building permanent structures. The early permanent structures, built out of stone or brick have survived several centuries of war and natural disaster, which has enabled the tracing of human development from cave dweller to city builder [1].

Human development has passed through several civilizations, some of which made improvements on the way things were done, while others hindered progress and brought whole periods of several centuries of stagnation or decline. The achievements of the ancient communities in the use of natural materials for building purposes are particularly impressive when one realizes that they had no theoretical bases for material testing, structural calculation and design. The method of design was simply trial and error, the accumulation of experience and the copying of successful precedents [2]. If a building collapsed, it was rebuilt with stronger parts and skills developed by harsh experience or observed during foreign travel were incorporated into craft-lore and handed down from master to apprentice. Sometimes, with the collapse of a civilization, ideas and practices disappeared - to be rediscovered elsewhere centuries later.

While stone was originally used as a building material in the form of undressed blocks, lack of technology and tools to extract and shape it into the appropriate shape limited its use. Furthermore, communities that were not endowed with deposits of suitable stones could not build with stone in the form of blocks. However, because of its appealing properties measures were taken to bind together small size stones into a building material. The Romans named this material "concretus" which signifies "growing together" - a concise description of the binding of loose particles into a single mass. The binding together of small stones into a solid mass required some cementing material, which the Romans made from lime mixed with a volcanic sand called pozzolana after Pozzuoli, where it was first found. Using this cementing material together with broken stone aggregates as well as broken bricks, the Romans managed to make reasonably strong concretes.

From the time of Industrial Revolution, however, concrete production has been tremendously improved. The advent of Portland Cement and means of accurately producing specific sizes of aggregates meant that specified strengths of concrete could be met. The focus of concrete production has to a large extent been directed towards the achievement of as high strength as possible. Studies on economic production of low strength concrete for low-cost housing need to be strengthened if concrete has to make a significant contribution in this area. While the importance of producing high strength concrete by strictly adhering to specified concrete constituents and proportions is acknowledged, this paper argues that production of concrete for low-cost housing does not require similar specifications and the use of locally available constituents could produce economic low-strength concrete that might contribute to the alleviation of the housing problem.

The paper proceeds by describing the cost components of building materials followed by the description of the constituents of concrete, and using a study carried out in Botswana it outlines how locally available materials could be used to replace the traditional concrete constituents when producing low-strength concretes.

Cost Components of Building Materials

The main components of building material costs are reflected in the energy used in the life cycle of the materials, which includes:

- Cost of procurement (extraction), manufacturing and processing.
- Cost of transporting the materials to the processing plant and/or to the building site.
- Cost of installing the materials in the building and of carrying out subsequent maintenance.

The relative costs of these three main areas will vary depending on the type of the material. It is tempting to attribute the different cost components to the different participants in a building project. Thus, extraction, manufacturing and processing costs are often attributed to material manufacturers; transport costs partly to building materials merchants and partly to the building owner; and the installation and maintenance costs to the building owner. However, in order to achieve an affordable building material, these costs should be considered simultaneously and cost reduction measures should aim at all the three components.

In most developing countries, labour costs, particularly of unskilled labour, are relatively low compared with the costs of materials, equipment and transport. Therefore, measures to use as many labour-intensive methods as possible in the extraction, manufacturing and processing of building materials should be aimed at. Similarly, small scale building materials production that can be done on site or close to the site should be aimed at, so as to minimize the transport costs of both raw materials and finished products. To reduce installation costs, the materials that are produced should be familiar to the intended builders, so as to eliminate the need to hire or train people in special skills that might be necessary for the installation of unfamiliar materials. In other words, the materials should match the traditional materials as closely as possible.

Both reinforced and plain concrete are excellent examples of the principles discussed above. Production of concrete provides the flexibility to cope with a wide variety of local situations with regard to available materials and cultural and climatic factors. The cost of reinforced concrete, for instance, could be lowered by replacing the reinforcing materials and cement with locally available alternatives. Organic materials such as sisal fibres have been found suitable to replace mild steel in reinforced concrete, and replacement of part of cement by fly ash or other cementatious material such as pozzolana have also been found effective [3-5]. The use of these alternative materials could be employed in small-scale production of concrete using labour-intensive methods. However, an important disadvantage of this type of production, is the difficulty of maintaining consistent quality of the concrete. Lack of consistency is more of a problem than is the achievement of absolute quality, as variability, caused by lack of control, makes the concrete un-predictable. Large inconsistencies in batching and mixing, for example, are widespread and result in waste of about 20% of cement production in some countries [3].

Constituents of Concrete

The manner in which the concrete mix is described can be by prescriptive specification in which the exact proportions of all ingredients, one to the other are spelt out; or by performance specification, which spells out the strength requirement often accompanied by other limiting factors, such as minimum cement content, maximum water-cement ratio and some limitations on aggregate sizes and properties.

The basic factors affecting the proportioning of concrete constituents are strength and durability. Generally, the largest maximum size of aggregate should be used in concrete mixtures to develop the optimum properties of strength, durability and shrinkage. This is primarily achieved because the larger aggregate enables the use of minimum unit water content, since it offers lesser surface area to be wetted and higher denseness.

However, this does not imply that smaller coarse aggregate is unsuitable for concrete. It only requires relatively more sand (fine aggregate) with consequent higher water demand with accompanying lower strength and durability. Thus, the type of aggregate used, while having an important effect on its performance, is generally governed by local sources of supply. Concrete of a reasonable quality may be made with almost any natural stone, rock or gravel, though the suitability of various types of aggregates may differ considerably according to the kind of concrete required. Generally, however, the performance of an aggregate depends less on the type of rock than on the nature of the particles, their shape, size and grading.

The strength of concrete may range from as low as 10 MPa for low strength concrete to as high as 130 MPa for very high strength concretes [6,7]. While the production of high strength concrete necessitates both the use of high quality constituent materials, specified mix design and strict adherence to concreting procedures, production of low-strength concrete makes allowance for the use of a range of alternative materials that need only meet the minimum requirements for durability. Such materials as periwinkle shells and crushed laterite rock, for instance, have been used as coarse aggregate in low strength concrete with satisfactory results [8,9]. This provides a possibility for other materials with similar general properties to be investigated as alternatives for aggregate.

LOW-STRENGTH CONCRETE FOR LOW-COST HOUSING IN BOTSWANA

A housing study in Botswana that was carried out in two major villages determined that the main problem with the indigenous houses, which are the main type of housing in the villages, is low durability of the walling material [10]. As a result of the low durability substantial maintenance is required every year after the rainy season, and in case of heavy rainfall, some of the houses collapse. A detailed study of the development of the indigenous building technology in these villages as well as of the culture of the people established that the technology could be improved by adopting appropriate modern building concepts for incorporation in the traditional technology [11].

To improve the walls, the sun-dried soil brick, which is the indigenous walling material, was improved by stabilizing it with specific amounts of cement, which was established to be the most effective stabilizing agent through laboratory tests [10]. Using the cement stabilized bricks, demonstration houses were built using the indigenous skills [12], and in the process it was found that the indigenous foundation used in these houses needed improvement.

Traditionally foundations are not regarded as important, mainly because the soil type allows rainwater to run-off quickly when it rains. The foundation, is therefore, constructed by excavating to about 150 mm deep with a width of about 350mm. and then the soil at this depth is compacted using hand rammers. Sometimes, small stones of about 50 mm diameter are put into the trenches at this depth in single layer before compaction is done. The walls are then built of sun-dried mud blocks on this compacted layer after placing a 30 mm thick layer of mud building mortar.

Traditionally, the floor of the houses is constructed by filling the area within the walls with selected soil, compacting it , and finishing with a layer of soil-cowdung mixture. The top layer of the floor has to be maintained frequently because it is easily worn out. Also, it may generate dust if it is not sprinkled with water frequently. In general, the floor is neither strong enough for normal use nor does it provide healthy living surroundings mainly because of the dust that may be generated and the possibility of harbouring harmful organisms.

Improvement of the Foundation and Floor Using Low-Cost Concrete

To improve the foundation and the floor without departing too much from the indigenous practice, the study investigated how the people of Tsabong, one of the villages, obtain the 50 mm diameter stones that they sometimes use to construct the foundations. It was established that stones of right size were collected by women and children from the building site and surrounding areas. Sometimes, the right size is produced by breaking larger stones with hammers. Apparently, there is a large deposit of stones in Tsabong. However, the fine aggregate that is specified for conventional concrete is not available in the village. In order to comply with building regulations, the river sand (fine aggregate) that is used to produce concrete for institutional buildings (formal sector buildings) has to be transported into the village from Kanye, which is about 300 km away. The cost of procuring this type of sand makes it impossible for potential house owners to use it. The study, therefore, investigated the possibility of using the local soil and "pit sand" (sand collected from pits dug along a seasonal stream that passes through the village) to produce low-strength concrete.

Sample Collection

Three samples were collected as follows.

1. Small size stones (approx. 50 mm diameter) that are sometimes used by the indigenous builders at the foundation level.

2. Soil similar to that used to produce cement stabilized bricks.

3. "Pit sand"

Atterberg limits and sedimentation tests determined that soil sample No. 2 contained 63% sand, 22.5 percent silt and 14.5 percent clay; while "pit sand" sample No.3 contained 34 percent sand, 56 percent silt and 10 percent clay.

Test Cubes

Six different types of cubes were made using the following specifications, which are elaborated on Table 1.

A1: Same cement content, but different proportions of stones and soil.

A2: Different cement contents, but same proportion of stones and soil

B1: Same cement content, but different proportions of stones and "pit sand"

B2: Different cement contents, but same proportion of stones and "pit sand"

As a control, cubes were also produced using the sand that is specified for institutional buildings. Similar stones to those used in the other tests were used.

C1: Same cement content but different proportions of stones and sand.

C2: Different cement contents, but same proportion of stones and sand.

Although volume-proportion is not as accurate as weight-proportion, the study adopted the former so as to follow the traditional practice as closely as possible. It is difficult for the people in the village to correct for bulking of the materials with moisture and to adopt weight-proportion because they lack the necessary equipment and insistence on its use will require very close supervision or training of the villagers on how to convert volumes into equivalent weights. Such measures are unlikely to succeed.

The constituents were mixed using a pan mixer and cubes were made according to standard procedures. Six cubes were produced in each case, making a total of 36. After keeping the cubes in the moulds for 48 hours, they were immersed in water for curing. However, on examining the cubes made using soil, it was noted that soil particles were being washed from the surface. It was, therefore, decided to remove all the cubes from the curing basin and cure them by sprinkling them with water and covering them with damp sacking instead. This was done daily for 28 days.

Failure Tests of the Cubes

Tests were only carried out after 28 days of curing because the primary aim of the study was to investigate whether the concrete produced using the local materials could develop adequate strength for the intended purpose. The results of crushing tests on the cubes are shown on Table 1.

Table 1 Mean compressive strengths with soil, "pit sand" and sand as fine aggregate

MIX PROPORTIONS	COMPRESSIVE STRENGTH (MPa)		
Cement: F/Aggregate: C/Aggregate	Soil (A)	Pit Sand (B)	River Sand (C)
1. Same Cement Content			
1 : 1 : 5	7	15	25
1 : 2 : 4	6	18	24
1 : 3 : 3	6	17	22
2. Different Cement Contents			
0.5 : 2 : 4	5	14	17
1 : 2 : 4	6	18	24
2 : 2 : 4	7	20	28

F/Aggregate – Fine aggregate, C/Aggregate – Coarse aggregate

The test results were compared with the compressive strength that is specified for foundations and floors of institutional buildings. The results of samples C1 and C2, (produced using river sand) closely matched the specified strength of 20 MPa for this type of construction. All average compressive strengths, except one, i.e., 17 MPa, were above this value.

An increase in the amount of cement resulted in an increase in strength. Similarly, a larger proportion of stones in the mix resulted in a higher strength.

This is in line with observations by Raina [13], that for low-strength concretes, strength increases with aggregate size and amount, while at the higher strength ranges, concretes containing the smaller maximum size aggregates (20 mm) generally develop greater strength.

The results of samples A1 and A2 (produced using soil) were far below the specified strength and no significant difference resulted from either varying the amount of cement in the sample or the proportion of soil to stones. Moreover, the failure mechanism was different from normal fracture, which occurs either along the cement paste or through the stones. In these cases, there was adhesion failure between the stones and the soil paste.

This means that the strength developed by stabilizing the soil with cement is higher than the adhesive strength between the stones and the soil paste. The concrete produced using soil was, therefore, considered unsuitable for foundation and floor construction.

The results of samples B1 and B2 (produced using pit sand) were much closer to the specified strength and an increase in the amount of cement when using this type of fine aggregate led to a significant increase in compressive strength, i.e., 14 MPa to 20 MPa. Similarly, a larger proportion of stones in the mix resulted in an increase in strength.

CONCLUSIONS

The objective of this study was to investigate the possibility of using local resources in Tsabong village as constituents of low-strength concrete to be used in the construction of foundations for low-cost housing.

Information about how people in the village mobilize building materials and how they build their houses was gathered, and also about the skills used in building these houses. It was decided to improve the strength of the indigenous foundation and floors using low-strength concrete that is produced using the familiar local materials. Therefore, concrete was produced using the soil, "pit sand" and the stones available in the village.

Tests on standards cubes of the concrete so produced indicated that concrete produced using soil has very low strength and is neither suitable for foundation nor floor, which require a strength of at least 10 MPa. Cubes produced using "pit sand" showed an average strength of 17 MPa, which is suitable for both foundation and floor of low-cost housing. Control cubes produced using sand brought in from Kanye together with locally sourced stone aggregate met the specified strength of 20 MPa and this meant that the local stone is acceptable for concrete production.

Since the test only determined the 28-day strength of concrete, further study is necessary to determine the growth in strength and the influence of curing conditions in order to be able to recommend to the people of the village when to load the foundation and how to cure the concrete.

REFERENCES

1. BEALL, C. Masonry Design and Detailing for Architects, Engineers and Contractors, Third edition, McGraw-Hill,1993.

2. MEAD, M. Continuities in Cultural Evolution, Yale University Press, New Haven, CT,1964.

3. SPENCE, R, J, S AND COOK, D, J. Building Materials in Developing Countries, Chichester, John Wiley, 1983.

4. DA SILVA GUIMARES, S. Experimental Mixing and Moulding with Vegetable Fibre Reinforced Cement Composites, in Ghavami, K, and M, Y, Fang (eds.) Envo, Low-cost and Energy Saving Construction Materials, Vol. 1, 1984, pp 68-82.

5. OKPALA, D, C. Some Engineering Properties of Sandcrete Blocks Containing Rice Husk Ash, Building and Environment, Vol. 28, No. 3, 1993, pp 235-241.

6. WEBB, J. High-strength Concrete: Economic, Design and Ductility, Concrete International, 8 (1), 1993, pp 27-32.

7. WALRAVEN, J. High Performance Concrete: Exploring a New Material, Structural Engineering International, 10 (3), 1995, pp 182-187.

8. THOMAS, K AND LISK, W, A, A. Investigation Into Suitability of Crushed Laterite Rocks for Use as Coarse Aggregate for Concrete, Proceedings of Conference on Concrete and Reinforced Concrete in Hot Countries, Haifa, Israel, 1971, pp 183-198.

9. FALADE, F. An Investigation of Periwinkle Shells as Coarse Aggregate in Concrete, Building and Environment, Vol. 30, No. 4, 1995, pp 573-577.

10. NGOWI, A, B. Improving the Traditional Earth Construction: A Case Study of Botswana, Construction and Building Materials, Vol. 11, No. 1, 1997, pp 1-7.

11. NGOWI, A, B. The Potential for Alleviation of Housing Needs Through Improvement of the Indigenous Construction Technology – A Case Study of Botswana, Unpublished PhD Thesis, University of the Witwatersrand, South Africa, 1998.

12. NGOWI, A, B. A Hybrid Approach to House Construction – A Case Study in Botswana, Building Research and Information, Vol. 25, No. 3, 1997, pp 142-147.

13. RAINA, V, K. Concrete for Construction: Facts and Practice, Tata McGraw-Hill, New Delhi, 1990.

AUTHOR INDEX

SUBJECT INDEX

This index has been compiled from the keywords assigned to the papers, edited and extended as appropriate. The page references are to the first page of the relevant paper